T0345061

TOWARD ANALYTICAL CHAOS IN NONLINEAR SYSTEMS

TOWARD ANALYTICAL CHAOS IN NONLINEAR SYSTEMS

Albert C. J. Luo
Southern Illinois University, USA

This edition first published 2014
© 2014 John Wiley & Sons Ltd

Registered office
John Wiley & Sons Ltd, The Atrium, Southern Gate, Chichester, West Sussex, PO19 8SQ, United Kingdom

For details of our global editorial offices, for customer services and for information about how to apply for permission to reuse the copyright material in this book please see our website at www.wiley.com.

The right of the author to be identified as the author of this work has been asserted in accordance with the Copyright, Designs and Patents Act 1988.

All rights reserved. No part of this publication may be reproduced, stored in a retrieval system, or transmitted, in any form or by any means, electronic, mechanical, photocopying, recording or otherwise, except as permitted by the UK Copyright, Designs and Patents Act 1988, without the prior permission of the publisher.

Wiley also publishes its books in a variety of electronic formats. Some content that appears in print may not be available in electronic books.

Designations used by companies to distinguish their products are often claimed as trademarks. All brand names and product names used in this book are trade names, service marks, trademarks or registered trademarks of their respective owners. The publisher is not associated with any product or vendor mentioned in this book.

Limit of Liability/Disclaimer of Warranty: While the publisher and author have used their best efforts in preparing this book, they make no representations or warranties with respect to the accuracy or completeness of the contents of this book and specifically disclaim any implied warranties of merchantability or fitness for a particular purpose. It is sold on the understanding that the publisher is not engaged in rendering professional services and neither the publisher nor the author shall be liable for damages arising herefrom. If professional advice or other expert assistance is required, the services of a competent professional should be sought.

Library of Congress Cataloging-in-Publication Data

Luo, Albert C. J.
Toward analytical chaos in nonlinear systems / Albert C. J. Luo.
pages cm
Includes bibliographical references and index.
ISBN 978-1-118-65861-1 (hardback)
1. Differentiable dynamical systems. 2. Nonlinear oscillations. 3. Chaotic behavior in systems. I. Title.
QA867.5.L86 2014
003′.857 – dc23

2014001972

A catalogue record for this book is available from the British Library.

Typeset in 10/12pt TimesLTStd by Laserwords Private Limited, Chennai, India
Printed and bound in Singapore by Markono Print Media Pte Ltd

1 2014

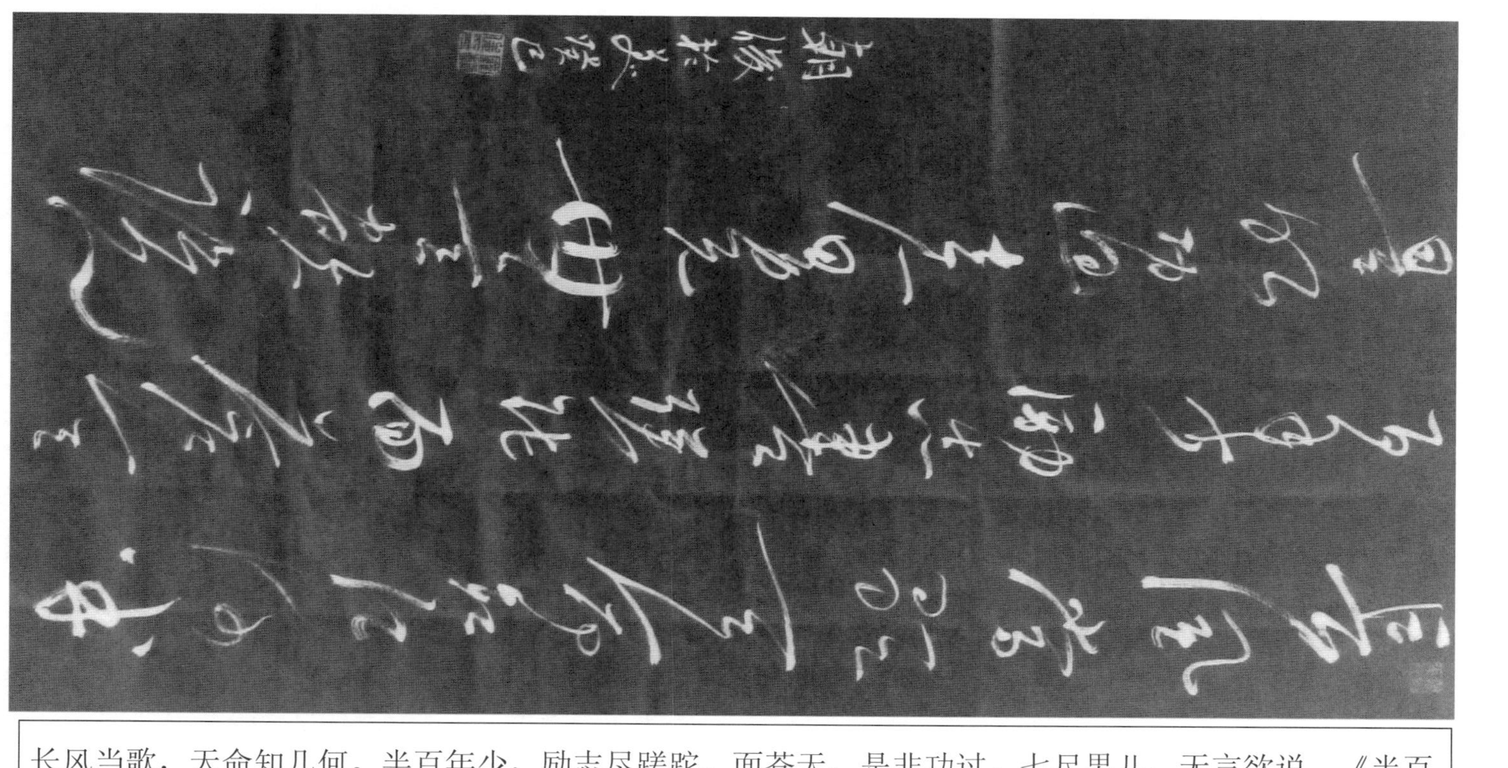

长风当歌，天命知几何。半百年少，励志尽蹉跎。面苍天，是非功过。七尺男儿，无言欲说。《半百人生》说 – 朝俊。Albert Luo's self-sketch poem of the 50th birthday

Contents

Preface

There has been interest in periodic solutions of dynamical systems for a few centuries. The periodic motion as a steady-state motion has attracted many scientists' attention. Until now, it still cannot be achieved analytically and accurately. In the second half of the twentieth century, using computer software, another steady-state motion (i.e., chaos) has been observed to be caused by bifurcations of periodic motions. Such a steady-state motion can be numerically simulated with computational errors. Other than that, it was not known how to obtain the bifurcation trees of periodic motions to chaos analytically in nonlinear dynamical systems, and the mathematical structures of solutions for chaos are still unknown. This is because it is not yet known how to obtain the exact solutions of periodic motions analytically. This book will try to solve this problem. Further, the title of this book is *Toward Analytical Chaos in Nonlinear Systems*. The author hopes this book can attract more attention to finding accurate analytical solutions of periodic motions to chaos, and the mathematical structures of chaos solutions can be achieved.

Since 1788, Lagrange has used the fundamental matrix of the linearized system as a moving coordinate transformation to obtain the Lagrange stand form, which is based on the variation of parameter procedure. From the Lagrange standard form, the method of averaging was used to investigate the gravitational three-body problem through a two-body problem with a perturbation. At the end of the nineteenth century, Poincare developed the perturbation method to investigate the periodic motion of the three-body problem. Since then, this cycle of perturbation method for periodic motions has remained, and it is hard to change such a cycle. Indeed, the perturbation method helps one understand some nonlinear phenomena and has given some reasonable explanations. However, based on the Lagrange standard form or perturbation methods, vector fields in dynamical systems have been changed, and the deformed vector fields cannot represent the original nonlinear dynamical systems. From the idea of the Lagrange standard form, the normal forms of nonlinear dynamical systems at a state of equilibrium cannot be used for periodic motions and chaos in the original nonlinear dynamical systems.

This book provides an analytical method for determining periodic flows and quasi-periodic flows in nonlinear dynamical systems with/without time-delay. From the analytical solutions of periodic motions, the bifurcation trees of periodic flows to chaos can be determined analytically. Further, one can achieve analytical solutions of chaos and understand the corresponding mathematical structures. The method presented in this book gives frequency-responses for nonlinear dynamical systems as the Laplace transformation for linear dynamical systems. This book has six chapters. Chapter 1 gives a brief history of the study of periodic motions in nonlinear dynamical systems. The stability, stability switching, and bifurcation of equilibriums in nonlinear continuous systems are introduced in Chapter 2, which is different from the traditional presentation. In Chapter 3, the analytical method for period-m flows in dynamical systems is presented, including autonomous and non-autonomous dynamical systems with/without time delay. In Chapter 4, the analytical determination of period-m to quasi-periodic flows is presented. In Chapter 5, the analytical bifurcation tree of periodic motion to chaos is presented through a periodically excited, quadratic nonlinear oscillator. In Chapter 6, the analytical solution for a time-delayed, quadratic nonlinear system is presented. The materials presented in this book will provide a different way to achieve analytical solutions of periodic flows to chaos in nonlinear dynamical systems.

Finally, I would like to thank my students (Bo Yu and Hanxiang Jin) for applying the recently developed analytical method to two nonlinear systems and completing numerical computations. Also, I would like to thank my wife (Sherry X. Huang) and my children (Yanyi Luo, Robin Ruo-Bing Luo, and Robert Zong-Yuan Luo) again for tolerance, patience, understanding, and continuous support.

Albert C.J. Luo
Edwardsville, Illinois, USA

1

Introduction

In this chapter, a brief literature survey of analytical solutions of periodic motions in nonlinear dynamical systems will be presented. The perturbation analysis has played an important role in such an approximate analysis of periodic motions in nonlinear systems. The perturbation method, method of averaging, harmonic balance, and generalized harmonic balance will be reviewed. The application of perturbation method in time-delayed systems will be discussed briefly.

1.1 Brief History

Since the seventeenth century, there has been interest in periodic motions in dynamical systems. The Fourier series theory shows that any periodic function can be expressed by a Fourier series expansion with different harmonics. In addition to simple oscillations, there has been interest in the motions of moon, earth, and sun in the three-body problem. The earliest approximation method is the method of averaging, and the idea of averaging originates from Lagrange (1788). At the end of the nineteenth century, Poincare (1890) provided the qualitative analysis of dynamical systems to determine periodic solutions and stability, and developed the perturbation theory for periodic solutions. In addition, Poincare (1899) discovered that the motion of a nonlinear coupled oscillator is sensitive to the initial condition, and qualitatively stated that the motion in the vicinity of unstable fixed points of nonlinear oscillation systems may be stochastic under regular applied forces. In the twentieth century, one followed Poincare's ideas to develop and apply the qualitative theory to investigate the complexity of motions in dynamical systems. With Poincare's influence, Birkhoff (1913) continued Poincare's work, and proof of Poincare's geometric theorem was given. Birkhoff (1927) showed that both stable and unstable fixed points of nonlinear oscillation systems with two degrees of freedom must exist whenever their frequency ratio (or called resonance) is rational. The sub-resonances in periodic motions of such systems change the topological structures of phase trajectories, and the island chains are obtained when the dynamical systems can be renormalized with fine scales. In such qualitative and quantitative analysis, the Taylor series expansion and the perturbation analysis play an important role. However, the Taylor series expansion analysis is valid in the small finite domain under certain convergent conditions, and the perturbation analysis

Toward Analytical Chaos in Nonlinear Systems, First Edition. Albert C. J. Luo.
© 2014 John Wiley & Sons, Ltd. Published 2014 by John Wiley & Sons, Ltd.

based on the small parameters, as an approximate estimate, is only acceptable for a very small domain with a short time period.

van der Pol (1920) used the averaging method to determine the periodic motions of self-excited systems in circuits, and the presence of natural entrainment frequencies in such a system was observed in van der Pol and van der Mark (1927). Cartwright and Littlewood (1945) discussed the periodic motions of the van der Pol equation and proved the existence of periodic motions. Cartwright and Littlewood (1947) discussed the periodic motions of a generalized nonlinear equation based on the similar Duffing equation. Levinson (1948) used a piecewise linear model to describe the van der Pol equation and determined the existence of periodic motions. Levinson (1949) further developed the structures of periodic solutions in such a second order differential equation through the piecewise linear model, and discovered that infinite periodic solutions exist in such a piecewise linear model. From the Levinson's results, Smale (1967) used the topological point to present the Smale horseshoe with discontinuous mappings to describe the existence of infinite periodic motions. Further, a differentiable dynamical system theory was developed. Such a theory has been extensively used to interpret the homoclinic tangle phenomenon in nonlinear dynamical systems. Smale found the infinite, many periodic motions, and a perfect minimal Cantor set near a homoclinic motion can be formed. Melnikov (1962) used the concept of Poincare (1892) to investigate the behavior of trajectories of perturbed systems near autonomous Hamiltonian systems. Melnikov (1963) further investigated the behavior of trajectories of perturbed Hamiltonian systems, and the width of the separatrix splitting was approximately estimated. The width gives the domain of the chaotic motion in the vicinity of the generic separatrix. Even if the width of the separatrix splitting was approximately estimated, the dynamics of the separatrix splitting was not developed.

Since the nonlinear phenomena was observed in engineering, Duffing (1918) used the hardening spring model to investigate the vibration of electro-magnetized vibrating beam, and after that, the Duffing oscillator has been extensively used in structural dynamics. In addition to determining the existence of periodic motions in nonlinear differential equations of the second order in mathematics, one has applied the Poincare perturbation methods for periodic motions in nonlinear dynamical systems. Fatou (1928) provided the first proof of asymptotic validity of the method of averaging through the existence of solutions of differential equations. Krylov and Bogolyubov (1935) systematically developed the method of averaging and the detailed discussion can be found in Bogoliubov and Mitropolsky (1961). The classic perturbation methods for nonlinear oscillators were presented (e.g., Stoker, 1950; Minorsky, 1962; Hayashi, 1964). Hayashi (1964) used the method of averaging and harmonic balance method to discuss the approximate periodic solutions of nonlinear systems and the corresponding stability. Nayfeh (1973) employed the multiple-scale perturbation method to develop approximate solutions of periodic motions in the Duffing oscillators. Holmes and Rand (1976) discussed the stability and bifurcation of periodic motions in the Duffing oscillator. Nayfeh and Mook (1979) applied the perturbation analysis to nonlinear structural vibrations via the Duffing oscillators, and Holmes (1979) demonstrated chaotic motions in nonlinear oscillators through the Duffing oscillator with a twin-well potential. Ueda (1980) numerically simulated chaos via period-doubling of periodic motions of Duffing oscillators.

Based on the work of Melnikov (1963), Greenspan (1981) extended the similar ideas to the dissipative dynamical systems (also see, Greenspan and Holmes, 1983; Guckenheimer and Holmes, 1983). Further, the Melnikov method was developed for the global transversality in dissipative nonlinear systems. Once the global transversality to the separatrix exists, one thought that the Smale horseshoe presented in Smale (1967) may exist, and furthermore chaos in such a nonlinear dynamical system may occur. However, from such a prediction based on the Melnikov method, one cannot observe the global transversality in nonlinear dynamical systems. The Smale horseshoe theory may not be adequate for nonlinear dynamical systems rather than the topological structure. From the perturbation analysis, the Melnikov function was obtained for Hamiltonian systems with a small perturbation. One used such a function to analytically predict global behaviors (e.g., chaos) in the Hamiltonian systems with a small perturbation. Because of the perturbation analysis, the Melnikov method can give a reasonable analysis of the global behavior only when the perturbation is very small and close to zero. However, the perturbation is very small to zero, chaos in nonlinear dynamical systems may not occur. So the Melnikov method may not help us understand the global behaviors of nonlinear dynamical systems. Luo (1995) used the Chirikov criterion to determine Hamiltonian chaos and applied the Melnikov function to investigate the global transversality (also see, Luo and Han, 1999; Luo, 2008, 2012a). The conclusion is that the Melnikov method cannot provide an adequate prediction of chaotic motions in the dissipative system. For a better understanding of the Melnikov method, the work of Melnikov (1963) should be revisited. Melnikov (1963) presented a perturbation analysis to estimate the width of the separatrix splitting. Indeed, the width of the separatrix can be approximately estimated, but it cannot be used for predicting the existence of chaos. The Melnikov function is an approximate energy increment during a certain time period, which can be found in references (e.g., Arnold, 1964; Chirikov, 1979; Luo and Han, 2001). If the Melnikov function is zero, from a physical point of view, the system energy is conserved during a certain time period. Such a zero value of the Melnikov function does not imply that the flow has any global transversality to the separatrix. One has difficulty finding a connection from periodic motions to chaos. Thus, one continues using the perturbation analysis to determine the approximate analytical solutions of periodic motions. Coppola and Rand (1990) determined limit cycles of nonlinear oscillators through elliptic functions in the averaging method. Wang *et al.* (1992) used the harmonic balance method and the Floquet theory to investigate the nonlinear behaviors of the Duffing oscillator with a bounded potential well (also see, Kao *et al.*, 1992). Luo and Han (1997) determined the stability and bifurcation conditions of periodic motions of the Duffing oscillator. However, only symmetric periodic motions of the Duffing oscillators were investigated. Luo and Han (1999) investigated the analytical prediction of chaos in nonlinear rods through the Duffing oscillator. Peng *et al.* (2008) presented the approximate symmetric solution of period-1 motions in the Duffing oscillator by the harmonic balance method with three harmonic terms. Luo (2012a) developed a generalized harmonic balance method to get the approximate analytical solutions of periodic motions and chaos in nonlinear dynamical systems. This method used the finite Fourier series to express periodic motions and the coefficients are time-varying. With averaging, a dynamical system of coefficients are obtained from which the steady-state solution are achieved and the corresponding stability and bifurcation are completed. Luo and Huang (2012a) used the generalized harmonic balance method with finite terms to obtain the analytical solution of period-1 motion of the Duffing oscillator with a twin-well potential. Luo and Huang (2012b) employed a generalized harmonic balance method to find analytical solutions of period-m motions in such a Duffing oscillator. The analytical bifurcation trees of periodic motions in the Duffing oscillator to chaos were obtained (also see, Luo and Huang, 2012c,d, 2013a,b,c, 2014). Such analytical bifurcation trees show the connection from periodic solution to chaos analytically. To better understand nonlinear behaviors in nonlinear dynamical systems, the analytical solutions for the bifurcation trees from period-1 motion to chaos in a periodically forced oscillator with quadratic nonlinearity were presented in Luo and Yu (2013a, b, c), and period-*m* motions in the periodically forced, van der Pol equation was presented in Luo and Laken (2013). The analytical solutions for the van der Pol oscillator can be used to verify the conclusions in Cartwright and Littlewood (1947) and Levinson (1949). The results for the quadratic nonlinear oscillator in Luo and Yu (2013a, b, c) analytically show the complicated period-1 motions and the corresponding bifurcation structures.

In recent years, time-delayed systems are of great interest since such systems extensively exist in engineering (e.g., Tlusty, 2000; Hu and Wang, 2002). The infinite dimensional state space causes a significant difficulty in understanding such a time-delayed problem. One tried to work on numerical methods to get the corresponding complicated behaviors. On the other hand, one is interested in the stability and bifurcation of equilibriums of the time-delayed systems (e.g., Stepan, 1989; Sun, 2009; Insperger and Stepan, 2011). In addition, one is also interested in periodic solutions in time-delayed dynamical systems. Perturbation methods have been used in recent years for such periodic motions in delayed dynamical systems. For instance, the approximate solutions of the time-delayed nonlinear oscillator were investigated by the method of multiple scales (e.g., Hu, Dowell, and Virgin 1998; Wang and Hu, 2006). The harmonic balance method was also used to determine approximate periodic solutions for delayed nonlinear oscillators (e.g., MacDonald, 1995; Liu and Kalmar-Nagy, 2010; Leung and Guo, 2012). However, such approximate solutions of periodic motions in the time-delayed oscillators are based on one or two harmonic terms, which are not accurate enough. In addition, the corresponding stability and bifurcation analysis of such approximate solutions may not be adequate. In this book, an alternative way of finding the accurate analytical solutions of periodic flows in time-delayed dynamical systems will be presented. This method is without any small-parameter requirement. In addition, this approach can also be applicable to the coefficient varying with time.

1.2 Book Layout

In this book, a new analytical method will be presented for analytical solutions of periodic motions in nonlinear dynamical systems with/ without time delay. The basic theory of nonlinear systems will be briefly introduced. The analytic method based on the generalized harmonic balance will be comprehensively discussed, and this method will be applied to nonlinear dynamical systems to find the periodic motions analytically and to determine the analytical bifurcation trees of periodic motions to chaos. The main body in this book is summarized as follows:

- In Chapter 2, the basic theory of nonlinear dynamical systems will be introduced. Local theory, global theory, and bifurcation theory of nonlinear dynamical systems will be briefly discussed. The stability switching and bifurcation on specific eigenvectors of the linearized system of a nonlinear system at equilibrium will be discussed. The higher-order singularity and stability for nonlinear systems on the specific eigenvectors will be developed.
- In Chapter 3, from Luo (2012a), the analytical dynamics of periodic flows and chaos in nonlinear dynamical systems will be presented. The analytical solutions of periodic flows

and chaos in autonomous systems will be discussed first, and the analytical dynamics of periodically forced nonlinear dynamical systems will be presented. The analytical solutions of periodic motions in free and periodically forced vibration systems will be presented. In a similar fashion, the analytical solutions of periodic flows for time-delayed nonlinear systems will be presented with/without periodic excitations, and time-delayed nonlinear vibration systems will be also discussed for engineering application. The analytical solutions of periodic flows and chaos are independent of the small parameters, which are different from the traditional perturbation methods. The methodology presented herein will end the history of chaos being numerically simulated only.

- In Chapter 4, from the idea of Luo (2012a, 2013), period-m flows to quasi-periodic flows in nonlinear dynamical systems will be presented. The analytical solutions of quasi-periodic flows in autonomous systems will be discussed, and the analytical solutions of quasi-periodic flows in periodically forced nonlinear dynamical systems will be presented. The analytical solutions of quasi-periodic motions in free and periodically forced vibration systems will be presented. The analytical solutions of quasi-periodic flows for time-delayed nonlinear systems will be presented with/without periodic excitations, and time-delayed nonlinear vibration systems will be discussed as well.
- In Chapter 5, analytical solutions for period-m motions in a periodically forced, quadratic nonlinear oscillator will be presented through the Fourier series solutions with finite harmonic terms, and the stability and bifurcation analyses of the corresponding period-1 motions will be carried out. There are many period-1 motions in such a nonlinear oscillator, and the parameter map for excitation amplitude and frequency will be developed for different period-1 motions. For each period-1 motion branch, analytical bifurcation trees of period-1 motions to chaos will be presented. For a better understanding of complex period-m motions in such a quadratic nonlinear oscillator, trajectories, and amplitude spectrums will be illustrated numerically.
- In Chapter 6, analytical solutions for period-m motions in a time-delayed, nonlinear oscillator will be presented through the Fourier series, and the stability and bifurcation analyses of the corresponding periodic motions will be presented through the eigenvalue analysis. Analytical bifurcation trees of periodic motions to chaos will be presented through the frequency-amplitude curves. Trajectories and amplitude spectrums of periodic motions in such a time-delayed nonlinear system will be illustrated numerically for a better understanding of time-delayed nonlinear dynamical systems.

2

Nonlinear Dynamical Systems

In this chapter, the basic theory of nonlinear dynamical systems will be introduced. Local theory, global theory, and bifurcation theory of nonlinear dynamical systems will be briefly discussed. The stability switching and bifurcation on specific eigenvectors of the linearized system at equilibrium will be discussed. The higher-order singularity and stability for nonlinear systems on the specific eigenvectors will be developed.

2.1 Continuous Systems

Definition 2.1 For $I \subseteq \mathscr{R}, \Omega \subseteq \mathscr{R}^n$, and $\Lambda \subseteq \mathscr{R}^m$, consider a vector function $\mathbf{f} : \Omega \times I \times \Lambda \to \mathscr{R}^n$ which is $C^r(r \geq 1)$-continuous, and there is an ordinary differential equation in a form of

$$\dot{\mathbf{x}} = \mathbf{f}(\mathbf{x}, t, \mathbf{p}) \text{ for } t \in I, \mathbf{x} \in \Omega \text{ and } \mathbf{p} \in \Lambda \tag{2.1}$$

where $\dot{\mathbf{x}} = d\mathbf{x}/dt$ is derivative with respect to time t, which is simply called the velocity vector of the state variables $\mathbf{x}$. With an initial condition of $\mathbf{x}(t_0) = \mathbf{x}_0$, the solution of Equation (2.1) is given by

$$\mathbf{x}(t) = \mathbf{\Phi}(\mathbf{x}_0, t - t_0, \mathbf{p}). \tag{2.2}$$

1. The ordinary differential equation with the initial condition is called a dynamical system.
2. The vector function $\mathbf{f}(\mathbf{x}, t, \mathbf{p})$ is called a vector field on domain Ω.
3. The solution $\mathbf{\Phi}(\mathbf{x}_0, t - t_0, \mathbf{p})$ is called the flow of dynamical system.
4. The projection of the solution $\mathbf{\Phi}(\mathbf{x}_0, t - t_0, \mathbf{p})$ on domain Ω is called the trajectory, phase curve, or orbit of dynamical system, which is defined as

$$\Gamma = \{\mathbf{x}(t) \in \Omega | \mathbf{x}(t) = \mathbf{\Phi}(\mathbf{x}_0, t - t_0, \mathbf{p}) \text{ for } t \in I\} \subset \Omega. \tag{2.3}$$

Definition 2.2 If the vector field of the dynamical system in Equation (2.1) is independent of time, such a system is called an autonomous dynamical system. Thus, Equation (2.1) becomes

$$\dot{\mathbf{x}} = \mathbf{f}(\mathbf{x}, \mathbf{p}) \text{ for } t \in I \subseteq \mathscr{R}, \mathbf{x} \in \Omega \subseteq \mathscr{R}^n \text{ and } \mathbf{p} \in \Lambda \subseteq \mathscr{R}^m \tag{2.4}$$

Otherwise, such a system is called non-autonomous dynamical systems if the vector field of the dynamical system in Equation (2.1) is dependent on time and state variables.

Toward Analytical Chaos in Nonlinear Systems, First Edition. Albert C. J. Luo.
© 2014 John Wiley & Sons, Ltd. Published 2014 by John Wiley & Sons, Ltd.

Definition 2.3 For a vector function $\mathbf{f} \in \mathscr{R}^n$ with $\mathbf{x} \in \mathscr{R}^n$, the operator norm of $\mathbf{f}$ is defined by

$$||\mathbf{f}|| = \sum_{i=1}^{n} \max_{||\mathbf{x}||\leq 1, t\in I} |f_i(\mathbf{x}, t)|. \tag{2.5}$$

For $\mathbf{f}(\mathbf{x}, \mathbf{p}) = \mathbf{A}\mathbf{x}$ with an $n \times n$ matrix $\mathbf{A} = (a_{ij})_{n\times n}$, the corresponding norm is defined by

$$||\mathbf{A}|| = \sum_{i,j=1}^{n} |a_{ij}|. \tag{2.6}$$

Definition 2.4 For a vector function $\mathbf{x}(t) = (x_1, x_2, \ldots, x_n)^{\mathrm{T}} \in \mathscr{R}^n$, the derivative and integral of $\mathbf{x}(t)$ are defined by

$$\begin{aligned} \frac{d\mathbf{x}(t)}{dt} &= \left(\frac{dx_1(t)}{dt}, \frac{dx_2(t)}{dt}, \ldots, \frac{dx_n(t)}{dt}\right)^{\mathrm{T}}, \\ \int \mathbf{x}(t)dt &= \left(\int x_1(t)\, dt, \int x_2(t)dt, \ldots, \int x_n(t)dt\right)^{\mathrm{T}}. \end{aligned} \tag{2.7}$$

For an $n \times n$ matrix $\mathbf{A} = (a_{ij})_{n\times n}$, the corresponding derivative and integral are defined by

$$\frac{d\mathbf{A}(t)}{dt} = \left(\frac{da_{ij}(t)}{dt}\right)_{n\times n} \text{ and } \int \mathbf{A}(t)dt = \left(\int a_{ij}(t)\, dt\right)_{n\times n}. \tag{2.8}$$

Definition 2.5 For $I \subseteq \mathscr{R}$, $\Omega \subseteq \mathscr{R}^n$, and $\Lambda \subseteq \mathscr{R}^m$, the vector function $\mathbf{f}(\mathbf{x}, t, \mathbf{p})$ with $\mathbf{f} : \Omega \times I \times \Lambda \to \mathscr{R}^n$ is differentiable at $\mathbf{x}_0 \in \Omega$ if

$$\left.\frac{\partial \mathbf{f}(\mathbf{x}, t, \mathbf{p})}{\partial \mathbf{x}}\right|_{(\mathbf{x}_0, t, \mathbf{p})} = \lim_{\Delta \mathbf{x} \to \mathbf{0}} \frac{\mathbf{f}(\mathbf{x}_0 + \Delta\mathbf{x}, t, \mathbf{p}) - \mathbf{f}(\mathbf{x}_0, t, \mathbf{p})}{\Delta \mathbf{x}}. \tag{2.9}$$

$\partial\mathbf{f}/\partial\mathbf{x}$ is called the spatial derivative of $\mathbf{f}(\mathbf{x}, t, \mathbf{p})$ at $\mathbf{x}_0$, and the derivative is given by the Jacobian matrix

$$\frac{\partial \mathbf{f}(\mathbf{x}, t, \mathbf{p})}{\partial \mathbf{x}} = (\partial f_i / \partial x_j)_{n\times n}. \tag{2.10}$$

Definition 2.6 For $I \subseteq \mathscr{R}$, $\Omega \subseteq \mathscr{R}^n$, and $\Lambda \subseteq \mathscr{R}^m$, consider a vector function $\mathbf{f}(\mathbf{x}, t, \mathbf{p})$ with $\mathbf{f} : \Omega \times I \times \Lambda \to \mathscr{R}^n$, $t \in I$, and $\mathbf{x} \in \Omega$ and $\mathbf{p} \in \Lambda$. The vector function $\mathbf{f}(\mathbf{x}, t, \mathbf{p})$ satisfies the Lipschitz condition with respect to $\mathbf{x}$ for $\Omega \times I \times \Lambda$,

$$||\mathbf{f}(\mathbf{x}_2, t, \mathbf{p}) - \mathbf{f}(\mathbf{x}_1, t, \mathbf{p})|| \leq L||\mathbf{x}_2 - \mathbf{x}_1|| \tag{2.11}$$

with $\mathbf{x}_1, \mathbf{x}_2 \in \Omega$ and L a constant. The constant L is called the Lipschitz constant.

Theorem 2.1 *Consider a dynamical system as*

$$\dot{\mathbf{x}} = \mathbf{f}(\mathbf{x}, t, \mathbf{p}) \text{ with } \mathbf{x}(t_0) = \mathbf{x}_0 \tag{2.12}$$

with $t_0, t \in I = [t_1, t_2]$, $\mathbf{x} \in \Omega = \{\mathbf{x} \,|\, ||\mathbf{x} - \mathbf{x}_0|| \leq d\}$ *and* $\mathbf{p} \in \Lambda$. *If the vector function* $\mathbf{f}(\mathbf{x}, t, \mathbf{p})$ *is* C^r*-continuous* ($r \geq 1$) *in* $G = \Omega \times I \times \Lambda$, *then the dynamical system in Equation (2.12) has one and only one solution* $\mathbf{\Phi}(\mathbf{x}_0, t - t_0, \mathbf{p})$ *for*

$$|t - t_0| \leq \min(t_2 - t_1, d/M) \text{ with } M = \max_{G} ||\mathbf{f}||. \tag{2.13}$$

Proof. The proof of this theorem can be referred to the book by Coddington and Levinson (1955). ∎

Theorem 2.2 *(Gronwall) Suppose there is a continuous real valued function $g(t) \geq 0$ to satisfy*

$$g(t) \leq \delta_1 \int_{t_0}^{t} g(\tau)d\tau + \delta_2 \tag{2.14}$$

for all $t \in [t_0, t_1]$ and δ_1 and δ_2 are positive constants. For $t \in [t_0, t_1]$, one obtains

$$g(t) \leq \delta_2 e^{\delta_1(t-t_0)}. \tag{2.15}$$

Proof. The proof can be referred to Luo (2012b). ∎

Theorem 2.3 *Consider a dynamical system as $\dot{\mathbf{x}} = \mathbf{f}(\mathbf{x}, t, \mathbf{p})$ with $\mathbf{x}(t_0) = \mathbf{x}_0$ in Equation (2.12) with $t_0, t \in I = [t_1, t_2]$, $\mathbf{x} \in \Omega = \{\mathbf{x} | \|\mathbf{x} - \mathbf{x}_0\| \leq d\}$ and $\mathbf{p} \in \Lambda$. The vector function $\mathbf{f}(\mathbf{x}, t, \mathbf{p})$ is C^r-continuous ($r \geq 1$) in $G = \Omega \times I \times \Lambda$. If the solution of $\dot{\mathbf{x}} = \mathbf{f}(\mathbf{x}, t, \mathbf{p})$ with $\mathbf{x}(t_0) = \mathbf{x}_0$ is $\mathbf{x}(t)$ on G and the solution of $\dot{\mathbf{y}} = \mathbf{f}(\mathbf{y}, t, \mathbf{p})$ with $\mathbf{y}(t_0) = \mathbf{y}_0$ is $\mathbf{y}(t)$ on G. For a given $\varepsilon > 0$, if $\|\mathbf{x}_0 - \mathbf{y}_0\| \leq \varepsilon$, then*

$$\|\mathbf{x}(t) - \mathbf{y}(t)\| \leq \varepsilon e^{L(t-t_0)} \, \text{on} \, I \times \Lambda \tag{2.16}$$

Proof. The proof can be referred to Luo (2012b). ∎

2.2 Equilibriums and Stability

Definition 2.7 Consider a metric space Ω and $\Omega_\alpha \subseteq \Omega$ ($\alpha = 1, 2, \ldots$).

1. A map $\mathbf{h}$ is called a homeomorphism of Ω_α onto Ω_β ($\alpha, \beta = 1, 2, \ldots$) if the map $\mathbf{h} : \Omega_\alpha \to \Omega_\beta$ is continuous and one-to-one, and $\mathbf{h}^{-1} : \Omega_\beta \to \Omega_\alpha$ is continuous.
2. Two set Ω_α and Ω_β are homeomorphic or topologically equivalent if there is a homeomorphism of Ω_α onto Ω_β.

Definition 2.8 A connected, metric space Ω with an open cover $\{\Omega_\alpha\}$ (i.e., $\Omega = \cup_\alpha \Omega_\alpha$) is called an n-dimensional, $C^r (r \geq 1)$ differentiable manifold if the following properties exist.

1. There is an open unit ball $B = \{\mathbf{x} \in \mathscr{R}^n | \|\mathbf{x}\| < 1\}$.
2. For all α, there is an homeomorphism $\mathbf{h}_\alpha : \Omega_\alpha \to B$.
3. If $\mathbf{h}_\alpha : \Omega_\alpha \to B$ and $\mathbf{h}_\beta : \Omega_\beta \to B$ are homeomorphisms for $\Omega_\alpha \cap \Omega_\beta \neq \emptyset$, then there is a C^r-differentiable map $\mathbf{h} = \mathbf{h}_\alpha \circ \mathbf{h}_\beta^{-1}$ for $\mathbf{h}_\alpha(\Omega_\alpha \cap \Omega_\beta) \subset \mathscr{R}^n$ and $\mathbf{h}_\beta(\Omega_\alpha \cap \Omega_\beta) \subset \mathscr{R}^n$ with

$$\mathbf{h} : \mathbf{h}_\beta(\Omega_\alpha \cap \Omega_\beta) \to \mathbf{h}_\alpha(\Omega_\alpha \cap \Omega_\beta), \tag{2.17}$$

and for all $\mathbf{x} \in \mathbf{h}_\beta(\Omega_\alpha \cap \Omega_\beta)$, the Jacobian determinant $\det D\mathbf{h}(\mathbf{x}) \neq 0$.

The manifold Ω is said to be analytic if the maps $\mathbf{h} = \mathbf{h}_\alpha \circ \mathbf{h}_\beta^{-1}$ are analytic.

Definition 2.9 Consider an autonomous, nonlinear dynamical system $\dot{\mathbf{x}} = \mathbf{f}(\mathbf{x}, \mathbf{p})$ in Equation (2.4). A point $\mathbf{x}^* \in \Omega$ is called an equilibrium point or critical point of a nonlinear system $\dot{\mathbf{x}} = \mathbf{f}(\mathbf{x}, \mathbf{p})$ if

$$\mathbf{f}(\mathbf{x}^*, \mathbf{p}) = \mathbf{0}. \tag{2.18}$$

The linearized system of the nonlinear system $\dot{\mathbf{x}} = \mathbf{f}(\mathbf{x}, \mathbf{p})$ in Equation (2.4) at the equilibrium point $\mathbf{x}^*$ is given by

$$\dot{\mathbf{y}} = D\mathbf{f}(\mathbf{x}^*, \mathbf{p})\mathbf{y} \text{ where } \mathbf{y} = \mathbf{x} - \mathbf{x}^*. \tag{2.19}$$

Definition 2.10 Consider an n-dimensional, autonomous, nonlinear dynamical system $\dot{\mathbf{x}} = \mathbf{f}(\mathbf{x}, \mathbf{p})$ in Equation (2.4) with an equilibrium point $\mathbf{x}^*$. The linearized system of the nonlinear system at the equilibrium point $\mathbf{x}^*$ is $\dot{\mathbf{y}} = D\mathbf{f}(\mathbf{x}^*, \mathbf{p})\mathbf{y}$ ($\mathbf{y} = \mathbf{x} - \mathbf{x}^*$) in Equation (2.19). The matrix $D\mathbf{f}(\mathbf{x}^*, \mathbf{p})$ possesses n eigenvalues λ_k ($k = 1, 2, \ldots n$). Set $N = \{1, 2, \ldots, n\}$, $N_i = \{i_1, i_2, \ldots, i_{n_i}\} \cup \emptyset$ with $i_j \in N$ ($j = 1, 2, \ldots, n_i$; $i = 1, 2, 3$) and $\Sigma_{i=1}^3 n_i = n$. $\cup_{i=1}^3 N_i = N$ and $N_i \cap N_l = \emptyset$ ($l \neq i$). $N_i = \emptyset$ if $n_i = 0$. The corresponding vectors for the negative, positive, and zero eigenvalues of $D\mathbf{f}(\mathbf{x}^*, \mathbf{p})$ are $\{\mathbf{u}_k\}$ ($k \in N_i$, $i = 1, 2, 3$), respectively. The stable, unstable, and invariant subspaces of the linearized nonlinear system in Equation (2.19) are defined as

$$\begin{aligned}
\mathscr{E}^{\mathrm{s}} &= span\{\mathbf{u}_k | (D\mathbf{f}(\mathbf{x}^*, \mathbf{p}) - \lambda_k \mathbf{I})\mathbf{u}_k = \mathbf{0}, \lambda_k < 0, k \in N_1 \subseteq N \cup \emptyset\};\\
\mathscr{E}^{\mathrm{u}} &= span\{\mathbf{u}_k | (D\mathbf{f}(\mathbf{x}^*, \mathbf{p}) - \lambda_k \mathbf{I})\mathbf{u}_k = \mathbf{0}, \lambda_k > 0, k \in N_2 \subseteq N \cup \emptyset\};\\
\mathscr{E}^{\mathrm{i}} &= span\{\mathbf{u}_k | (D\mathbf{f}(\mathbf{x}^*, \mathbf{p}) - \lambda_k \mathbf{I})\mathbf{u}_k = \mathbf{0}, \lambda_k = 0, k \in N_3 \subseteq N \cup \emptyset\}.
\end{aligned} \tag{2.20}$$

Definition 2.11 Consider a $2n$-dimensional, autonomous dynamical system $\dot{\mathbf{x}} = \mathbf{f}(\mathbf{x}, \mathbf{p})$ in Equation (2.4) with an equilibrium point $\mathbf{x}^*$. The linearized system of the nonlinear system at the equilibrium point $\mathbf{x}^*$ is $\dot{\mathbf{y}} = D\mathbf{f}(\mathbf{x}^*, \mathbf{p})\mathbf{y}$ ($\mathbf{y} = \mathbf{x} - \mathbf{x}^*$) in Equation (2.19). The matrix $D\mathbf{f}(\mathbf{x}^*, \mathbf{p})$ has complex eigenvalues $\alpha_k \pm \mathbf{i}\beta_k$ with eigenvectors $\mathbf{u}_k \pm \mathbf{i}\mathbf{v}_k$ ($k \in \{1, 2, \ldots, n\}$) and the base of vector is

$$\mathbf{B} = \{\mathbf{u}_1, \mathbf{v}_1, \ldots, \mathbf{u}_k, \mathbf{v}_k, \ldots, \mathbf{u}_n, \mathbf{v}_n\}. \tag{2.21}$$

The stable, unstable, center subspaces of Equation (2.19) are linear subspaces spanned by $\{\mathbf{u}_k, \mathbf{v}_k\}$ ($k \in N_i$, $i = 1, 2, 3$), respectively. $N_i = \{i_1, i_2, \ldots, i_{n_i}\} \cup \emptyset \subseteq N \cup \emptyset$ and $N = \{1, 2, \ldots, n\}$ with $i_j \in N$ ($j = 1, 2, \ldots, n_i$) and $\Sigma_{i=1}^3 n_i = n$. $\cup_{i=1}^3 N_i = N$ and $N_i \cap N_l = \emptyset$ ($l \neq i$). $N_i = \emptyset$ if $n_i = 0$. The stable, unstable, center subspaces of the linearized nonlinear system in Equation (2.19) are defined as

$$\begin{aligned}
\mathscr{E}^{\mathrm{s}} &= span\left\{ (\mathbf{u}_k, \mathbf{v}_k) \left| \begin{array}{l} \alpha_k < 0, \beta_k \neq 0, \\ (D\mathbf{f}(\mathbf{x}^*, \mathbf{p}) - (\alpha_k \pm \mathbf{i}\beta_k)\mathbf{I})(\mathbf{u}_k \pm \mathbf{i}\mathbf{v}_k) = \mathbf{0}, \\ k \in N_1 \subseteq \{1, 2, \ldots, n\} \cup \emptyset \end{array} \right. \right\};\\
\mathscr{E}^{\mathrm{u}} &= span\left\{ (\mathbf{u}_k, \mathbf{v}_k) \left| \begin{array}{l} \alpha_k > 0, \beta_k \neq 0, \\ (D\mathbf{f}(\mathbf{x}^*, \mathbf{p}) - (\alpha_k \pm \mathbf{i}\beta_k)\mathbf{I})(\mathbf{u}_k \pm \mathbf{i}\mathbf{v}_k) = \mathbf{0}, \\ k \in N_2 \subseteq \{1, 2, \ldots, n\} \cup \emptyset \end{array} \right. \right\};\\
\mathscr{E}^{\mathrm{c}} &= span\left\{ (\mathbf{u}_k, \mathbf{v}_k) \left| \begin{array}{l} \alpha_k = 0, \beta_k \neq 0, \\ (D\mathbf{f}(\mathbf{x}^*, \mathbf{p}) - (\alpha_k \pm \mathbf{i}\beta_k)\mathbf{I})(\mathbf{u}_k \pm \mathbf{i}\mathbf{v}_k) = \mathbf{0}, \\ k \in N_3 \subseteq \{1, 2, \ldots, n\} \cup \emptyset \end{array} \right. \right\}.
\end{aligned} \tag{2.22}$$

Theorem 2.4 *Consider an* n*-dimensional, autonomous, nonlinear dynamical system* $\dot{\mathbf{x}} = \mathbf{f}(\mathbf{x}, \mathbf{p})$ *in Equation (2.4) with an equilibrium point* $\mathbf{x}^*$. *The linearized system of the* *nonlinear system at the equilibrium point* $\mathbf{x}^*$ *is* $\dot{\mathbf{y}} = D\mathbf{f}(\mathbf{x}^*, \mathbf{p})\mathbf{y}$ $(\mathbf{y} = \mathbf{x} - \mathbf{x}^*)$ *in Equation (2.19). The eigenspace of* $D\mathbf{f}(\mathbf{x}^*, \mathbf{p})$ *(i.e.,* $\mathscr{E} \subseteq \mathscr{R}^n$*) in the linearized dynamical system is expressed by the direct sum of three subspaces*

$$\mathscr{E} = \mathscr{E}^s \oplus \mathscr{E}^u \oplus \mathscr{E}^c \tag{2.23}$$

where $\mathscr{E}^s$, $\mathscr{E}^u$ and $\mathscr{E}^c$ *are the stable, unstable, and center spaces* $\mathscr{E}^s$, $\mathscr{E}^u$, *and* $\mathscr{E}^c$ *respectively.*

Proof. This proof can be referred to Luo (2012b). ■

Definition 2.12 Consider an n-dimensional, autonomous, nonlinear dynamical system $\dot{\mathbf{x}} = \mathbf{f}(\mathbf{x}, \mathbf{p})$ in Equation (2.4) with an equilibrium point $\mathbf{x}^*$ and $\mathbf{f}(\mathbf{x}, \mathbf{p})$ is C^r $(r \geq 1)$-continuous in a neighborhood of the equilibrium $\mathbf{x}^*$. The corresponding solution is $\mathbf{x}(t) = \mathbf{\Phi}(\mathbf{x}_0, t - t_0, \mathbf{p}) = \mathbf{\Phi}_t(\mathbf{x}_0)$. The linearized system of the nonlinear system at the equilibrium point $\mathbf{x}^*$ is $\dot{\mathbf{y}} = D\mathbf{f}(\mathbf{x}^*, \mathbf{p})\mathbf{y}$ $(\mathbf{y} = \mathbf{x} - \mathbf{x}^*)$ in Equation (2.19). Suppose there is a neighborhood of the equilibrium $\mathbf{x}^*$ as $U(\mathbf{x}^*) \subset \Omega$, and in the neighborhood

$$\lim_{||\mathbf{y}|| \to 0} \frac{||\mathbf{f}(\mathbf{x}^* + \mathbf{y}, \mathbf{p}) - D\mathbf{f}(\mathbf{x}^*, \mathbf{p})\mathbf{y}||}{||\mathbf{y}||} = 0. \tag{2.24}$$

1. A C^r invariant manifold

$$\mathscr{S}_{loc}(\mathbf{x}, \mathbf{x}^*) = \left\{ \mathbf{x} \in U\left(\mathbf{x}^*\right) \mid \lim_{t \to \infty} \mathbf{x}(t) = \mathbf{x}^*, \mathbf{x}(t) \in U(\mathbf{x}^*) \text{ for all } t \geq 0 \right\} \tag{2.25}$$

   is called the local stable manifold of $\mathbf{x}^*$, and the corresponding global, stable manifold is defined as

$$\mathscr{S}(\mathbf{x}, \mathbf{x}^*) = \cup_{t \leq 0} \mathbf{\Phi}_t(\mathscr{S}_{loc}(\mathbf{x}, \mathbf{x}^*)). \tag{2.26}$$

2. A C^r invariant manifold

$$\mathscr{U}_{loc}(\mathbf{x}, \mathbf{x}^*) = \left\{ \mathbf{x} \in U\left(\mathbf{x}^*\right) \mid \lim_{t \to -\infty} \mathbf{x}(t) = \mathbf{x}^*, \mathbf{x}(t) \in U(\mathbf{x}^*) \text{ for all } t \leq 0 \right\} \tag{2.27}$$

   is called the local unstable manifold of $\mathbf{x}^*$, and the corresponding global, unstable manifold is defined as

$$\mathscr{U}(\mathbf{x}, \mathbf{x}^*) = \cup_{t \geq 0} \mathbf{\Phi}_t(\mathscr{U}_{loc}(\mathbf{x}, \mathbf{x}^*)). \tag{2.28}$$

3. A C^{r-1} invariant manifold $\mathscr{C}_{loc}(\mathbf{x}, \mathbf{x}^*)$ is called the center manifold of $\mathbf{x}^*$ if $\mathscr{C}_{loc}(\mathbf{x}, \mathbf{x}^*)$ possesses the same dimension of $\mathscr{E}^c$ for $\mathbf{x}^* \in \mathscr{S}(\mathbf{x}, \mathbf{x}^*)$, and the tangential space of $\mathscr{C}_{loc}(\mathbf{x}, \mathbf{x}^*)$ is identical to $\mathscr{E}^c$.

The stable and unstable manifolds are unique, but the center manifold is not unique. If the nonlinear vector field $\mathbf{f}$ is C^∞-continuous, then a C^r center manifold can be found for any $r < \infty$.

Theorem 2.5 *Consider an* n*-dimensional, autonomous, nonlinear dynamical system* $\dot{\mathbf{x}} = \mathbf{f}(\mathbf{x}, \mathbf{p})$ *in Equation (2.4) with a hyperbolic equilibrium point* $\mathbf{x}^*$ *and* $\mathbf{f}(\mathbf{x}, \mathbf{p})$ *is* C^r *$(r \geq 1)$-continuous in the neighborhood of the equilibrium* $\mathbf{x}^*$*. The corresponding solution is* $\mathbf{x}(t) = \mathbf{\Phi}(\mathbf{x}_0, t - t_0, \mathbf{p}) = \mathbf{\Phi}_t(\mathbf{x}_0)$*. The linearized system of the nonlinear system at the equilibrium point* $\mathbf{x}^*$ *is* $\dot{\mathbf{y}} = D\mathbf{f}(\mathbf{x}^*, \mathbf{p})\mathbf{y}$ $(\mathbf{y} = \mathbf{x} - \mathbf{x}^*)$ *in Equation (2.19). Suppose there is a* *neighborhood of the hyperbolic equilibrium* $\mathbf{x}^*$ *as* $U(\mathbf{x}^*) \subset \Omega$. *If the homeomorphism between the local invariant subspace* $E(\mathbf{x}, \mathbf{x}^*) \subset U(\mathbf{x}^*)$ *under the flow* $\mathbf{\Phi}(\mathbf{x}_0, t - t_0, \mathbf{p})$ *of* $\dot{\mathbf{x}} = \mathbf{f}(\mathbf{x}, \mathbf{p})$ *in Equation (2.4) and the eigenspace* $\mathscr{E}$ *of the linearized system exists with the condition in Equation (2.24), the local invariant subspace is decomposed by*

$$E(\mathbf{x}, \mathbf{x}^*) = \mathscr{S}_{loc}(\mathbf{x}, \mathbf{x}^*) \oplus \mathscr{U}_{loc}(\mathbf{x}, \mathbf{x}^*). \tag{2.29}$$

1. *The local stable invariant manifold* $\mathscr{S}_{loc}(\mathbf{x}, \mathbf{x}^*)$ *possesses the following properties:*
   a. *for* $\mathbf{x}^* \in \mathscr{S}_{loc}(\mathbf{x}, \mathbf{x}^*)$, $\mathscr{S}_{loc}(\mathbf{x}, \mathbf{x}^*)$ *possesses the same dimension of* $\mathscr{E}^s$ *and the tangential space of* $\mathscr{S}_{loc}(\mathbf{x}, \mathbf{x}^*)$ *is identical to* $\mathscr{E}^s$;
   b. *for* $\mathbf{x}_0 \in \mathscr{S}_{loc}(\mathbf{x}, \mathbf{x}^*)$, $\mathbf{x}(t) \in \mathscr{S}_{loc}(\mathbf{x}, \mathbf{x}^*)$ *for all time* $t \geq t_0$ *and* $\lim_{t\to\infty} \mathbf{x}(t) = \mathbf{x}^*$;
   c. *for* $\mathbf{x}_0 \notin \mathscr{S}_{loc}(\mathbf{x}, \mathbf{x}^*)$, $\|\mathbf{x} - \mathbf{x}^*\| \geq \delta$ *for* $\delta > 0$ *with* $t \geq t_1 \geq t_0$.
2. *The local unstable invariant manifold* $\mathscr{U}_{loc}(\mathbf{x}, \mathbf{x}^*)$ *possesses the following properties:*
   a. *for* $\mathbf{x}^* \in \mathscr{U}_{loc}(\mathbf{x}, \mathbf{x}^*)$, $\mathscr{U}_{loc}(\mathbf{x}, \mathbf{x}^*)$ *possesses the same dimension of* $\mathscr{E}^u$ *and the tangential space of* $\mathscr{U}_{loc}(\mathbf{x}, \mathbf{x}^*)$ *is identical to* $\mathscr{E}^u$;
   b. *for* $\mathbf{x}_0 \in \mathscr{U}_{loc}(\mathbf{x}, \mathbf{x}^*)$, $\mathbf{x}(t) \in \mathscr{U}_{loc}(\mathbf{x}, \mathbf{x}^*)$ *for all time* $t \leq t_0$ *and* $\lim_{t\to-\infty} \mathbf{x}(t) = \mathbf{x}^*$;
   c. *for* $\mathbf{x}_0 \notin \mathscr{U}_{loc}(\mathbf{x}, \mathbf{x}^*)$, $\|\mathbf{x} - \mathbf{x}^*\| \geq \delta$ *for* $\delta > 0$ *with* $t \leq t_1 \leq t_0$.

Proof. The proof for stable and unstable manifold can be found in Hartman (1964). The proof for center manifold can be referenced to Marsden and McCracken (1976) or Carr (1981). ■

Theorem 2.6 *Consider an* n*-dimensional, autonomous, nonlinear dynamical system* $\dot{\mathbf{x}} = \mathbf{f}(\mathbf{x}, \mathbf{p})$ *in Equation (2.4) with an equilibrium point* $\mathbf{x}^*$. *Suppose there is a neighborhood of the equilibrium* $\mathbf{x}^*$ *as* $U(\mathbf{x}^*) \subset \Omega$, *then* $\mathbf{f}(\mathbf{x}, \mathbf{p})$ *is* C^r $(r \geq 1)$*-continuous in a neighborhood of the equilibrium* $\mathbf{x}^*$. *The corresponding solution is* $\mathbf{x}(t) = \mathbf{\Phi}(\mathbf{x}_0, t - t_0, \mathbf{p})$. *The linearized system of the nonlinear system at the equilibrium point* $\mathbf{x}^*$ *is* $\dot{\mathbf{y}} = D\mathbf{f}(\mathbf{x}^*, \mathbf{p})\mathbf{y}$ $(\mathbf{y} = \mathbf{x} - \mathbf{x}^*)$ *in Equation (2.19). If the homeomorphism between the local invariant subspace* $E(\mathbf{x}, \mathbf{x}^*) \subset U(\mathbf{x}^*)$ *under the flow* $\mathbf{\Phi}(\mathbf{x}_0, t - t_0, \mathbf{p})$ *of* $\dot{\mathbf{x}} = \mathbf{f}(\mathbf{x}, \mathbf{p})$ *in Equation (2.4) and the eigenspace* $\mathscr{E}$ *of the linearized system exists with the condition in Equation (2.24), in addition to the local stable and unstable invariant manifolds, there is a* C^{r-1} *center manifold* $\mathscr{C}_{loc}(\mathbf{x}, \mathbf{x}^*)$. *The center manifold possesses the same dimension of* $\mathscr{E}^c$ *for* $\mathbf{x}^* \in \mathscr{C}_{loc}(\mathbf{x}, \mathbf{x}^*)$, *and the tangential space of* $\mathscr{C}_{loc}(\mathbf{x}, \mathbf{x}^*)$ *is identical to* $\mathscr{E}^c$. *Thus, the local invariant subspace is decomposed by*

$$E(\mathbf{x}, \mathbf{x}^*) = \mathscr{S}_{loc}(\mathbf{x}, \mathbf{x}^*) \oplus \mathscr{U}_{loc}(\mathbf{x}, \mathbf{x}^*) \oplus \mathscr{C}_{loc}(\mathbf{x}, \mathbf{x}^*). \tag{2.30}$$

Proof. The proof for stable and unstable manifold can be referenced to Hartman (1964). The proof for center manifold can be referenced to Marsden and McCracken (1976) or Carr (1981). ■

Definition 2.13 Consider an n-dimensional, autonomous, nonlinear dynamical system $\dot{\mathbf{x}} = \mathbf{f}(\mathbf{x}, \mathbf{p})$ in Equation (2.4) with an equilibrium point $\mathbf{x}^*$ and $\mathbf{f}(\mathbf{x}, \mathbf{p})$ is C^r $(r \geq 1)$-continuous in a neighborhood of the equilibrium $\mathbf{x}^*$.

1. The equilibrium $\mathbf{x}^*$ is stable if all $\varepsilon > 0$, there is a $\delta > 0$ such that for all $\mathbf{x}_0 \in U_\delta(\mathbf{x}^*)$ where $U_\delta(\mathbf{x}^*) = \{\mathbf{x} \mid \|\mathbf{x} - \mathbf{x}^*\| < \delta\}$ and $t \geq 0$,

$$\mathbf{\Phi}(\mathbf{x}_0, t - t_0, \mathbf{p}) \in U_\varepsilon(\mathbf{x}^*). \tag{2.31}$$

2. The equilibrium $\mathbf{x}^*$ is unstable if it is not stable or if all $\varepsilon > 0$, there is a $\delta > 0$ such that for all $\mathbf{x}_0 \in U_\delta(\mathbf{x}^*)$ where $U_\delta(\mathbf{x}^*) = \{\mathbf{x} | \|\mathbf{x} - \mathbf{x}^*\| < \delta\}$ and $t \geq t_1 > 0$,

$$\mathbf{\Phi}(\mathbf{x}_0, t - t_0, \mathbf{p}) \notin U_\varepsilon(\mathbf{x}^*). \tag{2.32}$$

3. The equilibrium $\mathbf{x}^*$ is asymptotically stable if all $\varepsilon > 0$, there is a $\delta > 0$ such that for all $\mathbf{x}_0 \in U_\delta(\mathbf{x}^*)$ where $U_\delta(\mathbf{x}^*) = \{\mathbf{x} | \|\mathbf{x} - \mathbf{x}^*\| < \delta\}$ and $t \geq 0$,

$$\lim_{t\to\infty} \mathbf{\Phi}(\mathbf{x}_0, t - t_0, \mathbf{p}) = \mathbf{x}^*. \tag{2.33}$$

4. The equilibrium $\mathbf{x}^*$ is asymptotically unstable if all $\varepsilon > 0$, there is a $\delta > 0$ such that for all $\mathbf{x}_0 \in U_\delta(\mathbf{x}^*)$ where $U_\delta(\mathbf{x}^*) = \{\mathbf{x} | \|\mathbf{x} - \mathbf{x}^*\| < \delta\}$ and $t \leq 0$,

$$\lim_{t\to-\infty} \mathbf{\Phi}(\mathbf{x}_0, t - t_0, \mathbf{p}) = \mathbf{x}^*. \tag{2.34}$$

Definition 2.14 Consider an n-dimensional, autonomous, nonlinear dynamical system $\dot{\mathbf{x}} = \mathbf{f}(\mathbf{x}, \mathbf{p})$ in Equation (2.4) with an equilibrium point $\mathbf{x}^*$. Suppose there is a neighborhood of the equilibrium $\mathbf{x}^*$ as $U(\mathbf{x}^*) \subset \Omega$, then $\mathbf{f}(\mathbf{x}, \mathbf{p})$ is C^r $(r \geq 1)$ continuous and Equation (2.24) holds in $U(\mathbf{x}^*) \subset \Omega$. The corresponding solution is $\mathbf{x}(t) = \mathbf{\Phi}(\mathbf{x}_0, t - t_0, \mathbf{p})$. For a linearized dynamical system in Equation (2.19), consider a real eigenvalue λ_k of matrix $D\mathbf{f}(\mathbf{x}^*, \mathbf{p})$ $(k \in N = \{1, 2, \ldots, n\})$ with an eigenvector $\mathbf{v}_k$. For $\mathbf{y}^{(k)} = c^{(k)}\mathbf{v}_k$, $\dot{\mathbf{y}}^{(k)} = \dot{c}^{(k)}\mathbf{v}_k = \lambda_k c^{(k)}\mathbf{v}_k$, thus $\dot{c}^{(k)} = \lambda_k c^{(k)}$.

1. $\mathbf{x}^{(k)}$ at the equilibrium $\mathbf{x}^*$ on the direction $\mathbf{v}_k$ is stable if

$$\lim_{t\to\infty} c^{(k)} = \lim_{t\to\infty} c_0^{(k)} e^{\lambda_k t} = 0 \text{ for } \lambda_k < 0. \tag{2.35}$$

2. $\mathbf{x}^{(k)}$ at the equilibrium $\mathbf{x}^*$ on the direction $\mathbf{v}_k$ is unstable if

$$\lim_{t\to\infty} |c^{(k)}| = \lim_{t\to\infty} |c_0^{(k)} e^{\lambda_k t}| = \infty \text{ for } \lambda_k > 0. \tag{2.36}$$

3. $\mathbf{x}^{(i)}$ at the equilibrium $\mathbf{x}^*$ on the direction $\mathbf{v}_k$ is uncertain (critical) if

$$\lim_{t\to\infty} c^{(k)} = \lim_{t\to\infty} e^{\lambda_k t} c_0^{(k)} = c_0^{(k)} \text{ for } \lambda_k = 0. \tag{2.37}$$

Definition 2.15 Consider a $2n$-dimensional, autonomous, nonlinear dynamical system $\dot{\mathbf{x}} = \mathbf{f}(\mathbf{x}, \mathbf{p})$ in Equation (2.4) with an equilibrium point $\mathbf{x}^*$. Suppose there is a neighborhood of the equilibrium $\mathbf{x}^*$ as $U(\mathbf{x}^*) \subset \Omega$, then $\mathbf{f}(\mathbf{x}, \mathbf{p})$ is $C^r(r \geq 1)$ continuous and Equation (2.24) holds in $U(\mathbf{x}^*) \subset \Omega$. The corresponding solution in $\mathbf{x}(t) = \mathbf{\Phi}(\mathbf{x}_0, t - t_0, \mathbf{p})$. For a linearized dynamical system in Equation (2.19), consider a pair of complex eigenvalue $\alpha_k \pm \mathbf{i}\beta_k$ $(k \in N = \{1, 2, \ldots, n\}, \mathbf{i} = \sqrt{-1})$ of matrix $D\mathbf{f}(\mathbf{x}^*, \mathbf{p})$ with a pair of eigenvectors $\mathbf{u}_k \pm \mathbf{i}\mathbf{v}_k$. On the invariant plane of $(\mathbf{u}_k, \mathbf{v}_k)$, consider $\mathbf{y}^{(k)} = \mathbf{y}_+^{(k)} + \mathbf{y}_-^{(k)}$ with

$$\mathbf{y}^{(k)} = c^{(k)}\mathbf{u}_k + d^{(k)}\mathbf{v}_k, \dot{\mathbf{y}}^{(k)} = \dot{c}^{(k)}\mathbf{u}_k + \dot{d}^{(k)}\mathbf{v}_k. \tag{2.38}$$

Thus, $\mathbf{c}^{(k)} = (c^{(k)}, d^{(k)})^{\mathrm{T}}$ with

$$\dot{\mathbf{c}}^{(k)} = \mathbf{E}_k \mathbf{c}^{(k)} \Rightarrow \mathbf{c}^{(k)} = e^{\alpha_k t}\mathbf{B}_k \mathbf{c}_0^{(k)} \tag{2.39}$$

where

$$\mathbf{E}_k = \begin{bmatrix} \alpha_k & \beta_k \\ -\beta_k & \alpha_k \end{bmatrix} \text{ and } \mathbf{B}_k = \begin{bmatrix} \cos\beta_k t & \sin\beta_k t \\ -\sin\beta_k t & \cos\beta_k t \end{bmatrix}. \tag{2.40}$$

1. $\mathbf{x}^{(k)}$ at the equilibrium $\mathbf{x}^*$ on the plane of $(\mathbf{u}_k, \mathbf{v}_k)$ is spirally stable if

$$\lim_{t\to\infty} \|\mathbf{c}^{(k)}\| = \lim_{t\to\infty} e^{\alpha_k t}\|\mathbf{B}_k\| \times \|\mathbf{c}_0^{(k)}\| = 0 \text{ for } \mathrm{Re}\lambda_k = \alpha_k < 0. \tag{2.41}$$

2. $\mathbf{x}^{(k)}$ at the equilibrium $\mathbf{x}^*$ on the plane of $(\mathbf{u}_k, \mathbf{v}_k)$ is spirally unstable if

$$\lim_{t\to\infty} \|\mathbf{c}^{(k)}\| = \lim_{t\to\infty} e^{\alpha_k t}\|\mathbf{B}_k\| \times \|\mathbf{c}_0^{(k)}\| = \infty \text{ for } \mathrm{Re}\lambda_k = \alpha_k > 0. \tag{2.42}$$

3. $\mathbf{x}^{(k)}$ at the equilibrium $\mathbf{x}^*$ on the plane of $(\mathbf{u}_k, \mathbf{v}_k)$ is on the invariant circle if

$$\lim_{t\to\infty} \|\mathbf{c}^{(k)}\| = \lim_{t\to\infty} e^{\alpha_k t}\|\mathbf{B}_k\| \times \|\mathbf{c}_0^{(k)}\| = \|\mathbf{c}_0^{(k)}\| \text{ for } \mathrm{Re}\lambda_k = \alpha_k = 0. \tag{2.43}$$

4. $\mathbf{x}^{(k)}$ at the equilibrium $\mathbf{x}^*$ on the plane of $(\mathbf{u}_k, \mathbf{v}_k)$ is degenerate in the direction of $\mathbf{u}_k$ if $\mathrm{Im}\lambda_k = 0$.

Definition 2.16 Consider an n-dimensional, autonomous, nonlinear dynamical system $\dot{\mathbf{x}} = \mathbf{f}(\mathbf{x}, \mathbf{p})$ in Equation (2.4) with an equilibrium point $\mathbf{x}^*$. Suppose there is a neighborhood of the equilibrium $\mathbf{x}^*$ as $U(\mathbf{x}^*) \subset \Omega$, and in the neighborhood, $\mathbf{f}(\mathbf{x}, \mathbf{p})$ is C^r $(r \geq 1)$-continuous and Equation (2.24) holds. The corresponding solution is $\mathbf{x}(t) = \mathbf{\Phi}(\mathbf{x}_0, t - t_0, \mathbf{p})$. The linearized system of the nonlinear system at the equilibrium point $\mathbf{x}^*$ is $\dot{\mathbf{y}} = D\mathbf{f}(\mathbf{x}^*, \mathbf{p})\mathbf{y}$ $(\mathbf{y} = \mathbf{x} - \mathbf{x}^*)$ in Equation (2.19).

1. The equilibrium $\mathbf{x}^*$ is said *a hyperbolic equilibrium* if none of eigenvalues of $D\mathbf{f}(\mathbf{x}^*, \mathbf{p})$ is zero real part (i.e., $\mathrm{Re}\lambda_k \neq 0$ $(k = 1, 2, \ldots, n)$).
2. The equilibrium $\mathbf{x}^*$ is said *a sink* if all of eigenvalues of $D\mathbf{f}(\mathbf{x}^*, \mathbf{p})$ have negative real parts (i.e., $\mathrm{Re}\lambda_k < 0$ $(k = 1, 2, \ldots, n)$).
3. The equilibrium $\mathbf{x}^*$ is said *a source* if all of eigenvalues of $D\mathbf{f}(\mathbf{x}^*, \mathbf{p})$ have positive real parts (i.e., $\mathrm{Re}\lambda_k > 0$ $(k = 1, 2, \ldots, n)$).
4. The equilibrium $\mathbf{x}^*$ is said *a saddle* if it is a hyperbolic equilibrium and $D\mathbf{f}(\mathbf{x}^*, \mathbf{p})$ have at least one eigenvalue with a positive real part (i.e., $\mathrm{Re}\lambda_j > 0$ $(j \in \{1, 2, \ldots, n\})$ and one with a negative real part (i.e., $\mathrm{Re}\lambda_k < 0$ $(k \in \{1, 2, \ldots, n\})$.
5. The equilibrium $\mathbf{x}^*$ is called a center if all of eigenvalues of $D\mathbf{f}(\mathbf{x}^*, \mathbf{p})$ have zero real parts (i.e., $\mathrm{Re}\lambda_j = 0$ $(j = 1, 2, \ldots, n)$) with distinct eigenvalues.
6. The equilibrium $\mathbf{x}^*$ is called a stable node if all of eigenvalues of $D\mathbf{f}(\mathbf{x}^*, \mathbf{p})$ are real $\lambda_k < 0$ $(k = 1, 2, \ldots, n)$.
7. The equilibrium $\mathbf{x}^*$ is called an unstable node if all of eigenvalues of $D\mathbf{f}(\mathbf{x}^*, \mathbf{p})$ are real $\lambda_k > 0$ $(k = 1, 2, \ldots, n)$.
8. The equilibrium $\mathbf{x}^*$ is called a degenerate case if all of eigenvalues of $D\mathbf{f}(\mathbf{x}^*, \mathbf{p})$ are zero $\lambda_k = 0$ $(k = 1, 2, \ldots, n)$.

As in Luo (2012a), the generalized stability and bifurcation of flows in linearized, nonlinear dynamical systems in Equation (2.4) will be discussed as follows.

Definition 2.17 Consider an n-dimensional, autonomous, nonlinear dynamical system $\dot{\mathbf{x}} = \mathbf{f}(\mathbf{x}, \mathbf{p})$ in Equation (2.4) with an equilibrium point $\mathbf{x}^*$. Suppose there is a neighborhood of the equilibrium $\mathbf{x}^*$ as $U(\mathbf{x}^*) \subset \Omega$, and in the neighborhood $\mathbf{f}(\mathbf{x}, \mathbf{p})$ is C^r $(r \geq 1)$-continuous and Equation (2.24) holds. The corresponding solution is $\mathbf{x}(t) = \mathbf{\Phi}(\mathbf{x}_0, t - t_0, \mathbf{p})$.

The matrix $D\mathbf{f}(\mathbf{x}^*, \mathbf{p})$ in Equation (2.19) possesses n eigenvalues λ_k $(k = 1, 2, \ldots, n)$. Set $N = \{1, 2, \ldots, m, m+1, \ldots, (n-m)/2\}$, $N_i = \{i_1, i_2, \ldots, i_{n_i}\} \cup \emptyset$ with $i_j \in N$ $(j = 1, 2, \ldots, n_i;$ $i = 1, 2, \ldots, 6)$, $\Sigma_{i=1}^{3} n_i = m$, and $2\Sigma_{i=4}^{6} n_i = n - m$. $\cup_{i=1}^{6} N_i = N$ with $N_i \cap N_l = \emptyset (l \neq i)$. $N_i = \emptyset$ if $n_i = 0$. The matrix $D\mathbf{f}(\mathbf{x}^*, \mathbf{p})$ possesses n_1-stable, n_2-unstable, and n_3-invariant real eigenvectors plus n_4-stable, n_5-unstable, and n_6-center pairs of complex eigenvectors. Without repeated complex eigenvalues of $\mathrm{Re}\lambda_k = 0\,(k \in N_3 \cup N_6)$, the flow $\mathbf{\Phi}(t)$ of the nonlinear system $\dot{\mathbf{x}} = \mathbf{f}(\mathbf{x}, \mathbf{p})$ is an $(n_1 : n_2 : [n_3; m_3] | n_4 : n_5 : n_6)$ flow in the neighborhood of $\mathbf{x}^*$. However, with repeated complex eigenvalues of $\mathrm{Re}\lambda_k = 0$ $(k \in N_3 \cup N_6)$, the flow $\mathbf{\Phi}(t)$ of the nonlinear system $\dot{\mathbf{x}} = \mathbf{f}(\mathbf{x}, \mathbf{p})$ is an $(n_1 : n_2 : [n_3; m_3] | n_4 : n_5 : [n_6, l; \mathbf{m}_6])$ flow in the neighborhood of $\mathbf{x}^*$. The meanings of notations in the aforementioned structures are defined as follows:

1. n_1 represents exponential sinks on n_1-directions of $\mathbf{v}_k$ if $\lambda_k < 0$ $(k \in N_1$ and $1 \leq n_1 \leq n)$ with distinct or repeated eigenvalues.
2. n_2 represents exponential sources on n_2-directions of $\mathbf{v}_k$ if $\lambda_k > 0$ $(k \in N_2$ and $1 \leq n_2 \leq n)$ with distinct or repeated eigenvalues.
3. $n_3 = 1$ represents an invariant center on 1-direction of $\mathbf{v}_k$ if $\lambda_k = 0$ $(k \in N_3$ and $n_3 = 1)$.
4. n_4 represents spiral sinks on n_4-pairs of $(\mathbf{u}_k, \mathbf{v}_k)$ if $\mathrm{Re}\lambda_k < 0$ and $\mathrm{Im}\lambda_k \neq 0$ $(k \in N_4$ and $1 \leq n_4 \leq n)$ with distinct or repeated eigenvalues.
5. n_5 represents spiral sources on n_5-pairs of $(\mathbf{u}_k, \mathbf{v}_k)$ if $\mathrm{Re}\lambda_k > 0$ and $\mathrm{Im}\lambda_k \neq 0$ $(k \in N_5$ and $1 \leq n_5 \leq n)$ with distinct or repeated eigenvalues.
6. n_6 represents invariant centers on n_6-pairs of $(\mathbf{u}_k, \mathbf{v}_k)$ if $\mathrm{Re}\lambda_k = 0$ and $\mathrm{Im}\lambda_k \neq 0$ $(k \in N_6$ and $1 \leq n_6 \leq n)$ with distinct eigenvalues.
7. $\emptyset$ represents empty or none if $n_i = 0$ $(i \in \{1, 2, \ldots, 6\})$.
8. $[n_3; m_3]$ represents invariant centers on $(n_3 - m_3)$-directions of $\mathbf{v}_{k_3}$ $(k_3 \in N_3)$ and sources in m_3-directions of $\mathbf{v}_{j_3}$ $(j_3 \in N_3$ and $j_3 \neq k_3)$ if $\lambda_k = 0$ $(k \in N_3$ and $n_3 \leq n)$ with the $(m_3 + 1)$th-order nilpotent matrix $\mathbf{N}_3^{m_3+1} = \mathbf{0}$ $(0 < m_3 \leq n_2 - 1)$.
9. $[n_3; \emptyset]$ represents invariant centers on n_3-directions of $\mathbf{v}_k$ if $\lambda_k = 0$ $(k \in N_3$ and $1 < n_3 \leq n)$ with a nilpotent matrix $\mathbf{N}_3 = \mathbf{0}$.
10. $[n_6, l; m_6]$ represents invariant centers on $(n_6 - m_6)$-pairs of $(\mathbf{u}_{k_6}, \mathbf{v}_{k_6})$ $(k_6 \in N_6)$, and sources in m_6-pairs of $(\mathbf{u}_{j_6}, \mathbf{v}_{j_6})$ $(j_6 \in N_6$ and $j_6 \neq k_6)$ if $\mathrm{Re}\lambda_k = 0$ and $\mathrm{Im}\lambda_k \neq 0$ $(k \in N_6$ and $n_6 \leq n)$ for $(l+1)$-pairs of repeated eigenvalues with the $(m_6 + 1)$th-order nilpotent matrix $\mathbf{N}_6^{m_6+1} = \mathbf{0}$ $(0 < m_6 \leq l)$.
11. $[n_6, l; \emptyset]$ represents invariant centers on n_6-pairs of $(\mathbf{u}_k, \mathbf{v}_k)$ if $\mathrm{Re}\lambda_k = 0$ and $\mathrm{Im}\lambda_k \neq 0$ $(k \in N_6$ and $1 \leq n_6 \leq n)$ for $(l+1)$ pairs of repeated eigenvalues with a nilpotent matrix $\mathbf{N}_6 = \mathbf{0}$.

Definition 2.18 Consider an n-dimensional, autonomous, nonlinear dynamical system $\dot{\mathbf{x}} = \mathbf{f}(\mathbf{x}, \mathbf{p})$ in Equation (2.4) with an equilibrium point $\mathbf{x}^*$. Suppose there is a neighborhood of the equilibrium $\mathbf{x}^*$ as $U(\mathbf{x}^*) \subset \Omega$, and in the neighborhood $\mathbf{f}(\mathbf{x}, \mathbf{p})$ is C^r $(r \geq 1)$-continuous and Equation (2.24) holds. The corresponding solution is $\mathbf{x}(t) = \mathbf{\Phi}(\mathbf{x}_0, t - t_0, \mathbf{p})$. The matrix $D\mathbf{f}(\mathbf{x}^*, \mathbf{p})$ in Equation (2.19) possesses n eigenvalues λ_k $(k = 1, 2, \ldots, n)$. Set $N = \{1, 2, \ldots, m, m+1, \ldots, (n-m)/2\}$, $N_i = \{i_1, i_2, \ldots, i_{n_i}\} \cup \emptyset$ with $i_j \in N$ $(j = 1, 2, \ldots, n_i;$ $i = 1, 2, \ldots, 6)$, $\Sigma_{i=1}^{3} n_i = m$, and $2\Sigma_{i=4}^{6} n_i = n - m$. $\cup_{i=1}^{6} N_i = N$ with $N_i \cap N_l = \emptyset (l \neq i)$. $N_i = \emptyset$ if $n_i = 0$. The matrix $D\mathbf{f}(\mathbf{x}^*, \mathbf{p})$ possesses n_1-stable, n_2-unstable, and n_3-invariant real eigenvectors plus n_4-stable, n_5-unstable, and n_6-center pairs of complex eigenvectors.

1. *Non-degenerate cases*
   a. The equilibrium point $\mathbf{x}^*$ is an $(n_1 : n_2 : \emptyset|n_4 : n_5 : \emptyset)$ hyperbolic point (or saddle) for the nonlinear system.
   b. The equilibrium point $\mathbf{x}^*$ is an $(n_1 : \emptyset : \emptyset|n_4 : \emptyset : \emptyset)$ sink for the nonlinear system.
   c. The equilibrium point $\mathbf{x}^*$ is an $(\emptyset : n_2 : \emptyset|\emptyset : n_5 : \emptyset)$ source for the nonlinear system.
   d. The equilibrium point $\mathbf{x}^*$ is an $(\emptyset : \emptyset : \emptyset|\emptyset : \emptyset : n/2)$ center for the nonlinear system.
   e. The equilibrium point $\mathbf{x}^*$ is an $(\emptyset : \emptyset : \emptyset|\emptyset : \emptyset : [n/2, l; \emptyset])$ center for the nonlinear system.
   f. The equilibrium point $\mathbf{x}^*$ is an $(\emptyset : \emptyset : \emptyset|\emptyset : \emptyset : [n/2, l; m])$ point for the nonlinear system.
   g. The equilibrium point $\mathbf{x}^*$ is an $(n_1 : \emptyset : \emptyset|n_4 : \emptyset : n_6)$ point for the nonlinear system.
   h. The equilibrium point $\mathbf{x}^*$ is an $(\emptyset : n_2 : \emptyset|\emptyset : n_5 : n_6)$ point for the nonlinear system.
   i. The equilibrium point $\mathbf{x}^*$ is an $(n_1 : n_2 : \emptyset|n_4 : n_5 : n_6)$ point for the nonlinear system.
2. *Simple degenerate cases*
   a. The equilibrium point $\mathbf{x}^*$ is an $(\emptyset : \emptyset : [n; \emptyset]|\emptyset : \emptyset : \emptyset)$-invariant (or static) center for the nonlinear system.
   b. The equilibrium point $\mathbf{x}^*$ is an $(\emptyset : \emptyset : [n; m_3]|\emptyset : \emptyset : \emptyset)$ point for the nonlinear system.
   c. The equilibrium point $\mathbf{x}^*$ is an $(\emptyset : \emptyset : [n_3; \emptyset]|\emptyset : \emptyset : n_6)$ point for the nonlinear system.
   d. The equilibrium point $\mathbf{x}^*$ is an $(\emptyset : \emptyset : [n_3; m_3]|\emptyset : \emptyset : n_6)$ point for the nonlinear system
   e. The equilibrium point $\mathbf{x}^*$ is an $(\emptyset : \emptyset : [n_3; \emptyset]|\emptyset : \emptyset : [n_6; \emptyset])$ point for the nonlinear system.
   f. The equilibrium point $\mathbf{x}^*$ is an $(\emptyset : \emptyset : [n_3; m_3]|\emptyset : \emptyset : [n_6; \emptyset])$ point for the nonlinear system.
   g. The equilibrium point $\mathbf{x}^*$ is an $(\emptyset : \emptyset : [n_3; \emptyset]|\emptyset : \emptyset : [n_6, l; m_6])$ point for the nonlinear system.
   h. The equilibrium point $\mathbf{x}^*$ is an $(\emptyset : \emptyset : [n_3; m_3]|\emptyset : \emptyset : [n_6, l; m_6])$ point for the nonlinear system.
3. *Complex degenerate cases*
   a. The equilibrium point $\mathbf{x}^*$ is an $(n_1 : \emptyset : [n_3; \emptyset]|n_4 : \emptyset : \emptyset)$ point for the nonlinear system.
   b. The equilibrium point $\mathbf{x}^*$ is an $(n_1 : \emptyset : [n_3; m_3]|n_4 : \emptyset : \emptyset)$ point for the nonlinear system.
   c. The equilibrium point $\mathbf{x}^*$ is an $(\emptyset : n_2 : [n_3; \emptyset]|\emptyset : n_5 : \emptyset)$ point for the nonlinear system.
   d. The equilibrium point $\mathbf{x}^*$ is an $(\emptyset : n_2 : [n_3; m_3]|\emptyset : n_5 : \emptyset)$ point for the nonlinear system.
   e. The equilibrium point $\mathbf{x}^*$ is an $(n_1 : \emptyset : [n_3; \emptyset]|n_4 : \emptyset : n_6)$ point for the nonlinear system.
   f. The equilibrium point $\mathbf{x}^*$ is an $(n_1 : \emptyset : [n_3; m_3]|n_4 : \emptyset : n_6)$ point for the nonlinear system.
   g. The equilibrium point $\mathbf{x}^*$ is an $(\emptyset : n_2 : [n_3; \emptyset]|\emptyset : n_5 : n_6)$ point for the nonlinear system.
   h. The equilibrium point $\mathbf{x}^*$ is an $(\emptyset : n_2 : [n_3; m_3]|\emptyset : n_5 : n_6)$ point for the nonlinear system.

2.3 Bifurcation and Stability Switching

The dynamical characteristics of equilibriums in nonlinear dynamical systems in Equation (2.4) are based on the given parameters. With varying parameters in dynamical systems, the corresponding dynamical behaviors will change qualitatively. The qualitative switching of dynamical behaviors in dynamical systems is called *bifurcation* and the corresponding parameter values are called *bifurcation values*. To understand the qualitative changes of dynamical behaviors of nonlinear systems with parameters in the neighborhood of equilibriums, the bifurcation theory for equilibrium of the nonlinear dynamical system in Equation (2.4) will be investigated. $D_{\mathbf{x}}() = \partial()/\partial\mathbf{x}$ and $D_{\mathbf{p}}() = \partial()/\partial\mathbf{p}$ will be adopted from now on. For no specific notice, $D \equiv D_{\mathbf{x}}$.

Definition 2.19 Consider an n-dimensional, autonomous, nonlinear dynamical system $\dot{\mathbf{x}} = \mathbf{f}(\mathbf{x}, \mathbf{p})$ in Equation (2.4) with an equilibrium point $(\mathbf{x}^*, \mathbf{p})$. Suppose there is a neighborhood of the equilibrium $\mathbf{x}^*$ as $U(\mathbf{x}^*) \subset \Omega$, and in the neighborhood, Equation (2.24) holds. The linearized system of the nonlinear system at the equilibrium point $(\mathbf{x}^*, \mathbf{p})$ is $\dot{\mathbf{y}} = D_{\mathbf{x}}\mathbf{f}(\mathbf{x}^*, \mathbf{p})\mathbf{y}$ $(\mathbf{y} = \mathbf{x} - \mathbf{x}^*)$ in Equation (2.19).

1. The equilibrium point $(\mathbf{x}_0^*, \mathbf{p}_0)$ is called *the switching point* of equilibrium solutions if $D_{\mathbf{x}}\mathbf{f}(\mathbf{x}^*, \mathbf{p})$ at $(\mathbf{x}_0^*, \mathbf{p}_0)$ possesses at least one more real eigenvalue (or one more pair of complex eigenvalues) with zero real part.
2. The value $\mathbf{p}_0$ in Equation (2.4) is called *a switching value* of $\mathbf{p}$ if the dynamical characteristics at point $(\mathbf{x}_0^*, \mathbf{p}_0)$ change from one state into another state.
3. The equilibrium point $(\mathbf{x}_0^*, \mathbf{p}_0)$ is called the *bifurcation point* of equilibrium solutions if $D_{\mathbf{x}}\mathbf{f}(\mathbf{x}^*, \mathbf{p})$ at $(\mathbf{x}_0^*, \mathbf{p}_0)$ possesses at least one more real eigenvalue (or one more pair of complex eigenvalues) with zero real part, and more than one branch of equilibrium solutions appear or disappear.
4. The value $\mathbf{p}_0$ in Equation (2.4) is called a bifurcation value of $\mathbf{p}$ if the dynamical characteristics at point $(\mathbf{x}_0^*, \mathbf{p}_0)$ change from one stable state into another unstable state.

2.3.1 Stability and Switching

To extend the idea of Definitions 2.14 and 2.15, a new function will be defined to determine the stability and the stability state switching.

Definition 2.20 Consider an n-dimensional, autonomous, nonlinear dynamical system $\dot{\mathbf{x}} = \mathbf{f}(\mathbf{x}, \mathbf{p})$ in Equation (2.4) with an equilibrium point $\mathbf{x}^*$ and $\mathbf{f}(\mathbf{x}, \mathbf{p})$ is C^r $(r \geq 1)$-continuous in a neighborhood of the equilibrium $\mathbf{x}^*$. The corresponding solution is $\mathbf{x}(t) = \mathbf{\Phi}(\mathbf{x}_0, t - t_0, \mathbf{p})$. Suppose $U(\mathbf{x}^*) \subset \Omega$ is a neighborhood of equilibrium $\mathbf{x}^*$, and there are n linearly independent vectors $\mathbf{v}_k$ $(k = 1, 2, \ldots, n)$. For a perturbation of equilibrium $\mathbf{y} = \mathbf{x} - \mathbf{x}^*$, let $\mathbf{y}^{(k)} = c_k\mathbf{v}_k$ and $\dot{\mathbf{y}}^{(k)} = \dot{c}_k\mathbf{v}_k$,

$$s_k = \mathbf{v}_k^{\mathrm{T}} \cdot \mathbf{y} = \mathbf{v}_k^{\mathrm{T}} \cdot (\mathbf{x} - \mathbf{x}^*) \tag{2.44}$$

where $s_k = c_k||\mathbf{v}_k||^2$. Define the following functions

$$G_k(\mathbf{x}, \mathbf{p}) = \mathbf{v}_k^{\mathrm{T}} \cdot \mathbf{f}(\mathbf{x}, \mathbf{p}) \tag{2.45}$$

and

$$\begin{aligned} G_{s_k}^{(1)}(\mathbf{x},\mathbf{p}) &= \mathbf{v}_k^{\mathrm{T}} \cdot D_{s_k}\mathbf{f}(\mathbf{x}(s_k),\mathbf{p}) \\ &= \mathbf{v}_k^{\mathrm{T}} \cdot D_{\mathbf{x}}\mathbf{f}(\mathbf{x}(s_k),\mathbf{p})\partial_{c_k}\mathbf{x}\partial_{s_k}c_k \\ &= \mathbf{v}_k^{\mathrm{T}} \cdot D_{\mathbf{x}}\mathbf{f}(\mathbf{x}(s_k),\mathbf{p})\mathbf{v}_k\|\mathbf{v}_k\|^{-2}. \end{aligned} \tag{2.46}$$

$$\begin{aligned} G_{s_k}^{(m)}(\mathbf{x},\mathbf{p}) &= \mathbf{v}_k^{\mathrm{T}} \cdot D_{s_k}^{(m)}\mathbf{f}(\mathbf{x}(s_k),\mathbf{p}) \\ &= \mathbf{v}_k^{\mathrm{T}} \cdot D_{s_k}(D_{s_k}^{(m-1)}\mathbf{f}(\mathbf{x}(s_k),\mathbf{p})) \end{aligned} \tag{2.47}$$

where $D_{s_k}(\cdot) = \partial(\cdot)/\partial s_k$ and $D_{s_k}^{(m)}(\cdot) = D_{s_k}(D_{s_k}^{(m-1)}(\cdot))$. $G_{s_k}^{(0)}(\mathbf{x},\mathbf{p}) = G_k(\mathbf{x},\mathbf{p})$ if $m = 0$.

Definition 2.21 Consider an n-dimensional, autonomous, nonlinear dynamical system $\dot{\mathbf{x}} = \mathbf{f}(\mathbf{x},\mathbf{p})$ in Equation (2.4) with an equilibrium point $\mathbf{x}^*$ and $\mathbf{f}(\mathbf{x},\mathbf{p})$ is C^r $(r \geq 1)$-continuous in a neighborhood of the equilibrium $\mathbf{x}^*$. The corresponding solution is $\mathbf{x}(t) = \mathbf{\Phi}(\mathbf{x}_0, t - t_0, \mathbf{p})$. Suppose $U(\mathbf{x}^*) \subset \Omega$ is a neighborhood of equilibrium $\mathbf{x}^*$, and there are n linearly-independent vectors $\mathbf{v}_k$ $(k = 1, 2, \ldots, n)$. For a perturbation of equilibrium $\mathbf{y} = \mathbf{x} - \mathbf{x}^*$, let $\mathbf{y}^{(k)} = c_k\mathbf{v}_k$ and $\dot{\mathbf{y}}^{(k)} = \dot{c}_k\mathbf{v}_k$.

1. $\mathbf{x}^{(k)}$ at the equilibrium $\mathbf{x}^*$ on the direction $\mathbf{v}_k$ is stable if

$$\begin{aligned} &\mathbf{v}_k^{\mathrm{T}} \cdot (\mathbf{x}(t+\varepsilon) - \mathbf{x}(t)) < 0 \ \text{ for } \mathbf{v}_k^{\mathrm{T}} \cdot (\mathbf{x}(t) - \mathbf{x}^*) > 0; \\ &\mathbf{v}_k^{\mathrm{T}} \cdot (\mathbf{x}(t+\varepsilon) - \mathbf{x}(t)) > 0 \ \text{ for } \mathbf{v}_k^{\mathrm{T}} \cdot (\mathbf{x}(t) - \mathbf{x}^*) < 0; \end{aligned} \tag{2.48}$$

   for all $\mathbf{x} \in U(\mathbf{x}^*) \subset \Omega$ and all $t \in [t_0, \infty)$. The equilibrium $\mathbf{x}^*$ is called the sink (or stable node) on the direction $\mathbf{v}_k$.
2. $\mathbf{x}^{(k)}$ at the equilibrium $\mathbf{x}^*$ on the direction $\mathbf{v}_k$ is unstable if

$$\begin{aligned} &\mathbf{v}_k^{\mathrm{T}} \cdot (\mathbf{x}(t+\varepsilon) - \mathbf{x}(t)) > 0 \ \text{ for } \mathbf{v}_k^{\mathrm{T}} \cdot (\mathbf{x}(t) - \mathbf{x}^*) > 0; \\ &\mathbf{v}_k^{\mathrm{T}} \cdot (\mathbf{x}(t+\varepsilon) - \mathbf{x}(t)) < 0 \ \text{ for } \mathbf{v}_k^{\mathrm{T}} \cdot (\mathbf{x}(t) - \mathbf{x}^*) < 0; \end{aligned} \tag{2.49}$$

   for all $\mathbf{x} \in U(\mathbf{x}^*) \subset \Omega$ and all $t \in [t_0, \infty)$. The equilibrium $\mathbf{x}^*$ is called the source (or unstable node) on the direction $\mathbf{v}_k$.
3. $\mathbf{x}^{(k)}$ at the equilibrium $\mathbf{x}^*$ on the direction $\mathbf{v}_k$ is increasingly unstable if

$$\begin{aligned} &\mathbf{v}_k^{\mathrm{T}} \cdot (\mathbf{x}(t+\varepsilon) - \mathbf{x}(t)) > 0 \ \text{ for } \mathbf{v}_k^{\mathrm{T}} \cdot (\mathbf{x}(t) - \mathbf{x}^*) > 0; \\ &\mathbf{v}_k^{\mathrm{T}} \cdot (\mathbf{x}(t+\varepsilon) - \mathbf{x}(t)) > 0 \ \text{ for } \mathbf{v}_k^{\mathrm{T}} \cdot (\mathbf{x}(t) - \mathbf{x}^*) < 0; \end{aligned} \tag{2.50}$$

   for all $\mathbf{x} \in U(\mathbf{x}^*) \subset \Omega$ and all $t \in [t_0, \infty)$. The equilibrium $\mathbf{x}^*$ is called the increasing saddle on the direction $\mathbf{v}_k$.
4. $\mathbf{x}^{(k)}$ at the equilibrium $\mathbf{x}^*$ on the direction $\mathbf{v}_k$ is decreasingly unstable if

$$\begin{aligned} &\mathbf{v}_k^{\mathrm{T}} \cdot (\mathbf{x}(t+\varepsilon) - \mathbf{x}(t)) < 0 \ \text{ for } \mathbf{v}_k^{\mathrm{T}} \cdot (\mathbf{x}(t) - \mathbf{x}^*) > 0; \\ &\mathbf{v}_k^{\mathrm{T}} \cdot (\mathbf{x}(t+\varepsilon) - \mathbf{x}(t)) < 0 \ \text{ for } \mathbf{v}_k^{\mathrm{T}} \cdot (\mathbf{x}(t) - \mathbf{x}^*) < 0; \end{aligned} \tag{2.51}$$

   for all $\mathbf{x} \in U(\mathbf{x}^*) \subset \Omega$ and all $t \in [t_0, \infty)$. The equilibrium $\mathbf{x}^*$ is called the decreasing saddle on the direction $\mathbf{v}_k$.

5. $\mathbf{x}^{(i)}$ at the equilibrium $\mathbf{x}^*$ on the direction $\mathbf{v}_k$ is invariant if

$$\mathbf{v}_k^{\mathrm{T}} \cdot (\mathbf{x}(t+\varepsilon) - \mathbf{x}(t)) = 0$$
$$\text{for } \mathbf{v}_k^{\mathrm{T}} \cdot (\mathbf{x}(t) - \mathbf{x}^*) \neq 0; \tag{2.52}$$

for all $\mathbf{x} \in U(\mathbf{x}^*) \subset \Omega$ and all $t \in [t_0, \infty)$. The equilibrium $\mathbf{x}^*$ is called to be degenerate on the direction $\mathbf{v}_k$.

Theorem 2.7 *Consider an n-dimensional, autonomous, nonlinear dynamical system* $\dot{\mathbf{x}} = \mathbf{f}(\mathbf{x}, \mathbf{p})$ *in Equation (2.4) with an equilibrium point* $\mathbf{x}^*$ *and* $\mathbf{f}(\mathbf{x}, \mathbf{p})$ *is* C^r $(r \geq 1)$*-continuous in a neighborhood of the equilibrium* $\mathbf{x}^*$ *(i.e.,* $U(\mathbf{x}^*) \subset \Omega$*). The corresponding solution is* $\mathbf{x}(t) = \mathbf{\Phi}(\mathbf{x}_0, t - t_0, \mathbf{p})$*. Suppose Equation (2.24) holds in* $U(\mathbf{x}^*) \subset \Omega$*. For a linearized dynamical system in Equation (2.19), consider a real eigenvalue* λ_k *of matrix* $D\mathbf{f}(\mathbf{x}^*, \mathbf{p})$ $(k \in N = \{1, 2, \ldots, n\})$ *with an eigenvector* $\mathbf{v}_k$*. Let* $\mathbf{y}^{(k)} = c_k \mathbf{v}_k$ *and* $\dot{\mathbf{y}}^{(k)} = \dot{c}_k \mathbf{v}_k$*,* $s_k = \mathbf{v}_k^{\mathrm{T}} \cdot \mathbf{y} = \mathbf{v}_k^{\mathrm{T}} \cdot (\mathbf{x} - \mathbf{x}^*)$ *in Equation (2.44) with* $s_k = c_k \|\mathbf{v}_k\|^2$*. Define*

$$\dot{s}_k = \mathbf{v}_k^{\mathrm{T}} \cdot \dot{\mathbf{y}} = \mathbf{v}_k^{\mathrm{T}} \cdot \dot{\mathbf{x}} = \mathbf{v}_k^{\mathrm{T}} \cdot \mathbf{f}(\mathbf{x}, \mathbf{p}). \tag{2.53}$$

1. $\mathbf{x}^{(k)}$ *at the equilibrium* $\mathbf{x}^*$ *on the direction* $\mathbf{v}_k$ *is stable if and only if*

$$G_k(\mathbf{x}, \mathbf{p}) = \mathbf{v}_k^{\mathrm{T}} \cdot \mathbf{f}(\mathbf{x}, \mathbf{p}) < 0 \ \text{ for } s_k = \mathbf{v}_k^{\mathrm{T}} \cdot (\mathbf{x}(t) - \mathbf{x}^*) > 0;$$
$$G_k(\mathbf{x}, \mathbf{p}) = \mathbf{v}_k^{\mathrm{T}} \cdot \mathbf{f}(\mathbf{x}, \mathbf{p}) > 0 \ \text{ for } s_k = \mathbf{v}_k^{\mathrm{T}} \cdot (\mathbf{x}(t) - \mathbf{x}^*) < 0 \tag{2.54}$$

for all $\mathbf{x} \in U(\mathbf{x}^*) \subset \Omega$ *and all* $t \in [t_0, \infty)$*.*

2. $\mathbf{x}^{(k)}$ *at the equilibrium* $\mathbf{x}^*$ *on the direction* $\mathbf{v}_k$ *is unstable if and only if*

$$G_k(\mathbf{x}, \mathbf{p}) = \mathbf{v}_k^{\mathrm{T}} \cdot \mathbf{f}(\mathbf{x}, \mathbf{p}) > 0 \ \text{ for } s_k = \mathbf{v}_k^{\mathrm{T}} \cdot (\mathbf{x}(t) - \mathbf{x}^*) > 0;$$
$$G_k(\mathbf{x}, \mathbf{p}) = \mathbf{v}_k^{\mathrm{T}} \cdot \mathbf{f}(\mathbf{x}, \mathbf{p}) < 0 \ \text{ for } s_k = \mathbf{v}_k^{\mathrm{T}} \cdot (\mathbf{x}(t) - \mathbf{x}^*) < 0 \tag{2.55}$$

for all $\mathbf{x} \in U(\mathbf{x}^*) \subset \Omega$ *and all* $t \in [t_0, \infty)$*.*

3. $\mathbf{x}^{(k)}$ *at the equilibrium* $\mathbf{x}^*$ *on the direction* $\mathbf{v}_k$ *is increasingly unstable if and only if*

$$G_k(\mathbf{x}, \mathbf{p}) = \mathbf{v}_k^{\mathrm{T}} \cdot \mathbf{f}(\mathbf{x}, \mathbf{p}) > 0 \ \text{ for } s_k = \mathbf{v}_k^{\mathrm{T}} \cdot (\mathbf{x}(t) - \mathbf{x}^*) > 0;$$
$$G_k(\mathbf{x}, \mathbf{p}) = \mathbf{v}_k^{\mathrm{T}} \cdot \mathbf{f}(\mathbf{x}, \mathbf{p}) > 0 \ \text{ for } s_k = \mathbf{v}_k^{\mathrm{T}} \cdot (\mathbf{x}(t) - \mathbf{x}^*) < 0 \tag{2.56}$$

for all $\mathbf{x} \in U(\mathbf{x}^*) \subset \Omega$ *and all* $t \in [t_0, \infty)$*.*

4. $\mathbf{x}^{(k)}$ *at the equilibrium* $\mathbf{x}^*$ *on the direction* $\mathbf{v}_k$ *is decreasingly unstable if and only if*

$$G_k(\mathbf{x}, \mathbf{p}) = \mathbf{v}_k^{\mathrm{T}} \cdot \mathbf{f}(\mathbf{x}, \mathbf{p}) < 0 \ \text{ for } s_k = \mathbf{v}_k^{\mathrm{T}} \cdot (\mathbf{x}(t) - \mathbf{x}^*) > 0;$$
$$G_k(\mathbf{x}, \mathbf{p}) = \mathbf{v}_k^{\mathrm{T}} \cdot \mathbf{f}(\mathbf{x}, \mathbf{p}) < 0 \ \text{ for } s_k = \mathbf{v}_k^{\mathrm{T}} \cdot (\mathbf{x}(t) - \mathbf{x}^*) < 0 \tag{2.57}$$

for all $\mathbf{x} \in U(\mathbf{x}^*) \subset \Omega$ *and all* $t \in [t_0, \infty)$*.*

5. $\mathbf{x}^{(i)}$ *at the equilibrium* $\mathbf{x}^*$ *on the direction* $\mathbf{v}_k$ *is invariant if*

$$G_k(\mathbf{x}, \mathbf{p}) = \mathbf{v}_k^{\mathrm{T}} \cdot \mathbf{f}(\mathbf{x}, \mathbf{p}) = 0 \tag{2.58}$$

for all $\mathbf{x} \in U(\mathbf{x}^*) \subset \Omega$ *and all* $t \in [t_0, \infty)$*.*

Proof. The proof can be referred to Luo (2012a). ■

Theorem 2.8 *Consider an n-dimensional, autonomous, nonlinear dynamical system* $\dot{\mathbf{x}} = \mathbf{f}(\mathbf{x}, \mathbf{p})$ *in Equation (2.4) with an equilibrium point* $\mathbf{x}^*$ *and* $\mathbf{f}(\mathbf{x}, \mathbf{p})$ *is* C^r $(r \geq 1)$*-continuous in a neighborhood of the equilibrium* $\mathbf{x}^*$ *(i.e.,* $U(\mathbf{x}^*) \subset \Omega$*). The corresponding solution is* $\mathbf{x}(t) = \mathbf{\Phi}(\mathbf{x}_0, t - t_0, \mathbf{p})$*. Suppose Equation (2.24) holds in* $U(\mathbf{x}^*) \subset \Omega$*. For a linearized dynamical system in Equation (2.19), consider a real eigenvalue* λ_k *of matrix* $D\mathbf{f}(\mathbf{x}^*, \mathbf{p})$ $(k \in N = \{1, 2, \ldots, n\})$ *with an eigenvector* $\mathbf{v}_k$*. Let* $\mathbf{y}^{(k)} = c_k \mathbf{v}_k$ *and* $\dot{\mathbf{y}}^{(k)} = \dot{c}_k \mathbf{v}_k$*,* $s_k = \mathbf{v}_k^{\mathrm{T}} \cdot \mathbf{y} = \mathbf{v}_k^{\mathrm{T}} \cdot (\mathbf{x} - \mathbf{x}^*)$ *in Equation (2.44) with* $s_k = c_k \|\mathbf{v}_k\|^2$*. Define* $\dot{s}_k = \mathbf{v}_k^{\mathrm{T}} \cdot \mathbf{f}(\mathbf{x}, \mathbf{p})$ *in Equation (2.53). Suppose* $\|G_k^{(2)}(\mathbf{x}^*, \mathbf{p})\| < \infty$*.*

1. $\mathbf{x}^{(k)}$ *at the equilibrium* $\mathbf{x}^*$ *on the direction* $\mathbf{v}_k$ *is stable if and only if*

$$G_{s_k}^{(1)}(\mathbf{x}^*, \mathbf{p}) = \lambda_k < 0 \tag{2.59}$$

   for all $\mathbf{x} \in U(\mathbf{x}^*) \subset \Omega$ *and all* $t \in [t_0, \infty)$*.*
2. $\mathbf{x}^{(k)}$ *at the equilibrium* $\mathbf{x}^*$ *on the direction* $\mathbf{v}_k$ *is unstable if and only if*

$$G_{s_k}^{(1)}(\mathbf{x}^*, \mathbf{p}) = \lambda_k > 0 \tag{2.60}$$

   for all $\mathbf{x} \in U(\mathbf{x}^*) \subset \Omega$ *and all* $t \in [t_0, \infty)$*.*
3. $\mathbf{x}^{(k)}$ *at the equilibrium* $\mathbf{x}^*$ *on the direction* $\mathbf{v}_k$ *is increasingly unstable if and only if*

$$G_{s_k}^{(1)}(\mathbf{x}^*, \mathbf{p}) = \lambda_k = 0, \text{and } G_{s_k}^{(2)}(\mathbf{x}^*, \mathbf{p}) > 0 \tag{2.61}$$

   for all $\mathbf{x} \in U(\mathbf{x}^*) \subset \Omega$ *and all* $t \in [t_0, \infty)$*.*
4. $\mathbf{x}^{(k)}$ *at the equilibrium* $\mathbf{x}^*$ *on the direction* $\mathbf{v}_k$ *is decreasingly unstable if and only if*

$$G_{s_k}^{(1)}(\mathbf{x}^*, \mathbf{p}) = \lambda_k = 0, \text{and } G_{s_k}^{(2)}(\mathbf{x}^*, \mathbf{p}) < 0 \tag{2.62}$$

   for all $\mathbf{x} \in U(\mathbf{x}^*) \subset \Omega$ *and all* $t \in [t_0, \infty)$*.*
5. $\mathbf{x}^{(i)}$ *at the equilibrium* $\mathbf{x}^*$ *on the direction* $\mathbf{v}_k$ *is invariant if and only if*

$$G_{s_k}^{(m)}(\mathbf{x}^*, \mathbf{p}) = 0 \quad (m = 0, 1, 2, \ldots) \tag{2.63}$$

   for all $\mathbf{x} \in U(\mathbf{x}^*) \subset \Omega$ *and all* $t \in [t_0, \infty)$*.*

Proof. The proof can be referred to Luo (2012a). ■

Definition 2.22 Consider an n-dimensional, autonomous, nonlinear dynamical system $\dot{\mathbf{x}} = \mathbf{f}(\mathbf{x}, \mathbf{p})$ in Equation (2.4) with an equilibrium point $\mathbf{x}^*$ and $\mathbf{f}(\mathbf{x}, \mathbf{p})$ is C^r $(r \geq 1)$-continuous in a neighborhood of the equilibrium $\mathbf{x}^*$ (i.e., $U(\mathbf{x}^*) \subset \Omega$). The corresponding solution is $\mathbf{x}(t) = \mathbf{\Phi}(\mathbf{x}_0, t - t_0, \mathbf{p})$. Suppose Equation (2.24) holds in $U(\mathbf{x}^*) \subset \Omega$. For a linearized dynamical system in Equation (2.19), consider a real eigenvalue λ_k of matrix $D\mathbf{f}(\mathbf{x}^*, \mathbf{p})$ $(k \in N = \{1, 2, \ldots, n\})$ with an eigenvector $\mathbf{v}_k$ and let $\mathbf{y}^{(k)} = c_k \mathbf{v}_k$.

1. $\mathbf{x}^{(k)}$ at the equilibrium $\mathbf{x}^*$ on the direction $\mathbf{v}_k$ is stable of the $(2m_k + 1)$th order if

$$\begin{aligned} &G_{s_k}^{(r_k)}(\mathbf{x}^*, \mathbf{p}) = 0, r_k = 0, 1, 2, \ldots, 2m_k; \\ &\mathbf{v}_k^{\mathrm{T}} \cdot (\mathbf{x}(t + \varepsilon) - \mathbf{x}(t)) < 0 \ \text{ for } \mathbf{v}_k^{\mathrm{T}} \cdot (\mathbf{x}(t) - \mathbf{x}^*) > 0; \\ &\mathbf{v}_k^{\mathrm{T}} \cdot (\mathbf{x}(t + \varepsilon) - \mathbf{x}(t)) > 0 \ \text{ for } \mathbf{v}_k^{\mathrm{T}} \cdot (\mathbf{x}(t) - \mathbf{x}^*) < 0 \end{aligned} \tag{2.64}$$

for all $\mathbf{x} \in U(\mathbf{x}^*) \subset \Omega$ and all $t \in [t_0, \infty)$. The equilibrium $\mathbf{x}^*$ is called the sink (or stable node) of the $(2m_k + 1)$th order on the direction $\mathbf{v}_k$.

2. $\mathbf{x}^{(k)}$ at the equilibrium $\mathbf{x}^*$ on the direction $\mathbf{v}_k$ is unstable of the $(2m_k + 1)$th order if

$$\begin{aligned} &G_{s_k}^{(r_k)}(\mathbf{x}^*, \mathbf{p}) = 0, r_k = 0, 1, 2, \ldots, 2m_k; \\ &\mathbf{v}_k^{\mathrm{T}} \cdot (\mathbf{x}(t+\varepsilon) - \mathbf{x}(t)) > 0 \ \text{ for } \mathbf{v}_k^{\mathrm{T}} \cdot (\mathbf{x}(t) - \mathbf{x}^*) > 0; \\ &\mathbf{v}_k^{\mathrm{T}} \cdot (\mathbf{x}(t+\varepsilon) - \mathbf{x}(t)) < 0 \ \text{ for } \mathbf{v}_k^{\mathrm{T}} \cdot (\mathbf{x}(t) - \mathbf{x}^*) < 0. \end{aligned} \tag{2.65}$$

for all $\mathbf{x} \in U(\mathbf{x}^*) \subset \Omega$ and all $t \in [t_0, \infty)$. The equilibrium $\mathbf{x}^*$ is called the source (or unstable node) of the $(2m_k + 1)$th order on the direction $\mathbf{v}_k$.

3. $\mathbf{x}^{(k)}$ at the equilibrium $\mathbf{x}^*$ on the direction $\mathbf{v}_k$ is increasingly unstable of the $(2m_k)$th order if

$$\begin{aligned} &G_{s_k}^{(r_k)}(\mathbf{x}^*, \mathbf{p}) = 0, r_k = 0, 1, 2, \ldots, 2m_k - 1; \\ &\mathbf{v}_k^{\mathrm{T}} \cdot (\mathbf{x}(t+\varepsilon) - \mathbf{x}(t)) > 0 \ \text{ for } \mathbf{v}_k^{\mathrm{T}} \cdot (\mathbf{x}(t) - \mathbf{x}^*) > 0; \\ &\mathbf{v}_k^{\mathrm{T}} \cdot (\mathbf{x}(t+\varepsilon) - \mathbf{x}(t)) > 0 \ \text{ for } \mathbf{v}_k^{\mathrm{T}} \cdot (\mathbf{x}(t) - \mathbf{x}^*) < 0 \end{aligned} \tag{2.66}$$

for all $\mathbf{x} \in U(\mathbf{x}^*) \subset \Omega$ and all $t \in [t_0, \infty)$. The equilibrium $\mathbf{x}^*$ is called the increasing saddle of the $(2m_k)$th order on the direction $\mathbf{v}_k$.

4. $\mathbf{x}^{(k)}$ at the equilibrium $\mathbf{x}^*$ on the direction $\mathbf{v}_k$ is decreasingly unstable of the $(2m_k)$th order if

$$\begin{aligned} &G_{s_k}^{(r_k)}(\mathbf{x}^*, \mathbf{p}) = 0, r_k = 0, 1, 2, \ldots, 2m_k - 1; \\ &\mathbf{v}_k^{\mathrm{T}} \cdot (\mathbf{x}(t+\varepsilon) - \mathbf{x}(t)) < 0 \ \text{ for } \mathbf{v}_k^{\mathrm{T}} \cdot (\mathbf{x}(t) - \mathbf{x}^*) > 0; \\ &\mathbf{v}_k^{\mathrm{T}} \cdot (\mathbf{x}(t+\varepsilon) - \mathbf{x}(t)) < 0 \ \text{ for } \mathbf{v}_k^{\mathrm{T}} \cdot (\mathbf{x}(t) - \mathbf{x}^*) < 0 \end{aligned} \tag{2.67}$$

for all $\mathbf{x} \in U(\mathbf{x}^*) \subset \Omega$ and all $t \in [t_0, \infty)$. The equilibrium $\mathbf{x}^*$ is called the decreasing saddle of the $(2m_k)$th order on the direction $\mathbf{v}_k$.

Theorem 2.9 *Consider an n-dimensional, autonomous, nonlinear dynamical system* $\dot{\mathbf{x}} = \mathbf{f}(\mathbf{x}, \mathbf{p})$ *in Equation (2.4) with an equilibrium point* $\mathbf{x}^*$ *and* $\mathbf{f}(\mathbf{x}, \mathbf{p})$ *is* C^r $(r \geq 1)$*-continuous in a neighborhood of the equilibrium* $\mathbf{x}^*$ *(i.e.,* $U(\mathbf{x}^*) \subset \Omega$*). The corresponding solution is* $\mathbf{x}(t) = \mathbf{\Phi}(\mathbf{x}_0, t - t_0, \mathbf{p})$. *Suppose Equation (2.24) holds in* $U(\mathbf{x}^*) \subset \Omega$. *For a linearized dynamical system in Equation (2.19), consider a real eigenvalue* λ_k *of matrix* $D\mathbf{f}(\mathbf{x}^*, \mathbf{p})$ $(k \in N = \{1, 2, \ldots, n\})$ *with an eigenvector* $\mathbf{v}_k$ *and let* $\mathbf{y}^{(k)} = c_k \mathbf{v}_k$.

1. $\mathbf{x}^{(k)}$ *at the equilibrium* $\mathbf{x}^*$ *on the direction* $\mathbf{v}_k$ *is stable of the* $(2m_k + 1)$th *order if and only if*

$$\begin{aligned} &G_{s_k}^{(r_k)}(\mathbf{x}^*, \mathbf{p}) = 0, r_k = 0, 1, 2, \ldots, 2m_k; \\ &G_{s_k}^{(2m_k+1)}(\mathbf{x}^*, \mathbf{p}) < 0 \end{aligned} \tag{2.68}$$

for all $\mathbf{x} \in U(\mathbf{x}^*) \subset \Omega$ *and all* $t \in [t_0, \infty)$.

2. $\mathbf{x}^{(k)}$ *at the equilibrium* $\mathbf{x}^*$ *on the direction* $\mathbf{v}_k$ *is unstable of the* $(2m_k + 1)$th *order if and only if*

$$\begin{aligned} &G_{s_k}^{(r_k)}(\mathbf{x}^*, \mathbf{p}) = 0, r_k = 0, 1, 2, \ldots, 2m_k; \\ &G_{s_k}^{(2m_k+1)}(\mathbf{x}^*, \mathbf{p}) > 0 \end{aligned} \tag{2.69}$$

for all $\mathbf{x} \in U(\mathbf{x}^*) \subset \Omega$ *and all* $t \in [t_0, \infty)$.

3. $\mathbf{x}^{(k)}$ *at the equilibrium* $\mathbf{x}^*$ *on the direction* $\mathbf{v}_k$ *is increasingly unstable of the* $(2m_k)$th *order if and only if*

$$G_{s_k}^{(r_k)}(\mathbf{x}^*, \mathbf{p}) = 0, r_k = 0, 1, 2, \ldots, 2m_k - 1;$$
$$G_{s_k}^{(2m_k)}(\mathbf{x}^*, \mathbf{p}) > 0 \tag{2.70}$$

for all $\mathbf{x} \in U(\mathbf{x}^*) \subset \Omega$ *and all* $t \in [t_0, \infty)$.

4. $\mathbf{x}^{(k)}$ *at the equilibrium* $\mathbf{x}^*$ *on the direction* $\mathbf{v}_k$ *is decreasingly unstable of the* $(2m_k)$th *order if and only if*

$$G_{s_k}^{(r_k)}(\mathbf{x}^*, \mathbf{p}) = 0, r_k = 0, 1, 2, \ldots, 2m_k - 1;$$
$$G_{s_k}^{(2m_k)}(\mathbf{x}^*, \mathbf{p}) < 0 \tag{2.71}$$

for all $\mathbf{x} \in U(\mathbf{x}^*) \subset \Omega$ *and all* $t \in [t_0, \infty)$.

Proof. The proof can be referred to Luo (2012a). ■

Definition 2.23 Consider an n-dimensional, autonomous, nonlinear dynamical system $\dot{\mathbf{x}} = \mathbf{f}(\mathbf{x}, \mathbf{p})$ in Equation (2.4) with an equilibrium point $\mathbf{x}^*$ and $\mathbf{f}(\mathbf{x}, \mathbf{p})$ is C^r $(r \geq 1)$-continuous in a neighborhood of the equilibrium $\mathbf{x}^*$ (i.e., $U(\mathbf{x}^*) \subset \Omega$). The corresponding solution is $\mathbf{x}(t) = \boldsymbol{\Phi}(\mathbf{x}_0, t - t_0, \mathbf{p})$. Suppose Equation (2.24) holds in $U(\mathbf{x}^*) \subset \Omega$. For a linearized dynamical system in Equation (2.19), consider a pair of complex eigenvalues $\alpha_k \pm \mathbf{i}\beta_k$ $(k \in N = \{1, 2, \ldots, n\}$, $\mathbf{i} = \sqrt{-1})$ of matrix $D\mathbf{f}(\mathbf{x}^*, \mathbf{p})$ with a pair of eigenvectors $\mathbf{u}_k \pm \mathbf{i}\mathbf{v}_k$. On the invariant plane of $(\mathbf{u}_k, \mathbf{v}_k)$, consider $\mathbf{r}_k = \mathbf{y}_k = \mathbf{y}_+^{(k)} + \mathbf{y}_-^{(k)}$ with

$$\mathbf{r}_k = c_k\mathbf{u}_k + d_k\mathbf{v}_k = r_k\mathbf{e}_{r_k},$$
$$\dot{\mathbf{r}}_k = \dot{c}_k\mathbf{u}_k + \dot{d}_k\mathbf{v}_k = \dot{r}_k\mathbf{e}_{r_k} + r_k\dot{\mathbf{e}}_{r_k} \tag{2.72}$$

and

$$c_k = \frac{1}{\Delta}[\Delta_2(\mathbf{u}_k^{\mathrm{T}} \cdot \mathbf{y}) - \Delta_{12}(\mathbf{v}_k^{\mathrm{T}} \cdot \mathbf{y})],$$
$$d_k = \frac{1}{\Delta}[\Delta_1(\mathbf{v}_k^{\mathrm{T}} \cdot \mathbf{y}) - \Delta_{12}(\mathbf{u}_k^{\mathrm{T}} \cdot \mathbf{y})],$$
$$\Delta_1 = ||\mathbf{u}_k||^2, \Delta_2 = ||\mathbf{v}_k||^2, \Delta_{12} = \mathbf{u}_k^{\mathrm{T}} \cdot \mathbf{v}_k,$$
$$\Delta = \Delta_1\Delta_2 - \Delta_{12}^2 \tag{2.73}$$

Consider a polar coordinate of (r_k, θ_k) defined by

$$c_k = r_k \cos\theta_k, \text{and } d_k = r_k \sin\theta_k;$$
$$r_k = \sqrt{c_k^2 + d_k^2}, \text{and } \theta_k = \arctan d_k/c_k; \tag{2.74}$$

$$\mathbf{e}_{r_k} = \cos\theta_k\mathbf{u}_k + \sin\theta_k\mathbf{v}_k \text{ and } \mathbf{e}_{\theta_k} = -\cos\theta_k\mathbf{u}_k^{\perp}\Delta_3 + \sin\theta_k\mathbf{v}_k^{\perp}\Delta_4$$
$$\Delta_3 = \mathbf{v}_k^{\mathrm{T}} \cdot \mathbf{u}_k^{\perp} \text{ and } \Delta_4 = \mathbf{u}_k^{\mathrm{T}} \cdot \mathbf{v}_k^{\perp} \tag{2.75}$$

where $\mathbf{u}_k^{\perp}$ and $\mathbf{v}_k^{\perp}$ are the normal vectors of $\mathbf{u}_k$ and $\mathbf{v}_k$, respectively.

$$\dot{c}_k = \frac{1}{\Delta}[\Delta_2 G_{c_k}(\mathbf{x},\mathbf{p}) - \Delta_{12} G_{d_k}(\mathbf{x},\mathbf{p})],$$
$$\dot{d}_k = \frac{1}{\Delta}[\Delta_1 G_{d_k}(\mathbf{x},\mathbf{p}) - \Delta_{12} G_{d_k}(\mathbf{x},\mathbf{p})] \tag{2.76}$$

where

$$G_{c_k}(\mathbf{x},\mathbf{p}) = \mathbf{u}_k^{\mathrm{T}} \cdot \mathbf{f}(\mathbf{x},\mathbf{p}) = \sum_{m=1}^{\infty} G_{c_k}^{(m)}(\mathbf{x}^*,\mathbf{p}) r_k^m,$$
$$G_{d_k}(\mathbf{x},\mathbf{p}) = \mathbf{v}_k^{\mathrm{T}} \cdot \mathbf{f}(\mathbf{x},\mathbf{p}) = \sum_{m=1}^{\infty} G_{d_k}^{(m)}(\mathbf{x}^*,\mathbf{p}) r_k^m; \tag{2.77}$$

$$G_{c_k}^{(m)}(\mathbf{x}^*,\mathbf{p}) = \mathbf{u}_k^{\mathrm{T}} \cdot \partial_{\mathbf{x}}^{(m)} \mathbf{f}(\mathbf{x},\mathbf{p})[\mathbf{u}_k \cos\theta_k + \mathbf{v}_k \sin\theta_k]^m|_{(\mathbf{x}^*,\mathbf{p})},$$
$$G_{d_k}^{(m)}(\mathbf{x}^*,\mathbf{p}) = \mathbf{v}_k^{\mathrm{T}} \cdot \partial_{\mathbf{x}}^{(m)} \mathbf{f}(\mathbf{x},\mathbf{p})[\mathbf{u}_k \cos\theta_k + \mathbf{v}_k \sin\theta_k]^m|_{(\mathbf{x}^*,\mathbf{p})}. \tag{2.78}$$

Thus

$$\dot{r}_k = \dot{c}_k \cos\theta_k + \dot{d}_k \sin\theta_k = \sum_{m=1}^{\infty} G_{r_k}^{(m)}(\theta_k) r_k^m,$$
$$\dot{\theta}_k = r_k^{-1}(\dot{d}_k \cos\theta_k - \dot{c}_k \sin\theta_k) = r_k^{-1} \sum_{m=1}^{\infty} G_{\theta_k}^{(m)}(\theta_k) r_k^{m-1}. \tag{2.79}$$

where

$$G_{r_k}^{(m)}(\theta_k) = \frac{1}{\Delta}[(\Delta_2 \cos\theta_k - \Delta_{12}\sin\theta_k)\mathbf{u}_k^{\mathrm{T}} + (\Delta_2 \sin\theta_k - \Delta_{12}\cos\theta_k)\mathbf{v}_k^{\mathrm{T}}]$$
$$\cdot \partial_{\mathbf{x}}^{(m)} \mathbf{f}(\mathbf{x},\mathbf{p})(\mathbf{u}_k \cos\theta_k + \mathbf{v}_k \sin\theta_k)^m|_{(\mathbf{x}^*,\mathbf{p})},$$
$$G_{\theta_k}^{(m)}(\theta_k) = -\frac{1}{\Delta}[(\Delta_2 \sin\theta_k + \Delta_{12}\cos\theta_k)\mathbf{u}_k^{\mathrm{T}} - (\Delta_1 \cos\theta_k - \Delta_{12}\sin\theta_k)\mathbf{v}_k^{\mathrm{T}}]$$
$$\cdot \partial_{\mathbf{x}}^{(m)} \mathbf{f}(\mathbf{x},\mathbf{p})(\mathbf{u}_k \cos\theta_k + \mathbf{v}_k \sin\theta_k)^m|_{(\mathbf{x}^*,\mathbf{p})}. \tag{2.80}$$

From the foregoing definition, consider the first order terms of G-function

$$G_{c_k}^{(1)}(\mathbf{x},\mathbf{p}) = G_{c_k 1}^{(1)}(\mathbf{x},\mathbf{p}) + G_{c_k 2}^{(1)}(\mathbf{x},\mathbf{p}),$$
$$G_{d_k}^{(1)}(\mathbf{x},\mathbf{p}) = G_{d_k 1}^{(1)}(\mathbf{x},\mathbf{p}) + G_{d_k 2}^{(1)}(\mathbf{x},\mathbf{p}) \tag{2.81}$$

where

$$G_{c_k 1}^{(1)}(\mathbf{x},\mathbf{p}) = \mathbf{u}_k^{\mathrm{T}} \cdot D_{\mathbf{x}} \mathbf{f}(\mathbf{x},\mathbf{p}) \partial_{c_k} \mathbf{x} = \mathbf{u}_k^{\mathrm{T}} \cdot D_{\mathbf{x}} \mathbf{f}(\mathbf{x},\mathbf{p}) \mathbf{u}_k$$
$$= \mathbf{u}_k^{\mathrm{T}} \cdot (-\beta_k \mathbf{v}_k + \alpha_k \mathbf{u}_k) = \alpha_k \Delta_1 - \beta_k \Delta_{12},$$
$$G_{c_k 2}^{(1)}(\mathbf{x},\mathbf{p}) = \mathbf{u}_k^{\mathrm{T}} \cdot D_{\mathbf{x}} \mathbf{f}(\mathbf{x},\mathbf{p}) \partial_{d_k} \mathbf{x} = \mathbf{u}_k^{\mathrm{T}} \cdot D_{\mathbf{x}} \mathbf{f}(\mathbf{x},\mathbf{p}) \mathbf{v}_k$$
$$= \mathbf{u}_k^{\mathrm{T}} \cdot (\beta_k \mathbf{u}_k + \alpha_k \mathbf{v}_k) = \alpha_k \Delta_{12} + \beta_k \Delta_1; \tag{2.82}$$

and

$$
\begin{aligned}
G^{(1)}_{d_k 1}(\mathbf{x},\mathbf{p}) &= \mathbf{v}_k^{\mathrm{T}} \cdot D_{\mathbf{x}}\mathbf{f}(\mathbf{x},\mathbf{p})\partial_{c_k}\mathbf{x} = \mathbf{v}_k^{\mathrm{T}} \cdot D_{\mathbf{x}}\mathbf{f}(\mathbf{x},\mathbf{p})\mathbf{u}_k \\
&= \mathbf{v}_k^{\mathrm{T}} \cdot (-\beta_k \mathbf{v}_k + \alpha_k \mathbf{u}_k) = -\beta_k \Delta_2 + \alpha_k \Delta_{12}, \\
G^{(1)}_{d_k 2}(\mathbf{x},\mathbf{p}) &= \mathbf{v}_k^{\mathrm{T}} \cdot D_{\mathbf{x}}\mathbf{f}(\mathbf{x},\mathbf{p})\partial_{d_k}\mathbf{x} = \mathbf{v}_k^{\mathrm{T}} \cdot D_{\mathbf{x}}\mathbf{f}(\mathbf{x},\mathbf{p})\mathbf{v}_k \\
&= \mathbf{v}_k^{\mathrm{T}} \cdot (\beta_k \mathbf{u}_k + \alpha_k \mathbf{v}_k) = \alpha_k \Delta_2 + \beta_k \Delta_{12}.
\end{aligned} \tag{2.83}
$$

Substitution of Equations (2.81)–(2.83) into Equation (2.78) gives

$$
\begin{aligned}
G^{(1)}_{c_k}(\mathbf{x},\mathbf{p}) &= G^{(1)}_{c_k 1}(\mathbf{x},\mathbf{p})\cos\theta_k + G^{(1)}_{c_k 2}(\mathbf{x},\mathbf{p})\sin\theta_k \\
&= (\alpha_k \Delta_1 - \beta_k \Delta_{12})\cos\theta_k + (\alpha_k \Delta_{12} + \beta_k \Delta_1)\sin\theta_k, \\
G^{(1)}_{d_k}(\mathbf{x},\mathbf{p}) &= G^{(1)}_{d_k 1}(\mathbf{x},\mathbf{p})\cos\theta_k + G^{(1)}_{d_k 2}(\mathbf{x},\mathbf{p})\sin\theta_k \\
&= (-\beta_k \Delta_2 + \alpha_k \Delta_{12})\cos\theta_k + (\alpha_k \Delta_2 + \beta_k \Delta_{12})\sin\theta_k.
\end{aligned} \tag{2.84}
$$

From Equation (2.80), we have

$$
\begin{aligned}
G^{(1)}_{r_k}(\theta_k) &= \frac{1}{\Delta}[(G^{(1)}_{c_k}\Delta_2 - G^{(1)}_{d_k}\Delta_{12})\cos\theta_k + (G^{(1)}_{d_k}\Delta_1 - G^{(1)}_{c_k}\Delta_{12})\sin\theta_k] = \alpha_k; \\
G^{(1)}_{\theta_k}(\theta_k) &= \frac{1}{\Delta}[(G^{(1)}_{d_k}\Delta_1 - G^{(1)}_{c_k}\Delta_{12})\cos\theta_k - (G^{(1)}_{c_k}\Delta_2 - G^{(1)}_{d_k}\Delta_{12})\sin\theta_k] = -\beta_k.
\end{aligned} \tag{2.85}
$$

Furthermore, Equation (2.79) gives

$$
\dot{r}_k = \alpha_k r_k + o(r_k) \text{ and } \dot{\theta}_k r_k = -\beta_k r_k + o(r_k). \tag{2.86}
$$

As $r_k << 1$ and $r_k \to 0$, we have

$$
\dot{r}_k = \alpha_k r_k \text{ and } \dot{\theta}_k = -\beta_k. \tag{2.87}
$$

With an initial condition of $r_k = r_k^0$ and $\theta_k = \theta_k^0$, the corresponding solution of Equation (2.87) is

$$
r_k = r_k^0 e^{\alpha_k t} \text{ and } \theta_k = -\beta_k t + \theta_k^0. \tag{2.88}
$$

and

$$
\begin{aligned}
c_k &= r_k^0 e^{\alpha_k t}\cos(-\beta_k t + \theta_k^0) = e^{\alpha_k t}[\cos(\beta_k t)c_k^0 + \sin(\beta_k t)d_k^0]; \\
d_k &= r_k^0 e^{\alpha_k t}\sin(-\beta_k t + \theta_k^0) = e^{\alpha_k t}[-\sin(\beta_k t)c_k^0 + \cos(\beta_k t)d_k^0].
\end{aligned} \tag{2.89}
$$

Letting $\mathbf{c}^{(k)} = (c^{(k)}, d^{(k)})^{\mathrm{T}}$, we have

$$
\dot{\mathbf{c}}^{(k)} = \mathbf{E}_k \mathbf{c}^{(k)} \Rightarrow \mathbf{c}^{(k)} = e^{\alpha_k t}\mathbf{B}_k \mathbf{c}_0^{(k)} \tag{2.90}
$$

where

$$
\mathbf{E}_i = \begin{bmatrix} \alpha_k & \beta_k \\ -\beta_k & \alpha_k \end{bmatrix} \text{ and } \mathbf{B}_k = \begin{bmatrix} \cos\beta_k t & \sin\beta_k t \\ -\sin\beta_k t & \cos\beta_k t \end{bmatrix}. \tag{2.91}
$$

If $G_{r_k}^{(m)}(\theta_k)$ and $G_{\theta_k}^{(m)}(\theta_k)$ are dependent on θ_k, Equation (2.79) gives the dynamical systems based on the polar coordinates on the invariant plane of $(\mathbf{u}_k, \mathbf{v}_k)$ of matrix $D\mathbf{f}(\mathbf{x}^*, \mathbf{p})$ with a pair of eigenvectors $\mathbf{u}_k \pm \mathbf{i}\mathbf{v}_k$. If $G_{r_k}^{(m)}(\theta_k)$ and $G_{\theta_k}^{(m)}(\theta_k)$ are independent of θ_k, the deformed dynamical system on the plane of $(\mathbf{u}_k, \mathbf{v}_k)$ is dependent on r_k, then the G-functions can be used to determine the stability of $\mathbf{x}^{(k)}$ at the equilibrium $\mathbf{x}^*$ on the plane of $(\mathbf{u}_k, \mathbf{v}_k)$.

Definition 2.24 Consider an n-dimensional, autonomous, nonlinear dynamical system $\dot{\mathbf{x}} = \mathbf{f}(\mathbf{x}, \mathbf{p})$ in Equation (2.4) with an equilibrium point $\mathbf{x}^*$ and $\mathbf{f}(\mathbf{x}, \mathbf{p})$ is C^r $(r \geq 1)$-continuous in a neighborhood of the equilibrium $\mathbf{x}^*$. The corresponding solution is $\mathbf{x}(t) = \mathbf{\Phi}(\mathbf{x}_0, t - t_0, \mathbf{p})$. Suppose $U(\mathbf{x}^*) \subset \Omega$ is a neighborhood of equilibrium $\mathbf{x}^*$, and there are n linearly independent vectors $\mathbf{v}_k$ $(k = 1, 2, \ldots, n)$. For a linearized dynamical system in Equation (2.19), consider a pair of complex eigenvalues $\alpha_k \pm \mathbf{i}\beta_k$ $(k \in N = \{1, 2, \ldots, n\}, \mathbf{i} = \sqrt{-1})$ of matrix $D\mathbf{f}(\mathbf{x}^*, \mathbf{p})$ with a pair of eigenvectors $\mathbf{u}_k \pm \mathbf{i}\mathbf{v}_k$. On the invariant plane of $(\mathbf{u}_k, \mathbf{v}_k)$, consider $\mathbf{y}^{(k)} = \mathbf{y}_+^{(k)} + \mathbf{y}_-^{(k)}$ with Equations (2.72) and (2.74). For any arbitrarily small $\varepsilon > 0$, the stability of the equilibrium $\mathbf{x}^*$ on the invariant plane of $(\mathbf{u}_k, \mathbf{v}_k)$ can be determined.

1. $\mathbf{x}^{(k)}$ at the equilibrium $\mathbf{x}^*$ on the plane of $(\mathbf{u}_k, \mathbf{v}_k)$ is spirally stable if

$$r_k(t + \varepsilon) - r_k(t) < 0. \tag{2.92}$$

2. $\mathbf{x}^{(k)}$ at the equilibrium $\mathbf{x}^*$ on the plane of $(\mathbf{u}_k, \mathbf{v}_k)$ is spirally unstable if

$$r_k(t + \varepsilon) - r_k(t) > 0. \tag{2.93}$$

3. $\mathbf{x}^{(k)}$ at the equilibrium $\mathbf{x}^*$ on the plane of $(\mathbf{u}_k, \mathbf{v}_k)$ is spirally stable with the m_kth-order singularity if for $\theta_k \in [0, 2\pi]$

$$\begin{aligned} G_{r_k}^{(s_k)}(\theta_k) = 0 \ \text{ for } s_k = 0, 1, 2, \ldots, m_k - 1, \\ r_k(t + \varepsilon) - r_k(t) < 0. \end{aligned} \tag{2.94}$$

4. $\mathbf{x}^{(k)}$ at the equilibrium $\mathbf{x}^*$ on the plane of $(\mathbf{u}_k, \mathbf{v}_k)$ is spirally unstable with the m_kth-order singularity if for $\theta_k \in [0, 2\pi]$

$$\begin{aligned} G_{r_k}^{(s_k)}(\theta_k) = 0 \ \text{ for } s_k = 0, 1, 2, \ldots, m_k - 1, \\ r_k(t + \varepsilon) - r_k(t) > 0. \end{aligned} \tag{2.95}$$

5. $\mathbf{x}^{(k)}$ at the equilibrium $\mathbf{x}^*$ on the plane of $(\mathbf{u}_k, \mathbf{v}_k)$ is circular if for $\theta_k \in [0, 2\pi]$

$$r_k(t + \varepsilon) - r_k(t) = 0. \tag{2.96}$$

6. $\mathbf{x}^{(k)}$ at the equilibrium $\mathbf{x}^*$ on the plane of $(\mathbf{u}_k, \mathbf{v}_k)$ is degenerate in the direction of $\mathbf{u}_k$ if

$$\beta_k = 0 \text{ and } \theta_k(t + \varepsilon) - \theta_k(t) = 0. \tag{2.97}$$

Theorem 2.10 *Consider an n-dimensional, autonomous, nonlinear dynamical system* $\dot{\mathbf{x}} = \mathbf{f}(\mathbf{x}, \mathbf{p})$ *in Equation (2.4) with an equilibrium point* $\mathbf{x}^*$ *and* $\mathbf{f}(\mathbf{x}, \mathbf{p})$ *is* C^r $(r \geq 1)$*-continuous in a neighborhood of the equilibrium* $\mathbf{x}^*$*. The corresponding solution is* $\mathbf{x}(t) = \mathbf{\Phi}(\mathbf{x}_0, t - t_0, \mathbf{p})$.

Suppose $U(\mathbf{x}^*) \subset \Omega$ is a neighborhood of equilibrium $\mathbf{x}^*$, and there are n linearly independent vectors $\mathbf{v}_k$ ($k = 1, 2, \ldots, n$). For a linearized dynamical system in Equation (2.19), consider a pair of complex eigenvalues $\alpha_k \pm \mathbf{i}\beta_k$ ($k \in N = \{1, 2, \ldots, n\}, \mathbf{i} = \sqrt{-1}$) of matrix $D\mathbf{f}(\mathbf{x}^*, \mathbf{p})$ with a pair of eigenvectors $\mathbf{u}_k \pm \mathbf{i}\mathbf{v}_k$. On the invariant plane of $(\mathbf{u}_k, \mathbf{v}_k)$, consider $\mathbf{y}^{(k)} = \mathbf{y}_+^{(k)} + \mathbf{y}_-^{(k)}$ with Equations (2.72) and (2.74) with $G_{r_k}^{(s_k)}(\theta_k) = const$. For any arbitrarily small $\varepsilon > 0$, the stability of the equilibrium $\mathbf{x}^*$ on the invariant plane of $(\mathbf{u}_k, \mathbf{v}_k)$ can be determined.

1. *$\mathbf{x}^{(k)}$ at the equilibrium $\mathbf{x}^*$ on the plane of $(\mathbf{u}_k, \mathbf{v}_k)$ is spirally stable if and only if*

$$G_{r_k}^{(1)}(\theta_k) = \alpha_k < 0. \tag{2.98}$$

2. *$\mathbf{x}^{(k)}$ at the equilibrium $\mathbf{x}^*$ on the plane of $(\mathbf{u}_k, \mathbf{v}_k)$ is spirally unstable if and only if*

$$G_{r_k}^{(1)}(\theta_k) = \alpha_k > 0. \tag{2.99}$$

3. *$\mathbf{x}^{(k)}$ at the equilibrium $\mathbf{x}^*$ on the plane of $(\mathbf{u}_k, \mathbf{v}_k)$ is spirally stable with the m_kth-order singularity if and only if for $\theta_k \in [0, 2\pi]$*

$$\begin{aligned} G_{r_k}^{(s_k)}(\theta_k) &= 0 \ \text{ for } s_k = 1, 2, \ldots, m_k - 1, \\ \text{and } G_{r_k}^{(m_k)}(\theta_k) &< 0. \end{aligned} \tag{2.100}$$

4. *$\mathbf{x}^{(k)}$ at the equilibrium $\mathbf{x}^*$ on the plane of $(\mathbf{u}_k, \mathbf{v}_k)$ is spirally unstable with the m_kth-order singularity if and only if for $\theta_k \in [0, 2\pi]$*

$$\begin{aligned} G_{r_k}^{(s_k)}(\theta_k) &= 0 \ \text{ for } s_k = 1, 2, \ldots, m_k - 1, \\ \text{and } G_{r_k}^{(m_k)}(\theta_k) &> 0. \end{aligned} \tag{2.101}$$

5. *$\mathbf{x}^{(k)}$ at the equilibrium $\mathbf{x}^*$ on the plane of $(\mathbf{u}_k, \mathbf{v}_k)$ is circular if and only if for $\theta_k \in [0, 2\pi]$*

$$G_{r_k}^{(s_k)}(\theta_k) = 0 \ \text{ for } s_k = 1, 2, \ldots. \tag{2.102}$$

6. *$\mathbf{x}^{(k)}$ at the equilibrium $\mathbf{x}^*$ on the plane of $(\mathbf{u}_k, \mathbf{v}_k)$ is degenerate in the direction of $\mathbf{u}_k$ if and only if*

$$\text{Im}\lambda_k = \beta_k = 0 \text{ and } G_{\theta_k}^{(s_k)}(\theta_k) = 0 \ \text{ for } s_k = 2, 3, \ldots. \tag{2.103}$$

Proof. The proof can be referred to Luo (2012a). ■

Note that $G_{r_k}^{(s_k)}(\theta_k) = const$ requires $s_k = 2m_k - 1$ and one obtains $G_{r_k}^{(s_k)}(\theta_k) = 0$ for $s_k = 2m_k$.

2.3.2 *Bifurcations*

Definition 2.25 Consider an n-dimensional, autonomous, nonlinear dynamical system $\dot{\mathbf{x}} = \mathbf{f}(\mathbf{x}, \mathbf{p})$ in Equation (2.4) with an equilibrium point $\mathbf{x}^*$ and $\mathbf{f}(\mathbf{x}, \mathbf{p})$ is C^r ($r \geq 1$)-continuous in a neighborhood of the equilibrium $\mathbf{x}^*$ (i.e., $U(\mathbf{x}^*) \subset \Omega$). The corresponding solution is $\mathbf{x}(t) = \mathbf{\Phi}(\mathbf{x}_0, t - t_0, \mathbf{p})$. Suppose Equation (2.24) holds in $U(\mathbf{x}^*) \subset \Omega$. For a linearized dynamical system in Equation (2.19), consider a real eigenvalue λ_k of matrix $D\mathbf{f}(\mathbf{x}^*, \mathbf{p}^*)$

$(k \in N = \{1, 2, \ldots, n\})$ with an eigenvector $\mathbf{v}_k$. Suppose one of n independent solutions $\mathbf{y} = c_k \mathbf{v}_k$ and $\dot{\mathbf{y}} = \dot{c}_k \mathbf{v}_k$,

$$s_k = \mathbf{v}_k^{\mathrm{T}} \cdot \mathbf{y} = \mathbf{v}_k^{\mathrm{T}} \cdot (\mathbf{x} - \mathbf{x}^*) \tag{2.104}$$

where $s_k = c_k \|\mathbf{v}_k\|^2$.

$$\dot{s}_k = \mathbf{v}_k^{\mathrm{T}} \cdot \dot{\mathbf{y}} = \mathbf{v}_k^{\mathrm{T}} \cdot \dot{\mathbf{x}} = \mathbf{v}_k^{\mathrm{T}} \cdot \mathbf{f}(\mathbf{x}, \mathbf{p}). \tag{2.105}$$

In the vicinity of point $(\mathbf{x}_0^*, \mathbf{p}_0)$, $\mathbf{v}_k^{\mathrm{T}} \cdot \mathbf{f}(\mathbf{x}, \mathbf{p})$ can be expanded for $(0 < \theta < 1)$ as

$$\begin{aligned} \mathbf{v}_k^{\mathrm{T}} \cdot \mathbf{f}(\mathbf{x}, \mathbf{p}) &= a_k (s_k - s_{k0}^*) + \mathbf{b}_k^{\mathrm{T}} \cdot (\mathbf{p} - \mathbf{p}_0) \\ &+ \sum_{r=0}^{m>1} C_m^r \mathbf{a}_k^{(m-r,r)} (s_k - s_{k0}^*)^{m-r} (\mathbf{p} - \mathbf{p}_0)^r \\ &+ [(s_k - s_{k0}^*)\partial_{s_k} + (\mathbf{p} - \mathbf{p}_0)\partial_{\mathbf{p}}]^{m+1} (\mathbf{v}_k^{\mathrm{T}} \cdot \mathbf{f}(\mathbf{x}_0^* + \theta \Delta \mathbf{x}, \mathbf{p}_0 + \theta \Delta \mathbf{p})) \end{aligned} \tag{2.106}$$

where

$$\begin{aligned} a_k &= \mathbf{v}_k^{\mathrm{T}} \cdot \partial_{s_k} \mathbf{f}(\mathbf{x}, \mathbf{p})|_{(\mathbf{x}_0^*, \mathbf{p}_0)}, \\ \mathbf{b}_k^{\mathrm{T}} &= \mathbf{v}_k^{\mathrm{T}} \cdot \partial_{\mathbf{p}} \mathbf{f}(\mathbf{x}, \mathbf{p})|_{(\mathbf{x}_0^*, \mathbf{p}_0)}, \\ \mathbf{a}_k^{(r,s)} &= \mathbf{v}_k^{\mathrm{T}} \cdot \partial_{s_k}^{(r)} \partial_{\mathbf{p}}^{(s)} \mathbf{f}(\mathbf{x}, \mathbf{p})|_{(\mathbf{x}_0^*, \mathbf{p}_0)}. \end{aligned} \tag{2.107}$$

If $a_k = 0$ and $\mathbf{p} = \mathbf{p}_0$, the stability of current equilibrium $\mathbf{x}^*$ on an eigenvector $\mathbf{v}_k$ changes from stable to unstable state (or from unstable to stable state). The bifurcation manifold in the direction of $\mathbf{v}_k$ is determined by

$$\mathbf{b}_k^{\mathrm{T}} \cdot (\mathbf{p} - \mathbf{p}_0) + \sum_{r=0}^{m>1} C_m^r \mathbf{a}_k^{(m-r,r)} (s_k^* - s_{k0}^*)^{m-r} (\mathbf{p} - \mathbf{p}_0)^r = 0. \tag{2.108}$$

In the neighborhood of $(\mathbf{x}_0^*, \mathbf{p}_0)$, when other components of equilibrium $\mathbf{x}^*$ on the eigenvector of $\mathbf{v}_j$ for all $j \neq k$, $(j, k \in N)$ do not change their stability states, Equation (2.108) possesses l-branch solutions of equilibrium s_k^* $(0 < l \le m)$ with l_1-stable and l_2-unstable solutions $(l_1, l_2 \in \{0, 1, 2, \ldots, l\})$. Such l-branch solutions are called the bifurcation solutions of equilibrium $\mathbf{x}^*$ on the eigenvector of $\mathbf{v}_k$ in the neighborhood of $(\mathbf{x}_0^*, \mathbf{p}_0)$. Such a bifurcation at point $(\mathbf{x}_0^*, \mathbf{p}_0)$ is called the hyperbolic bifurcation of mth-order on the eigenvector of $\mathbf{v}_k$. Three special cases are defined as

1. If

$$\begin{aligned} \mathbf{a}_k^{(1,1)} &= \mathbf{0}, \\ \mathbf{b}_k^{\mathrm{T}} \cdot (\mathbf{p} - \mathbf{p}_0) + \frac{1}{2!} a_k^{(2,0)} (s_k^* - s_{k0}^*)^2 &= 0 \end{aligned} \tag{2.109}$$

where

$$\begin{aligned} a_k^{(2,0)} &= \mathbf{v}_k^{\mathrm{T}} \cdot \partial_{s_k}^{(2)} \partial_{\mathbf{p}}^{(0)} \mathbf{f}(\mathbf{x}, \mathbf{p})|_{(\mathbf{x}_0^*, \mathbf{p}_0)} = \mathbf{v}_k^{\mathrm{T}} \cdot \partial_{s_k}^{(2)} \mathbf{f}(\mathbf{x}, \mathbf{p})|_{(\mathbf{x}_0^*, \mathbf{p}_0)} \\ &= \mathbf{v}_k^{\mathrm{T}} \cdot \partial_{\mathbf{x}}^{(2)} \mathbf{f}(\mathbf{x}, \mathbf{p})(\mathbf{v}_k \mathbf{v}_k)|_{(\mathbf{x}_0^*, \mathbf{p}_0)} = G_k^{(2)}(\mathbf{x}_0^*, \mathbf{p}_0) \neq 0, \\ \mathbf{b}_k^{\mathrm{T}} &= \mathbf{v}_k^{\mathrm{T}} \cdot \partial_{\mathbf{p}} \mathbf{f}(\mathbf{x}, \mathbf{p})|_{(\mathbf{x}_0^*, \mathbf{p}_0)} \neq \mathbf{0}, \end{aligned} \tag{2.110}$$

$$a_k^{(2,0)} \times [\mathbf{b}_k^{\mathrm{T}} \cdot (\mathbf{p} - \mathbf{p}_0)] < 0, \tag{2.111}$$

such a bifurcation at point $(\mathbf{x}_0^*, \mathbf{p}_0)$ is called the *saddle-node* bifurcation on the eigenvector of $\mathbf{v}_k$.

2. If

$$\begin{gathered} \mathbf{b}_\alpha^{\mathrm{T}} \cdot (\mathbf{p} - \mathbf{p}_0) = 0, \\ \mathbf{a}_k^{(1,1)} \cdot (\mathbf{p} - \mathbf{p}_0)(s_k^* - s_{k0}^*) + \frac{1}{2!} a_k^{(2,0)} (s_k^* - s_{k0}^*)^2 = 0 \end{gathered} \tag{2.112}$$

where

$$\begin{aligned} a_k^{(2,0)} &= \mathbf{v}_k^{\mathrm{T}} \cdot \partial_{s_k}^{(2)} \partial_{\mathbf{p}}^{(0)} \mathbf{f}(\mathbf{x}, \mathbf{p})|_{(\mathbf{x}_0^*, \mathbf{p}_0)} = \mathbf{v}_k^{\mathrm{T}} \cdot \partial_{s_k}^{(2)} \mathbf{f}(\mathbf{x}, \mathbf{p})|_{(\mathbf{x}_0^*, \mathbf{p}_0)} \\ &= \mathbf{v}_k^{\mathrm{T}} \cdot \partial_{\mathbf{x}}^{(2)} \mathbf{f}(\mathbf{x}, \mathbf{p})(\mathbf{v}_k \mathbf{v}_k)|_{(\mathbf{x}_0^*, \mathbf{p}_0)} = G_k^{(2)}(\mathbf{x}_0^*, \mathbf{p}_0) \neq 0, \\ \mathbf{a}_k^{(1,1)} &= \mathbf{v}_k^{\mathrm{T}} \cdot \partial_{s_k}^{(1)} \partial_{\mathbf{p}}^{(1)} \mathbf{f}(\mathbf{x}, \mathbf{p})|_{(\mathbf{x}_0^*, \mathbf{p}_0)} = \mathbf{v}_k^{\mathrm{T}} \cdot \partial_{s_k} \partial_{\mathbf{p}} \mathbf{f}(\mathbf{x}, \mathbf{p})|_{(\mathbf{x}_0^*, \mathbf{p}_0)} \\ &= \mathbf{v}_k^{\mathrm{T}} \cdot \partial_{\mathbf{x}} \partial_{\mathbf{p}} \mathbf{f}(\mathbf{x}, \mathbf{p}) \mathbf{v}_k|_{(\mathbf{x}_0^*, \mathbf{p}_0)} \neq \mathbf{0}, \end{aligned} \tag{2.113}$$

$$a_k^{(2,0)} \times [\mathbf{a}_k^{(1,1)} \cdot (\mathbf{p} - \mathbf{p}_0)] < 0, \tag{2.114}$$

such a bifurcation at point $(\mathbf{x}_0^*, \mathbf{p}_0)$ is called the *transcritical* bifurcation on the eigenvector of $\mathbf{v}_k$.

3. If

$$\begin{gathered} \mathbf{b}_\alpha^{\mathrm{T}} \cdot (\mathbf{p} - \mathbf{p}_0) = 0, \\ a_k^{(2,0)} = 0, \mathbf{a}_k^{(2,1)} = 0, \mathbf{a}_k^{(1,2)} = 0, \\ \mathbf{a}_k^{(1,1)} \cdot (\mathbf{p} - \mathbf{p}_0)(s_k^* - s_{k0}^*) + \frac{1}{3!} a_k^{(3,0)} (s_k^* - s_{k0}^*)^3 = 0 \end{gathered} \tag{2.115}$$

where

$$\begin{aligned} a_k^{(3,0)} &= \mathbf{v}_k^{\mathrm{T}} \cdot \partial_{s_k}^{(3)} \partial_{\mathbf{p}}^{(0)} \mathbf{f}(\mathbf{x}, \mathbf{p})|_{(\mathbf{x}_0^*, \mathbf{p}_0)} = \mathbf{v}_k^{\mathrm{T}} \cdot \partial_{s_k}^{(3)} \mathbf{f}(\mathbf{x}, \mathbf{p})|_{(\mathbf{x}_0^*, \mathbf{p}_0)} \\ &= \mathbf{v}_k^{\mathrm{T}} \cdot \partial_{\mathbf{x}}^{(3)} \mathbf{f}(\mathbf{x}, \mathbf{p})(\mathbf{v}_k \mathbf{v}_k \mathbf{v}_k)|_{(\mathbf{x}_0^*, \mathbf{p}_0)} = G_k^{(3)}(\mathbf{x}_0^*, \mathbf{p}_0) \neq 0, \\ \mathbf{a}_k^{(1,1)} &= \mathbf{v}_k^{\mathrm{T}} \cdot \partial_{s_k}^{(1)} \partial_{\mathbf{p}}^{(1)} \mathbf{f}(\mathbf{x}, \mathbf{p})|_{(\mathbf{x}_0^*, \mathbf{p}_0)} = \mathbf{v}_k^{\mathrm{T}} \cdot \partial_{s_k} \partial_{\mathbf{p}} \mathbf{f}(\mathbf{x}, \mathbf{p})|_{(\mathbf{x}_0^*, \mathbf{p}_0)} \\ &= \mathbf{v}_k^{\mathrm{T}} \cdot \partial_{\mathbf{x}} \partial_{\mathbf{p}} \mathbf{f}(\mathbf{x}, \mathbf{p}) \mathbf{v}_k|_{(\mathbf{x}_0^*, \mathbf{p}_0)} \neq \mathbf{0}, \end{aligned} \tag{2.116}$$

$$a_k^{(3,0)} \times [\mathbf{a}_k^{(1,1)} \cdot (\mathbf{p} - \mathbf{p}_0)] < 0, \tag{2.117}$$

such a bifurcation at point $(\mathbf{x}_0^*, \mathbf{p}_0)$ is called the *pitchfork* bifurcation on the eigenvector of $\mathbf{v}_k$.

From the analysis, the bifurcation points possess the higher-order singularity of the flow in dynamical system. For the saddle-node bifurcation, the $(2m)$th order singularity of the flow at the bifurcation point exists as a saddle of the $(2m)$th order. For the transcritical bifurcation, the $(2m)$th order singularity of the flow at the bifurcation point exists as a saddle of the $(2m)$th order. However, for the stable pitchfork bifurcation, the $(2m+1)$th order singularity of the flow at the bifurcation point exists as a sink of the $(2m+1)$th order. For the unstable pitchfork bifurcation, the $(2m+1)$th order singularity of the flow at the bifurcation point exists as a source of the $(2m+1)$th order.

Definition 2.26 Consider an n-dimensional, autonomous, nonlinear dynamical system $\dot{\mathbf{x}} = \mathbf{f}(\mathbf{x},\mathbf{p})$ in Equation (2.4) with an equilibrium point $\mathbf{x}^*$ and $\mathbf{f}(\mathbf{x},\mathbf{p})$ is C^r $(r \geq 1)$-continuous in a neighborhood of the equilibrium $\mathbf{x}^*$. The corresponding solution is $\mathbf{x}(t) = \boldsymbol{\Phi}(\mathbf{x}_0, t-t_0, \mathbf{p})$. Suppose $U(\mathbf{x}^*) \subset \Omega$ is a neighborhood of equilibrium $\mathbf{x}^*$, and there are n linearly independent vectors $\mathbf{v}_k$ $(k = 1, 2, \ldots, n)$. For a linearized dynamical system in Equation (2.19), consider a pair of complex eigenvalues $\alpha_k \pm \mathbf{i}\beta_k$ $(k \in N = \{1, 2, \ldots, n\}, \mathbf{i} = \sqrt{-1})$ of matrix $D\mathbf{f}(\mathbf{x}^*, \mathbf{p})$ with a pair of eigenvectors $\mathbf{u}_k \pm \mathbf{i}\mathbf{v}_k$. On the invariant plane of $(\mathbf{u}_k, \mathbf{v}_k)$, consider $\mathbf{r}_k = \mathbf{y}_+^{(k)} + \mathbf{y}_-^{(k)}$ with

$$\begin{aligned} \mathbf{r}_k &= c_k\mathbf{u}_k + d_k\mathbf{v}_k = r_k\mathbf{e}_{r_k}, \\ \dot{\mathbf{r}}_k &= \dot{c}_k\mathbf{u}_k + \dot{d}_k\mathbf{v}_k = \dot{r}_k\mathbf{e}_{r_k} + r_k\dot{\mathbf{e}}_{r_k} \end{aligned} \tag{2.118}$$

and

$$\begin{aligned} c_k &= \frac{1}{\Delta}[\Delta_2(\mathbf{u}_k^{\mathrm{T}}\cdot\mathbf{y}) - \Delta_{12}(\mathbf{v}_k^{\mathrm{T}}\cdot\mathbf{y})] \text{ and } d_k = \frac{1}{\Delta}[\Delta_1(\mathbf{v}_k^{\mathrm{T}}\cdot\mathbf{y}) - \Delta_{12}(\mathbf{u}_k^{\mathrm{T}}\cdot\mathbf{y})], \\ \Delta_1 &= \|\mathbf{u}_k\|^2, \Delta_2 = \|\mathbf{v}_k\|^2, \Delta_{12} = \mathbf{u}_k^{\mathrm{T}}\cdot\mathbf{v}_k \text{ and } \Delta = \Delta_1\Delta_2 - \Delta_{12}^2. \end{aligned} \tag{2.119}$$

Consider a polar coordinate of (r_k, θ_k) defined by

$$\begin{aligned} c_k &= r_k\cos\theta_k, \text{and } d_k = r_k\sin\theta_k; \\ r_k &= \sqrt{c_k^2 + d_k^2}, \text{and } \theta_k = \arctan d_k/c_k; \\ \mathbf{e}_{r_k} &= \cos\theta_k\mathbf{u}_k + \sin\theta_k\mathbf{v}_k \text{ and} \\ \mathbf{e}_{\theta_k} &= -\cos\theta_k\mathbf{u}_k^{\perp}\Delta_3 + \sin\theta_k\mathbf{v}_k^{\perp}\Delta_4, \\ \Delta_3 &= \mathbf{v}_k^{\mathrm{T}}\cdot\mathbf{u}_k^{\perp} \text{ and } \Delta_4 = \mathbf{u}_k^{\mathrm{T}}\cdot\mathbf{v}_k^{\perp}. \end{aligned} \tag{2.120}$$

Thus

$$\begin{aligned} \dot{c}_k &= \frac{1}{\Delta}[\Delta_2 G_{c_k}(\mathbf{x},\mathbf{p}) - \Delta_{12}G_{d_k}(\mathbf{x},\mathbf{p})], \\ \dot{d}_k &= \frac{1}{\Delta}[\Delta_1 G_{d_k}(\mathbf{x},\mathbf{p}) - \Delta_{12}G_{d_k}(\mathbf{x},\mathbf{p})] \end{aligned} \tag{2.121}$$

where

$$\begin{aligned} G_{c_k}(\mathbf{x},\mathbf{p}) &= \mathbf{u}_k^{\mathrm{T}}\cdot\mathbf{f}(\mathbf{x},\mathbf{p}) = \mathbf{a}_k^{\mathrm{T}}\cdot(\mathbf{p}-\mathbf{p}_0) + a_{k11}(c_k - c_{k0}^*) + a_{k12}(d_k - d_{k0}^*) \\ &+ \sum_{r=0}^{m>1} C_m^r \mathbf{G}_{c_k}^{(m-r,r)}(\mathbf{x}^*,\mathbf{p}_0)(\mathbf{p}-\mathbf{p}_0)^r r_k^{m-r} \\ &+ [(c_k - c_{k0}^*)\partial_{c_k} + (d_k - d_{k0}^*)\partial_{d_k} + (\mathbf{p}-\mathbf{p}_0)\partial_{\mathbf{p}}]^{m+1} \\ &\times (\mathbf{u}_k^{\mathrm{T}}\cdot\mathbf{f}(\mathbf{x}_0^* + \theta\Delta\mathbf{x}, \mathbf{p}_0 + \theta\Delta\mathbf{p})), \end{aligned}$$

$$
\begin{aligned}
G_{d_k}(\mathbf{x},\mathbf{p}) = \mathbf{v}_k^{\mathrm{T}}\cdot\mathbf{f}(\mathbf{x},\mathbf{p}) &= \mathbf{b}_k^{\mathrm{T}}\cdot(\mathbf{p}-\mathbf{p}_0) + a_{k21}(c_k - c_{k0}^*) + a_{k22}(d_k - d_{k0}^*) \\
&+ \sum_{r=0}^{m>1} C_m^r \mathbf{G}_{d_k}^{(m-r,r)}(\mathbf{x}^*,\mathbf{p}_0)(\mathbf{p}-\mathbf{p}_0)^r r_k^{m-r} \\
&+ [(c_k - c_{k0}^*)\partial_{c_k} + (d_k - d_{k0}^*)\partial_{d_k} + (\mathbf{p}-\mathbf{p}_0)\partial_{\mathbf{p}}]^{m+1} \\
&\times(\mathbf{v}_k^{\mathrm{T}}\cdot\mathbf{f}(\mathbf{x}_0^* + \theta\Delta\mathbf{x}, \mathbf{p}_0 + \theta\Delta\mathbf{p}))
\end{aligned} \tag{2.122}
$$

and

$$
\begin{aligned}
\mathbf{G}_{c_k}^{(s,r)}(\mathbf{x}^*,\mathbf{p}) &= \mathbf{u}_k^{\mathrm{T}}\cdot[\partial_{\mathbf{x}}()\mathbf{u}_k\cos\theta_k + \partial_{\mathbf{x}}()\mathbf{v}_k\sin\theta_k]^s\partial_{\mathbf{p}}^{(r)}\mathbf{f}(\mathbf{x},\mathbf{p})|_{(\mathbf{x}^*,\mathbf{p})}, \\
\mathbf{G}_{d_k}^{(s,r)}(\mathbf{x}^*,\mathbf{p}) &= \mathbf{v}_k^{\mathrm{T}}\cdot[\partial_{\mathbf{x}}()\mathbf{u}_k\cos\theta_k + \partial_{\mathbf{x}}()\mathbf{v}_k\sin\theta_k]^s\partial_{\mathbf{p}}^{(r)}\mathbf{f}(\mathbf{x},\mathbf{p})|_{(\mathbf{x}^*,\mathbf{p})};
\end{aligned} \tag{2.123}
$$

$$
\begin{aligned}
\mathbf{a}_k^{\mathrm{T}} &= \mathbf{u}_k^{\mathrm{T}}\cdot\partial_{\mathbf{p}}\mathbf{f}(\mathbf{x},\mathbf{p}),\ \ \mathbf{b}_k^{\mathrm{T}} = \mathbf{v}_k^{\mathrm{T}}\cdot\partial_{\mathbf{p}}\mathbf{f}(\mathbf{x},\mathbf{p}); \\
a_{k11} &= \mathbf{u}_k^{\mathrm{T}}\cdot\partial_{\mathbf{x}}\mathbf{f}(\mathbf{x},\mathbf{p})\mathbf{u}_k,\ \ a_{k12} = \mathbf{u}_k^{\mathrm{T}}\cdot\partial_{\mathbf{x}}\mathbf{f}(\mathbf{x},\mathbf{p})\mathbf{v}_k; \\
a_{k21} &= \mathbf{v}_k^{\mathrm{T}}\cdot\partial_{\mathbf{x}}\mathbf{f}(\mathbf{x},\mathbf{p})\mathbf{u}_k,\ \ a_{k22} = \mathbf{v}_k^{\mathrm{T}}\cdot\partial_{\mathbf{x}}\mathbf{f}(\mathbf{x},\mathbf{p})\mathbf{v}_k.
\end{aligned} \tag{2.124}
$$

Thus

$$
\begin{aligned}
\dot{r}_k &= \dot{c}_k\cos\theta_k + \dot{d}_k\sin\theta_k \\
&= \sum_{r=0}^{m} C_m^r \mathbf{G}_{r_k}^{(m-r,r)}(\theta_k,\mathbf{p}_0)(\mathbf{p}-\mathbf{p}_0)^{m-r} r_k^r \\
\dot{\theta}_k &= r_k^{-1}(\dot{d}_k\cos\theta_k - \dot{c}_k\sin\theta_k) \\
&= \sum_{r=0}^{m} C_m^r \mathbf{G}_{\theta_k}^{(m-r,r)}(\theta_k,\mathbf{p}_0)(\mathbf{p}-\mathbf{p}_0)^{m-r} r_k^{r-1}.
\end{aligned} \tag{2.125}
$$

where

$$
\begin{aligned}
\mathbf{G}_{r_k}^{(m-r,r)}(\theta_k,\mathbf{p}_0) &= \frac{1}{\Delta}[(\Delta_2\cos\theta_k - \Delta_{12}\sin\theta_k)\mathbf{G}_{c_k}^{(m-r,r)}(\mathbf{x}^*,\mathbf{p}_0) \\
&\quad + (\Delta_2\sin\theta_k - \Delta_{12}\cos\theta_k)\mathbf{G}_{d_k}^{(m-r,r)}(\mathbf{x}^*,\mathbf{p}_0)], \\
\mathbf{G}_{\theta_k}^{(m-r,r)}(\theta_k,\mathbf{p}_0) &= -\frac{1}{\Delta}[(\Delta_2\sin\theta_k + \Delta_{12}\cos\theta_k)\mathbf{G}_{c_k}^{(m-r,r)}(\mathbf{x}^*,\mathbf{p}_0) \\
&\quad - (\Delta_1\cos\theta_k - \Delta_{12}\sin\theta_k)\mathbf{G}_{d_k}^{(m-r,r)}(\mathbf{x}^*,\mathbf{p}_0)].
\end{aligned} \tag{2.126}
$$

Suppose

$$
\mathbf{a}_k^{\mathrm{T}}\cdot(\mathbf{p}-\mathbf{p}_0) = 0 \text{ and } \mathbf{b}_k^{\mathrm{T}}\cdot(\mathbf{p}-\mathbf{p}_0) = 0 \tag{2.127}
$$

then

$$
\begin{aligned}
\dot{r}_k &= (\alpha_k + \mathbf{G}_{r_k}^{(1,1)}(\theta_k,\mathbf{p}_0)\cdot(\mathbf{p}-\mathbf{p}_0))r_k + G_{r_k}^{(3,0)}(\theta_k,\mathbf{p}_0)r_k^3 + o(r_k^3), \\
\dot{\theta}_k &= \beta_k + \mathbf{G}_{\theta_k}^{(1,1)}(\theta_k,\mathbf{p}_0)\cdot(\mathbf{p}-\mathbf{p}_0) + \mathrm{G}_{\theta_k}^{(3,0)}(\theta_k,\mathbf{p}_0)r_k^2 + o(r_k^2)
\end{aligned} \tag{2.128}
$$

where

$$
\begin{aligned}
\mathbf{G}_{r_k}^{(1,1)}(\theta_k,\mathbf{p}_0) &= \mathbf{G}_{r_k}^{(1,1)}(\mathbf{p}_0) \text{ and } G_{r_k}^{(3,0)}(\theta_k,\mathbf{p}_0) = G_{r_k}^{(3,0)}(\mathbf{p}_0), \\
\mathbf{G}_{\theta_k}^{(1,1)}(\theta_k,\mathbf{p}_0) &= \mathbf{G}_{\theta_k}^{(1,1)}(\mathbf{p}_0) \text{ and } G_{\theta_k}^{(3,0)}(\theta_k,\mathbf{p}_0) = G_{\theta_k}^{(3,0)}(\mathbf{p}_0)
\end{aligned} \tag{2.129}
$$

If $\alpha_k = 0$ and $\mathbf{p} = \mathbf{p}_0$, the stability of current equilibrium $\mathbf{x}^*$ on an eigenvector plane of changes from stable to unstable state (or from unstable to stable state). The bifurcation manifold on the plane of $(\mathbf{u}_k, \mathbf{v}_k)$ is determined by

$$
\begin{aligned}
&(\alpha_{k0} + \mathbf{G}_{r_k}^{(1,1)}(\theta_k, \mathbf{p}_0) \cdot (\mathbf{p} - \mathbf{p}_0))r_k + \frac{1}{3!}G_{r_k}^{(3,0)}(\theta_k, \mathbf{p}_0)r_k^3 = 0,\\
&\beta_{k0} + \mathbf{G}_{\theta_k}^{(1,1)}(\theta_k, \mathbf{p}_0) \cdot (\mathbf{p} - \mathbf{p}_0) + \frac{1}{3!}\mathrm{G}_{\theta_k}^{(3,0)}(\theta_k, \mathbf{p}_0)r_k^2 = 0
\end{aligned} \tag{2.130}
$$

where

$$
\begin{aligned}
&\mathbf{G}_{r_k}^{(1,1)}(\theta_k, \mathbf{p}_0) = \partial_{\mathbf{p}}\alpha_k|_{(\mathbf{x}_0^*, \mathbf{p}_0)} \neq \mathbf{0},\\
&[\mathbf{G}_{r_k}^{(1,1)}(\theta_k, \mathbf{p}_0) \cdot (\mathbf{p} - \mathbf{p}_0)] \times G_{r_k}^{(3,0)}(\theta_k, \mathbf{p}_0) < 0
\end{aligned} \tag{2.131}
$$

Such a bifurcation at point $(\mathbf{x}_0^*, \mathbf{p}_0)$ is called the Hopf bifurcation on the eigenvector plane of $(\mathbf{u}_k, \mathbf{v}_k)$.

For the repeated eigenvalues of $D\mathbf{f}(\mathbf{x}^*, \mathbf{p})$, the bifurcation of equilibrium can be similarly discussed in the foregoing two Theorems 2.9 and 2.10. Herein, such a procedure will not be repeated.

In Luo (2012a), the Hopf bifurcation points possess the higher-order singularity of the flow in the dynamical system in the corresponding radial direction. For the stable Hopf bifurcation, the mth order singularity of the flow at the bifurcation point exists as a sink of the mth order in the radial direction. For the unstable Hopf bifurcation, the mth order singularity of the flow at the bifurcation point exists as a source of the mth order in the radial direction.

The stability and bifurcation of equilibrium for a 2-D nonlinear dynamic system are summarized with $\det(D\mathbf{f}) = \det(D\mathbf{f}(\mathbf{x}_0^*, \mathbf{p}_0))$ and $\mathrm{tr}(D\mathbf{f}) = \mathrm{tr}(D\mathbf{f}(\mathbf{x}_0^*, \mathbf{p}_0))$ in Figure 2.1. The thick

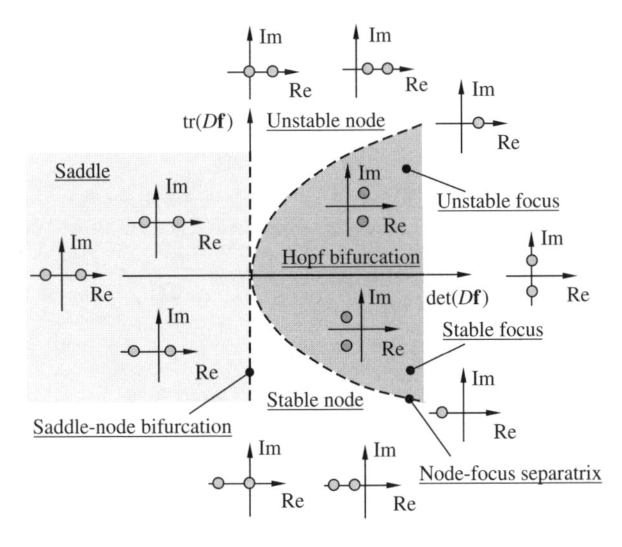

Figure 2.1 Stability and bifurcation diagrams through the complex plane of eigenvalues for 2D dynamical systems

dashed lines are bifurcation lines. The stability of equilibriums is given by the eigenvalues in complex plane. The stability of equilibriums for higher dimensional systems can be identified by using a naming of stability for linear dynamical systems in Luo (2012a). The saddle-node bifurcation possesses stable saddle-node bifurcation (critical) and unstable saddle-node bifurcation (degenerate).

3

An Analytical Method for Periodic Flows

In this chapter, from Luo (2012a), the analytical dynamics of periodic flows and chaos in nonlinear dynamical systems will be presented. The analytical solutions of periodic flows and chaos in autonomous systems will be discussed first, and the analytical dynamics of non-autonomous nonlinear dynamical systems will be presented. The analytical solutions of periodic motions in free and periodically excited vibration systems will be presented. In a similar fashion, the analytical solutions of periodic flows for time-delayed nonlinear systems will be presented with/without periodic excitations, and time-delayed nonlinear vibration systems will also be discussed for engineering applications. The analytical solutions of periodic flows and chaos are independent of the small parameters, which are different from the traditional perturbation methods. The methodology presented herein will end the history of chaos being numerically simulated only.

3.1 Nonlinear Dynamical Systems

In this section, analytical periodic flows in autonomous and periodically forced, nonlinear dynamical systems will be discussed. A generalized harmonic balance method with the Fourier series expressions will be presented for such analytical, periodic flows, and chaos in nonlinear dynamical systems. The local stability and bifurcation theory of equilibriums in nonlinear autonomous systems of coefficients in the Fourier series solutions will be employed to classify analytical solutions of periodic flows and chaos in nonlinear dynamical systems.

3.1.1 Autonomous Nonlinear Systems

Periodic flows in autonomous dynamical systems will be presented herein. If an autonomous nonlinear system has a periodic flow with period $T = 2\pi/\Omega$, then such a periodic flow can be expressed by a generalized coordinate transformation based on the Fourier series. The method is stated through the following theorem.

Theorem 3.1 *Consider a nonlinear dynamical system as*

$$\dot{\mathbf{x}} = \mathbf{f}(\mathbf{x}, \mathbf{p}) \in \mathscr{R}^n \tag{3.1}$$

Toward Analytical Chaos in Nonlinear Systems, First Edition. Albert C. J. Luo.
© 2014 John Wiley & Sons, Ltd. Published 2014 by John Wiley & Sons, Ltd.

where $\mathbf{f}(\mathbf{x}, \mathbf{p})$ *is a* C^r*-continuous nonlinear function vector* $(r \geq 1)$*. If such a dynamical system has a periodic flow* $\mathbf{x}(t)$ *with finite norm* $\|\mathbf{x}\|$ *and period* $T = 2\pi/\Omega$*, there is a generalized coordinate transformation with* $\theta = \Omega t$ *for the periodic flow of Equation (3.1) in a form of*

$$\mathbf{x}(t) = \mathbf{a}_0(t) + \sum_{k=1}^{\infty} \mathbf{b}_k(t)\cos(k\theta) + \mathbf{c}_k(t)\sin(k\theta) \tag{3.2}$$

with

$$\begin{aligned} \mathbf{a}_0 &= (a_{01}, a_{02}, \ldots, a_{0n})^{\mathrm{T}}, \\ \mathbf{b}_k &= (b_{k1}, b_{k2}, \ldots, b_{kn})^{\mathrm{T}}, \\ \mathbf{c}_k &= (c_{k1}, c_{k2}, \ldots, c_{kn})^{\mathrm{T}} \end{aligned} \tag{3.3}$$

and

$$\begin{gathered} \|\mathbf{x}\| = \|\mathbf{a}_0\| + \sum_{k=1}^{\infty} \|\mathbf{A}_k\|, \text{ and } \lim_{k\to\infty} \|\mathbf{A}_k\| = 0 \text{ but not uniform} \\ \text{with } \mathbf{A}_k = (A_{k1}, A_{k2}, \ldots, A_{kn})^{\mathrm{T}} \text{ and } A_{kj} = \sqrt{b_{kj}^2 + c_{kj}^2} \; (j = 1, 2, \ldots, n). \end{gathered} \tag{3.4}$$

For $\|\mathbf{x}(t) - \mathbf{x}^*(t)\| < \varepsilon$ *with a prescribed small positive* $\varepsilon > 0$*, the infinite term transformation of the periodic flow* $\mathbf{x}(t)$ *of Equation (3.1), given by Equation (3.2), can be approximated by a finite term transformation* $\mathbf{x}^*(t)$ *as*

$$\mathbf{x}^*(t) = \mathbf{a}_0(t) + \sum_{k=1}^{N} \mathbf{b}_k(t)\cos(k\theta) + \mathbf{c}_k(t)\sin(k\theta) \tag{3.5}$$

and the generalized coordinates are determined by

$$\begin{aligned} \dot{\mathbf{a}}_0 &= \mathbf{F}_0(\mathbf{a}_0, \mathbf{b}, \mathbf{c}), \\ \dot{\mathbf{b}} &= -\Omega \mathbf{k}_1 \mathbf{c} + \mathbf{F}_1(\mathbf{a}_0, \mathbf{b}, \mathbf{c}), \\ \dot{\mathbf{c}} &= \Omega \mathbf{k}_1 \mathbf{b} + \mathbf{F}_2(\mathbf{a}_0, \mathbf{b}, \mathbf{c}) \end{aligned} \tag{3.6}$$

where

$$\begin{aligned} &\mathbf{k}_1 = diag(\mathbf{I}_{n\times n}, 2\mathbf{I}_{n\times n}, \ldots, N\mathbf{I}_{n\times n}), \\ &\mathbf{b} = (\mathbf{b}_1, \mathbf{b}_2, \ldots, \mathbf{b}_N)^{\mathrm{T}}, \\ &\mathbf{c} = (\mathbf{c}_1, \mathbf{c}_2, \ldots, \mathbf{c}_N)^{\mathrm{T}}, \\ &\mathbf{F}_1 = (\mathbf{F}_{11}, \mathbf{F}_{12}, \ldots, \mathbf{F}_{1N})^{\mathrm{T}}, \\ &\mathbf{F}_2 = (\mathbf{F}_{21}, \mathbf{F}_{22}, \ldots, \mathbf{F}_{2N})^{\mathrm{T}} \\ &\text{for } N = 1, 2, \ldots, \infty \end{aligned} \tag{3.7}$$

and

$$\begin{aligned} \mathbf{F}_0(\mathbf{a}_0, \mathbf{b}, \mathbf{c}) &= \frac{1}{2\pi}\int_0^{2\pi} \mathbf{f}(\mathbf{x}^*, \mathbf{p})d\theta; \\ \mathbf{F}_{1k}(\mathbf{a}_0, \mathbf{b}, \mathbf{c}) &= \frac{1}{\pi}\int_0^{2\pi} \mathbf{f}(\mathbf{x}^*, \mathbf{p})\cos(k\theta)d\theta, \end{aligned}$$

$$\mathbf{F}_{2k}(\mathbf{a}_0, \mathbf{b}, \mathbf{c}) = \frac{1}{\pi}\int_0^{2\pi} \mathbf{f}(\mathbf{x}^*, \mathbf{p}) \sin(k\theta) d\theta$$

$$\text{for } k = 1, 2, \ldots, N. \tag{3.8}$$

Equation (3.6) becomes

$$\dot{\mathbf{z}} = \mathbf{f}(\mathbf{z}) \tag{3.9}$$

where

$$\begin{aligned} \mathbf{z} &= (\mathbf{a}_0, \mathbf{b}, \mathbf{c})^{\mathrm{T}}, \\ \mathbf{f} &= (\mathbf{F}_0, -\Omega \mathbf{k}_1 \mathbf{c} + \mathbf{F}_1, \Omega \mathbf{k}_1 \mathbf{b} + \mathbf{F}_2)^{\mathrm{T}}. \end{aligned} \tag{3.10}$$

If equilibrium $\mathbf{z}^*$ of Equation (3.6) (i.e., $\mathbf{f}(\mathbf{z}^*) = \mathbf{0}$) exists, then the approximate solution of periodic flow exists as in Equation (3.5). In vicinity of equilibrium $\mathbf{z}^*$, with $\mathbf{z} = \mathbf{z}^* + \Delta\mathbf{z}$, the linearized equation of Equation (3.4) is

$$\Delta\dot{\mathbf{z}} = D\mathbf{f}(\mathbf{z}^*)\Delta\mathbf{z} \tag{3.11}$$

and the eigenvalue analysis of equilibrium $\mathbf{z}^*$ is given by

$$|D\mathbf{f}(\mathbf{z}^*) - \lambda \mathbf{I}_{n(2N+1)\times n(2N+1)}| = 0 \tag{3.12}$$

where $D\mathbf{f}(\mathbf{z}^*) = \partial\mathbf{f}(\mathbf{z})/\partial\mathbf{z}|_{\mathbf{z}^*}$. Thus, the stability and bifurcation of periodic flow can be classified by the eigenvalues of $D\mathbf{f}(\mathbf{z}^*)$ with

$$(n_1, n_2, n_3 \mid n_4, n_5, n_6) \tag{3.13}$$

where n_1 is the total number of negative real eigenvalues, n_2 is the total number of positive real eigenvalues, n_3 is the total number of zero real eigenvalues; n_4 is the total pair number of complex eigenvalues with negative real parts, n_5 is the total pair number of complex eigenvalues with positive real parts, n_6 is the total pair number of complex eigenvalues with zero real parts.

1. *If all eigenvalues of the equilibrium possess negative real parts, the approximate periodic solution is stable.*
2. *If at least one eigenvalue of the equilibrium possesses positive real part, the approximate periodic solution is unstable.*
3. *The boundaries between stable and unstable equilibriums with higher order singularity give bifurcation and stability conditions with higher order singularity.*

Proof. The proof is similar to Luo (2012a). If $\mathbf{f}(\mathbf{x}, \mathbf{p})$ is a C^r-continuous nonlinear function vector ($r \geq 1$), then the velocity $\dot{\mathbf{x}}$ should be C^r-continuous ($r \geq 1$), and the acceleration $\ddot{\mathbf{x}}$ should be bounded (i.e., $||\ddot{\mathbf{x}}|| \leq K$). From Equation (3.2), the norms of the periodic flows are defined by

$$||\mathbf{x}|| = ||\mathbf{a}_0|| + \sum_{k=1}^{\infty} ||\mathbf{A}_k||$$

where

$$\begin{aligned} \mathbf{A}_k &= (A_{k1}, A_{k2}, \ldots, A_{kn})^{\mathrm{T}}, \\ A_{kj} &= \sqrt{b_{kj}^2 + c_{kj}^2} \; (j = 1, 2, \ldots, n), \end{aligned}$$

Because the periodic flow in Equation (3.1) is bounded (i.e., $||\mathbf{x}(t)|| < C$), we have

$$\lim_{k\to\infty} ||\mathbf{A}_k|| = 0 \text{ but not uniform.}$$

Thus, the Fourier series transformation of periodic flow as in Equation (3.2) is convergent. From Equation (3.3), using Equations (3.2) and (3.4) gives

$$||\mathbf{x}(t) - \mathbf{x}^*(t)|| = \left\| \sum_{k=N+1}^{\infty} \mathbf{b}_k(t)\cos(k\theta) + \mathbf{c}_k(t)\sin(k\theta) \right\| = \sum_{k=N+1}^{\infty} ||\mathbf{A}_k||.$$

For the prescribed small positive $\varepsilon > 0$, if $||\mathbf{x}(t) - \mathbf{x}^*(t)|| < \varepsilon$ exists, then we have

$$\sum_{k=N+1}^{\infty} ||\mathbf{A}_k|| < \varepsilon.$$

Therefore, the convergent infinite term transformation in Equation (3.2) can be approximated by a finite term transformation in Equation (3.5) in the sense of ε.

Taking the derivative of Equation (3.5) with respect to time generates

$$\dot{\mathbf{x}}^*(t) = \dot{\mathbf{a}}_0 + \sum_{k=1}^{N} [\dot{\mathbf{b}}_k + k\Omega\mathbf{c}_k]\cos(k\theta) + [\dot{\mathbf{c}}_k - k\Omega\mathbf{b}_k]\sin(k\theta).$$

Substitution of the foregoing equation into the nonlinear system in Equation (3.1), and application of the virtual work principle for a basis of constant, $\cos(k\theta)$ and $\sin(k\theta)$ ($k = 1, 2, \ldots$) as a set of virtual displacements gives

$$\frac{1}{2\pi}\int_0^{2\pi} [\dot{\mathbf{x}} - \mathbf{f}(\mathbf{x}, \mathbf{p})]d\theta = 0,$$

$$\frac{1}{\pi}\int_0^{2\pi} [\dot{\mathbf{x}} - \mathbf{f}(\mathbf{x}, \mathbf{p})]\cos(k\theta)d\theta = 0,$$

$$\frac{1}{\pi}\int_0^{2\pi} [\dot{\mathbf{x}} - \mathbf{f}(\mathbf{x}, \mathbf{p})]\sin(k\theta)d\theta = 0.$$

Under $||\mathbf{x} - \mathbf{x}^*|| < \varepsilon$ with continuity $||\ddot{\mathbf{x}}|| < K$ (constant) and small $\varepsilon > 0$, for $k = 1, 2, \ldots, N$, the foregoing equation gives

$$\frac{1}{2\pi}\int_0^{2\pi} [\dot{\mathbf{x}}^* - \mathbf{f}(\mathbf{x}^*, \mathbf{p})]d\theta + O(\delta) = 0,$$

$$\frac{1}{\pi}\int_0^{2\pi} [\dot{\mathbf{x}}^* - \mathbf{f}(\mathbf{x}^*, \mathbf{p})]\cos(k\theta)d\theta + O(\delta) = 0,$$

$$\frac{1}{\pi}\int_0^{2\pi} [\dot{\mathbf{x}}^* - \mathbf{f}(\mathbf{x}^*, \mathbf{p})]\sin(k\theta)d\theta + O(\delta) = 0$$

where $\delta = \max(\varepsilon, \varepsilon_t)$ and $||\dot{\mathbf{x}} - \dot{\mathbf{x}}^*|| < \varepsilon_t$ with small $\varepsilon_t > 0$. The foregoing equation generates

$$\dot{\mathbf{a}}_0 = \mathbf{F}_0(\mathbf{z}),$$

$$\dot{\mathbf{b}}_k = -\Omega k\mathbf{c}_k + \mathbf{F}_{1k}(\mathbf{z}),$$

$$\dot{\mathbf{c}}_k = \Omega k\mathbf{b}_k + \mathbf{F}_{2k}(\mathbf{z})$$

where for $k = 1, 2, \ldots, N$

$$\mathbf{F}_0(\mathbf{z}) = \frac{1}{2\pi}\int_0^{2\pi} \mathbf{f}(\mathbf{x}^*, \mathbf{p})d\theta;$$

$$\mathbf{F}_{1k}(\mathbf{z}) = \frac{1}{\pi}\int_0^{2\pi} \mathbf{f}(\mathbf{x}^*, \mathbf{p})\cos(k\theta)d\theta,$$

$$\mathbf{F}_{2k}(\mathbf{z}) = \frac{1}{\pi}\int_0^{2\pi} \mathbf{f}(\mathbf{x}^*, \mathbf{p})\sin(k\theta)d\theta;$$

and for $N = 1, 2, \ldots, \infty$

$$\begin{aligned}
\mathbf{k}_1 &= diag(\mathbf{I}_{n\times n}, 2\mathbf{I}_{n\times n}, \ldots, N\mathbf{I}_{n\times n}),\\
\mathbf{b} &= (\mathbf{b}_1, \mathbf{b}_2, \ldots, \mathbf{b}_N)^{\mathrm{T}},\\
\mathbf{c} &= (\mathbf{c}_1, \mathbf{c}_2, \ldots, \mathbf{c}_N)^{\mathrm{T}};\\
\mathbf{F}_1 &= (\mathbf{F}_{11}, \mathbf{F}_{12}, \ldots, \mathbf{F}_{1N})^{\mathrm{T}},\\
\mathbf{F}_2 &= (\mathbf{F}_{21}, \mathbf{F}_{22}, \ldots, \mathbf{F}_{2N})^{\mathrm{T}};\\
\mathbf{z} &= (\mathbf{a}_0, \mathbf{b}, \mathbf{c})^{\mathrm{T}}.
\end{aligned}$$

Rearranging the foregoing equation gives Equation (3.6), that is,

$$\begin{aligned}
\dot{\mathbf{a}}_0 &= \mathbf{F}_0(\mathbf{a}_0, \mathbf{b}, \mathbf{c}),\\
\dot{\mathbf{b}} &= -\Omega\mathbf{k}_1\mathbf{c} + \mathbf{F}_1(\mathbf{a}_0, \mathbf{b}, \mathbf{c}),\\
\dot{\mathbf{c}} &= \Omega\mathbf{k}_1\mathbf{b} + \mathbf{F}_2(\mathbf{a}_0, \mathbf{b}, \mathbf{c}).
\end{aligned}$$

Introduce

$$\mathbf{f} = (\mathbf{F}_0, -\Omega\mathbf{k}_1\mathbf{c} + \mathbf{F}_1, \Omega\mathbf{k}_1\mathbf{b} + \mathbf{F}_2)^{\mathrm{T}}.$$

The equation in Equation (3.6) becomes

$$\dot{\mathbf{z}} = \mathbf{f}(\mathbf{z}, \mathbf{z}^{\tau}).$$

Consider equilibriums of the foregoing equation (i.e., $\mathbf{f}(\mathbf{z}^*) = \mathbf{0}$) by

$$\begin{aligned}
\mathbf{0} &= \mathbf{F}_0(\mathbf{a}_0^*, \mathbf{b}^*, \mathbf{c}^*),\\
\mathbf{0} &= -\Omega\mathbf{k}_1\mathbf{c}^* + \mathbf{F}_1(\mathbf{a}_0^*, \mathbf{b}^*, \mathbf{c}^*),\\
\mathbf{0} &= \Omega\mathbf{k}_1\mathbf{b}^* + \mathbf{F}_2(\mathbf{a}_0^*, \mathbf{b}^*, \mathbf{c}^*).
\end{aligned}$$

Thus, the solutions of the forgoing equation are the existence conditions of the periodic solutions for the nonlinear dynamical systems. The foregoing equation gives equilibrium $\mathbf{z}^*$. In vicinity of $\mathbf{z}^*$, with $\mathbf{z} = \mathbf{z}^* + \Delta\mathbf{z}$, the linearized equation of $\dot{\mathbf{z}} = \mathbf{f}(\mathbf{z})$ is

$$\Delta\dot{\mathbf{z}} = D_{\mathbf{z}}\mathbf{f}(\mathbf{z}^*)\Delta\mathbf{z}$$

and the eigenvalue analysis of equilibrium $\mathbf{z}^*$ is completed via

$$|D_{\mathbf{z}}\mathbf{f}(\mathbf{z}^*) - \lambda\mathbf{I}_{n(2N+1)\times n(2N+1)}| = 0$$

where

$$D_{\mathbf{z}}\mathbf{f}(\mathbf{z}^{*}) = \frac{\partial \mathbf{f}(\mathbf{z})}{\partial \mathbf{z}}|_{\mathbf{z}^{*}}.$$

Therefore, as discussed before, the eigenvalues of $D\mathbf{f}(\mathbf{z}^{*})$ are classified by

$$(n_1, n_2, n_3 \mid n_4, n_5, n_6).$$

From the stability and bifurcation theory of dynamical systems at equilibrium, the stability, and bifurcation of the periodic solutions can be classified as stated in the theorem. This theorem is proved. ■

If the Hopf bifurcation of equilibriums of Equation (3.6) occurs, there is a periodic solution of generalized coordinates in Equation (3.5) with a frequency ω. Thus, the periodic solution of the generalized coordinates can be expressed as a new generalized coordinate transformation with $\vartheta = \omega t$

$$\mathbf{a}_0(t) = \mathbf{A}_{00}(t) + \sum_{m=1}^{\infty} \mathbf{A}_{0m}^{(1)}(t)\cos(m\vartheta) + \mathbf{A}_{0m}^{(2)}(t)\sin(m\vartheta).$$

$$\mathbf{b}_k(t) = \mathbf{B}_{k0}(t) + \sum_{m=1}^{\infty} \mathbf{B}_{km}^{(1)}(t)\cos(m\vartheta) + \mathbf{B}_{km}^{(2)}(t)\sin(m\vartheta),$$

$$\mathbf{c}_k(t) = \mathbf{C}_{k0}(t) + \sum_{m=1}^{\infty} \mathbf{C}_{km}^{(1)}(t)\cos(m\vartheta) + \mathbf{C}_{km}^{(2)}(t)\sin(m\vartheta), \tag{3.14}$$

Substitution of Equation (3.14) into Equation (3.5) yields

$$\begin{aligned}\mathbf{x}(t) = {} & \mathbf{A}_{00}(t) + \sum_{m=1}^{\infty} \mathbf{A}_{0m}^{(1)}(t)\cos(m\vartheta) + \mathbf{A}_{0m}^{(2)}(t)\sin(m\vartheta) \\ & + \sum_{k=1}^{\infty} \mathbf{B}_{k0}(t)\cos(k\theta) + \mathbf{C}_{k0}(t)\sin(k\theta) \\ & + \sum_{k=1}^{\infty}\sum_{m=1}^{\infty} \mathbf{B}_{km}^{(1)}(t)\cos(m\vartheta)\cos(k\theta) + \mathbf{B}_{km}^{(2)}(t)\sin(m\vartheta)\cos(k\theta) \\ & + \sum_{k=1}^{\infty}\sum_{m=1}^{\infty} \mathbf{C}_{km}^{(1)}(t)\cos(m\vartheta)\sin(k\theta) + \mathbf{C}_{km}^{(2)}(t)\sin(m\vartheta)\sin(k\theta).\end{aligned} \tag{3.15}$$

If the new solution is still periodic with excitation period $T = 2\pi/\Omega$, then for specific m, we have

$$m\vartheta = k\theta \Rightarrow m\omega = k\Omega. \tag{3.16}$$

For this case, $k = 1$ should be inserted because $k > 1$ terms are already included in the Fourier series expression. Thus,

$$m\vartheta = \theta \Rightarrow m\omega = \Omega. \tag{3.17}$$

For $m = 1$, the period-1 flow is obtained, and Equation (3.15) will become Equation (3.5). For the period-m flow, we have a new generalized coordinate transformation as

$$\mathbf{x}^{(m)}(t) = \mathbf{a}_0^{(m)}(t) + \sum_{k=1}^{\infty} \mathbf{b}_{k/m}(t)\cos\left(\frac{k}{m}\theta\right) + \mathbf{c}_{k/m}(t)\sin\left(\frac{k}{m}\theta\right). \tag{3.18}$$

If $\|\mathbf{x}^{(m)}(t) - \mathbf{x}^{(m)*}(t)\| < \varepsilon$ with a prescribed small $\varepsilon > 0$, the solution of period-m flow in Equation (3.18) can be approximated by a finite term transformation as

$$\mathbf{x}^{(m)*}(t) = \mathbf{a}_0^{(m)}(t) + \sum_{k=1}^{N} \mathbf{b}_{k/m}(t)\cos\left(\frac{k}{m}\theta\right) + \mathbf{c}_{k/m}(t)\sin\left(\frac{k}{m}\theta\right). \tag{3.19}$$

If $m\omega \neq k\Omega$ for any m and k, the solution will be quasi-periodic or chaotic instead of periodic. Herein, this case will not be discussed. If period-1 motion possesses at least N_1 harmonic vector terms, then the total harmonic vector terms for period-m flow should be $N \geq mN_1$. Similarly, the period-m flows in nonlinear dynamical systems can be discussed.

Theorem 3.2 *Consider a nonlinear dynamical system in Equation (3.1). If such a dynamical system has a period-*m *flow* $\mathbf{x}^{(m)}(t)$ *with finite norm* $\|\mathbf{x}^{(m)}\|$ *and period* $T = 2\pi/\Omega$, *there is a generalized coordinate transformation with* $\theta = \Omega t$ *for the period-*m *flow of Equation (3.1) in the form of*

$$\mathbf{x}^{(m)}(t) = \mathbf{a}_0^{(m)}(t) + \sum_{k=1}^{\infty} \mathbf{b}_{k/m}(t)\cos\left(\frac{k}{m}\theta\right) + \mathbf{c}_{k/m}(t)\sin\left(\frac{k}{m}\theta\right) \tag{3.20}$$

with

$$\begin{aligned}
\mathbf{a}_0^{(m)} &= (a_{01}^{(m)}, a_{02}^{(m)}, \ldots, a_{0n}^{(m)})^{\mathrm{T}},\\
\mathbf{b}_{k/m} &= (b_{k/m1}, b_{k/m2}, \ldots, b_{k/mn})^{\mathrm{T}},\\
\mathbf{c}_{k/m} &= (c_{k/m1}, c_{k/m2}, \ldots, c_{k/mn})^{\mathrm{T}}
\end{aligned} \tag{3.21}$$

and

$$\begin{aligned}
&\|\mathbf{x}^{(m)}\| = \|\mathbf{a}_0^{(m)}\| + \sum_{k=1}^{\infty} \|\mathbf{A}_{k/m}\|, \text{ and } \lim_{k\to\infty} \|\mathbf{A}_{k/m}\| = 0 \text{ but not uniform}\\
&\text{with } \mathbf{A}_{k/m} = (A_{k/m1}, A_{k/m2} \ldots, A_{k/mn})^{\mathrm{T}}\\
&\text{and } A_{k/mj} = \sqrt{b_{k/mj}^2 + c_{k/mj}^2} \quad (j = 1, 2, \ldots, n).
\end{aligned} \tag{3.22}$$

For $\|\mathbf{x}^{(m)}(t) - \mathbf{x}^{(m)*}(t)\| < \varepsilon$ *with a prescribed small* $\varepsilon > 0$, *the infinite term transformation* $\mathbf{x}^{(m)}(t)$ *of the period-*m *flow of Equation (3.1), given by Equation (3.20), can be approximated by a finite term transformation* $\mathbf{x}^{(m)*}(t)$ *as*

$$\mathbf{x}^{(m)*}(t) = \mathbf{a}_0^{(m)}(t) + \sum_{k=1}^{N} \mathbf{b}_{k/m}(t)\cos\left(\frac{k}{m}\theta\right) + \mathbf{c}_{k/m}(t)\sin\left(\frac{k}{m}\theta\right) \tag{3.23}$$

and the generalized coordinates are determined by

$$\begin{aligned}
\dot{\mathbf{a}}_0^{(m)} &= \mathbf{F}_0^{(m)}(\mathbf{a}_0^{(m)}, \mathbf{b}^{(m)}, \mathbf{c}^{(m)}),\\
\dot{\mathbf{b}}^{(m)} &= -\frac{\Omega}{m}\mathbf{k}_1\mathbf{c}^{(m)} + \mathbf{F}_1^{(m)}(\mathbf{a}_0^{(m)}, \mathbf{b}^{(m)}, \mathbf{c}^{(m)}),\\
\dot{\mathbf{c}}^{(m)} &= \frac{\Omega}{m}\mathbf{k}_1\mathbf{b}^{(m)} + \mathbf{F}_2^{(m)}(\mathbf{a}_0^{(m)}, \mathbf{b}^{(m)}, \mathbf{c}^{(m)})
\end{aligned} \tag{3.24}$$

where

$$
\begin{aligned}
&\mathbf{k}_1 = diag(\mathbf{I}_{n\times n}, 2\mathbf{I}_{n\times n}, \ldots, N\mathbf{I}_{n\times n}),\\
&\mathbf{b}^{(m)} = (\mathbf{b}_{1/m}, \mathbf{b}_{2/m}, \ldots, \mathbf{b}_{N/m})^{\mathrm{T}},\\
&\mathbf{c}^{(m)} = (\mathbf{c}_{1/m}, \mathbf{c}_{2/m}, \ldots, \mathbf{c}_{N/m})^{\mathrm{T}},\\
&\mathbf{F}_1^{(m)} = (\mathbf{F}_{11}^{(m)}, \mathbf{F}_{12}^{(m)}, \ldots, \mathbf{F}_{1N}^{(m)})^{\mathrm{T}},\\
&\mathbf{F}_2^{(m)} = (\mathbf{F}_{21}^{(m)}, \mathbf{F}_{22}^{(m)}, \ldots, \mathbf{F}_{2N}^{(m)})^{\mathrm{T}}\\
&\text{for } N = 1, 2, \ldots, \infty;
\end{aligned}
\tag{3.25}
$$

and

$$
\begin{aligned}
&\mathbf{F}_0^{(m)}(\mathbf{a}_0^{(m)}, \mathbf{b}^{(m)}, \mathbf{c}^{(m)}) = \frac{1}{2m\pi}\int_0^{2m\pi} \mathbf{f}(\mathbf{x}^{(m)*}, \mathbf{p})d\theta;\\
&\mathbf{F}_{1k}^{(m)}(\mathbf{a}_0^{(m)}, \mathbf{b}^{(m)}, \mathbf{c}^{(m)}) = \frac{1}{m\pi}\int_0^{2m\pi} \mathbf{f}(\mathbf{x}^{(m)*}, \mathbf{p})\cos\left(\frac{k}{m}\theta\right)d\theta,\\
&\mathbf{F}_{2k}^{(m)}(\mathbf{a}_0^{(m)}, \mathbf{b}^{(m)}, \mathbf{c}^{(m)}) = \frac{1}{m\pi}\int_0^{2m\pi} \mathbf{f}(\mathbf{x}^{(m)*}, \mathbf{p})\sin\left(\frac{k}{m}\theta\right)d\theta\\
&\text{for } k = 1, 2, \ldots, N.
\end{aligned}
\tag{3.26}
$$

Equation (3.24) becomes

$$\dot{\mathbf{z}}^{(m)} = \mathbf{f}^{(m)}(\mathbf{z}^{(m)}) \tag{3.27}$$

where

$$
\begin{aligned}
&\mathbf{z}^{(m)} = (\mathbf{a}_0^{(m)}, \mathbf{b}^{(m)}, \mathbf{c}^{(m)})^{\mathrm{T}}\\
&\mathbf{f}^{(m)} = (\mathbf{F}_0^{(m)}, -\Omega\mathbf{k}_1\mathbf{c}^{(m)}/m + \mathbf{F}_1^{(m)}, \Omega\mathbf{k}_1\mathbf{b}^{(m)}/m + \mathbf{F}_2^{(m)})^{\mathrm{T}}
\end{aligned}
\tag{3.28}
$$

*If equilibrium $\mathbf{z}^{(m)*}$ of Equation (3.27) (i.e., $\mathbf{f}^{(m)}(\mathbf{z}^{(m)*}) = \mathbf{0}$) exists, then the approximate solution of period-*m *flow exists as in Equation (3.23). In vicinity of equilibrium $\mathbf{z}^{(m)*}$, with $\mathbf{z}^{(m)} = \mathbf{z}^{(m)*} + \Delta\mathbf{z}^{(m)}$, the linearized equation of Equation (3.27) is*

$$\Delta\dot{\mathbf{z}}^{(m)} = D\mathbf{f}^{(m)}(\mathbf{z}^{(m)*})\Delta\mathbf{z}^{(m)} \tag{3.29}$$

and the eigenvalue analysis of equilibrium $\mathbf{z}^*$ is given by

$$|D\mathbf{f}^{(m)}(\mathbf{z}^{(m)*}) - \lambda\mathbf{I}_{n(2N+1)\times n(2N+1)}| = 0 \tag{3.30}$$

*where $D\mathbf{f}^{(m)}(\mathbf{z}^{(m)*}) = \partial\mathbf{f}^{(m)}(\mathbf{z}^{(m)})/\partial\mathbf{z}^{(m)}|_{\mathbf{z}^{(m)*}}$. The stability and bifurcation of such a period-*m *flow can be classified by the eigenvalues of $D\mathbf{f}^{(m)}(\mathbf{z}^{(m)*})$ with*

$$(n_1, n_2, n_3 \mid n_4, n_5, n_6) \tag{3.31}$$

where n_1 is the total number of negative real eigenvalues, n_2 is the total number of positive real eigenvalues, n_3 is the total number of zero real eigenvalues; n_4 is the total pair number of complex eigenvalues with negative real parts, n_5 is the total pair number of complex eigenvalues with positive real parts, n_6 is the total pair number of complex eigenvalues with zero real parts.

1. *If all eigenvalues of the equilibrium possess negative real parts, the approximate periodic solution is stable.*
2. *If at least one eigenvalue of the equilibrium possesses positive real part, the approximate periodic solution is unstable.*
3. *The boundaries between stable and unstable equilibriums with higher order singularity give bifurcation and stability conditions with higher order singularity.*

Proof. The proof is similar to Luo (2012a). Since $\mathbf{f}(\mathbf{x}, \mathbf{p})$ is a C^r-continuous nonlinear function vector ($r \geq 1$), the velocity $\dot{\mathbf{x}}$ should be C^r-continuous ($r \geq 1$), and then the acceleration $\ddot{\mathbf{x}}$ should be bounded (i.e., $||\ddot{\mathbf{x}}|| \leq K$). From Equation (3.20) the norms of the periodic flows are defined by

$$||\mathbf{x}^{(m)}|| = ||\mathbf{a}_0^{(m)}|| + \sum_{k=1}^{\infty} ||\mathbf{A}_{k/m}||,$$

where

$$\mathbf{A}_{k/m} = (A_{k/m1}, A_{k/m2} \ldots, A_{k/mn})^{\mathrm{T}},$$
$$A_{k/mj} = \sqrt{b_{k/mj}^2 + c_{k/mj}^2} \quad (j = 1, 2, \ldots, n).$$

Because the periodic flow in Equation (3.1) is bounded (i.e., $||\mathbf{x}(t)|| < C$), we have

$$\lim_{k \to \infty} ||\mathbf{A}_{k/m}|| = 0 \text{ but not uniform.}$$

Thus, the Fourier series transformation of periodic flow as in Equation (3.1) is convergent. From Equation (3.22), using Equations (3.20) and (3.23) gives

$$\begin{aligned} ||\mathbf{x}^{(m)}(t) - \mathbf{x}^{(m)*}(t)|| &= \left\| \sum_{k=N+1}^{\infty} \mathbf{b}_{k/m}(t) \cos\left(\frac{k}{m}\theta\right) + \mathbf{c}_{k/m}(t) \sin\left(\frac{k}{m}\theta\right) \right\| \\ &= \sum_{k=N+1}^{\infty} ||\mathbf{A}_{k/m}||. \end{aligned}$$

For the prescribed small positive $\varepsilon > 0$, if $||\mathbf{x}^{(m)}(t) - \mathbf{x}^{(m)*}(t)|| < \varepsilon$ exists, then

$$\sum_{k=N+1}^{\infty} ||\mathbf{A}_{k/m}|| < \varepsilon.$$

Therefore, the convergent infinite term transformation in Equation (3.20) can be approximated by a finite term transformation in Equation (3.23) in the sense of ε.

Taking the derivative of Equation (3.23) with respect to time gives

$$\begin{aligned} \dot{\mathbf{x}}^{(m)*}(t) = \dot{\mathbf{a}}_0^{(m)} + \sum_{k=1}^{N} &\left[\dot{\mathbf{b}}_{k/m} + \frac{k}{m}\Omega\mathbf{c}_{k/m}\right] \cos\left(\frac{k}{m}\theta\right) \\ &+ \left[\dot{\mathbf{c}}_{k/m} - \frac{k}{m}\Omega\mathbf{b}_{k/m}\right] \sin\left(\frac{k}{m}\theta\right). \end{aligned}$$

Substitution of the foregoing equation into the nonlinear system in Equation (3.1) and application of the virtual work principle for a basis of constant, $\cos(k\theta/m)$ and $\sin(k\theta/m)$ $(k = 1, 2, \ldots)$ as a set of virtual displacements gives

$$\frac{1}{2m\pi}\int_0^{2m\pi}[\dot{\mathbf{x}}^{(m)} - \mathbf{f}(\mathbf{x}^{(m)}, \mathbf{p})]d\theta = 0,$$

$$\frac{1}{m\pi}\int_0^{2m\pi}[\dot{\mathbf{x}}^{(m)} - \mathbf{f}(\mathbf{x}^{(m)}, \mathbf{p})]\cos\left(\frac{k}{m}\theta\right)d\theta = 0,$$

$$\frac{1}{m\pi}\int_0^{2m\pi}[\dot{\mathbf{x}}^{(m)} - \mathbf{f}(\mathbf{x}^{(m)}, \mathbf{p})]\sin\left(\frac{k}{m}\theta\right)d\theta = 0.$$

Under $\|\mathbf{x}^{(m)} - \mathbf{x}^{(m)*}\| < \varepsilon$ with continuity $\|\ddot{\mathbf{x}}^{(m)}\| < K$ (constant) and small $\varepsilon > 0$, for $k = 1, 2, \ldots, N$, the foregoing equation gives

$$\frac{1}{2m\pi}\int_0^{2m\pi}[\dot{\mathbf{x}}^{(m)*} - \mathbf{f}(\mathbf{x}^{(m)*}, \mathbf{p})]d\theta + O(\delta) = 0,$$

$$\frac{1}{m\pi}\int_0^{2m\pi}[\dot{\mathbf{x}}^{(m)*} - \mathbf{f}(\mathbf{x}^{(m)*}, \mathbf{p})]\cos\left(\frac{k}{m}\theta\right)d\theta + O(\delta) = 0,$$

$$\frac{1}{m\pi}\int_0^{2m\pi}[\dot{\mathbf{x}}^{(m)*} - \mathbf{f}(\mathbf{x}^{(m)*}, \mathbf{p})]\sin\left(\frac{k}{m}\theta\right)d\theta + O(\delta) = 0$$

where $\delta = \max(\varepsilon, \varepsilon_t)$ and $\|\dot{\mathbf{x}} - \dot{\mathbf{x}}^*\| < \varepsilon_t$ with small $\varepsilon_t > 0$. From the foregoing equation, we have

$$\dot{\mathbf{a}}_0^{(m)} = \mathbf{F}_0(\mathbf{z}^{(m)}),$$

$$\dot{\mathbf{b}}_{k/m} + \frac{\Omega}{m}k\mathbf{c}_{k/m} = \mathbf{F}_{1k}^{(m)}(\mathbf{z}^{(m)}),$$

$$\dot{\mathbf{c}}_{k/m} - \frac{\Omega}{m}k\mathbf{b}_{k/m} = \mathbf{F}_{2k}^{(m)}(\mathbf{z}^{(m)})$$

where

$$\mathbf{F}_0^{(m)}(\mathbf{z}^{(m)}) = \frac{1}{2m\pi}\int_0^{2m\pi}\mathbf{f}(\mathbf{x}^{(m)*}, \mathbf{p})d\theta;$$

$$\mathbf{F}_{1k}^{(m)}(\mathbf{z}^{(m)}) = \frac{1}{m\pi}\int_0^{2m\pi}\mathbf{f}(\mathbf{x}^{(m)*}, \mathbf{p})\cos\left(\frac{k}{m}\theta\right)d\theta,$$

$$\mathbf{F}_{2k}^{(m)}(\mathbf{z}^{(m)}) = \frac{1}{m\pi}\int_0^{2m\pi}\mathbf{f}(\mathbf{x}^{(m)*}, \mathbf{p})\sin\left(\frac{k}{m}\theta\right)d\theta$$

for $k = 1, 2, \ldots, N$; and

$$\mathbf{k}_1 = diag(\mathbf{I}_{n\times n}, 2\mathbf{I}_{n\times n}, \ldots, N\mathbf{I}_{n\times n}),$$

$$\mathbf{b}^{(m)} = (\mathbf{b}_{1/m}, \mathbf{b}_{2/m}, \ldots, \mathbf{b}_{N/m})^{\mathrm{T}},$$

$$\mathbf{c}^{(m)} = (\mathbf{c}_{1/m}, \mathbf{c}_{2/m}, \ldots, \mathbf{c}_{N/m})^{\mathrm{T}};$$

$$\mathbf{F}_1^{(m)} = (\mathbf{F}_{11}^{(m)}, \mathbf{F}_{12}^{(m)}, \ldots, \mathbf{F}_{1N}^{(m)})^{\mathrm{T}},$$
$$\mathbf{F}_2^{(m)} = (\mathbf{F}_{21}^{(m)}, \mathbf{F}_{22}^{(m)}, \ldots, \mathbf{F}_{2N}^{(m)})^{\mathrm{T}};$$
$$\mathbf{z}^{(m)} = (\mathbf{a}_0^{(m)}, \mathbf{b}^{(m)}, \mathbf{c}^{(m)})^{\mathrm{T}}$$
$$\text{for } N = 1, 2, \ldots, \infty.$$

Rearranging the foregoing equation gives Equation (3.24), that is,

$$\dot{\mathbf{a}}_0^{(m)} = \mathbf{F}_0^{(m)}(\mathbf{z}^{(m)}),$$
$$\dot{\mathbf{b}}^{(m)} = -\frac{\Omega}{m}\mathbf{k}_1\mathbf{c}^{(m)} + \mathbf{F}_1^{(m)}(\mathbf{z}^{(m)}),$$
$$\dot{\mathbf{c}}^{(m)} = \frac{\Omega}{m}\mathbf{k}_1\mathbf{b}^{(m)} + \mathbf{F}_2^{(m)}(\mathbf{z}^{(m)}).$$

Introducing

$$\mathbf{f}^{(m)} = (\mathbf{F}_0^{(m)}, -\Omega\mathbf{k}_1\mathbf{c}^{(m)}/m + \mathbf{F}_1^{(m)}, \Omega\mathbf{k}_1\mathbf{b}^{(m)}/m + \mathbf{F}_2^{(m)})^{\mathrm{T}},$$

the standard form of equation in Equation (3.24) becomes

$$\dot{\mathbf{z}}^{(m)} = \mathbf{f}^{(m)}(\mathbf{z}^{(m)}).$$

Consider the equilibrium solution of the foregoing equation with $\mathbf{z}^{(m)*}$ (i.e., $\mathbf{f}^{(m)}(\mathbf{z}^{(m)*}) = \mathbf{0}$) by

$$\mathbf{0} = \mathbf{F}_0^{(m)}(\mathbf{a}_0^{(m)*}, \mathbf{b}^{(m)*}, \mathbf{c}^{(m)*}),$$
$$\mathbf{0} = -\frac{\Omega}{m}\mathbf{k}_1\mathbf{c}^{(m)*} + \mathbf{F}_1^{(m)}(\mathbf{a}_0^{(m)*}, \mathbf{b}^{(m)*}, \mathbf{c}^{(m)*}),$$
$$\mathbf{0} = \frac{\Omega}{m}\mathbf{k}_1\mathbf{b}^{(m)*} + \mathbf{F}_2^{(m)}(\mathbf{a}_0^{(m)*}, \mathbf{b}^{(m)*}, \mathbf{c}^{(m)*}).$$

Thus, solutions of the foregoing equation are the existence conditions of periodic solutions for nonlinear dynamical systems. Thus, the foregoing equation gives $\mathbf{z}^{(m)*}$. In the vicinity of $\mathbf{z}^{(m)*}$, with $\mathbf{z}^{(m)} = \mathbf{z}^{(m)*} + \Delta\mathbf{z}^{(m)}$ the linearized equation of $\dot{\mathbf{z}}^{(m)} = \mathbf{f}^{(m)}(\mathbf{z}^{(m)})$ is

$$\Delta\dot{\mathbf{z}}^{(m)} = D\mathbf{f}^{(m)}(\mathbf{z}^{(m)*})\Delta\mathbf{z}^{(m)}$$

and the eigenvalue analysis of equilibrium $\mathbf{z}^{(m)*}$ is completed via

$$|D\mathbf{f}^{(m)}(\mathbf{z}^{(m)*}) - \lambda\mathbf{I}_{n(2N+1)\times n(2N+1)}| = 0$$

where

$$D\mathbf{f}^{(m)}(\mathbf{z}^{(m)*}) = \frac{\partial\mathbf{f}^{(m)}(\mathbf{z}^{(m)})}{\partial\mathbf{z}^{(m)}}\bigg|_{\mathbf{z}^{(m)*}}.$$

Thus, the corresponding eigenvalues are classified by

$$(n_1, n_2, n_3 \mid n_4, n_5, n_6).$$

From the stability and bifurcation theory of dynamical systems at equilibrium, the stability, and bifurcation of the periodic solutions can be classified as stated in the theorem. This theorem is proved. ■

If $m \to \infty$, Equation (3.20) will give the analytical expression of chaos in dynamical systems in Equation (3.1), which can be approximated by Equation (3.23) under the condition of $||\mathbf{x}^{(m)}(t) - \mathbf{x}^{(m)*}(t)|| < \varepsilon$. The route from the periodic flow to chaos is through the Hopf bifurcation.

3.1.2 Non-Autonomous Nonlinear Systems

Periodic flows in periodically excited nonlinear dynamical systems will be presented herein. If a periodically excited nonlinear system with an external period $T = 2\pi/\Omega$ has a periodic flow, then such a periodic flow can be expressed by a Fourier series. The methodology is presented through the following theorems.

Theorem 3.3 *Consider a non-autonomous nonlinear dynamical system as*

$$\dot{\mathbf{x}} = \mathbf{F}(\mathbf{x}, t, \mathbf{p}) \in \mathscr{R}^n \tag{3.32}$$

where $\mathbf{F}(\mathbf{x}, t, \mathbf{p})$ *is a* C^r*-continuous nonlinear function vector* ($r \geq 1$) *with an excitation period* $T = 2\pi/\Omega$. *If such a dynamical system has a periodic flow* $\mathbf{x}(t)$ *with finite norm* $||\mathbf{x}||$, *there is a generalized coordinate transformation with* $\theta = \Omega t$ *for the periodic flow of Equation (3.32) in a form of*

$$\mathbf{x}(t) = \mathbf{a}_0(t) + \sum_{k=1}^{\infty} \mathbf{b}_k(t)\cos(k\theta) + \mathbf{c}_k(t)\sin(k\theta) \tag{3.33}$$

with

$$\begin{aligned}
\mathbf{a}_0 &= (a_{01}, a_{02}, \ldots, a_{0n})^{\mathrm{T}},\\
\mathbf{b}_k &= (b_{k1}, b_{k2}, \ldots, b_{kn})^{\mathrm{T}},\\
\mathbf{c}_k &= (c_{k1}, c_{k2}, \ldots, c_{kn})^{\mathrm{T}};
\end{aligned} \tag{3.34}$$

and

$$\begin{gathered}
||\mathbf{x}|| = ||\mathbf{a}_0|| + \sum_{k=1}^{\infty} ||\mathbf{A}_k||, \text{ and } \lim_{k\to\infty} ||\mathbf{A}_k|| = 0 \text{ but not uniform}\\
\text{with } \mathbf{A}_k = (A_{k1}, A_{k2}, \ldots, A_{kn})^{\mathrm{T}} \text{ and } A_{kj} = \sqrt{b_{kj}^2 + c_{kj}^2}\ \ (j = 1, 2, \ldots, n).
\end{gathered} \tag{3.35}$$

For $||\mathbf{x}(t) - \mathbf{x}^*(t)|| < \varepsilon$ *with a prescribed small positive* $\varepsilon > 0$, *the infinite term transformation of the periodic flow* $\mathbf{x}(t)$ *of Equation (3.32), given by Equation (3.33), can be approximated by a finite term transformation* $\mathbf{x}^*(t)$ *as*

$$\mathbf{x}^*(t) = \mathbf{a}_0(t) + \sum_{k=1}^{N} \mathbf{b}_k(t)\cos(k\theta) + \mathbf{c}_k(t)\sin(k\theta) \tag{3.36}$$

and the generalized coordinates are determined by

$$\dot{\mathbf{a}}_0 = \mathbf{F}_0(\mathbf{a}_0, \mathbf{b}, \mathbf{c}),$$

$$\begin{aligned}\dot{\mathbf{b}} &= -\Omega \mathbf{k}_1 \mathbf{c} + \mathbf{F}_1(\mathbf{a}_0, \mathbf{b}, \mathbf{c}),\\ \dot{\mathbf{c}} &= \Omega \mathbf{k}_1 \mathbf{b} + \mathbf{F}_2(\mathbf{a}_0, \mathbf{b}, \mathbf{c});\end{aligned} \tag{3.37}$$

where

$$\begin{aligned}\mathbf{k}_1 &= diag(\mathbf{I}_{n\times n}, 2\mathbf{I}_{n\times n}, \ldots, N\mathbf{I}_{n\times n}),\\ \mathbf{b} &= (\mathbf{b}_1, \mathbf{b}_2, \ldots, \mathbf{b}_N)^{\mathrm{T}},\\ \mathbf{c} &= (\mathbf{c}_1, \mathbf{c}_2, \ldots, \mathbf{c}_N)^{\mathrm{T}},\\ \mathbf{F}_1 &= (\mathbf{F}_{11}, \mathbf{F}_{12}, \ldots, \mathbf{F}_{1N})^{\mathrm{T}},\\ \mathbf{F}_2 &= (\mathbf{F}_{21}, \mathbf{F}_{22}, \ldots, \mathbf{F}_{2N})^{\mathrm{T}}\\ &\text{for } N = 1, 2, \ldots, \infty;\end{aligned} \tag{3.38}$$

and for $k = 1, 2, \ldots, N$

$$\begin{aligned}\mathbf{F}_0(\mathbf{a}_0, \mathbf{b}, \mathbf{c}) &= \frac{1}{2\pi}\int_0^{2\pi} \mathbf{F}(\mathbf{x}^*, t, \mathbf{p}) d\theta;\\ \mathbf{F}_{1k}(\mathbf{a}_0, \mathbf{b}, \mathbf{c}) &= \frac{1}{\pi}\int_0^{2\pi} \mathbf{F}(\mathbf{x}^*, t, \mathbf{p}) \cos(k\theta) d\theta,\\ \mathbf{F}_{2k}(\mathbf{a}_0, \mathbf{b}, \mathbf{c}) &= \frac{1}{\pi}\int_0^{2\pi} \mathbf{F}(\mathbf{x}^*, t, \mathbf{p}) \sin(k\theta) d\theta.\end{aligned} \tag{3.39}$$

Equation (3.37) becomes

$$\dot{\mathbf{z}} = \mathbf{f}(\mathbf{z}) \tag{3.40}$$

where

$$\begin{aligned}\mathbf{z} &= (\mathbf{a}_0, \mathbf{b}, \mathbf{c})^{\mathrm{T}},\\ \mathbf{f} &= (\mathbf{F}_0, -\Omega \mathbf{k}_1 \mathbf{c} + \mathbf{F}_1, \Omega \mathbf{k}_1 \mathbf{b} + \mathbf{F}_2)^{\mathrm{T}}.\end{aligned} \tag{3.41}$$

If equilibrium $\mathbf{z}^*$ *of Equation (3.40) (i.e.,* $\mathbf{f}(\mathbf{z}^*) = \mathbf{0}$*) exists, then the approximate solution of periodic flow exists as in Equation (3.36). In vicinity of equilibrium* $\mathbf{z}^*$*, with* $\mathbf{z} = \mathbf{z}^* + \Delta\mathbf{z}$ *the linearized equation of Equation (3.40) is*

$$\Delta\dot{\mathbf{z}} = D\mathbf{f}(\mathbf{z}^*)\Delta\mathbf{z} \tag{3.42}$$

and the eigenvalue analysis of equilibrium $\mathbf{z}^*$ *is given by*

$$|D\mathbf{f}(\mathbf{z}^*) - \lambda \mathbf{I}_{n(2N+1)\times n(2N+1)}| = 0 \tag{3.43}$$

where $D\mathbf{f}(\mathbf{z}^*) = \partial\mathbf{f}(\mathbf{z})/\partial\mathbf{z}|_{\mathbf{z}^*}$*. Thus, the stability and bifurcation of periodic flow can be classified by the eigenvalues of* $D\mathbf{f}(\mathbf{z}^*)$ *with*

$$(n_1, n_2, n_3 \mid n_4, n_5, n_6). \tag{3.44}$$

1. *If all eigenvalues of the equilibrium possess negative real parts, the approximate periodic solution is stable.*

2. *If at least one of the eigenvalues of the equilibrium possesses positive real part, the approximate periodic solution is unstable.*
3. *The boundaries between stable and unstable equilibriums with higher order singularity give bifurcation and stability conditions with higher order singularity.*

Proof. The proof can be referenced to Luo (2012a) or similar to the proof of Theorem 3.1. ■

Because they are similar to the autonomous nonlinear system, period-m flows in the periodically excited, nonlinear dynamical system in Equation (3.32) can be discussed.

Theorem 3.4 *Consider an autonomous nonlinear dynamical system in Equation (3.32) with an excitation period $T = 2\pi/\Omega$. If such a dynamical system has a period-*m *flow $\mathbf{x}^{(m)}(t)$ with finite norm $\|\mathbf{x}^{(m)}\|$, there is a generalized coordinate transformation with $\theta = \Omega t$ for the periodic flow of Equation (3.32) in the form of*

$$\mathbf{x}^{(m)}(t) = \mathbf{a}_0^{(m)}(t) + \sum_{k=1}^{\infty} \mathbf{b}_{k/m}(t)\cos\left(\frac{k}{m}\theta\right) + \mathbf{c}_{k/m}(t)\sin\left(\frac{k}{m}\theta\right) \tag{3.45}$$

with

$$\begin{aligned} \mathbf{a}_0^{(m)} &= (a_{01}^{(m)}, a_{02}^{(m)}, \ldots, a_{0n}^{(m)})^{\mathrm{T}}, \\ \mathbf{b}_{k/m} &= (b_{k/m1}, b_{k/m2}, \ldots, b_{k/mn})^{\mathrm{T}}, \\ \mathbf{c}_{k/m} &= (c_{k/m1}, c_{k/m2}, \ldots, c_{k/mn})^{\mathrm{T}} \end{aligned} \tag{3.46}$$

and

$$\begin{aligned} &\|\mathbf{x}^{(m)}\| = \|\mathbf{a}_0^{(m)}\| + \sum_{k=1}^{\infty} \|\mathbf{A}_{k/m}\|, \text{ and } \lim_{k\to\infty} \|\mathbf{A}_{k/m}\| = 0 \text{ but not uniform} \\ &\text{with } \mathbf{A}_{k/m} = (A_{k/m1}, A_{k/m2} \ldots, A_{k/mn})^{\mathrm{T}} \\ &\text{and } A_{k/mj} = \sqrt{b_{k/mj}^2 + c_{k/mj}^2} \quad (j = 1, 2, \ldots, n). \end{aligned} \tag{3.47}$$

*For $\|\mathbf{x}^{(m)}(t) - \mathbf{x}^{(m)*}(t)\| < \varepsilon$ with a prescribed small $\varepsilon > 0$, the infinite term transformation $\mathbf{x}^{(m)}(t)$ of the period-*m *flow of Equation (3.32), given by Equation (3.45), can be approximated by a finite term transformation $\mathbf{x}^{(m)*}(t)$ as*

$$\mathbf{x}^*(t) = \mathbf{a}_0^{(m)}(t) + \sum_{k=1}^{N} \mathbf{b}_{k/m}(t)\cos\left(\frac{k}{m}\theta\right) + \mathbf{c}_{k/m}(t)\sin\left(\frac{k}{m}\theta\right) \tag{3.48}$$

and the generalized coordinates are determined by

$$\begin{aligned} \dot{\mathbf{a}}_0^{(m)} &= \mathbf{F}_0^{(m)}(\mathbf{a}_0^{(m)}, \mathbf{b}^{(m)}, \mathbf{c}^{(m)}), \\ \dot{\mathbf{b}}^{(m)} &= -\frac{\Omega}{m}\mathbf{k}_1\mathbf{c}^{(m)} + \mathbf{F}_1^{(m)}(\mathbf{a}_0^{(m)}, \mathbf{b}^{(m)}, \mathbf{c}^{(m)}), \\ \dot{\mathbf{c}}^{(m)} &= \frac{\Omega}{m}\mathbf{k}_1\mathbf{b}^{(m)} + \mathbf{F}_2^{(m)}(\mathbf{a}_0^{(m)}, \mathbf{b}^{(m)}, \mathbf{c}^{(m)}) \end{aligned} \tag{3.49}$$

where for $N = 1, 2, \ldots, \infty$

$$\begin{aligned}
\mathbf{k}_1 &= diag(\mathbf{I}_{n\times n}, 2\mathbf{I}_{n\times n}, \ldots, N\mathbf{I}_{n\times n}),\\
\mathbf{b}^{(m)} &= (\mathbf{b}_{1/m}, \mathbf{b}_{2/m}, \ldots, \mathbf{b}_{N/m})^{\mathrm{T}},\\
\mathbf{c}^{(m)} &= (\mathbf{c}_{1/m}, \mathbf{c}_{2/m}, \ldots, \mathbf{c}_{N/m})^{\mathrm{T}},\\
\mathbf{F}_1^{(m)} &= (\mathbf{F}_{11}^{(m)}, \mathbf{F}_{12}^{(m)}, \ldots, \mathbf{F}_{1N}^{(m)})^{\mathrm{T}},\\
\mathbf{F}_2^{(m)} &= (\mathbf{F}_{21}^{(m)}, \mathbf{F}_{22}^{(m)}, \ldots, \mathbf{F}_{2N}^{(m)})^{\mathrm{T}}
\end{aligned} \tag{3.50}$$

and

$$\begin{aligned}
\mathbf{F}_0^{(m)}(\mathbf{a}_0^{(m)}, \mathbf{b}^{(m)}, \mathbf{c}^{(m)}) &= \frac{1}{2m\pi}\int_0^{2m\pi} \mathbf{F}(\mathbf{x}^{(m)*}, t, \mathbf{p})d\theta;\\
\mathbf{F}_{1k}^{(m)}(\mathbf{a}_0^{(m)}, \mathbf{b}^{(m)}, \mathbf{c}^{(m)}) &= \frac{1}{m\pi}\int_0^{2m\pi} \mathbf{F}(\mathbf{x}^{(m)*}, t, \mathbf{p})\cos\left(\frac{k}{m}\theta\right)d\theta,\\
\mathbf{F}_{2k}^{(m)}(\mathbf{a}_0^{(m)}, \mathbf{b}^{(m)}, \mathbf{c}^{(m)}) &= \frac{1}{m\pi}\int_0^{2m\pi} \mathbf{F}(\mathbf{x}^{(m)*}, t, \mathbf{p})\sin\left(\frac{k}{m}\theta\right)d\theta\\
&\text{for } k = 1, 2, \ldots, N.
\end{aligned} \tag{3.51}$$

Equation (3.49) becomes

$$\dot{\mathbf{z}}^{(m)} = \mathbf{f}^{(m)}(\mathbf{z}^{(m)}) \tag{3.52}$$

where

$$\begin{aligned}
\mathbf{z}^{(m)} &= (\mathbf{a}_0^{(m)}, \mathbf{b}^{(m)}, \mathbf{c}^{(m)})^{\mathrm{T}},\\
\mathbf{f}^{(m)} &= (\mathbf{F}_0^{(m)}, -\Omega\mathbf{k}_1\mathbf{c}^{(m)}/m + \mathbf{F}_1^{(m)}, \Omega\mathbf{k}_1\mathbf{b}^{(m)}/m + \mathbf{F}_2^{(m)})^{\mathrm{T}}.
\end{aligned} \tag{3.53}$$

If equilibrium $\mathbf{z}^{(m)*}$ *of Equation (3.52) (i.e.,* $\mathbf{f}^{(m)}(\mathbf{z}^{(m)*}) = \mathbf{0}$*) exists, then the approximate solution of period-*m *flow exists as in Equation (3.48). In the vicinity of equilibrium* $\mathbf{z}^{(m)*}$*, with* $\mathbf{z}^{(m)} = \mathbf{z}^{(m)*} + \Delta\mathbf{z}^{(m)}$*, the linearized equation of Equation (3.52) is*

$$\Delta\dot{\mathbf{z}}^{(m)} = D\mathbf{f}^{(m)}(\mathbf{z}^{(m)*})\Delta\mathbf{z}^{(m)} \tag{3.54}$$

and the eigenvalue analysis of equilibrium $\mathbf{z}^*$ *is given by*

$$|D\mathbf{f}^{(m)}(\mathbf{z}^{(m)*}) - \lambda\mathbf{I}_{n(2N+1)\times n(2N+1)}| = 0 \tag{3.55}$$

where $D\mathbf{f}^{(m)}(\mathbf{z}^{(m)*}) = \partial\mathbf{f}^{(m)}(\mathbf{z}^{(m)})/\partial\mathbf{z}^{(m)}|_{\mathbf{z}^{(m)*}}$*. The stability and bifurcation of period-*m *flow can be classified by the eigenvalues of* $D\mathbf{f}^{(m)}(\mathbf{z}^{(m)*})$ *with*

$$(n_1, n_2, n_3 \mid n_4, n_5, n_6). \tag{3.56}$$

1. *If all eigenvalues of the equilibrium possess negative real parts, the approximate periodic solution is stable.*
2. *If at least one of the eigenvalues of the equilibrium possesses positive real part, the approximate periodic solution is unstable.*

3. The boundaries between stable and unstable equilibriums with higher order singularity give bifurcation and stability conditions with higher order singularity.

Proof. The proof can be referred to Luo (2012a) or similar to the proof of Theorem 3.2. ■

If $m \to \infty$, Equation (3.45) will give the analytical expression of chaos in periodically excited, nonlinear dynamical systems in Equation (3.32), which can be approximated by Equation (3.48) under the condition of $||\mathbf{x}^{(m)}(t) - \mathbf{x}^{(m)*}(t)|| < \varepsilon$ as $N \to \infty$.

3.2 Nonlinear Vibration Systems

In this section, the analytical solutions of periodic motions for nonlinear vibration systems are presented owing to extensive application, and the local stability and bifurcation theory will be applied to determine the stability and bifurcation of approximate solutions of periodic motions in nonlinear vibration systems.

3.2.1 Free Vibration Systems

Periodic motions in nonlinear vibration systems will be presented herein. If such a vibration system has periodic motions with period $T = 2\pi/\Omega$, then such periodic motions can be expressed by the Fourier series.

Theorem 3.5 *Consider a nonlinear vibration system as*

$$\ddot{\mathbf{x}} = \mathbf{f}(\mathbf{x}, \dot{\mathbf{x}}, \mathbf{p}) \in \mathscr{R}^n \tag{3.57}$$

where $\mathbf{f}(\mathbf{x}, \dot{\mathbf{x}}, \mathbf{p})$ *is a* C^r*-continuous nonlinear function vector* ($r \geq 1$). *If such a dynamical system has a periodic motion* $\mathbf{x}(t)$ *with finite norm* $||\mathbf{x}||$ *and period* $T = 2\pi/\Omega$, *there is a generalized coordinate transformation with* $\theta = \Omega t$ *for the periodic motion of Equation (3.57) in a form of*

$$\mathbf{x}(t) = \mathbf{a}_0(t) + \sum_{k=1}^{\infty} \mathbf{b}_k(t)\cos(k\theta) + \mathbf{c}_k(t)\sin(k\theta) \tag{3.58}$$

with

$$\begin{aligned}
\mathbf{a}_0 &= (a_{01}, a_{02}, \ldots, a_{0n})^{\mathrm{T}},\\
\mathbf{b}_k &= (b_{k1}, b_{k2}, \ldots, b_{kn})^{\mathrm{T}},\\
\mathbf{c}_k &= (c_{k1}, c_{k2}, \ldots, c_{kn})^{\mathrm{T}}
\end{aligned} \tag{3.59}$$

and

$$\begin{gathered}
||\mathbf{x}|| = ||\mathbf{a}_0|| + \sum_{k=1}^{\infty} ||\mathbf{A}_k||, \text{ and } \lim_{k\to\infty} ||\mathbf{A}_k|| = 0 \text{ but not uniform}\\
\text{with } \mathbf{A}_k = (A_{k1}, A_{k2}, \ldots, A_{kn})^{\mathrm{T}} \text{ and } A_{kj} = \sqrt{b_{kj}^2 + c_{kj}^2}\ (j = 1, 2, \ldots, n).
\end{gathered} \tag{3.60}$$

For $\|\mathbf{x}(t) - \mathbf{x}^*(t)\| < \varepsilon$ with a prescribed small positive $\varepsilon > 0$, the infinite term transformation of the periodic motion $\mathbf{x}(t)$ of Equation (3.57), given by Equation (3.58), can be approximated by a finite term transformation $\mathbf{x}^*(t)$ as

$$\mathbf{x}^*(t) = \mathbf{a}_0(t) + \sum_{k=1}^{N} \mathbf{b}_k(t)\cos(k\theta) + \mathbf{c}_k(t)\sin(k\theta) \tag{3.61}$$

and the generalized coordinates are determined by

$$\begin{aligned}
\ddot{\mathbf{a}}_0 &= \mathbf{F}_0(\mathbf{a}_0, \mathbf{b}, \mathbf{c}, \dot{\mathbf{a}}_0, \dot{\mathbf{b}}, \dot{\mathbf{c}}),\\
\ddot{\mathbf{b}} &= -2\Omega\mathbf{k}_1\dot{\mathbf{c}} + \Omega^2\mathbf{k}_2\mathbf{b} + \mathbf{F}_1(\mathbf{a}_0, \mathbf{b}, \mathbf{c}, \dot{\mathbf{a}}_0, \dot{\mathbf{b}}, \dot{\mathbf{c}}),\\
\ddot{\mathbf{c}} &= 2\Omega\mathbf{k}_1\dot{\mathbf{b}} + \Omega^2\mathbf{k}_2\mathbf{c} + \mathbf{F}_2(\mathbf{a}_0, \mathbf{b}, \mathbf{c}, \dot{\mathbf{a}}_0, \dot{\mathbf{b}}, \dot{\mathbf{c}})
\end{aligned} \tag{3.62}$$

where

$$\begin{aligned}
&\mathbf{k}_1 = diag(\mathbf{I}_{n\times n}, 2\mathbf{I}_{n\times n}, \ldots, N\mathbf{I}_{n\times n}),\\
&\mathbf{k}_2 = diag(\mathbf{I}_{n\times n}, 2^2\mathbf{I}_{n\times n}, \ldots, N^2\mathbf{I}_{n\times n}),\\
&\mathbf{b} = (\mathbf{b}_1, \mathbf{b}_2, \ldots, \mathbf{b}_N)^{\mathrm{T}},\\
&\mathbf{c} = (\mathbf{c}_1, \mathbf{c}_2, \ldots, \mathbf{c}_N)^{\mathrm{T}},\\
&\mathbf{F}_1 = (\mathbf{F}_{11}, \mathbf{F}_{12}, \ldots, \mathbf{F}_{1N})^{\mathrm{T}},\\
&\mathbf{F}_2 = (\mathbf{F}_{21}, \mathbf{F}_{22}, \ldots, \mathbf{F}_{2N})^{\mathrm{T}}\\
&\text{for } N = 1, 2, \ldots, \infty
\end{aligned} \tag{3.63}$$

and for $k = 1, 2, \ldots, N$

$$\begin{aligned}
\mathbf{F}_0(\mathbf{a}_0, \mathbf{b}, \mathbf{c}, \dot{\mathbf{a}}_0, \dot{\mathbf{b}}, \dot{\mathbf{c}}) &= \frac{1}{2\pi}\int_0^{2\pi} \mathbf{f}(\mathbf{x}^*, \dot{\mathbf{x}}^*, \mathbf{p})d\theta;\\
\mathbf{F}_{1k}(\mathbf{a}_0, \mathbf{b}, \mathbf{c}, \dot{\mathbf{a}}_0, \dot{\mathbf{b}}, \dot{\mathbf{c}}) &= \frac{1}{\pi}\int_0^{2\pi} \mathbf{f}(\mathbf{x}^*, \dot{\mathbf{x}}^*, \mathbf{p})\cos(k\theta)d\theta,\\
\mathbf{F}_{2k}(\mathbf{a}_0, \mathbf{b}, \mathbf{c}, \dot{\mathbf{a}}_0, \dot{\mathbf{b}}, \dot{\mathbf{c}}) &= \frac{1}{\pi}\int_0^{2\pi} \mathbf{f}(\mathbf{x}^*, \dot{\mathbf{x}}^*, \mathbf{p})\sin(k\theta)d\theta.
\end{aligned} \tag{3.64}$$

The state-space form of Equation (3.62) is

$$\dot{\mathbf{z}} = \mathbf{z}_1 \text{ and } \dot{\mathbf{z}}_1 = \mathbf{g}(\mathbf{z}, \mathbf{z}_1) \tag{3.65}$$

where

$$\begin{aligned}
&\mathbf{z} = (\mathbf{a}_0, \mathbf{b}, \mathbf{c})^{\mathrm{T}}, \dot{\mathbf{z}} = \mathbf{z}_1,\\
&\mathbf{g} = (\mathbf{F}_0, -2\Omega\mathbf{k}_1\dot{\mathbf{c}} + \Omega^2\mathbf{k}_2\mathbf{b} + \mathbf{F}_1, 2\Omega\mathbf{k}_1\dot{\mathbf{b}} + \Omega^2\mathbf{k}_2\mathbf{c} + \mathbf{F}_2)^{\mathrm{T}}.
\end{aligned} \tag{3.66}$$

An equivalent system of Equation (3.65) is

$$\dot{\mathbf{y}} = \mathbf{f}(\mathbf{y}) \tag{3.67}$$

where

$$\mathbf{y} = (\mathbf{z}, \mathbf{z}_1)^{\mathrm{T}} \text{ and } \mathbf{f} = (\mathbf{z}_1, \mathbf{g})^{\mathrm{T}}. \tag{3.68}$$

If equilibrium $\mathbf{y}^*$ *of Equation (3.67) (i.e.,* $\mathbf{f}(\mathbf{y}^*) = \mathbf{0}$*) exists, then the approximate solution of periodic motion exists as in Equation (3.61). In the vicinity of equilibrium* $\mathbf{y}^*$*, with* $\mathbf{y} = \mathbf{y}^* + \Delta\mathbf{y}$ *the linearized equation of Equation (3.67) is*

$$\Delta\dot{\mathbf{y}} = D\mathbf{f}(\mathbf{y}^*)\Delta\mathbf{y} \tag{3.69}$$

and the eigenvalue analysis of equilibrium $\mathbf{y}^*$ *is given by*

$$|D\mathbf{f}(\mathbf{y}^*) - \lambda\mathbf{I}_{2n(2N+1)\times 2n(2N+1)}| = 0 \tag{3.70}$$

where $D\mathbf{f}(\mathbf{y}^*) = \partial\mathbf{f}(\mathbf{y})/\partial\mathbf{y}|_{\mathbf{y}^*}$. *Thus, the stability and bifurcation of periodic motions can be classified by the eigenvalues of* $D\mathbf{f}(\mathbf{y}^*)$ *with*

$$(n_1, n_2, n_3 \mid n_4, n_5, n_6). \tag{3.71}$$

1. *If all eigenvalues of the equilibrium possess negative real parts, the approximate periodic solution is stable.*
2. *If at least one of the eigenvalues of the equilibrium possesses positive real part, the approximate periodic solution is unstable.*
3. *The boundaries between stable and unstable equilibriums with higher order singularity give bifurcation and stability conditions with higher order singularity.*

Proof. The proof is similar to Luo (2012a). Since $\mathbf{f}(\mathbf{x}, \dot{\mathbf{x}}; \mathbf{x}^\tau, \dot{\mathbf{x}}^\tau, \mathbf{p})$ is a C^r-continuous nonlinear function vector ($r \geq 1$), the velocity $\dot{\mathbf{x}}$ should be C^r-continuous ($r \geq 1$), and then the acceleration $\ddot{\mathbf{x}}$ should be bounded (i.e., $||\ddot{\mathbf{x}}|| \leq K$). From Equation (3.58), the norms of the periodic motion are defined by

$$||\mathbf{x}|| = ||\mathbf{a}_0|| + \sum_{k=1}^{\infty} ||\mathbf{A}_k||$$

where

$$\mathbf{A}_k = (A_{k1}, A_{k2}, \ldots, A_{kn})^{\mathrm{T}},$$
$$A_{kj} = \sqrt{b_{kj}^2 + c_{kj}^2} \; (j = 1, 2, \ldots, n).$$

Because the periodic motion in Equation (3.57) is bounded (i.e., $||\mathbf{x}(t)|| < C$), we have

$$\lim_{k\to\infty} ||\mathbf{A}_k|| = 0 \text{ but not uniform.}$$

Thus, the Fourier series transformation of periodic motion as in Equation (3.58) is convergent. From Equation (3.60), using Equations (3.58) and (3.61) gives

$$\begin{aligned} ||\mathbf{x}(t) - \mathbf{x}^*(t)|| &= \left\| \sum_{k=N+1}^{\infty} \mathbf{b}_k(t)\cos(k\theta) + \mathbf{c}_k(t)\sin(k\theta) \right\| \\ &= \sum_{k=N+1}^{\infty} ||\mathbf{A}_k||. \end{aligned}$$

For the prescribed small positive $\varepsilon > 0$, if $||\mathbf{x}(t) - \mathbf{x}^*(t)|| < \varepsilon$ exist, then we have

$$\sum_{k=N+1}^{\infty} ||\mathbf{A}_k|| < \varepsilon.$$

Therefore, the convergent infinite term transformation in Equation (3.58) can be approximated by a finite term transformation in Equation (3.61) in a sense of ε.

Taking the derivative of Equation (3.61) with respect to time gives

$$\dot{\mathbf{x}}^*(t) = \dot{\mathbf{a}}_0 + \sum_{k=1}^{N} (\dot{\mathbf{b}}_k + k\Omega\mathbf{c}_k)\cos(k\theta) + (\dot{\mathbf{c}}_k - k\Omega\mathbf{b}_k)\sin(k\theta),$$

$$\ddot{\mathbf{x}}^*(t) = \ddot{\mathbf{a}}_0 + \sum_{k=1}^{N} (\ddot{\mathbf{b}}_k + 2k\Omega\dot{\mathbf{c}}_k - k^2\Omega^2\mathbf{b}_k)\cos(k\theta)$$
$$+ (\ddot{\mathbf{c}}_k - 2k\Omega\dot{\mathbf{b}}_k - k^2\Omega^2\mathbf{c}_k)\sin(k\theta).$$

Substitution of the foregoing equation into Equation (3.57), and application of the virtual work principle for constant, $\cos(k\theta)$ and $\sin(k\theta)$ $(k = 1, 2, \ldots)$ as a set of virtual displacements gives,

$$\frac{1}{2\pi}\int_0^{2\pi} [\ddot{\mathbf{x}} - \mathbf{f}(\mathbf{x}, \dot{\mathbf{x}}, \mathbf{p})]d\theta = 0,$$

$$\frac{1}{\pi}\int_0^{2\pi} [\ddot{\mathbf{x}} - \mathbf{f}(\mathbf{x}, \dot{\mathbf{x}}, \mathbf{p})]\cos(k\theta)d\theta = 0,$$

$$\frac{1}{\pi}\int_0^{2\pi} [\ddot{\mathbf{x}} - \mathbf{f}(\mathbf{x}, \dot{\mathbf{x}}, \mathbf{p})]\sin(k\theta)d\theta = 0.$$

Under $||\mathbf{x}^{(m)} - \mathbf{x}^{(m)*}|| < \varepsilon$ with continuity $||\ddot{\mathbf{x}}^{(m)}|| < K$ (constant) and small $\varepsilon > 0$, for $k = 1, 2, \ldots, N$, the foregoing equation gives,

$$\frac{1}{2\pi}\int_0^{2\pi} [\ddot{\mathbf{x}}^* - \mathbf{f}(\mathbf{x}^*, \dot{\mathbf{x}}^*, \mathbf{p})]d\theta + O(\delta) = 0,$$

$$\frac{1}{\pi}\int_0^{2\pi} [\ddot{\mathbf{x}}^* - \mathbf{f}(\mathbf{x}^*, \dot{\mathbf{x}}^*, \mathbf{p})]\cos(k\theta)d\theta + O(\delta) = 0,$$

$$\frac{1}{\pi}\int_0^{2\pi} [\ddot{\mathbf{x}}^* - \mathbf{f}(\mathbf{x}^*, \dot{\mathbf{x}}^*, \mathbf{p})]\sin(k\theta)d\theta + O(\delta) = 0$$

where $\delta = \max(\varepsilon, \varepsilon_t, \varepsilon_{tt})$ with $||\dot{\mathbf{x}} - \dot{\mathbf{x}}^*|| < \varepsilon_t$, $||\ddot{\mathbf{x}} - \ddot{\mathbf{x}}^*|| < \varepsilon_{tt}$ for small $\{\varepsilon_t, \varepsilon_{tt}\} > 0$. The foregoing equation generates,

$$\ddot{\mathbf{a}}_0 = \mathbf{F}_0(\mathbf{z}, \mathbf{z}_1),$$
$$\ddot{\mathbf{b}}_k = -2\Omega k\dot{\mathbf{c}}_k + \Omega^2 k\mathbf{b}_k + \mathbf{F}_{1k}(\mathbf{z}, \mathbf{z}_1),$$
$$\ddot{\mathbf{c}}_k = 2\Omega k\dot{\mathbf{b}}_k + \Omega^2 k\mathbf{c}_k + \mathbf{F}_{2k}(\mathbf{z}, \mathbf{z}_1)$$

where

$$\mathbf{F}_0(\mathbf{z},\mathbf{z}_1) = \frac{1}{2\pi}\int_0^{2\pi} \mathbf{f}(\mathbf{x}^*,\dot{\mathbf{x}}^*\mathbf{p})d\theta;$$

$$\mathbf{F}_{1k}(\mathbf{z},\mathbf{z}_1) = \frac{1}{\pi}\int_0^{2\pi} \mathbf{f}(\mathbf{x}^*,\dot{\mathbf{x}}^*,\mathbf{p})\cos(k\theta)d\theta,$$

$$\mathbf{F}_{2k}(\mathbf{z},\mathbf{z}_1) = \frac{1}{\pi}\int_0^{2\pi} \mathbf{f}(\mathbf{x}^*,\dot{\mathbf{x}}^*,\mathbf{p})\sin(k\theta)d\theta.$$

Introduce

$$\begin{aligned}
\mathbf{k}_1 &= diag(\mathbf{I}_{n\times n}, 2\mathbf{I}_{n\times n}, \ldots, N\mathbf{I}_{n\times n}),\\
\mathbf{k}_2 &= diag(\mathbf{I}_{n\times n}, 2^2\mathbf{I}_{n\times n}, \ldots, N^2\mathbf{I}_{n\times n})\\
\mathbf{b} &= (\mathbf{b}_1, \mathbf{b}_2, \ldots, \mathbf{b}_N)^{\mathrm{T}},\\
\mathbf{c} &= (\mathbf{c}_1, \mathbf{c}_2, \ldots, \mathbf{c}_N)^{\mathrm{T}};\\
\mathbf{F}_1 &= (\mathbf{F}_{11}, \mathbf{F}_{12}, \ldots, \mathbf{F}_{1N})^{\mathrm{T}},\\
\mathbf{F}_2 &= (\mathbf{F}_{21}, \mathbf{F}_{22}, \ldots, \mathbf{F}_{2N})^{\mathrm{T}};\\
\mathbf{z} &= (\mathbf{a}_0, \mathbf{b}, \mathbf{c})^{\mathrm{T}}, \dot{\mathbf{z}} = \mathbf{z}_1.
\end{aligned}$$

Rearranging the foregoing equation gives Equation (3.62), that is,

$$\begin{aligned}
\ddot{\mathbf{a}}_0 &= \mathbf{F}_0(\mathbf{z},\mathbf{z}_1),\\
\ddot{\mathbf{b}} &= -2\Omega\mathbf{k}_1\dot{\mathbf{c}} + \Omega^2\mathbf{k}_2\mathbf{b} + \mathbf{F}_1(\mathbf{z},\mathbf{z}_1),\\
\ddot{\mathbf{c}} &= 2\Omega\mathbf{k}_1\dot{\mathbf{b}} + \Omega^2\mathbf{k}_2\mathbf{c} + \mathbf{F}_2(\mathbf{z},\mathbf{z}_1);.
\end{aligned}$$

Let

$$\begin{aligned}
\mathbf{z} &= (\mathbf{a}_0, \mathbf{b}, \mathbf{c})^{\mathrm{T}}, \dot{\mathbf{z}} = \mathbf{z}_1,\\
\mathbf{g} &= (\mathbf{F}_0, -2\Omega\mathbf{k}_1\dot{\mathbf{c}} + \Omega^2\mathbf{k}_2\mathbf{b} + \mathbf{F}_1, 2\Omega\mathbf{k}_1\dot{\mathbf{b}} + \Omega^2\mathbf{k}_2\mathbf{c} + \mathbf{F}_2)^{\mathrm{T}}.
\end{aligned}$$

The equation in Equation (3.62) becomes

$$\dot{\mathbf{z}} = \mathbf{z}_1 \text{ and } \dot{\mathbf{z}}_1 = \mathbf{g}(\mathbf{z},\mathbf{z}_1).$$

Letting $\mathbf{y} = (\mathbf{z},\mathbf{z}_1)^{\mathrm{T}}$ and $\mathbf{f} = (\mathbf{z}_1,\mathbf{g})^{\mathrm{T}}$, we have

$$\dot{\mathbf{y}} = \mathbf{f}(\mathbf{y})$$

Consider the equilibrium solution of the foregoing equation (i.e., $\mathbf{f}(\mathbf{y}^*) = \mathbf{0}$), that is,

$$\begin{aligned}
\mathbf{0} &= \mathbf{F}_0(\mathbf{a}_0^*, \mathbf{b}^*, \mathbf{c}^*, \mathbf{0}, \mathbf{0}, \mathbf{0}),\\
\mathbf{0} &= \Omega^2\mathbf{k}_2\mathbf{b}^* + \mathbf{F}_1(\mathbf{a}_0^*, \mathbf{b}^*, \mathbf{c}^*, \mathbf{0}, \mathbf{0}, \mathbf{0}),\\
\mathbf{0} &= \Omega^2\mathbf{k}_2\mathbf{c}^* + \mathbf{F}_2(\mathbf{a}_0^*, \mathbf{b}^*, \mathbf{c}^*, \mathbf{0}, \mathbf{0}, \mathbf{0}).
\end{aligned}$$

Thus, the solutions of the foregoing equation are the existence conditions of the periodic solutions for nonlinear vibration systems. If the foregoing equation gives the equilibrium $\mathbf{y}^*$. In vicinity of $\mathbf{y}^*$, with $\mathbf{y} = \mathbf{y}^* + \Delta\mathbf{y}$, the linearized equation of $\dot{\mathbf{y}} = \mathbf{f}(\mathbf{y})$ is

$$\Delta\dot{\mathbf{y}} = D_{\mathbf{y}}\mathbf{f}(\mathbf{y}^*)\Delta\mathbf{y}$$

and the eigenvalue analysis of equilibrium $\mathbf{y}^*$ is given by

$$|D_{\mathbf{y}}\mathbf{f}(\mathbf{y}^*) - \lambda\mathbf{I}_{2n(2N+1)\times 2n(2N+1)}| = 0$$

where

$$D_{\mathbf{y}}\mathbf{f}(\mathbf{y}^*) = \frac{\partial \mathbf{f}(\mathbf{y})}{\partial \mathbf{y}}|_{\mathbf{y}^*}.$$

Thus, the stability and bifurcation of periodic solution can be classified by the eigenvalues of Equation (3.71) at equilibrium $\mathbf{y}^*$ with

$$(n_1, n_2, n_3 \mid n_4, n_5, n_6).$$

From the stability and bifurcation theory of dynamical systems at equilibrium, the stability, and bifurcation of the periodic solutions can be classified as stated in the theorem. This theorem is proved. ■

If the Hopf bifurcation of equilibriums of Equation (3.67) occurs, there is a periodic solution of coefficients in Equation (3.62) with a frequency ω. As discussed from Equation (3.14) to Equation (3.19), there is a period-m motion as in Equation (3.18). If $m\omega \neq k\Omega$ for any m and k, the solution will be quasi-periodic or chaotic instead of periodic. Herein, the period-m motions in vibration systems will be discussed only.

Theorem 3.6 *Consider a nonlinear vibration system in Equation (3.57). If such a dynamical system has a period-*m *motion $\mathbf{x}^{(m)}(t)$ with finite norm $\|\mathbf{x}^{(m)}\|$ and period $T = 2\pi/\Omega$, there is a generalized coordinate transformation with $\theta = \Omega t$ for the periodic motion of Equation (3.57) in the form of*

$$\mathbf{x}^{(m)}(t) = \mathbf{a}_0^{(m)}(t) + \sum_{k=1}^{\infty} \mathbf{b}_{k/m}(t)\cos\left(\frac{k}{m}\theta\right) + \mathbf{c}_{k/m}(t)\sin\left(\frac{k}{m}\theta\right) \tag{3.72}$$

with

$$\begin{aligned} \mathbf{a}_0^{(m)} &= (a_{01}^{(m)}, a_{02}^{(m)}, \ldots, a_{0n}^{(m)})^{\mathrm{T}}, \\ \mathbf{b}_{k/m} &= (b_{k/m1}, b_{k/m2}, \ldots, b_{k/mn})^{\mathrm{T}}, \\ \mathbf{c}_{k/m} &= (c_{k/m1}, c_{k/m2}, \ldots, c_{k/mn})^{\mathrm{T}} \end{aligned} \tag{3.73}$$

and

$$\begin{aligned} &\|\mathbf{x}^{(m)}\| = \|\mathbf{a}_0^{(m)}\| + \sum_{k=1}^{\infty} \|\mathbf{A}_{k/m}\|, \text{ and } \lim_{k\to\infty} \|\mathbf{A}_{k/m}\| = 0 \text{ but not uniform} \\ &\text{with } \mathbf{A}_{k/m} = (A_{k/m1}, A_{k/m2} \cdots, A_{k/mn})^{\mathrm{T}} \\ &\text{and } A_{k/mj} = \sqrt{b_{k/mj}^2 + c_{k/mj}^2} \;\; (j = 1, 2, \ldots, n). \end{aligned} \tag{3.74}$$

For $\|\mathbf{x}^{(m)}(t) - \mathbf{x}^{(m)*}(t)\| < \varepsilon$ *with a prescribed small* $\varepsilon > 0$*, the infinite term transformation* $\mathbf{x}^{(m)}(t)$ *of period-*m *motion of Equation (3.57), given by Equation (3.72), can be approximated by a finite term transformation* $\mathbf{x}^{(m)*}(t)$ *as*

$$\mathbf{x}^{(m)*}(t) = \mathbf{a}_0^{(m)}(t) + \sum_{k=1}^{N} \mathbf{b}_{k/m}(t)\cos\left(\frac{k}{m}\theta\right) + \mathbf{c}_{k/m}(t)\sin\left(\frac{k}{m}\theta\right) \tag{3.75}$$

and the generalized coordinates are determined by

$$\begin{aligned}
\ddot{\mathbf{a}}_0^{(m)} &= \mathbf{F}_0^{(m)}(\mathbf{a}_0^{(m)}, \mathbf{b}^{(m)}, \mathbf{c}^{(m)}, \dot{\mathbf{a}}_0^{(m)}, \dot{\mathbf{b}}^{(m)}, \dot{\mathbf{c}}^{(m)}),\\
\ddot{\mathbf{b}}^{(m)} &= -2\frac{\Omega}{m}\mathbf{k}_1\dot{\mathbf{c}}^{(m)} + \frac{\Omega^2}{m^2}\mathbf{k}_2\mathbf{b}^{(m)} + \mathbf{F}_1^{(m)}(\mathbf{a}_0^{(m)}, \mathbf{b}^{(m)}, \mathbf{c}^{(m)}, \dot{\mathbf{a}}_0^{(m)}, \dot{\mathbf{b}}^{(m)}, \dot{\mathbf{c}}^{(m)}),\\
\ddot{\mathbf{c}}^{(m)} &= 2\frac{\Omega}{m}\mathbf{k}_1\dot{\mathbf{b}}^{(m)} + \frac{\Omega^2}{m^2}\mathbf{k}_2\mathbf{c}^{(m)} + \mathbf{F}_2^{(m)}(\mathbf{a}_0^{(m)}, \mathbf{b}^{(m)}, \mathbf{c}^{(m)}, \dot{\mathbf{a}}_0^{(m)}, \dot{\mathbf{b}}^{(m)}, \dot{\mathbf{c}}^{(m)})
\end{aligned} \tag{3.76}$$

where

$$\begin{aligned}
\mathbf{k}_1 &= diag(\mathbf{I}_{n\times n}, 2\mathbf{I}_{n\times n}, \ldots, N\mathbf{I}_{n\times n}),\\
\mathbf{k}_2 &= diag(\mathbf{I}_{n\times n}, 2^2\mathbf{I}_{n\times n}, \ldots, N^2\mathbf{I}_{n\times n}),\\
\mathbf{b}^{(m)} &= (\mathbf{b}_{1/m}, \mathbf{b}_{2/m}, \ldots, \mathbf{b}_{N/m})^{\mathrm{T}},\\
\mathbf{c}^{(m)} &= (\mathbf{c}_{1/m}, \mathbf{c}_{2/m}, \ldots, \mathbf{c}_{N/m})^{\mathrm{T}},\\
\mathbf{F}_1^{(m)} &= (\mathbf{F}_{11}^{(m)}, \mathbf{F}_{12}^{(m)}, \ldots, \mathbf{F}_{1N}^{(m)})^{\mathrm{T}},\\
\mathbf{F}_2^{(m)} &= (\mathbf{F}_{21}^{(m)}, \mathbf{F}_{22}^{(m)}, \ldots, \mathbf{F}_{2N}^{(m)})^{\mathrm{T}}\\
&\text{for } N = 1, 2, \ldots, \infty.
\end{aligned} \tag{3.77}$$

and for $k = 1, 2, \ldots, N$

$$\begin{aligned}
&\mathbf{F}_0^{(m)}(\mathbf{a}_0^{(m)}, \mathbf{b}^{(m)}, \mathbf{c}^{(m)}, \dot{\mathbf{a}}_0^{(m)}, \dot{\mathbf{b}}^{(m)}, \dot{\mathbf{c}}^{(m)})\\
&\quad = \frac{1}{2m\pi}\int_0^{2m\pi} \mathbf{f}(\mathbf{x}^{(m)*}, \dot{\mathbf{x}}^{(m)*}, \mathbf{p})d\theta;\\
&\mathbf{F}_{1k}^{(m)}(\mathbf{a}_0^{(m)}, \mathbf{b}^{(m)}, \mathbf{c}^{(m)}, \dot{\mathbf{a}}_0^{(m)}, \dot{\mathbf{b}}^{(m)}, \dot{\mathbf{c}}^{(m)})\\
&\quad = \frac{1}{m\pi}\int_0^{2m\pi} \mathbf{f}(\mathbf{x}^{(m)*}, \dot{\mathbf{x}}^{(m)*}, \mathbf{p})\cos\left(\frac{k}{m}\theta\right)d\theta,\\
&\mathbf{F}_{2k}^{(m)}(\mathbf{a}_0^{(m)}, \mathbf{b}^{(m)}, \mathbf{c}^{(m)}, \dot{\mathbf{a}}_0^{(m)}, \dot{\mathbf{b}}^{(m)}, \dot{\mathbf{c}}^{(m)})\\
&\quad = \frac{1}{m\pi}\int_0^{2m\pi} \mathbf{f}(\mathbf{x}^{(m)*}, \dot{\mathbf{x}}^{(m)*}, \mathbf{p})\sin\left(\frac{k}{m}\theta\right)d\theta.
\end{aligned} \tag{3.78}$$

The state-space form of Equation (3.76) is

$$\dot{\mathbf{z}}^{(m)} = \mathbf{z}_1^{(m)} \text{ and } \dot{\mathbf{z}}_1^{(m)} = \mathbf{g}^{(m)}(\mathbf{z}^{(m)}, \mathbf{z}_1^{(m)}) \tag{3.79}$$

where

$$\mathbf{z}^{(m)} = (\mathbf{a}_0^{(m)}, \mathbf{b}^{(m)}, \mathbf{c}^{(m)})^{\mathrm{T}}, \dot{\mathbf{z}}^{(m)} = \mathbf{z}_1^{(m)}$$

$$\mathbf{g}^{(m)} = \left(\mathbf{F}_0^{(m)}, -2\frac{\Omega}{m}\mathbf{k}_1\dot{\mathbf{c}}^{(m)} + \frac{\Omega^2}{m^2}\mathbf{k}_2\mathbf{b}^{(m)} + \mathbf{F}_1^{(m)}, \right.$$

$$\left. 2\frac{\Omega}{m}\mathbf{k}_1\dot{\mathbf{b}}^{(m)} + \frac{\Omega^2}{m^2}\mathbf{k}_2\mathbf{c}^{(m)} + \mathbf{F}_2^{(m)} \right)^{\mathrm{T}}. \tag{3.80}$$

An equivalent system of Equation (3.79) is

$$\dot{\mathbf{y}}^{(m)} = \mathbf{f}^{(m)}(\mathbf{y}^{(m)}) \tag{3.81}$$

where

$$\mathbf{y}^{(m)} = (\mathbf{z}^{(m)}, \mathbf{z}_1^{(m)})^{\mathrm{T}} \text{ and } \mathbf{f}^{(m)} = (\mathbf{z}_1^{(m)}, \mathbf{g}^{(m)})^{\mathrm{T}}. \tag{3.82}$$

*If equilibrium $\mathbf{y}^{(m)*}$ of Equation (3.81) (i.e., $\mathbf{f}^{(m)}(\mathbf{y}^{(m)*}) = \mathbf{0}$) exists, then the approximate solution of period-*m *motion exists as in Equation (3.75). In the vicinity of equilibrium $\mathbf{y}^{(m)*}$, with $\mathbf{y}^{(m)} = \mathbf{y}^{(m)*} + \Delta\mathbf{y}^{(m)}$ the linearized equation of Equation (3.81) is*

$$\Delta\dot{\mathbf{y}}^{(m)} = D\mathbf{f}^{(m)}(\mathbf{y}^{(m)*})\Delta\mathbf{y}^{(m)} \tag{3.83}$$

and the eigenvalue analysis of the equilibrium $\mathbf{y}^*$ is given by

$$|D\mathbf{f}^{(m)}(\mathbf{y}^{(m)*}) - \lambda\mathbf{I}_{2n(2N+1)\times 2n(2N+1)}| = 0 \tag{3.84}$$

*where $D\mathbf{f}^{(m)}(\mathbf{y}^{(m)*}) = \partial\mathbf{f}^{(m)}(\mathbf{y}^{(m)})/\partial\mathbf{y}^{(m)}|_{\mathbf{y}^{(m)*}}$. Thus, the stability and bifurcation of period-*m *motions can be classified by the eigenvalues of $D\mathbf{f}^{(m)}(\mathbf{y}^{(m)*})$ with*

$$(n_1, n_2, n_3 \mid n_4, n_5, n_6). \tag{3.85}$$

1. *If all eigenvalues of the equilibrium possess negative real parts, the approximate periodic solution is stable.*
2. *If at least one of the eigenvalues of the equilibrium possesses positive real part, the approximate periodic solution is unstable.*
3. *The boundaries between stable and unstable equilibriums with higher order singularity give bifurcation and stability conditions with higher order singularity.*

Proof. The proof is similar to Luo (2012a). Since $\mathbf{f}(\mathbf{x}, \dot{\mathbf{x}}, \mathbf{p})$ is a C^r-continuous nonlinear function vector ($r \geq 1$), the velocity $\dot{\mathbf{x}}$ should be C^r-continuous ($r \geq 1$), and then the acceleration $\ddot{\mathbf{x}}$ should be bounded (i.e., $||\ddot{\mathbf{x}}|| < K$). From Equation (3.72), the norms of the periodic motion are defined by

$$||\mathbf{x}^{(m)}|| = ||\mathbf{a}_0^{(m)}|| + \sum_{k=1}^{\infty} ||\mathbf{A}_{k/m}||$$

where

$$\mathbf{A}_{k/m} = (A_{k/m1}, A_{k/m2}, \ldots, A_{k/mn})^{\mathrm{T}},$$

$$A_{k/mj} = \sqrt{b_{k/mj}^2 + c_{k/mj}^2} \quad (j = 1, 2, \ldots, n).$$

Because the periodic motion in Equation (3.57) is bounded (i.e., $\|\mathbf{x}(t)\| < C$), we have

$$\lim_{k\to\infty} \|\mathbf{A}_{k/m}\| = 0 \text{ and } \lim_{k\to\infty} \|\mathbf{A}_{k/m}^{\tau}\| = 0 \text{ but not uniform.}$$

Thus, the Fourier series transformation of periodic motion as in Equation (3.72) is convergent. From Equation (3.74), using Equations (3.72) and (3.75) gives

$$\begin{aligned}
&\|\mathbf{x}^{(m)}(t) - \mathbf{x}^{(m)*}(t)\| \\
&= \left\| \sum_{k=N+1}^{\infty} \mathbf{b}_{k/m}(t)\cos\left(\frac{k}{m}\theta\right) + \mathbf{c}_{k/m}(t)\sin\left(\frac{k}{m}\theta\right) \right\| \\
&= \sum_{k=N+1}^{\infty} \|\mathbf{A}_{k/m}\|.
\end{aligned}$$

For the prescribed small positive $\varepsilon > 0$, if $\|\mathbf{x}^{(m)}(t) - \mathbf{x}^{(m)*}(t)\| < \varepsilon$ exists, then we have

$$\sum_{k=N+1}^{\infty} \|\mathbf{A}_{k/m}\| < \varepsilon.$$

Therefore, the convergent infinite term transformation in Equation (3.72) can be approximated by a finite term transformation in Equation (3.75) in the sense of ε.

Taking the derivatives of Equation (3.75) gives

$$\begin{aligned}
\dot{\mathbf{x}}^{(m)*}(t) = \dot{\mathbf{a}}_0^{(m)} + \sum_{k=1}^{N} &\left(\dot{\mathbf{b}}_{k/m} + \frac{k}{m}\Omega\mathbf{c}_{k/m}\right)\cos\left(\frac{k}{m}\theta\right) \\
&+ \left(\dot{\mathbf{c}}_{k/m} - \frac{k}{m}\Omega\mathbf{b}_{k/m}\right)\sin\left(\frac{k}{m}\theta\right).
\end{aligned}$$

Substitution of the foregoing equation into Equation (3.57), and application of the virtual work principle for constant, $\cos(k\theta/m)$ and $\sin(k\theta/m)$ $(k = 1, 2, \ldots)$ as a set of virtual displacements, gives

$$\begin{aligned}
&\frac{1}{2m\pi}\int_0^{2m\pi} [\ddot{\mathbf{x}}^{(m)} - \mathbf{f}(\mathbf{x}^{(m)}, \dot{\mathbf{x}}^{(m)}, \mathbf{p})]d\theta = 0, \\
&\frac{1}{m\pi}\int_0^{2m\pi} [\ddot{\mathbf{x}}^{(m)} - \mathbf{f}(\mathbf{x}^{(m)}, \dot{\mathbf{x}}^{(m)}, \mathbf{p})]\cos\left(\frac{k}{m}\theta\right)d\theta = 0, \\
&\frac{1}{m\pi}\int_0^{2m\pi} [\ddot{\mathbf{x}}^{(m)} - \mathbf{f}(\mathbf{x}^{(m)}, \dot{\mathbf{x}}^{(m)}, \mathbf{p})]\sin\left(\frac{k}{m}\theta\right)d\theta = 0.
\end{aligned}$$

Under $\|\mathbf{x}^{(m)} - \mathbf{x}^{(m)*}\| < \varepsilon$ with continuity $\|\ddot{\mathbf{x}}^{(m)}\| < K$ (constant) and small $\varepsilon > 0$, for $k = 1, 2, \ldots, N$, the foregoing equation becomes

$$\begin{aligned}
&\frac{1}{2m\pi}\int_0^{2m\pi} [\ddot{\mathbf{x}}^{(m)*} - \mathbf{f}(\mathbf{x}^{(m)*}, \dot{\mathbf{x}}^{(m)*}, \mathbf{p})]d\theta + O(\delta) = 0, \\
&\frac{1}{m\pi}\int_0^{2m\pi} [\ddot{\mathbf{x}}^{(m)*} - \mathbf{f}(\mathbf{x}^{(m)*}, \dot{\mathbf{x}}^{(m)*}, \mathbf{p})]\cos\left(\frac{k}{m}\theta\right)d\theta + O(\delta) = 0, \\
&\frac{1}{m\pi}\int_0^{2m\pi} [\ddot{\mathbf{x}}^{(m)*} - \mathbf{f}(\mathbf{x}^{(m)*}, \dot{\mathbf{x}}^{(m)*}, \mathbf{p})]\sin\left(\frac{k}{m}\theta\right)d\theta + O(\delta) = 0
\end{aligned}$$

where $\delta = \max(\varepsilon, \varepsilon_t, \varepsilon_{tt})$ with $\|\dot{\mathbf{x}}^{(m)} - \dot{\mathbf{x}}^{(m)*}\| < \varepsilon_t$, $\|\ddot{\mathbf{x}}^{(m)} - \ddot{\mathbf{x}}^{(m)*}\| < \varepsilon_{tt}$ for small $\{\varepsilon_t, \varepsilon_{tt}\} > 0$. The foregoing equation gives

$$\begin{aligned}
\ddot{\mathbf{a}}_0^{(m)} &= \mathbf{F}_0^{(m)}(\mathbf{z}^{(m)}, \mathbf{z}_1^{(m)}),\\
\ddot{\mathbf{b}}_{k/m} &= -2\frac{\Omega}{m}k\dot{\mathbf{c}}_{k/m} + \frac{\Omega^2}{m^2}k^2\mathbf{b}_{k/m} + \mathbf{F}_{1k}^{(m)}(\mathbf{z}^{(m)}, \mathbf{z}_1^{(m)}),\\
\ddot{\mathbf{c}}_{k/m} &= 2\frac{\Omega}{m}k\dot{\mathbf{b}}_{k/m} + \frac{\Omega^2}{m^2}k^2\mathbf{c}_{k/m} + \mathbf{F}_{2k}^{(m)}(\mathbf{z}^{(m)}, \mathbf{z}_1^{(m)})
\end{aligned}$$

where for $k = 1, 2, \ldots, N$

$$\begin{aligned}
&\mathbf{F}_0^{(m)}(\mathbf{z}^{(m)}, \mathbf{z}_1^{(m)})\\
&\quad = \frac{1}{2m\pi}\int_0^{2m\pi} \mathbf{f}(\mathbf{x}^{(m)*}, \dot{\mathbf{x}}^{(m)*}, \mathbf{p})d\theta;\\
&\mathbf{F}_{1k}^{(m)}(\mathbf{z}^{(m)}, \mathbf{z}_1^{(m)}; \mathbf{z}^{\tau(m)}, \mathbf{z}_1^{\tau(m)})\\
&\quad = \frac{1}{m\pi}\int_0^{2m\pi} \mathbf{f}(\mathbf{x}^{(m)*}, \dot{\mathbf{x}}^{(m)*}, \mathbf{p})\cos\left(\frac{k}{m}\theta\right)d\theta,\\
&\mathbf{F}_{2k}^{(m)}(\mathbf{z}^{(m)}, \mathbf{z}_1^{(m)}; \mathbf{z}^{\tau(m)}, \mathbf{z}_1^{\tau(m)})\\
&\quad = \frac{1}{m\pi}\int_0^{2m\pi} \mathbf{f}(\mathbf{x}^{(m)*}, \dot{\mathbf{x}}^{(m)*}, \mathbf{p})\sin\left(\frac{k}{m}\theta\right)d\theta;
\end{aligned}$$

and

$$\begin{aligned}
\mathbf{k}_1 &= diag(\mathbf{I}_{n\times n}, 2\mathbf{I}_{n\times n}, \ldots, N\mathbf{I}_{n\times n}),\\
\mathbf{k}_2 &= diag(\mathbf{I}_{n\times n}, 2^2\mathbf{I}_{n\times n}, \ldots, N^2\mathbf{I}_{n\times n});\\
\mathbf{b}^{(m)} &= (\mathbf{b}_{1/m}, \mathbf{b}_{2/m}, \ldots, \mathbf{b}_{N/m})^{\mathrm{T}},\\
\mathbf{c}^{(m)} &= (\mathbf{c}_{1/m}, \mathbf{c}_{2/m}, \ldots, \mathbf{c}_{N/m})^{\mathrm{T}};\\
\mathbf{F}_1^{(m)} &= (\mathbf{F}_{11}^{(m)}, \mathbf{F}_{12}^{(m)}, \ldots, \mathbf{F}_{1N}^{(m)})^{\mathrm{T}},\\
\mathbf{F}_2^{(m)} &= (\mathbf{F}_{21}^{(m)}, \mathbf{F}_{22}^{(m)}, \ldots, \mathbf{F}_{2N}^{(m)})^{\mathrm{T}};\\
\mathbf{z}^{(m)} &= (\mathbf{a}_0^{(m)}, \mathbf{b}^{(m)}, \mathbf{c}^{(m)})^{\mathrm{T}}, \dot{\mathbf{z}}^{(m)} = \mathbf{z}_1^{(m)}\\
&\text{for } N = 1, 2, \ldots, \infty.
\end{aligned}$$

Rearranging the foregoing equation gives Equation (3.76), that is,

$$\begin{aligned}
\ddot{\mathbf{a}}_0^{(m)} &= \mathbf{F}_0^{(m)}(\mathbf{z}^{(m)}, \mathbf{z}_1^{(m)}),\\
\ddot{\mathbf{b}}^{(m)} &= -2\frac{\Omega}{m}\mathbf{k}_1\dot{\mathbf{c}}^{(m)} + \frac{\Omega^2}{m^2}\mathbf{k}_2\mathbf{b}^{(m)} + \mathbf{F}_1^{(m)}(\mathbf{z}^{(m)}, \mathbf{z}_1^{(m)}),\\
\ddot{\mathbf{c}}^{(m)} &= 2\frac{\Omega}{m}\mathbf{k}_1\dot{\mathbf{b}}^{(m)} + \frac{\Omega^2}{m^2}\mathbf{k}_2\mathbf{c}^{(m)} + \mathbf{F}_2^{(m)}(\mathbf{z}^{(m)}, \mathbf{z}_1^{(m)})
\end{aligned}$$

Introduce

$$\mathbf{g}^{(m)} = \left(\mathbf{F}_0^{(m)}, -2\frac{\Omega}{m}\mathbf{k}_1\dot{\mathbf{c}}^{(m)} + \frac{\Omega^2}{m^2}\mathbf{k}_2\mathbf{b}^{(m)} + \mathbf{F}_1^{(m)},\right.$$
$$\left. 2\frac{\Omega}{m}\mathbf{k}_1\dot{\mathbf{b}}^{(m)} + \frac{\Omega^2}{m^2}\mathbf{k}_2\mathbf{c}^{(m)} + \mathbf{F}_2^{(m)}\right)^{\mathrm{T}}.$$

The equation in Equation (3.76) becomes

$$\dot{\mathbf{z}}^{(m)} = \mathbf{z}_1^{(m)} \text{ and } \dot{\mathbf{z}}_1^{(m)} = \mathbf{g}^{(m)}(\mathbf{z}^{(m)}, \mathbf{z}_1^{(m)}).$$

Letting

$$\mathbf{y}^{(m)} = (\mathbf{z}^{(m)}, \mathbf{z}_1^{(m)})^{\mathrm{T}}, \mathbf{f}^{(m)} = (\mathbf{z}_1^{(m)}, \mathbf{g}^{(m)})^{\mathrm{T}}$$

we have

$$\dot{\mathbf{y}}^{(m)} = \mathbf{f}^{(m)}(\mathbf{y}^{(m)}).$$

Consider equilibriums of the foregoing equation (i.e., $\mathbf{f}^{(m)}(\mathbf{y}^{(m)*}) = \mathbf{0}$), that is,

$$\mathbf{0} = \mathbf{F}_0(\mathbf{a}_0^*, \mathbf{b}^*, \mathbf{c}^*, \mathbf{0}, \mathbf{0}, \mathbf{0}),$$
$$\mathbf{0} = \Omega^2\mathbf{k}_2\mathbf{b}^* + \mathbf{F}_1(\mathbf{a}_0^*, \mathbf{b}^*, \mathbf{c}^*, \mathbf{0}, \mathbf{0}, \mathbf{0}),$$
$$\mathbf{0} = \Omega^2\mathbf{k}_2\mathbf{c}^* + \mathbf{F}_2(\mathbf{a}_0^*, \mathbf{b}^*, \mathbf{c}^*, \mathbf{0}, \mathbf{0}, \mathbf{0}).$$

Thus, the solutions of the foregoing equation are the existence conditions of the periodic solutions for nonlinear vibration systems. Thus, the foregoing equation gives equilibrium $\mathbf{y}^{(m)*}$. In the vicinity of $\mathbf{y}^{(m)*}$, with $\mathbf{y}^{(m)} = \mathbf{y}^{(m)*} + \Delta\mathbf{y}^{(m)}$ the linearized equation of $\dot{\mathbf{y}}^{(m)} = \mathbf{f}^{(m)}(\mathbf{y}^{(m)}, \mathbf{y}^{\tau(m)})$ is

$$\Delta\dot{\mathbf{y}}^{(m)} = D\mathbf{f}^{(m)}(\mathbf{y}^{(m)*}, \mathbf{y}^{\tau(m)*})\Delta\mathbf{y}^{(m)}$$

and the eigenvalue analysis of equilibrium $\mathbf{y}^*$ is given by

$$|D\mathbf{f}^{(m)}(\mathbf{y}^{(m)*}) - \lambda\mathbf{I}_{2n(2N+1)\times 2n(2N+1)}| = 0$$

where

$$D\mathbf{f}^{(m)}(\mathbf{y}^{(m)*}) = \frac{\partial\mathbf{f}^{(m)}(\mathbf{y}^{(m)})}{\partial\mathbf{y}^{(m)}}\Big|_{\mathbf{y}^{(m)*}}.$$

The stability and bifurcation of the period-m motion can be classified by the eigenvalues of Equation (3.85) at equilibrium $\mathbf{y}^{(m)*} = \mathbf{y}^{\tau(m)*}$ with

$$(n_1, n_2, n_3 \mid n_4, n_5, n_6).$$

From the stability and bifurcation theory of dynamical systems at such equilibrium, the stability and bifurcation of the periodic motion can be classified as stated in the theorem. This theorem is proved. ■

If the Hopf bifurcation of period-m motion occurs, a generalized coordinate transformation with $\theta = \Omega t$ for the period-doubling solution of period-m motion can be expressed by

$$\mathbf{x}^{(2m)}(t) = \mathbf{a}_0^{(2m)}(t) + \sum_{k=1}^{\infty} \mathbf{b}_{k/2m}(t)\cos\left(\frac{k}{2m}\theta\right) + \mathbf{c}_{k/2m}(t)\sin\left(\frac{k}{2m}\theta\right). \tag{3.86}$$

If $\|\mathbf{x}^{(2m)}(t) - \mathbf{x}^{(2m)*}(t)\| < \varepsilon$ with a prescribed small $\varepsilon > 0$, a finite term transformation for the solution of period-$2m$ motion in Equation (3.86) can be expressed by

$$\mathbf{x}^{(2m)*}(t) = \mathbf{a}_0^{(2m)}(t) + \sum_{k=1}^{N} \mathbf{b}_{k/2m}(t)\cos\left(\frac{k}{2m}\theta\right) + \mathbf{c}_{k/2m}(t)\sin\left(\frac{k}{2m}\theta\right) \tag{3.87}$$

where the generalized coordinates are computed for $k = 1, 2, \ldots, N$

$$\begin{aligned}
\ddot{\mathbf{a}}_0^{(2m)} &= \mathbf{F}_0^{(2m)}(\dot{\mathbf{a}}_0^{(2m)}, \dot{\mathbf{b}}^{(2m)}, \dot{\mathbf{c}}^{(2m)}, \mathbf{a}_0^{(2m)}, \mathbf{b}^{(2m)}, \mathbf{c}^{(2m)}),\\
\ddot{\mathbf{b}}^{(2m)} &+ 2\frac{\Omega}{(2m)}\mathbf{k}_1\dot{\mathbf{c}}^{(2m)} - \frac{\Omega^2}{(2m)^2}\mathbf{k}_2\mathbf{b}^{(2m)}\\
&= \mathbf{F}_{1k}^{(m)}(\dot{\mathbf{a}}_0^{(2m)}, \dot{\mathbf{b}}^{(2m)}, \dot{\mathbf{c}}^{(2m)}, \mathbf{a}_0^{(2m)}, \mathbf{b}^{(2m)}, \mathbf{c}^{(2m)}),\\
\ddot{\mathbf{c}}^{(2m)} &- 2\frac{\Omega}{(2m)}\mathbf{k}_1\dot{\mathbf{b}}^{(2m)} - \frac{\Omega^2}{(2m)^2}\mathbf{k}_2\mathbf{c}^{(2m)}\\
&= \mathbf{F}_{2k}^{(m)}(\dot{\mathbf{a}}_0^{(2m)}, \dot{\mathbf{b}}^{(2m)}, \dot{\mathbf{c}}^{(2m)}, \mathbf{a}_0^{(2m)}, \mathbf{b}^{(2m)}, \mathbf{c}^{(2m)})
\end{aligned} \tag{3.88}$$

where for $N = 1, 2, \ldots, \infty$

$$\begin{aligned}
\mathbf{k}_1 &= diag(\mathbf{I}_{n\times n}, 2\mathbf{I}_{n\times n}, \ldots, N\mathbf{I}_{n\times n}),\\
\mathbf{k}_2 &= diag(\mathbf{I}_{n\times n}, 2^2\mathbf{I}_{n\times n}, \ldots, N^2\mathbf{I}_{n\times n});\\
\mathbf{b}^{(2m)} &= (\mathbf{b}_{1/2m}, \mathbf{b}_{2/2m}, \ldots, \mathbf{b}_{N/2m})^{\mathrm{T}},\\
\mathbf{c}^{(2m)} &= (\mathbf{c}_{1/2m}, \mathbf{c}_{2/2m}, \ldots, \mathbf{c}_{N/2m})^{\mathrm{T}};\\
\mathbf{F}_1^{(2m)} &= (\mathbf{F}_{11}^{(2m)}, \mathbf{F}_{12}^{(2m)}, \ldots, \mathbf{F}_{1N}^{(2m)})^{\mathrm{T}},\\
\mathbf{F}_2^{(2m)} &= (\mathbf{F}_{21}^{(2m)}, \mathbf{F}_{22}^{(2m)}, \ldots, \mathbf{F}_{2N}^{(2m)})^{\mathrm{T}}
\end{aligned} \tag{3.89}$$

and for $k = 1, 2, \ldots, N$

$$\begin{aligned}
&\mathbf{F}_0^{(2m)}(\mathbf{a}_0^{(2m)}, \mathbf{b}^{(2m)}, \mathbf{c}^{(2m)}, \dot{\mathbf{a}}_0^{(2m)}, \dot{\mathbf{b}}^{(2m)}, \dot{\mathbf{c}}^{(2m)})\\
&\quad = \frac{1}{2(2m\pi)}\int_0^{2(2m\pi)} \mathbf{f}(\mathbf{x}^{(2m)*}, \dot{\mathbf{x}}^{(2m)*}, \mathbf{p})d\theta;\\
&\mathbf{F}_{1k}^{(2m)}(\mathbf{a}_0^{(2m)}, \mathbf{b}^{(2m)}, \mathbf{c}^{(2m)}, \dot{\mathbf{a}}_0^{(2m)}, \dot{\mathbf{b}}^{(2m)}, \dot{\mathbf{c}}^{(2m)})\\
&\quad = \frac{1}{2m\pi}\int_0^{2(2m\pi)} \mathbf{f}(\mathbf{x}^{(2m)*}, \dot{\mathbf{x}}^{(2m)*}, \mathbf{p})\cos\left(\frac{k}{2m}\theta\right)d\theta,\\
&\mathbf{F}_{2k}^{(2m)}(\mathbf{a}_0^{(2m)}, \mathbf{b}^{(2m)}, \mathbf{c}^{(2m)}, \dot{\mathbf{a}}_0^{(2m)}, \dot{\mathbf{b}}^{(2m)}, \dot{\mathbf{c}}^{(2m)})\\
&\quad = \frac{1}{2m\pi}\int_0^{2(2m\pi)} \mathbf{f}(\mathbf{x}^{(2m)*}, \dot{\mathbf{x}}^{(2m)*}, \mathbf{p})\sin\left(\frac{k}{2m}\theta\right)d\theta.
\end{aligned} \tag{3.90}$$

If the Hopf bifurcation of period-$2m$ motion occurs again and again, a generalized coordinate transformation with $\theta = \Omega t$ for a period-$2^l m$ motion can be expressed by

$$\mathbf{x}^{(2^l m)}(t) = \mathbf{a}_0^{(2^l m)}(t) + \sum_{k=1}^{\infty} \mathbf{b}_{k/2^l m}(t)\cos\left(\frac{k}{2^l m}\theta\right) + \mathbf{c}_{k/2^l m}(t)\sin\left(\frac{k}{2^l m}\theta\right). \tag{3.91}$$

If $\|\mathbf{x}^{(2^l m)}(t) - \mathbf{x}^{(2^l m)*}(t)\| < \varepsilon$ with a prescribed small $\varepsilon > 0$, a finite term transformation for the solution of period-$2^l m$ motion in Equation (3.91) can be approximated by

$$\mathbf{x}^{(2^l m)*}(t) = \mathbf{a}_0^{(2^l m)}(t) + \sum_{k=1}^{N} \mathbf{b}_{k/2^l m}(t)\cos\left(\frac{k}{2^l m}\theta\right) + \mathbf{c}_{k/2^l m}(t)\sin\left(\frac{k}{2^l m}\theta\right) \tag{3.92}$$

where the generalized coordinates are computed for $k = 1, 2, \ldots, N$

$$\begin{aligned}
&\ddot{\mathbf{a}}_0^{(2^l m)} = \mathbf{F}_0^{(2^l m)}(\dot{\mathbf{a}}_0^{(2^l m)}, \dot{\mathbf{b}}^{(2^l m)}, \dot{\mathbf{c}}^{(2^l m)}, \mathbf{a}_0^{(2^l m)}, \mathbf{b}^{(2^l m)}, \mathbf{c}^{(2^l m)}),\\
&\ddot{\mathbf{b}}^{(2^l m)} + 2\frac{\Omega}{2^l m}\mathbf{k}_1\dot{\mathbf{c}}^{(2^l m)} - \frac{\Omega^2}{(2^l m)^2}\mathbf{k}_2\mathbf{b}^{(2^l m)}\\
&\quad = \mathbf{F}_{1k}^{(2^l m)}(\dot{\mathbf{a}}_0^{(2^l m)}, \dot{\mathbf{b}}^{(2^l m)}, \dot{\mathbf{c}}^{(2^l m)}, \mathbf{a}_0^{(2^l m)}, \mathbf{b}^{(2^l m)}, \mathbf{c}^{(2^l m)}),\\
&\ddot{\mathbf{c}}^{(2^l m)} - 2\frac{\Omega}{2^l m}\mathbf{k}_1\dot{\mathbf{b}}^{(2^l m)} - \frac{\Omega^2}{(2^l m)^2}\mathbf{k}_2\mathbf{c}^{(2^l m)}\\
&\quad = \mathbf{F}_{2k}^{(2^l m)}(\dot{\mathbf{a}}_0^{(2^l m)}, \dot{\mathbf{b}}^{(2^l m)}, \dot{\mathbf{c}}^{(2^l m)}, \mathbf{a}_0^{(2^l m)}, \mathbf{b}^{(2^l m)}, \mathbf{c}^{(2^l m)});
\end{aligned} \tag{3.93}$$

and for $N = 1, 2, \ldots, \infty$

$$\begin{aligned}
&\mathbf{k}_1 = diag(\mathbf{I}_{n\times n}, 2\mathbf{I}_{n\times n}, \ldots, N\mathbf{I}_{n\times n}),\\
&\mathbf{k}_2 = diag(\mathbf{I}_{n\times n}, 2^2\mathbf{I}_{n\times n}, \ldots, N^2\mathbf{I}_{n\times n}),\\
&\mathbf{b}^{(2^l m)} = (\mathbf{b}_{1/2^l m}, \mathbf{b}_{2/2^l m}, \ldots, \mathbf{b}_{N/2^l m})^{\mathrm{T}},\\
&\mathbf{c}^{(2^l m)} = (\mathbf{c}_{1/2^l m}, \mathbf{c}_{2/2^l m}, \ldots, \mathbf{c}_{N/2^l m})^{\mathrm{T}},\\
&\mathbf{F}_1^{(2^l m)} = (\mathbf{F}_{11}^{(2^l m)}, \mathbf{F}_{12}^{(2^l m)}, \ldots, \mathbf{F}_{1N}^{(2^l m)})^{\mathrm{T}},\\
&\mathbf{F}_2^{(2^l m)} = (\mathbf{F}_{21}^{(2^l m)}, \mathbf{F}_{22}^{(2^l m)}, \ldots, \mathbf{F}_{2N}^{(2^l m)})^{\mathrm{T}};
\end{aligned} \tag{3.94}$$

$$\begin{aligned}
&\mathbf{F}_0^{(2^l m)}(\mathbf{a}_0^{(2^l m)}, \mathbf{b}^{(2^l m)}, \mathbf{c}^{(2^l m)}, \dot{\mathbf{a}}_0^{(2^l m)}, \dot{\mathbf{b}}^{(2^l m)}, \dot{\mathbf{c}}^{(2^l m)})\\
&\quad = \frac{1}{2^l(2m\pi)}\int_0^{2^l(2m\pi)} \mathbf{f}(\mathbf{x}^{(2^l m)*}, \dot{\mathbf{x}}^{(2^l m)*}, \mathbf{p})d\theta;\\
&\mathbf{F}_{1k}^{(2^l m)}(\mathbf{a}_0^{(2^l m)}, \mathbf{b}^{(2^l m)}, \mathbf{c}^{(2^l m)}, \dot{\mathbf{a}}_0^{(2^l m)}, \dot{\mathbf{b}}^{(2^l m)}, \dot{\mathbf{c}}^{(2^l m)})\\
&\quad = \frac{1}{2^{l-1}(2m\pi)}\int_0^{2^l(2m\pi)} \mathbf{f}(\mathbf{x}^{(2^l m)*}, \dot{\mathbf{x}}^{(2^l m)*}, \mathbf{p})\cos\left(\frac{k}{2^l m}\theta\right)d\theta,\\
&\mathbf{F}_{2k}^{(2^l m)}(\mathbf{a}_0^{(2^l m)}, \mathbf{b}^{(2^l m)}, \mathbf{c}^{(2^l m)}, \dot{\mathbf{a}}_0^{(2^l m)}, \dot{\mathbf{b}}^{(2^l m)}, \dot{\mathbf{c}}^{(2^l m)})\\
&\quad = \frac{1}{2^{l-1}(2m\pi)}\int_0^{2^l(2m\pi)} \mathbf{f}(\mathbf{x}^{(2^l m)*}, \dot{\mathbf{x}}^{(2^l m)*}, \mathbf{p})\sin\left(\frac{k}{2^l m}\theta\right)d\theta.
\end{aligned} \tag{3.95}$$

The solution of period-$2^l m$ motion can be determined by the equilibrium of the coefficient dynamical system in Equation (3.93), and the corresponding stability and bifurcation can be done. As $l \to \infty$, the stable and unstable chaos with $(n_1, n_2, n_3 \mid n_4, n_5, n_6)$ in Equation (3.85) can be obtained where

$$\sum_{i=1}^{3} n_i + 2\sum_{i=4}^{6} n_i = 2n(1+2N). \tag{3.96}$$

With increasing l, the Fourier truncated number $N >> 2^l m$ will dramatically increase. If period-1 motion possesses at least N_1 harmonic vector terms, then the total harmonic vector terms for period-$2^l m$ motion should be $N \geq 2^l m N_1$. The chaotic motion classifications with specific cases are given as follows.

1. For the chaotic motion of $(n_1, 0, 0 \mid 0, 0, 0)$ with $n_1 = 2n(1+2N)$, the chaotic motion is called the hyperbolic stable chaos.
2. For the chaotic motion of $(0, 0, 0 \mid n_4, 0, 0)$ with $n_4 = n(1+2N)$, the chaotic motion is called the spiral stable chaos.
3. For the chaotic motion of $(n_1, 0, 0 \mid n_4, 0, 0)$ with $n_1 + 2n_4 = 2n(1+2N)$, the chaotic motion is called the hyperbolic-spiral stable chaos.
4. For the chaotic motion of $(0, n_2, 0 \mid 0, 0, 0)$ with $n_2 = 2n(1+2N)$, the chaotic motion is called the hyperbolic unstable chaos.
5. For the chaotic motion of $(0, 0, 0 \mid 0, n_5, 0)$ with $n_5 = n(1+2N)$, the chaotic motion is called the spiral unstable chaos.
6. For the chaotic motion of $(0, n_2, 0 \mid 0, n_5, 0)$ with $n_2 + 2n_5 = 2n(1+2N)$, the chaotic motion is called the hyperbolic-spiral unstable chaos.
7. For the chaotic motion of $(n_1, n_2, 0 \mid 0, 0, 0)$ with $n_1 + n_2 = 2n(1+2N)$, the chaotic motion is called the saddle unstable chaos.
8. For the chaotic motion of $(n_1, n_2, 0 \mid n_4, n_5, 0)$ with $n_1 + n_2 + 2n_4 + 2n_5 = 2n(1+2N)$, the chaotic motion is called the spiral saddle unstable chaos.

Because m is an arbitrary positive integer number, it includes $(2^l m_1)$ for period-$2^l m_1$ motion. Thus, the expression for period-m motion can be used for any periodic motions. The expression in Equation (3.72) can be used to express the solution for chaotic motion as $m \to \infty$, which can be approximated by Equation (3.75) under the condition of $\|\mathbf{x}^{(m)}(t) - \mathbf{x}^{(m)*}(t)\| < \varepsilon$. The chaotic solutions can be classified as discussed for period-$2^l m_1$ motion.

3.2.2 Periodically Excited Vibration Systems

If a periodically excited, nonlinear vibration system possesses a periodic motion with period $T = 2\pi/\Omega$, then such a periodic motion can be expressed by the Fourier series, discussed as follows.

Theorem 3.7 *Consider a periodically excited, nonlinear vibration system as*

$$\ddot{\mathbf{x}} = \mathbf{F}(\mathbf{x}, \dot{\mathbf{x}}, t, \mathbf{p}) \in \mathscr{R}^n \tag{3.97}$$

where $\mathbf{F}(\mathbf{x}, \dot{\mathbf{x}}, t, \mathbf{p})$ is a C^r-continuous nonlinear function vector ($r \geq 1$) with an excitation period $T = 2\pi/\Omega$. If such a vibration system has a periodic motion $\mathbf{x}(t)$ with finite norm $\|\mathbf{x}\|$, there is a generalized coordinate transformation with $\theta = \Omega t$ for the periodic motion of Equation (3.97) in a form of

$$\mathbf{x}(t) = \mathbf{a}_0(t) + \sum_{k=1}^{\infty} \mathbf{b}_k(t)\cos(k\theta) + \mathbf{c}_k(t)\sin(k\theta) \tag{3.98}$$

with

$$\mathbf{a}_0 = (a_{01}, a_{02}, \ldots, a_{0n})^{\mathrm{T}},$$
$$\mathbf{b}_k = (b_{k1}, b_{k2}, \ldots, b_{kn})^{\mathrm{T}},$$
$$\mathbf{c}_k = (c_{k1}, c_{k2}, \ldots, c_{kn})^{\mathrm{T}} \tag{3.99}$$

and

$$\|\mathbf{x}\| = \|\mathbf{a}_0\| + \sum_{k=1}^{\infty} \|\mathbf{A}_k\|, \text{ and } \lim_{k\to\infty} \|\mathbf{A}_k\| = 0 \text{ but not uniform}$$
$$\text{with } \mathbf{A}_k = (A_{k1}, A_{k2}, \ldots, A_{kn})^{\mathrm{T}} \text{ and } A_{kj} = \sqrt{b_{kj}^2 + c_{kj}^2} \ (j = 1, 2, \ldots, n). \tag{3.100}$$

For $\|\mathbf{x}(t) - \mathbf{x}^*(t)\| < \varepsilon$ with a prescribed small positive $\varepsilon > 0$, the infinite term transformation of periodic motion $\mathbf{x}(t)$ of Equation (3.97), given by Equation (3.98), can be approximated by a finite term transformation $\mathbf{x}^*(t)$ as

$$\mathbf{x}^*(t) = \mathbf{a}_0(t) + \sum_{k=1}^{N} \mathbf{b}_k(t)\cos(k\theta) + \mathbf{c}_k(t)\sin(k\theta) \tag{3.101}$$

and the generalized coordinates are determined by

$$\ddot{\mathbf{a}}_0 = \mathbf{F}_0(\mathbf{a}_0, \mathbf{b}, \mathbf{c}, \dot{\mathbf{a}}_0, \dot{\mathbf{b}}, \dot{\mathbf{c}}),$$
$$\ddot{\mathbf{b}} = -2\Omega\mathbf{k}_1\dot{\mathbf{c}} + \Omega^2\mathbf{k}_2\mathbf{b} + \mathbf{F}_1(\mathbf{a}_0, \mathbf{b}, \mathbf{c}, \dot{\mathbf{a}}_0, \dot{\mathbf{b}}, \dot{\mathbf{c}}),$$
$$\ddot{\mathbf{c}} = 2\Omega\mathbf{k}_1\dot{\mathbf{b}} + \Omega^2\mathbf{k}_2\mathbf{c} + \mathbf{F}_2(\mathbf{a}_0, \mathbf{b}, \mathbf{c}, \dot{\mathbf{a}}_0, \dot{\mathbf{b}}, \dot{\mathbf{c}}); \tag{3.102}$$

where for $N = 1, 2, \ldots, \infty$

$$\mathbf{k}_1 = diag(\mathbf{I}_{n\times n}, 2\mathbf{I}_{n\times n}, \ldots, N\mathbf{I}_{n\times n}),$$
$$\mathbf{k}_2 = diag(\mathbf{I}_{n\times n}, 2^2\mathbf{I}_{n\times n}, \ldots, N^2\mathbf{I}_{n\times n}),$$
$$\mathbf{b} = (\mathbf{b}_1, \mathbf{b}_2, \ldots, \mathbf{b}_N)^{\mathrm{T}},$$
$$\mathbf{c} = (\mathbf{c}_1, \mathbf{c}_2, \ldots, \mathbf{c}_N)^{\mathrm{T}},$$
$$\mathbf{F}_1 = (\mathbf{F}_{11}, \mathbf{F}_{12}, \ldots, \mathbf{F}_{1N})^{\mathrm{T}},$$
$$\mathbf{F}_2 = (\mathbf{F}_{21}, \mathbf{F}_{22}, \ldots, \mathbf{F}_{2N})^{\mathrm{T}} \tag{3.103}$$

and for $k = 1, 2, \ldots, N$

$$\mathbf{F}_0(\mathbf{a}_0, \mathbf{b}, \mathbf{c}, \dot{\mathbf{a}}_0, \dot{\mathbf{b}}, \dot{\mathbf{c}}) = \frac{1}{2\pi}\int_0^{2\pi} \mathbf{F}(\mathbf{x}^*, \dot{\mathbf{x}}^*, t, \mathbf{p})d\theta;$$
$$\mathbf{F}_{1k}(\mathbf{a}_0, \mathbf{b}, \mathbf{c}, \dot{\mathbf{a}}_0, \dot{\mathbf{b}}, \dot{\mathbf{c}}) = \frac{1}{\pi}\int_0^{2\pi} \mathbf{F}(\mathbf{x}^*, \dot{\mathbf{x}}^*, t, \mathbf{p})\cos(k\theta)d\theta,$$
$$\mathbf{F}_{2k}(\mathbf{a}_0, \mathbf{b}, \mathbf{c}, \dot{\mathbf{a}}_0, \dot{\mathbf{b}}, \dot{\mathbf{c}}) = \frac{1}{\pi}\int_0^{2\pi} \mathbf{F}(\mathbf{x}^*, \dot{\mathbf{x}}^*, t, \mathbf{p})\sin(k\theta)d\theta. \tag{3.104}$$

The state-space form of Equation (3.102) is

$$\dot{\mathbf{z}} = \mathbf{z}_1 \text{ and } \dot{\mathbf{z}}_1 = \mathbf{g}(\mathbf{z}, \mathbf{z}_1) \tag{3.105}$$

where

$$\begin{aligned} &\mathbf{z} = (\mathbf{a}_0, \mathbf{b}, \mathbf{c})^{\mathrm{T}}, \dot{\mathbf{z}} = \mathbf{z}_1 \\ &\mathbf{g} = (\mathbf{F}_0, -2\Omega\mathbf{k}_1\dot{\mathbf{c}} + \Omega^2\mathbf{k}_2\mathbf{b} + \mathbf{F}_1, 2\Omega\mathbf{k}_1\dot{\mathbf{b}} + \Omega^2\mathbf{k}_2\mathbf{c} + \mathbf{F}_2)^{\mathrm{T}}. \end{aligned} \tag{3.106}$$

An equivalent system of Equation (3.105) is

$$\dot{\mathbf{y}} = \mathbf{f}(\mathbf{y}) \tag{3.107}$$

where

$$\mathbf{y} = (\mathbf{z}, \mathbf{z}_1)^{\mathrm{T}} \text{ and } \mathbf{f} = (\mathbf{z}_1, \mathbf{g})^{\mathrm{T}}. \tag{3.108}$$

If equilibrium $\mathbf{y}^*$ *of Equation (3.107) (i.e.,* $\mathbf{f}(\mathbf{y}^*) = \mathbf{0}$*) exists, then the approximate solution of periodic motion exists as in Equation (3.101). In the vicinity of equilibrium* $\mathbf{y}^*$*, with* $\mathbf{y} = \mathbf{y}^* + \Delta\mathbf{y}$ *the linearized equation of Equation (3.107) is*

$$\Delta\dot{\mathbf{y}} = D\mathbf{f}(\mathbf{y}^*)\Delta\mathbf{y} \tag{3.109}$$

and the eigenvalue analysis of equilibrium $\mathbf{y}^*$ *is given by*

$$|D\mathbf{f}(\mathbf{y}^*) - \lambda\mathbf{I}_{2n(2N+1)\times 2n(2N+1)}| = 0 \tag{3.110}$$

where $D\mathbf{f}(\mathbf{y}^*) = \partial\mathbf{f}(\mathbf{y})/\partial\mathbf{y}|_{\mathbf{y}^*}$*. Thus, the stability and bifurcation of periodic motions can be classified by the eigenvalues of* $D\mathbf{f}(\mathbf{y}^*)$ *with*

$$(n_1, n_2, n_3 \,|\, n_4, n_5, n_6). \tag{3.111}$$

1. *If all eigenvalues of the equilibrium possess negative real parts, the approximate periodic solution is stable.*
2. *If at least one of the eigenvalues of the equilibrium possesses positive real part, the approximate periodic solution is unstable.*
3. *The boundaries between stable and unstable equilibriums with higher order singularity give bifurcation and stability conditions with higher order singularity.*

Proof. The proof can be referenced to Luo (2012a) or similar to the proof of Theorem 3.5. ■

Similarly, period-m motions in periodically excited, nonlinear vibration systems will be discussed.

Theorem 3.8 *Consider a periodically excited, nonlinear dynamical system in Equation (3.97) with an excitation period* $T = 2\pi/\Omega$*. If such a vibration system has a period-*m *motion* $\mathbf{x}^{(m)}(t)$ *with finite norm* $\|\mathbf{x}^{(m)}\|$ *and period* $T = 2\pi/\Omega$*, there is a generalized coordinate transformation with* $\theta = \Omega t$ *for the periodic motion of Equation (3.97) in the form of*

$$\mathbf{x}^{(m)}(t) = \mathbf{a}_0^{(m)}(t) + \sum_{k=1}^{\infty} \mathbf{b}_{k/m}(t)\cos\left(\frac{k}{m}\theta\right) + \mathbf{c}_{k/m}(t)\sin\left(\frac{k}{m}\theta\right) \tag{3.112}$$

with

$$\begin{aligned}
\mathbf{a}_0^{(m)} &= (a_{01}^{(m)}, a_{02}^{(m)}, \ldots, a_{0n}^{(m)})^{\mathrm{T}},\\
\mathbf{b}_{k/m} &= (b_{k/m1}, b_{k/m2}, \ldots, b_{k/mn})^{\mathrm{T}},\\
\mathbf{c}_{k/m} &= (c_{k/m1}, c_{k/m2}, \ldots, c_{k/mn})^{\mathrm{T}}
\end{aligned} \tag{3.113}$$

and

$$\begin{aligned}
&\|\mathbf{x}^{(m)}\| = \|\mathbf{a}_0^{(m)}\| + \sum_{k=1}^{\infty} \|\mathbf{A}_{k/m}\|, \text{ and } \lim_{k\to\infty} \|\mathbf{A}_{k/m}\| = 0 \text{ but not uniform}\\
&\text{with } \mathbf{A}_{k/m} = (A_{k/m1}, A_{k/m2} \ldots, A_{k/mn})^{\mathrm{T}}\\
&\text{and } A_{k/mj} = \sqrt{b_{k/mj}^2 + c_{k/mj}^2}\ (j = 1, 2, \ldots, n).
\end{aligned} \tag{3.114}$$

For $\|\mathbf{x}^{(m)}(t) - \mathbf{x}^{(m)*}(t)\| < \varepsilon$ *with a prescribed small* $\varepsilon > 0$, *the infinite term transformation* $\mathbf{x}^{(m)}(t)$ *of period-*m *motion of Equation (3.97), given by Equation (3.112), can be approximated by a finite term transformation* $\mathbf{x}^{(m)*}(t)$ *as*

$$\mathbf{x}^{(m)*}(t) = \mathbf{a}_0^{(m)}(t) + \sum_{k=1}^{N} \mathbf{b}_{k/m}(t)\cos\left(\frac{k}{m}\theta\right) + \mathbf{c}_{k/m}(t)\sin\left(\frac{k}{m}\theta\right) \tag{3.115}$$

and the generalized coordinates are determined by

$$\begin{aligned}
\ddot{\mathbf{a}}_0^{(m)} &= \mathbf{F}_0^{(m)}(\mathbf{a}_0^{(m)}, \mathbf{b}^{(m)}, \mathbf{c}^{(m)}, \dot{\mathbf{a}}_0^{(m)}, \dot{\mathbf{b}}^{(m)}, \dot{\mathbf{c}}^{(m)}),\\
\ddot{\mathbf{b}}^{(m)} &= -2\frac{\Omega}{m}\mathbf{k}_1\dot{\mathbf{c}}^{(m)} + \frac{\Omega^2}{m^2}\mathbf{k}_2\mathbf{b}^{(m)} + \mathbf{F}_1^{(m)}(\mathbf{a}_0^{(m)}, \mathbf{b}^{(m)}, \mathbf{c}^{(m)}, \dot{\mathbf{a}}_0^{(m)}, \dot{\mathbf{b}}^{(m)}, \dot{\mathbf{c}}^{(m)}),\\
\ddot{\mathbf{c}}^{(m)} &= 2\frac{\Omega}{m}\mathbf{k}_1\dot{\mathbf{b}}^{(m)} + \frac{\Omega^2}{m^2}\mathbf{k}_2\mathbf{c}^{(m)} + \mathbf{F}_2^{(m)}(\mathbf{a}_0^{(m)}, \mathbf{b}^{(m)}, \mathbf{c}^{(m)}, \dot{\mathbf{a}}_0^{(m)}, \dot{\mathbf{b}}^{(m)}, \dot{\mathbf{c}}^{(m)})
\end{aligned} \tag{3.116}$$

where for $N = 1, 2, \ldots, \infty$

$$\begin{aligned}
\mathbf{k}_1 &= diag(\mathbf{I}_{n\times n}, 2\mathbf{I}_{n\times n}, \ldots, N\mathbf{I}_{n\times n}),\\
\mathbf{k}_2 &= diag(\mathbf{I}_{n\times n}, 2^2\mathbf{I}_{n\times n}, \ldots, N^2\mathbf{I}_{n\times n}),\\
\mathbf{b}^{(m)} &= (\mathbf{b}_{1/m}, \mathbf{b}_{2/m}, \ldots, \mathbf{b}_{N/m})^{\mathrm{T}},\\
\mathbf{c}^{(m)} &= (\mathbf{c}_{1/m}, \mathbf{c}_{2/m}, \ldots, \mathbf{c}_{N/m})^{\mathrm{T}},\\
\mathbf{F}_1^{(m)} &= (\mathbf{F}_{11}^{(m)}, \mathbf{F}_{12}^{(m)}, \ldots, \mathbf{F}_{1N}^{(m)})^{\mathrm{T}}\\
\mathbf{F}_2^{(m)} &= (\mathbf{F}_{21}^{(m)}, \mathbf{F}_{22}^{(m)}, \ldots, \mathbf{F}_{2N}^{(m)})^{\mathrm{T}};
\end{aligned} \tag{3.117}$$

and for $k = 1, 2, \ldots, N$

$$\begin{aligned}
&\mathbf{F}_0^{(m)}(\mathbf{a}_0^{(m)}, \mathbf{b}^{(m)}, \mathbf{c}^{(m)}, \dot{\mathbf{a}}_0^{(m)}, \dot{\mathbf{b}}^{(m)}, \dot{\mathbf{c}}^{(m)})\\
&\quad = \frac{1}{2m\pi}\int_0^{2m\pi} \mathbf{F}(\mathbf{x}^{(m)*}, \dot{\mathbf{x}}^{(m)*}, t, \mathbf{p})d\theta;
\end{aligned}$$

$$
\begin{aligned}
&\mathbf{F}_{1k}^{(m)}(\mathbf{a}_0^{(m)}, \mathbf{b}^{(m)}, \mathbf{c}^{(m)}, \dot{\mathbf{a}}_0^{(m)}, \dot{\mathbf{b}}^{(m)}, \dot{\mathbf{c}}^{(m)}) \\
&= \frac{1}{m\pi} \int_0^{2m\pi} \mathbf{F}(\mathbf{x}^{(m)*}, \dot{\mathbf{x}}^{(m)*}, t, \mathbf{p}) \cos\left(\frac{k}{m}\theta\right) d\theta, \\
&\mathbf{F}_{2k}^{(m)}(\mathbf{a}_0^{(m)}, \mathbf{b}^{(m)}, \mathbf{c}^{(m)}, \dot{\mathbf{a}}_0^{(m)}, \dot{\mathbf{b}}^{(m)}, \dot{\mathbf{c}}^{(m)}) \\
&= \frac{1}{m\pi} \int_0^{2m\pi} \mathbf{F}(\mathbf{x}^{(m)*}, \dot{\mathbf{x}}^{(m)*}, t, \mathbf{p}) \sin\left(\frac{k}{m}\theta\right) d\theta.
\end{aligned} \tag{3.118}
$$

The state-space form of Equation (3.116) is

$$
\dot{\mathbf{z}}^{(m)} = \mathbf{z}_1^{(m)} \text{ and } \dot{\mathbf{z}}_1^{(m)} = \mathbf{g}^{(m)}(\mathbf{z}^{(m)}, \mathbf{z}_1^{(m)}) \tag{3.119}
$$

where

$$
\begin{aligned}
\mathbf{z}^{(m)} &= (\mathbf{a}_0^{(m)}, \mathbf{b}^{(m)}, \mathbf{c}^{(m)})^{\mathrm{T}}, \dot{\mathbf{z}}^{(m)} = \mathbf{z}_1^{(m)} \\
\mathbf{g}^{(m)} &= (\mathbf{F}_0^{(m)}, -2\frac{\Omega}{m}\mathbf{k}_1\dot{\mathbf{c}}^{(m)} + \frac{\Omega^2}{m^2}\mathbf{k}_2\mathbf{b}^{(m)} + \mathbf{F}_1^{(m)}, \\
&\quad 2\frac{\Omega}{m}\mathbf{k}_1\dot{\mathbf{b}}^{(m)} + \frac{\Omega^2}{m^2}\mathbf{k}_2\mathbf{c}^{(m)} + \mathbf{F}_2^{(m)})^{\mathrm{T}}.
\end{aligned} \tag{3.120}
$$

An equivalent system of Equation (3.119) is

$$
\dot{\mathbf{y}}^{(m)} = \mathbf{f}^{(m)}(\mathbf{y}^{(m)}) \tag{3.121}
$$

where

$$
\mathbf{y}^{(m)} = (\mathbf{z}^{(m)}, \mathbf{z}_1^{(m)})^{\mathrm{T}} \text{ and } \mathbf{f}^{(m)} = (\mathbf{z}_1^{(m)}, \mathbf{g}^{(m)})^{\mathrm{T}}. \tag{3.122}
$$

*If equilibrium $\mathbf{y}^{(m)*}$ of Equation (3.121) exists (i.e., $\mathbf{f}^{(m)}(\mathbf{y}^{(m)*}) = \mathbf{0}$), then the approximate solution of period-*m *motion exists as in Equation (3.115). In the vicinity of equilibrium $\mathbf{y}^{(m)*}$, with $\mathbf{y}^{(m)} = \mathbf{y}^{(m)*} + \Delta\mathbf{y}^{(m)}$ the linearized equation of Equation (3.121) is*

$$
\Delta\dot{\mathbf{y}}^{(m)} = D\mathbf{f}^{(m)}(\mathbf{y}^{(m)*})\Delta\mathbf{y}^{(m)} \tag{3.123}
$$

and the eigenvalue analysis of equilibrium $\mathbf{y}^*$ is given by

$$
|D\mathbf{f}^{(m)}(\mathbf{y}^{(m)*}) - \lambda\mathbf{I}_{2n(2N+1)\times 2n(2N+1)}| = 0 \tag{3.124}
$$

*where $D\mathbf{f}^{(m)}(\mathbf{y}^{(m)*}) = \partial\mathbf{f}^{(m)}(\mathbf{y}^{(m)})/\partial\mathbf{y}^{(m)}|_{\mathbf{y}^{(m)*}}$. The stability and bifurcation of period-*m *motions can be classified by eigenvalues of $D\mathbf{f}^{(m)}(\mathbf{y}^{(m)*})$ with*

$$
(n_1, n_2, n_3 \mid n_4, n_5, n_6). \tag{3.125}
$$

1. *If all eigenvalues of the equilibrium possess negative real parts, the approximate periodic solution is stable.*
2. *If at least one of the eigenvalues of the equilibrium possesses positive real part, the approximate periodic solution is unstable.*
3. *The boundaries between stable and unstable equilibriums with higher order singularity give bifurcation and stability conditions with higher order singularity.*

Proof. The proof can be referred to Luo (2012a) or similar to the proof of Theorem 3.6. ■

3.3 Time-Delayed Nonlinear Systems

In this section, analytical periodic flows in autonomous, time-delayed, nonlinear dynamical systems will be discussed, and the local stability and bifurcation theory of equilibriums in a time-delayed, nonlinear system of coefficients will be used to classify analytical solutions of periodic flows and chaos in time-delayed, nonlinear dynamical systems. A generalized harmonic balance method will be employed for periodic flows and chaos in time-delayed, nonlinear dynamical systems.

3.3.1 Autonomous Time-Delayed Nonlinear Systems

Periodic flows in autonomous, time-delayed, dynamical systems will be discussed first. If such a time-delayed system has a periodic flow with a period of $T = 2\pi/\Omega$, then such a periodic flow can be expressed by the Fourier series.

Theorem 3.9 *Consider a nonlinear, time-delayed dynamical system as*

$$\dot{\mathbf{x}} = \mathbf{f}(\mathbf{x}, \mathbf{x}^{\tau}, \mathbf{p}) \in \mathscr{R}^{n} \tag{3.126}$$

where $\mathbf{f}(\mathbf{x}, \mathbf{x}^{\tau}, \mathbf{p})$ is a C^r-continuous nonlinear function vector ($r \geq 1$). If such a dynamical system has a periodic flow $\mathbf{x}(t)$ with finite norm $\|\mathbf{x}\|$ and period $T = 2\pi/\Omega$, there is a generalized coordinate transformation with $\theta = \Omega t$ for the periodic flow of Equation (3.126) in the form of

$$\begin{aligned}
&\mathbf{x} \equiv \mathbf{x}(t) = \mathbf{a}_0(t) + \sum_{k=1}^{\infty} \mathbf{b}_k(t)\cos(k\theta) + \mathbf{c}_k(t)\sin(k\theta);\\
&\mathbf{x}^{\tau} \equiv \mathbf{x}(t-\tau) = \mathbf{a}_0^{\tau}(t) + \sum_{k=1}^{\infty} \mathbf{b}_k^{\tau}(t)\cos[k(\theta-\theta^{\tau})] + \mathbf{c}_k^{\tau}(t)\sin[k(\theta-\theta^{\tau})]
\end{aligned} \tag{3.127}$$

with $\mathbf{a}_0^{\tau} = \mathbf{a}_0(t-\tau)$, $\mathbf{b}_k^{\tau} = \mathbf{b}_k(t-\tau)$, $\mathbf{c}_k^{\tau} = \mathbf{c}_k(t-\tau)$, $\theta^{\tau} = \Omega\tau$ and

$$\begin{aligned}
\mathbf{a}_0 &= (a_{01}, a_{02}, \dots, a_{0n})^{\mathrm{T}},\\
\mathbf{b}_k &= (b_{k1}, b_{k2}, \dots, b_{kn})^{\mathrm{T}},\\
\mathbf{c}_k &= (c_{k1}, c_{k2}, \dots, c_{kn})^{\mathrm{T}};\\
\mathbf{a}_0^{\tau} &= (a_{01}^{\tau}, a_{02}^{\tau}, \dots, a_{0n}^{\tau})^{\mathrm{T}},\\
\mathbf{b}_k^{\tau} &= (b_{k1}^{\tau}, b_{k2}^{\tau}, \dots, b_{kn}^{\tau})^{\mathrm{T}},\\
\mathbf{c}_k^{\tau} &= (c_{k1}^{\tau}, c_{k2}^{\tau}, \dots, c_{kn}^{\tau})^{\mathrm{T}}
\end{aligned} \tag{3.128}$$

and

$$\begin{aligned}
&\|\mathbf{x}\| = \|\mathbf{a}_0\| + \sum_{k=1}^{\infty} \|\mathbf{A}_k\|, \text{ and } \lim_{k\to\infty} \|\mathbf{A}_k\| = 0 \text{ but not uniform}\\
&\text{with } \mathbf{A}_k = (A_{k1}, A_{k2}, \dots, A_{kn})^{\mathrm{T}} \text{ and } A_{kj} = \sqrt{b_{kj}^2 + c_{kj}^2}\ (j = 1, 2, \dots, n).
\end{aligned} \tag{3.129}$$

$$\|\mathbf{x}^{\tau}\| = \|\mathbf{a}_0^{\tau}\| + \sum_{k=1}^{\infty} \|\mathbf{A}_k^{\tau}\|, \text{ and } \lim_{k\to\infty} \|\mathbf{A}_k^{\tau}\| = 0 \text{ but not uniform}$$

$$\text{with } \mathbf{A}_k^{\tau} = (A_{k1}^{\tau}, A_{k2}^{\tau}, \ldots, A_{kn}^{\tau})^{\mathrm{T}} \text{ and } A_{kj}^{\tau} = \sqrt{(b_{kj}^{\tau})^2 + (c_{kj}^{\tau})^2}\ (j = 1, 2, \ldots, n). \tag{3.130}$$

For $\|\mathbf{x}(t) - \mathbf{x}^*(t)\| < \varepsilon$ *and* $\|\mathbf{x}^{\tau}(t) - \mathbf{x}^{\tau*}(t)\| < \varepsilon^{\tau}$ *with prescribed small positive* $\varepsilon > 0$ *and* $\varepsilon^{\tau} > 0$, *the infinite term transformation of periodic flow* $\mathbf{x}(t)$ *of Equation (3.126), given by Equation (3.127), can be approximated by a finite term transformation* $\mathbf{x}^*(t)$ *as*

$$\mathbf{x}^* \equiv \mathbf{x}^*(t) = \mathbf{a}_0(t) + \sum_{k=1}^{N} \mathbf{b}_k(t)\cos(k\theta) + \mathbf{c}_k(t)\sin(k\theta),$$

$$\mathbf{x}^{\tau*} \equiv \mathbf{x}^*(t-\tau) = \mathbf{a}_0^{\tau}(t) + \sum_{k=1}^{N} \mathbf{b}_k^{\tau}(t)\cos[k(\theta-\theta^{\tau})] + \mathbf{c}_k^{\tau}(t)\sin[k(\theta-\theta^{\tau})] \tag{3.131}$$

and the generalized coordinates are determined by

$$\begin{aligned} \dot{\mathbf{a}}_0 &= \mathbf{F}_0(\mathbf{z}, \mathbf{z}^{\tau}), \\ \dot{\mathbf{b}} &= -\Omega\mathbf{k}_1\mathbf{c} + \mathbf{F}_1(\mathbf{z}, \mathbf{z}^{\tau}), \\ \dot{\mathbf{c}} &= \Omega\mathbf{k}_1\mathbf{b} + \mathbf{F}_2(\mathbf{z}, \mathbf{z}^{\tau}) \end{aligned} \tag{3.132}$$

where

$$\begin{aligned} \mathbf{k}_1 &= diag(\mathbf{I}_{n\times n}, 2\mathbf{I}_{n\times n}, \ldots, N\mathbf{I}_{n\times n}), \\ \mathbf{b} &= (\mathbf{b}_1, \mathbf{b}_2, \ldots, \mathbf{b}_N)^{\mathrm{T}}, \\ \mathbf{c} &= (\mathbf{c}_1, \mathbf{c}_2, \ldots, \mathbf{c}_N)^{\mathrm{T}}; \\ \mathbf{b}^{\tau} &= (\mathbf{b}_1^{\tau}, \mathbf{b}_2^{\tau}, \ldots, \mathbf{b}_N^{\tau})^{\mathrm{T}}, \\ \mathbf{c}^{\tau} &= (\mathbf{c}_1^{\tau}, \mathbf{c}_2^{\tau}, \ldots, \mathbf{c}_N^{\tau})^{\mathrm{T}}; \\ \mathbf{F}_1 &= (\mathbf{F}_{11}, \mathbf{F}_{12}, \ldots, \mathbf{F}_{1N})^{\mathrm{T}}, \\ \mathbf{F}_2 &= (\mathbf{F}_{21}, \mathbf{F}_{22}, \ldots, \mathbf{F}_{2N})^{\mathrm{T}}; \\ \mathbf{z} &= (\mathbf{a}_0, \mathbf{b}, \mathbf{c})^{\mathrm{T}}, \\ \mathbf{z}^{\tau} &= (\mathbf{a}_0^{\tau}, \mathbf{b}^{\tau}, \mathbf{c}^{\tau})^{\mathrm{T}} \\ &\text{for } N = 1, 2, \ldots, \infty; \end{aligned} \tag{3.133}$$

and for $k = 1, 2, \ldots, N$

$$\begin{aligned} \mathbf{F}_0(\mathbf{z}, \mathbf{z}^{\tau}) &= \frac{1}{2\pi}\int_0^{2\pi} \mathbf{f}(\mathbf{x}^*, \mathbf{x}^{\tau*}, \mathbf{p})d\theta; \\ \mathbf{F}_{1k}(\mathbf{z}, \mathbf{z}^{\tau}) &= \frac{1}{\pi}\int_0^{2\pi} \mathbf{f}(\mathbf{x}^*, \mathbf{x}^{\tau*}, \mathbf{p})\cos(k\theta)d\theta, \\ \mathbf{F}_{2k}(\mathbf{z}, \mathbf{z}^{\tau}) &= \frac{2}{\pi}\int_0^{2\pi} \mathbf{f}(\mathbf{x}^*, \mathbf{x}^{\tau*}, \mathbf{p})\sin(k\theta)d\theta. \end{aligned} \tag{3.134}$$

Equation (3.132) becomes

$$\dot{\mathbf{z}} = \mathbf{f}(\mathbf{z}, \mathbf{z}^{\tau}) \tag{3.135}$$

where

$$\mathbf{f} = (\mathbf{F}_0, -\Omega\mathbf{k}_1\mathbf{c} + \mathbf{F}_1, \Omega\mathbf{k}_1\mathbf{b} + \mathbf{F}_2)^{\mathrm{T}}. \tag{3.136}$$

If equilibrium $\mathbf{z}^* = \mathbf{z}^{\tau *}$ *of Equation (3.135) (i.e.,* $\mathbf{f}(\mathbf{z}^*, \mathbf{z}^{*\tau}) = \mathbf{0}$*) exists, then the approximate solution of periodic flow exists as in Equation (3.131). In the vicinity of equilibrium* $\mathbf{z}^* = \mathbf{z}^{\tau *}$*, with* $\mathbf{z} = \mathbf{z}^* + \Delta\mathbf{z}$ *and* $\mathbf{z}^{\tau} = \mathbf{z}^{\tau *} + \Delta\mathbf{z}^{\tau}$*, the linearized equation of Equation (3.135) is*

$$\Delta\dot{\mathbf{z}} = D_{\mathbf{z}}\mathbf{f}(\mathbf{z}^*, \mathbf{z}^{*\tau})\Delta\mathbf{z} + D_{\mathbf{z}^{\tau}}\mathbf{f}(\mathbf{z}^*, \mathbf{z}^{*\tau})\Delta\mathbf{z}^{\tau} \tag{3.137}$$

and the eigenvalue analysis of equilibrium $\mathbf{z}^* = \mathbf{z}^{\tau *}$ *is given by*

$$|D_{\mathbf{z}}\mathbf{f}(\mathbf{z}^*, \mathbf{z}^{\tau *}) - \lambda\mathbf{I}_{n(2N+1)\times n(2N+1)} + D_{\mathbf{z}}\mathbf{f}(\mathbf{z}^*, \mathbf{z}^{\tau *})e^{-\lambda\tau}| = 0 \tag{3.138}$$

where

$$D_{\mathbf{z}}\mathbf{f}(\mathbf{z}^*, \mathbf{z}^{\tau *}) = \frac{\partial\mathbf{f}(\mathbf{z}, \mathbf{z}^{\tau})}{\partial\mathbf{z}}|_{(\mathbf{z}^*, \mathbf{z}^{\tau *})},$$
$$D_{\mathbf{z}^{\tau}}\mathbf{f}(\mathbf{z}^*, \mathbf{z}^{\tau *}) = \frac{\partial\mathbf{f}(\mathbf{z}, \mathbf{z}^{\tau})}{\partial\mathbf{z}^{\tau}}|_{(\mathbf{z}^*, \mathbf{z}^{\tau *})}. \tag{3.139}$$

Thus, the stability and bifurcation of periodic flow can be classified by the eigenvalues of Equation (3.137) at equilibrium $\mathbf{z}^* = \mathbf{z}^{\tau *}$ *with*

$$(n_1, n_2, n_3 \mid n_4, n_5, n_6). \tag{3.140}$$

1. *If all eigenvalues of the equilibrium possess negative real parts, the approximate periodic solution is stable.*
2. *If at least one of the eigenvalues of the equilibrium possesses positive real part, the approximate periodic solution is unstable.*
3. *The boundaries between stable and unstable equilibriums with higher order singularity give bifurcation and stability conditions with higher order singularity.*

Proof. If $\mathbf{f}(\mathbf{x}, \mathbf{x}^{\tau}, \mathbf{p})$ is a C^r-continuous nonlinear function vector ($r \geq 1$), then the velocity $\dot{\mathbf{x}}$ should be C^r-continuous ($r \geq 1$), and the acceleration $\ddot{\mathbf{x}}$ should be bounded (i.e., $||\ddot{\mathbf{x}}|| \leq K$). From Equation (3.127), the norms of the periodic flows are defined by

$$||\mathbf{x}|| = ||\mathbf{a}_0|| + \sum_{k=1}^{\infty}||\mathbf{A}_k|| \text{ and } ||\mathbf{x}^{\tau}|| = ||\mathbf{a}_0^{\tau}|| + \sum_{k=1}^{\infty}||\mathbf{A}_k^{\tau}||$$

where

$$\mathbf{A}_k = (A_{k1}, A_{k2}, \ldots, A_{kn})^{\mathrm{T}},$$
$$A_{kj} = \sqrt{b_{kj}^2 + c_{kj}^2} \quad (j = 1, 2, \ldots, n),$$

$$\mathbf{A}_k^{\tau} = (A_{k1}^{\tau}, A_{k2}^{\tau}, \ldots, A_{kn}^{\tau})^{\mathrm{T}},$$
$$A_{kj}^{\tau} = \sqrt{(b_{kj}^{\tau})^2 + (c_{kj}^{\tau})^2} \quad (j = 1, 2, \ldots, n).$$

Because the periodic flow in Equation (3.126) is bounded (i.e., $||\mathbf{x}(t)|| < C$), we have

$$\lim_{k\to\infty} ||\mathbf{A}_k|| = 0 \text{ and } \lim_{k\to\infty} ||\mathbf{A}_k^\tau|| = 0 \text{ but not uniform.}$$

Thus, the Fourier series transformation of periodic flow as in Equation (3.127) is convergent. From Equations (3.129) and (3.130), using Equations (3.127) and (3.131) gives

$$||\mathbf{x}(t) - \mathbf{x}^*(t)|| = \left\| \sum_{k=N+1}^{\infty} \mathbf{b}_k(t)\cos(k\theta) + \mathbf{c}_k(t)\sin(k\theta) \right\| = \sum_{k=N+1}^{\infty} ||\mathbf{A}_k||,$$

$$||\mathbf{x}^\tau(t) - \mathbf{x}^{\tau *}(t)|| = \left\| \sum_{k=N+1}^{\infty} \mathbf{b}_k^\tau(t)\cos[k(\theta-\theta^\tau)] + \mathbf{c}_k^\tau(t)\sin[k(\theta-\theta^\tau)] \right\|$$

$$= \sum_{k=N+1}^{\infty} ||\mathbf{A}_k^\tau||.$$

For the prescribed small positive $\varepsilon > 0$ and $\varepsilon^\tau > 0$, if $||\mathbf{x}(t) - \mathbf{x}^*(t)|| < \varepsilon$ and $||\mathbf{x}^\tau(t) - \mathbf{x}^{\tau *}(t)|| < \varepsilon^\tau$ exist, then we have

$$\sum_{k=N+1}^{\infty} ||\mathbf{A}_k|| < \varepsilon \text{ and } \sum_{k=N+1}^{\infty} ||\mathbf{A}_k|| < \varepsilon^\tau.$$

Therefore, the convergent infinite term transformation in Equation (3.127) can be approximated by a finite term transformation in Equation (3.131) in the sense of ε.

Taking the derivative of Equation (3.131) with respect to time generates

$$\dot{\mathbf{x}}^*(t) = \dot{\mathbf{a}}_0 + \sum_{k=1}^{N} [\dot{\mathbf{b}}_k + k\Omega\mathbf{c}_k]\cos(k\theta) + [\dot{\mathbf{c}}_k - k\Omega\mathbf{b}_k]\sin(k\theta),$$

$$\dot{\mathbf{x}}^{\tau *}(t) = \dot{\mathbf{a}}_0^\tau + \sum_{k=1}^{N} [\dot{\mathbf{b}}_k^\tau + k\Omega\mathbf{c}_k^\tau]\cos[k(\theta-\theta^\tau)] + [\dot{\mathbf{c}}_k^\tau - k\Omega\mathbf{b}_k^\tau]\sin[k(\theta-\theta^\tau)]$$

where $\mathbf{a}_0^\tau = \mathbf{a}_0(t-\tau)$, $\mathbf{b}_k^\tau = \mathbf{b}_k(t-\tau)$, $\mathbf{c}_k^\tau = \mathbf{c}_k(t-\tau)$, $\theta^\tau = \Omega\tau$. Substitution of the foregoing equation into the time-delayed system in Equation (3.126), and application of the virtual work principle for a basis of constant, $\cos(k\theta)$ and $\sin(k\theta)$ ($k = 1, 2, \ldots$) as a set of virtual displacements, that is,

$$\frac{1}{2\pi}\int_0^{2\pi} [\dot{\mathbf{x}} - \mathbf{f}(\mathbf{x}, \mathbf{x}^\tau, \mathbf{p})]d\theta = 0,$$

$$\frac{1}{\pi}\int_0^{2\pi} [\dot{\mathbf{x}} - \mathbf{f}(\mathbf{x}, \mathbf{x}^\tau, \mathbf{p})]\cos(k\theta)d\theta = 0,$$

$$\frac{1}{\pi}\int_0^{2\pi} [\dot{\mathbf{x}} - \mathbf{f}(\mathbf{x}, \mathbf{x}^\tau, \mathbf{p})]\sin(k\theta)d\theta = 0.$$

Under $||\mathbf{x} - \mathbf{x}^*|| < \varepsilon$ and $||\mathbf{x}^\tau - \mathbf{x}^{\tau *}|| < \varepsilon^\tau$ with continuity $||\dot{\mathbf{x}}|| < K$ (constant) and small $\varepsilon > 0$ and $\varepsilon^\tau > 0$, for $k = 1, 2, \ldots, N$, the foregoing equation gives

$$\frac{1}{2\pi}\int_0^{2\pi} [\dot{\mathbf{x}}^* - \mathbf{f}(\mathbf{x}^*, \mathbf{x}^{\tau *}, \mathbf{p})]d\theta + O(\delta) = 0,$$

$$\frac{1}{\pi}\int_0^{2\pi}[\dot{\mathbf{x}}^* - \mathbf{f}(\mathbf{x}^*, \mathbf{x}^{\tau*}, \mathbf{p})]\cos(k\theta)d\theta + O(\delta) = 0,$$
$$\frac{1}{\pi}\int_0^{2\pi}[\dot{\mathbf{x}}^* - \mathbf{f}(\mathbf{x}^*, \mathbf{x}^{\tau*}, \mathbf{p})]\sin(k\theta)d\theta + O(\delta) = 0$$

where $\delta = \max(\varepsilon, \varepsilon^\tau, \varepsilon_t)$ and $||\dot{\mathbf{x}} - \dot{\mathbf{x}}^*|| < \varepsilon_t$ with small $\varepsilon_t > 0$. The foregoing equation gives

$$\begin{aligned}
\dot{\mathbf{a}}_0 &= \mathbf{F}_0(\mathbf{z}, \mathbf{z}^\tau),\\
\dot{\mathbf{b}}_k &= -\Omega k\mathbf{c}_k + \mathbf{F}_{1k}(\mathbf{z}, \mathbf{z}^\tau),\\
\dot{\mathbf{c}}_k &= \Omega k\mathbf{b}_k + \mathbf{F}_{2k}(\mathbf{z}, \mathbf{z}^\tau)
\end{aligned}$$

where for $k = 1, 2, \ldots, N$

$$\begin{aligned}
\mathbf{F}_0(\mathbf{z}, \mathbf{z}^\tau) &= \frac{1}{2\pi}\int_0^{2\pi}\mathbf{f}(\mathbf{x}^*, \mathbf{x}^{\tau*}, \mathbf{p})d\theta;\\
\mathbf{F}_{1k}(\mathbf{z}, \mathbf{z}^\tau) &= \frac{1}{\pi}\int_0^{2\pi}\mathbf{f}(\mathbf{x}^*, \mathbf{x}^{\tau*}, \mathbf{p})\cos(k\theta)d\theta,\\
\mathbf{F}_{2k}(\mathbf{z}, \mathbf{z}^\tau) &= \frac{1}{\pi}\int_0^{2\pi}\mathbf{f}(\mathbf{x}^*, \mathbf{x}^{\tau*}, \mathbf{p})\sin(k\theta)d\theta;
\end{aligned}$$

and for $N = 1, 2, \ldots, \infty$

$$\begin{aligned}
\mathbf{k}_1 &= diag(\mathbf{I}_{n\times n}, 2\mathbf{I}_{n\times n}, \ldots, N\mathbf{I}_{n\times n}),\\
\mathbf{b} &= (\mathbf{b}_1, \mathbf{b}_2, \ldots, \mathbf{b}_N)^{\mathrm{T}},\\
\mathbf{c} &= (\mathbf{c}_1, \mathbf{c}_2, \ldots, \mathbf{c}_N)^{\mathrm{T}};\\
\mathbf{b}^\tau &= (\mathbf{b}_1^\tau, \mathbf{b}_2^\tau, \ldots, \mathbf{b}_N^\tau)^{\mathrm{T}},\\
\mathbf{c}^\tau &= (\mathbf{c}_1^\tau, \mathbf{c}_2^\tau, \ldots, \mathbf{c}_N^\tau)^{\mathrm{T}};\\
\mathbf{F}_1 &= (\mathbf{F}_{11}, \mathbf{F}_{12}, \ldots, \mathbf{F}_{1N})^{\mathrm{T}},\\
\mathbf{F}_2 &= (\mathbf{F}_{21}, \mathbf{F}_{22}, \ldots, \mathbf{F}_{2N})^{\mathrm{T}};\\
\mathbf{z} &= (\mathbf{a}_0, \mathbf{b}, \mathbf{c})^{\mathrm{T}},\\
\mathbf{z}^\tau &= (\mathbf{a}_0^\tau, \mathbf{b}^\tau, \mathbf{c}^\tau)^{\mathrm{T}}.
\end{aligned}$$

Rearranging the foregoing equation gives Equation (3.132), that is,

$$\begin{aligned}
\dot{\mathbf{a}}_0 &= \mathbf{F}_0(\mathbf{a}_0, \mathbf{b}, \mathbf{c}; \mathbf{a}_0^\tau, \mathbf{b}^\tau, \mathbf{c}^\tau),\\
\dot{\mathbf{b}} &= -\Omega\mathbf{k}_1\mathbf{c} + \mathbf{F}_1(\mathbf{a}_0, \mathbf{b}, \mathbf{c}; \mathbf{a}_0^\tau, \mathbf{b}^\tau, \mathbf{c}^\tau),\\
\dot{\mathbf{c}} &= \Omega\mathbf{k}_1\mathbf{b} + \mathbf{F}_2(\mathbf{a}_0, \mathbf{b}, \mathbf{c}; \mathbf{a}_0^\tau, \mathbf{b}^\tau, \mathbf{c}^\tau).
\end{aligned}$$

Introduce

$$\mathbf{f} = (\mathbf{F}_0, -\Omega\mathbf{k}_1\mathbf{c} + \mathbf{F}_1, \Omega\mathbf{k}_1\mathbf{b} + \mathbf{F}_2)^{\mathrm{T}}.$$

The equation in Equation (3.132) becomes

$$\dot{\mathbf{z}} = \mathbf{f}(\mathbf{z}, \mathbf{z}^{\tau}).$$

Consider equilibriums of the foregoing equation (i.e., $\mathbf{f}(\mathbf{z}^*, \mathbf{z}^{\tau*}) = \mathbf{0}$) by

$$\begin{aligned}
\mathbf{0} &= \mathbf{F}_0(\mathbf{a}_0^*, \mathbf{b}^*, \mathbf{c}^*, \mathbf{a}_0^{\tau*}, \mathbf{b}^{\tau*}, \mathbf{c}^{\tau*}),\\
\mathbf{0} &= -\Omega\mathbf{k}_1\mathbf{c}^* + \mathbf{F}_1(\mathbf{a}_0^*, \mathbf{b}^*, \mathbf{c}^*, \mathbf{a}_0^{\tau*}, \mathbf{b}^{\tau*}, \mathbf{c}^{\tau*}),\\
\mathbf{0} &= \Omega\mathbf{k}_1\mathbf{b}^* + \mathbf{F}_2(\mathbf{a}_0^*, \mathbf{b}^*, \mathbf{c}^*, \mathbf{a}_0^{\tau*}, \mathbf{b}^{\tau*}, \mathbf{c}^{\tau*})
\end{aligned}$$

with

$$\mathbf{a}_0^* = \mathbf{a}_0^{\tau*}, \mathbf{b}^* = \mathbf{b}^{\tau*}, \mathbf{c}^* = \mathbf{c}^{\tau*}.$$

Thus, the solutions of the foregoing equation are the existence conditions of the periodic solutions for time-delayed, nonlinear dynamical systems. The foregoing equation gives equilibrium $\mathbf{z}^* = \mathbf{z}^{\tau*}$. In the vicinity of $\mathbf{z}^*$ and $\mathbf{z}^{\tau*}$, with $\mathbf{z} = \mathbf{z}^* + \Delta\mathbf{z}$ and $\mathbf{z}^{\tau} = \mathbf{z}^{\tau*} + \Delta\mathbf{z}^{\tau}$, the linearized equation of $\dot{\mathbf{z}} = \mathbf{f}(\mathbf{z}, \mathbf{z}^{\tau})$ is

$$\Delta\dot{\mathbf{z}} = D_{\mathbf{z}}\mathbf{f}(\mathbf{z}^*, \mathbf{z}^{\tau*})\Delta\mathbf{z} + D_{\mathbf{z}^{\tau}}\mathbf{f}(\mathbf{z}^*, \mathbf{z}^{\tau*})\Delta\mathbf{z}^{\tau}$$

and the eigenvalue analysis of equilibrium $\mathbf{z}^* = \mathbf{z}^{\tau*}$ is completed via

$$|D_{\mathbf{z}}\mathbf{f}(\mathbf{z}^*, \mathbf{z}^{\tau*}) - \lambda\mathbf{I}_{n(2N+1)\times n(2N+1)} + D_{\mathbf{z}^{\tau}}\mathbf{f}(\mathbf{z}^*, \mathbf{z}^{\tau*})e^{-\lambda\tau}| = 0$$

where

$$\begin{aligned}
D_{\mathbf{z}}\mathbf{f}(\mathbf{z}^*, \mathbf{z}^{\tau*}) &= \frac{\partial\mathbf{f}(\mathbf{z}, \mathbf{z}^{\tau})}{\partial\mathbf{z}}|_{(\mathbf{z}^*, \mathbf{z}^{\tau*})},\\
D_{\mathbf{z}^{\tau}}\mathbf{f}(\mathbf{z}^*, \mathbf{z}^{\tau*}) &= \frac{\partial\mathbf{f}(\mathbf{z}, \mathbf{z}^{\tau})}{\partial\mathbf{z}^{\tau}}|_{(\mathbf{z}^*, \mathbf{z}^{\tau*})}.
\end{aligned}$$

Therefore, as discussed before, the eigenvalues of Equation (3.137) at equilibrium $\mathbf{z}^* = \mathbf{z}^{\tau*}$ are classified by

$$(n_1, n_2, n_3 \mid n_4, n_5, n_6).$$

From the stability and bifurcation theory of dynamical systems at equilibrium, the stability, and bifurcation of the periodic solutions can be classified as stated in the theorem. This theorem is proved. ■

If the Hopf bifurcation of equilibriums of Equation (3.132) occurs, there is a periodic solution of coefficients in Equation (3.127) with a frequency ω. Thus, the coefficients solution can be expressed as

$$\mathbf{a}_0(t) = \mathbf{A}_{00}(t) + \sum_{m=1}^{\infty} \mathbf{A}_{0m}^{(1)}(t)\cos(m\vartheta) + \mathbf{A}_{0m}^{(2)}(t)\sin(m\vartheta),$$

$$\mathbf{b}_k(t) = \mathbf{B}_{k0}(t) + \sum_{m=1}^{\infty} \mathbf{B}_{km}^{(1)}(t)\cos(m\vartheta) + \mathbf{B}_{km}^{(2)}(t)\sin(m\vartheta),$$

$$\mathbf{c}_k(t) = \mathbf{C}_{k0}(t) + \sum_{m=1}^{\infty} \mathbf{C}_{km}^{(1)}(t)\cos(m\vartheta) + \mathbf{C}_{km}^{(2)}(t)\sin(m\vartheta);$$

$$\mathbf{a}_0^{\tau}(t) = \mathbf{A}_{00}^{\tau}(t) + \sum_{m=1}^{\infty} \mathbf{A}_{0m}^{\tau(1)}(t)\cos[m(\vartheta - \vartheta^{\tau})] + \mathbf{A}_{0m}^{\tau(2)}(t)\sin[m(\vartheta - \vartheta^{\tau})],$$

$$\mathbf{b}_k^{\tau}(t) = \mathbf{B}_{k0}^{\tau}(t) + \sum_{m=1}^{\infty} \mathbf{B}_{km}^{\tau(1)}(t)\cos[m(\vartheta - \vartheta^{\tau})] + \mathbf{B}_{km}^{\tau(2)}(t)\sin[m(\vartheta - \vartheta^{\tau})],$$

$$\mathbf{c}_k^{\tau}(t) = \mathbf{C}_{k0}^{\tau}(t) + \sum_{m=1}^{\infty} \mathbf{C}_{km}^{\tau(1)}(t)\cos[m(\vartheta - \vartheta^{\tau})] + \mathbf{C}_{km}^{\tau(2)}(t)\sin[m(\vartheta - \vartheta^{\tau})]. \tag{3.141}$$

Substitution of Equation (3.141) into Equation (3.127) gives

$$\begin{aligned}
\mathbf{x}(t) = {} & \mathbf{A}_{00}(t) + \sum_{m=1}^{\infty} \mathbf{A}_{0m}^{(1)}(t)\cos(m\vartheta) + \mathbf{A}_{0m}^{(2)}(t)\sin(m\vartheta) \\
& + \sum_{k=1}^{\infty} \mathbf{B}_{k0}(t)\cos(k\theta) + \mathbf{C}_{k0}(t)\sin(k\theta) \\
& + \sum_{k=1}^{\infty}\sum_{m=1}^{\infty} \mathbf{B}_{km}^{(1)}(t)\cos(m\vartheta)\cos(k\theta) + \mathbf{B}_{km}^{(2)}(t)\sin(m\vartheta)\cos(k\theta) \\
& + \sum_{k=1}^{\infty}\sum_{m=1}^{\infty} \mathbf{C}_{km}^{(1)}(t)\cos(m\vartheta)\sin(k\theta) + \mathbf{C}_{km}^{(2)}(t)\sin(m\vartheta)\sin(k\theta);
\end{aligned} \tag{3.142}$$

$$\begin{aligned}
\mathbf{x}^{\tau}(t) = {} & \mathbf{A}_{00}^{\tau}(t) + \sum_{m=1}^{\infty} \mathbf{A}_{0m}^{\tau(1)}(t)\cos[m(\vartheta - \vartheta^{\tau}))] + \mathbf{A}_{0m}^{(2)}(t)\sin[m(\vartheta - \vartheta^{\tau})] \\
& + \sum_{k=1}^{\infty} \mathbf{B}_{k0}^{\tau}(t)\cos[k(\theta - \theta^{\tau})] + \mathbf{C}_{k0}^{\tau}(t)\sin[k(\theta - \theta^{\tau})] \\
& + \sum_{k=1}^{\infty}\sum_{m=1}^{\infty} \{\mathbf{B}_{km}^{\tau(1)}(t)\cos[m(\vartheta - \vartheta^{\tau})]\cos[k(\theta - \theta^{\tau})] \\
& + \mathbf{B}_{km}^{\tau(2)}(t)\sin[m(\vartheta - \vartheta^{\tau})]\cos[k(\theta - \theta^{\tau})]\} \\
& + \sum_{k=1}^{\infty}\sum_{m=1}^{\infty} \{\mathbf{C}_{km}^{\tau(1)}(t)\cos[m(\vartheta - \vartheta^{\tau})]\sin[k(\theta - \theta^{\tau})] \\
& + \mathbf{C}_{km}^{\tau(2)}(t)\sin[m(\vartheta - \vartheta^{\tau})]\sin[k(\theta - \theta^{\tau})]\}.
\end{aligned} \tag{3.143}$$

If the new solution is still periodic with excitation period $T = 2\pi/\Omega$, then for specific m, the following relation should be satisfied.

$$m(\vartheta - \vartheta^{\tau}) = k(\theta - \theta^{\tau}) \quad \Rightarrow m\omega = k\Omega. \tag{3.144}$$

For this case, $k = 1$ should be inserted because $k > 1$ terms are already included in the Fourier series expression. Thus, as in dynamical systems without delay,

$$m(\vartheta - \vartheta^{\tau}) = (\theta - \theta^{\tau}) \quad \Rightarrow m\omega = \Omega. \tag{3.145}$$

For $m = 1$, the period-1 flow is obtained and Equations (3.142) and (3.143) will become Equation (3.127). For the period-m flow, we have

$$\begin{aligned}
\mathbf{x}^{(m)}(t) &= \mathbf{a}_0^{(m)}(t) + \sum_{k=1}^{\infty} \mathbf{b}_{k/m}(t)\cos\left(\frac{k}{m}\theta\right) + \mathbf{c}_{k/m}(t)\sin\left(\frac{k}{m}\theta\right);\\
\mathbf{x}^{\tau(m)}(t) &= \mathbf{a}_0^{\tau(m)}(t) + \sum_{k=1}^{\infty} \mathbf{b}_{k/m}^{\tau}(t)\cos\left[\frac{k}{m}(\theta - \theta^{\tau})\right] + \mathbf{c}_{k/m}^{\tau}(t)\sin\left[\frac{k}{m}(\theta - \theta^{\tau})\right].
\end{aligned} \tag{3.146}$$

For $\|\mathbf{x}^{(m)}(t) - \mathbf{x}^{(m)*}(t)\| < \varepsilon$ and $\|\mathbf{x}^{\tau(m)}(t) - \mathbf{x}^{\tau(m)*}(t)\| < \varepsilon^{\tau}$ with prescribed small $\varepsilon > 0$ and $\varepsilon^{\tau} > 0$, the solution of period-m flow in Equation (3.146) can be approximated by

$$\begin{aligned}
\mathbf{x}^{(m)*}(t) &= \mathbf{a}_0^{(m)}(t) + \sum_{k=1}^{N} \mathbf{b}_{k/m}(t)\cos\left(\frac{k}{m}\theta\right) + \mathbf{c}_{k/m}(t)\sin\left(\frac{k}{m}\theta\right),\\
\mathbf{x}^{\tau(m)*}(t) &= \mathbf{a}_0^{\tau(m)}(t) + \sum_{k=1}^{N} \mathbf{b}_{k/m}^{\tau}(t)\cos\left[\frac{k}{m}(\theta - \theta^{\tau})\right] + \mathbf{c}_{k/m}^{\tau}(t)\sin\left[\frac{k}{m}(\theta - \theta^{\tau})\right].
\end{aligned} \tag{3.147}$$

If $m\omega \neq k\Omega$ for any m and k, the solutions will be quasi-periodic or chaotic instead of periodic in the time-delayed nonlinear dynamical systems. However, we will not discuss it herein. If period-1 flow possesses at least N_1 harmonic vector terms, then the total harmonic vector terms for period-m flow should be $N \geq mN_1$. The period-m flow in a time-delayed, nonlinear dynamical system will be discussed as follows.

Theorem 3.10 *Consider a time-delayed, nonlinear dynamical system in Equation (3.126). If such a time-delayed dynamical system has a period-*m *flow* $\mathbf{x}^{(m)}(t)$ *with finite norm* $\|\mathbf{x}^{(m)}\|$ *and period* $T = 2\pi/\Omega$, *there is a generalized coordinate transformation for the period-*m *flow of Equation (3.126) in the form of*

$$\begin{aligned}
\mathbf{x}^{(m)}(t) &= \mathbf{a}_0^{(m)}(t) + \sum_{k=1}^{\infty} \mathbf{b}_{k/m}(t)\cos\left(\frac{k}{m}\theta\right) + \mathbf{c}_{k/m}(t)\sin\left(\frac{k}{m}\theta\right);\\
\mathbf{x}^{\tau(m)}(t) &= \mathbf{a}_0^{\tau(m)}(t) + \sum_{k=1}^{\infty} \mathbf{b}_{k/m}^{\tau}(t)\cos\left[\frac{k}{m}(\theta - \theta^{\tau})\right] + \mathbf{c}_{k/m}^{\tau}(t)\sin\left[\frac{k}{m}(\theta - \theta^{\tau})\right]
\end{aligned} \tag{3.148}$$

with $\mathbf{a}_0^{\tau(m)} = \mathbf{a}_0^{(m)}(t-\tau), \mathbf{b}_k^{\tau(m)} = \mathbf{b}_k^{(m)}(t-\tau), \mathbf{c}_k^{\tau(m)} = \mathbf{c}_k^{(m)}(t-\tau), \theta^\tau = \Omega\tau$ *and*

$$
\begin{aligned}
\mathbf{a}_0^{(m)} &= (a_{01}^{(m)}, a_{02}^{(m)}, \ldots, a_{0n}^{(m)})^{\mathrm{T}},\\
\mathbf{b}_{k/m} &= (b_{k/m1}, b_{k/m2}, \ldots, b_{k/mn})^{\mathrm{T}},\\
\mathbf{c}_{k/m} &= (c_{k/m1}, c_{k/m2}, \ldots, c_{k/mn})^{\mathrm{T}};\\
\mathbf{a}_0^{\tau(m)} &= (a_{01}^{\tau(m)}, a_{02}^{\tau(m)}, \ldots, a_{0n}^{\tau(m)})^{\mathrm{T}},\\
\mathbf{b}_{k/m}^{\tau} &= (b_{k/m1}^{\tau}, b_{k/m2}^{\tau}, \ldots, b_{k/mn}^{\tau})^{\mathrm{T}},\\
\mathbf{c}_{k/m}^{\tau} &= (c_{k/m1}^{\tau}, c_{k/m2}^{\tau}, \ldots, c_{k/mn}^{\tau})^{\mathrm{T}}
\end{aligned}
\tag{3.149}
$$

and

$$
\begin{aligned}
&\|\mathbf{x}^{(m)}\| = \|\mathbf{a}_0^{(m)}\| + \sum_{k=1}^{\infty}\|\mathbf{A}_{k/m}\|, \text{ and } \lim_{k\to\infty}\|\mathbf{A}_{k/m}\| = 0 \text{ but not uniform}\\
&\text{with } \mathbf{A}_{k/m} = (A_{k/m1}, A_{k/m2} \cdots, A_{k/mn})^{\mathrm{T}}\\
&\text{and } A_{k/mj} = \sqrt{b_{k/mj}^2 + c_{k/mj}^2}\ (j = 1, 2, \ldots, n);
\end{aligned}
\tag{3.150}
$$

$$
\begin{aligned}
&\|\mathbf{x}^{\tau(m)}\| = \|\mathbf{a}_0^{\tau(m)}\| + \sum_{k=1}^{\infty}\|\mathbf{A}_{k/m}^{\tau}\|, \text{ and } \lim_{k\to\infty}\|\mathbf{A}_{k/m}^{\tau}\| = 0 \text{ but not uniform}\\
&\text{with } \mathbf{A}_{k/m}^{\tau} = (A_{k/m1}^{\tau}, A_{k/m2}^{\tau} \cdots, A_{k/mn}^{\tau})^{\mathrm{T}}\\
&\text{and } A_{k/mj}^{\tau} = \sqrt{(b_{k/mj}^{\tau})^2 + (c_{k/mj}^{\tau})^2}\ (j = 1, 2, \ldots, n).
\end{aligned}
\tag{3.151}
$$

For $\|\mathbf{x}^{(m)}(t) - \mathbf{x}^{(m)*}(t)\| < \varepsilon$ *and* $\|\mathbf{x}^{\tau(m)}(t) - \mathbf{x}^{\tau(m)*}(t)\| < \varepsilon^\tau$ *with prescribed small* $\varepsilon > 0$ *and* $\varepsilon^\tau > 0$, *the infinite term transformation of period-*m *flow* $\mathbf{x}^{(m)}(t)$ *of Equation (3.126), given by Equation (3.148), can be approximated by a finite term transformation* $\mathbf{x}^{(m)*}(t)$ *as*

$$
\begin{aligned}
\mathbf{x}^{(m)*}(t) &= \mathbf{a}_0^{(m)}(t) + \sum_{k=1}^{N}\mathbf{b}_{k/m}(t)\cos\left(\frac{k}{m}\theta\right) + \mathbf{c}_{k/m}(t)\sin\left(\frac{k}{m}\theta\right);\\
\mathbf{x}^{\tau(m)*}(t) &= \mathbf{a}_0^{\tau(m)}(t) + \sum_{k=1}^{N}\mathbf{b}_{k/m}^{\tau}(t)\cos\left[\frac{k}{m}(\theta-\theta^\tau)\right] + \mathbf{c}_{k/m}^{\tau}(t)\sin\left[\frac{k}{m}(\theta-\theta^\tau)\right]
\end{aligned}
\tag{3.152}
$$

and the generalized coordinates are determined by

$$
\begin{aligned}
\dot{\mathbf{a}}_0^{(m)} &= \mathbf{F}_0^{(m)}(\mathbf{z}^{(m)}, \mathbf{z}^{\tau(m)}),\\
\dot{\mathbf{b}}^{(m)} &= -\frac{\Omega}{m}\mathbf{k}_1\mathbf{c}^{(m)} + \mathbf{F}_1^{(m)}(\mathbf{z}^{(m)}, \mathbf{z}^{\tau(m)}),\\
\dot{\mathbf{c}}^{(m)} &= \frac{\Omega}{m}\mathbf{k}_1\mathbf{b}^{(m)} + \mathbf{F}_2^{(m)}(\mathbf{z}^{(m)}, \mathbf{z}^{\tau(m)})
\end{aligned}
\tag{3.153}
$$

where

$$
\begin{aligned}
\mathbf{k}_1 &= diag(\mathbf{I}_{n\times n}, 2\mathbf{I}_{n\times n}, \dots, N\mathbf{I}_{n\times n}),\\
\mathbf{b}^{(m)} &= (\mathbf{b}_{1/m}, \mathbf{b}_{2/m}, \dots, \mathbf{b}_{N/m})^{\mathrm{T}},\\
\mathbf{c}^{(m)} &= (\mathbf{c}_{1/m}, \mathbf{c}_{2/m}, \dots, \mathbf{c}_{N/m})^{\mathrm{T}};\\
\mathbf{b}^{\tau(m)} &= (\mathbf{b}^{\tau}_{1/m}, \mathbf{b}^{\tau}_{2/m}, \dots, \mathbf{b}^{\tau}_{N/m})^{\mathrm{T}},\\
\mathbf{c}^{\tau(m)} &= (\mathbf{c}^{\tau}_{1/m}, \mathbf{c}^{\tau}_{2/m}, \dots, \mathbf{c}^{\tau}_{N/m})^{\mathrm{T}};\\
\mathbf{F}^{(m)}_1 &= (\mathbf{F}^{(m)}_{11}, \mathbf{F}^{(m)}_{12}, \dots, \mathbf{F}^{(m)}_{1N})^{\mathrm{T}},\\
\mathbf{F}^{(m)}_2 &= (\mathbf{F}^{(m)}_{21}, \mathbf{F}^{(m)}_{22}, \dots, \mathbf{F}^{(m)}_{2N})^{\mathrm{T}};\\
\mathbf{z}^{(m)} &= (\mathbf{a}^{(m)}_0, \mathbf{b}^{(m)}, \mathbf{c}^{(m)})^{\mathrm{T}},\\
\mathbf{z}^{\tau(m)} &= (\mathbf{a}^{\tau(m)}_0, \mathbf{b}^{\tau(m)}, \mathbf{c}^{\tau(m)})^{\mathrm{T}}\\
&\text{for } N = 1, 2, \dots, \infty;
\end{aligned}
\tag{3.154}
$$

and

$$
\begin{aligned}
\mathbf{F}^{(m)}_0(\mathbf{z}^{(m)}, \mathbf{z}^{\tau(m)}) &= \frac{1}{2m\pi}\int_0^{2m\pi} \mathbf{f}(\mathbf{x}^{(m)*}, \mathbf{x}^{\tau(m)*}, \mathbf{p})d\theta;\\
\mathbf{F}^{(m)}_{1k}(\mathbf{z}^{(m)}, \mathbf{z}^{\tau(m)}) &= \frac{1}{m\pi}\int_0^{2m\pi} \mathbf{f}(\mathbf{x}^{(m)*}, \mathbf{x}^{\tau(m)*}, \mathbf{p})\cos\left(\frac{k}{m}\theta\right)d\theta,\\
\mathbf{F}^{(m)}_{2k}(\mathbf{z}^{(m)}, \mathbf{z}^{\tau(m)}) &= \frac{1}{m\pi}\int_0^{2m\pi} \mathbf{f}(\mathbf{x}^{(m)*}, \mathbf{x}^{\tau(m)*}, \mathbf{p})\sin\left(\frac{k}{m}\theta\right)d\theta\\
&\text{for } k = 1, 2, \dots, N.
\end{aligned}
\tag{3.155}
$$

Equation (3.153) becomes

$$\dot{\mathbf{z}}^{(m)} = \mathbf{f}^{(m)}(\mathbf{z}^{(m)}, \mathbf{z}^{\tau(m)}) \tag{3.156}$$

where

$$\mathbf{f}^{(m)} = \left(\mathbf{F}^{(m)}_0, -\frac{\Omega}{m}\mathbf{k}_1\mathbf{c}^{(m)} + \mathbf{F}^{(m)}_1, \frac{\Omega}{m}\mathbf{k}_1\mathbf{b}^{(m)} + \mathbf{F}^{(m)}_2\right)^{\mathrm{T}}. \tag{3.157}$$

*If equilibrium ($\mathbf{z}^{(m)*} = \mathbf{z}^{\tau(m)*}$) of Equation (3.156) (i.e., $\mathbf{f}^{(m)}(\mathbf{z}^{(m)*}, \mathbf{z}^{\tau(m)*}) = \mathbf{0}$) exists, then the approximate solution of period-*m *flow exists as in Equation (3.152). In the vicinity of equilibrium $\mathbf{z}^{(m)*} = \mathbf{z}^{\tau(m)*}$, with $\mathbf{z}^{(m)} = \mathbf{z}^{(m)*} + \Delta\mathbf{z}^{(m)}$ and $\mathbf{z}^{\tau(m)} = \mathbf{z}^{\tau(m)*} + \Delta\mathbf{z}^{\tau(m)}$, the linearized equation of Equation (3.156) is*

$$\Delta\dot{\mathbf{z}}^{(m)} = D_{\mathbf{z}^{(m)}}\mathbf{f}^{(m)}(\mathbf{z}^{(m)*}, \mathbf{z}^{\tau(m)*})\Delta\mathbf{z}^{(m)} + D_{\mathbf{z}^{\tau(m)}}\mathbf{f}^{(m)}(\mathbf{z}^{(m)*}, \mathbf{z}^{\tau(m)*})\Delta\mathbf{z}^{\tau(m)} \tag{3.158}$$

and the eigenvalue analysis of equilibrium $\mathbf{z}^*$ is given by

$$|D_{\mathbf{z}^{(m)}}\mathbf{f}^{(m)}(\mathbf{z}^{(m)*}, \mathbf{z}^{\tau(m)*}) - \lambda\mathbf{I}_{n(2N+1)\times n(2N+1)} + D_{\mathbf{z}^{\tau(m)}}\mathbf{f}^{(m)}(\mathbf{z}^{(m)*}, \mathbf{z}^{\tau(m)*})e^{-\lambda\tau}| = 0 \tag{3.159}$$

where

$$D_{\mathbf{z}^{(m)}}\mathbf{f}^{(m)}(\mathbf{z}^{(m)*},\mathbf{z}^{\tau(m)*}) = \frac{\partial \mathbf{f}^{(m)}(\mathbf{z}^{(m)},\mathbf{z}^{\tau(m)})}{\partial \mathbf{z}^{(m)}}\Big|_{(\mathbf{z}^{(m)*},\mathbf{z}^{\tau(m)*})},$$

$$D_{\mathbf{z}^{\tau(m)}}\mathbf{f}^{(m)}(\mathbf{z}^{(m)*},\mathbf{z}^{\tau(m)*}) = \frac{\partial \mathbf{f}^{(m)}(\mathbf{z}^{(m)},\mathbf{z}^{\tau(m)})}{\partial \mathbf{z}^{\tau(m)}}\Big|_{(\mathbf{z}^{(m)*},\mathbf{z}^{\tau(m)*})}. \tag{3.160}$$

The stability and bifurcation of periodic flow can be classified by the eigenvalues of Equation (3.156) at equilibrium $\mathbf{z}^{(m)*} = \mathbf{z}^{\tau(m)*}$ with

$$(n_1, n_2, n_3 \mid n_4, n_5, n_6). \tag{3.161}$$

1. *If all eigenvalues of the equilibrium possess negative real parts, the periodic solution is stable.*
2. *If at least one of the eigenvalues of the equilibrium possesses a positive real part, the approximate steady-state solution is unstable.*
3. *The boundary between the stable and unstable equilibriums with higher order singularity gives the bifurcation conditions and stability with higher order singularity.*

Proof. The proof is similar to the proof of Theorem 3.9. Since $\mathbf{f}(\mathbf{x},\mathbf{x}^{\tau},\mathbf{p})$ is a C^r-continuous nonlinear function vector ($r \geq 1$)), the velocity $\dot{\mathbf{x}}$ should be C^r-continuous ($r \geq 1$), and then the acceleration $\ddot{\mathbf{x}}$ should be bounded (i.e., $\|\ddot{\mathbf{x}}\| \leq K$). From Equation (3.148), the norms of the periodic flows are defined by

$$\|\mathbf{x}^{(m)}\| = \|\mathbf{a}_0^{(m)}\| + \sum_{k=1}^{\infty}\|\mathbf{A}_{k/m}\|,$$

$$\|\mathbf{x}^{\tau(m)}\| = \|\mathbf{a}_0^{\tau(m)}\| + \sum_{k=1}^{\infty}\|\mathbf{A}_{k/m}^{\tau}\|,$$

where

$$\mathbf{A}_{k/m} = (A_{k/m1}, A_{k/m2} \ldots, A_{k/mn})^{\mathrm{T}},$$

$$A_{k/mj} = \sqrt{b_{k/mj}^2 + c_{k/mj}^2}\ \ (j = 1, 2, \ldots, n),$$

$$\mathbf{A}_{k/m}^{\tau} = (A_{k/m1}^{\tau}, A_{k/m2}^{\tau}, \ldots, A_{k/mn}^{\tau})^{\mathrm{T}},$$

$$A_{k/mj}^{\tau} = \sqrt{(b_{k/mj}^{\tau})^2 + (c_{k/mj}^{\tau})^2}\ \ (j = 1, 2, \ldots, n).$$

Because the periodic flow in Equation (3.126) is bounded (i.e., $\|\mathbf{x}(t)\| < C$), we have

$$\lim_{k\to\infty}\|\mathbf{A}_{k/m}\| = 0 \text{ and } \lim_{k\to\infty}\|\mathbf{A}_{k/m}^{\tau}\| = 0 \text{ but not uniform.}$$

Thus, the Fourier series transformation of periodic flow as in Equation (3.148) is convergent. From Equations (3.150) and (3.151), using Equations (3.148) and (3.152) gives

$$\|\mathbf{x}^{(m)}(t) - \mathbf{x}^{(m)*}(t)\| = \left\| \sum_{k=N+1}^{\infty} \mathbf{b}_{k/m}(t)\cos\left(\frac{k}{m}\theta\right) + \mathbf{c}_{k/m}(t)\sin\left(\frac{k}{m}\theta\right) \right\|$$

$$= \sum_{k=N+1}^{\infty} \|\mathbf{A}_{k/m}\|,$$

$$\begin{aligned}
&\|\mathbf{x}^{\tau(m)}(t) - \mathbf{x}^{\tau(m)*}(t)\| \\
&= \left\| \sum_{k=N+1}^{\infty} \mathbf{b}^{\tau}_{k/m}(t) \cos\left[\frac{k}{m}(\theta - \theta^{\tau})\right] + \mathbf{c}^{\tau}_{k/m}(t) \sin\left[\frac{k}{m}(\theta - \theta^{\tau})\right] \right\| \\
&= \sum_{k=N+1}^{\infty} \|\mathbf{A}^{\tau}_{k/m}\|.
\end{aligned}$$

For the prescribed small positive $\varepsilon > 0$ and $\varepsilon^{\tau} > 0$, if $\|\mathbf{x}^{(m)}(t) - \mathbf{x}^{(m)*}(t)\| < \varepsilon$ and $\|\mathbf{x}^{\tau(m)}(t) - \mathbf{x}^{\tau(m)*}(t)\| < \varepsilon^{\tau}$ exist, then we have

$$\sum_{k=N+1}^{\infty} \|\mathbf{A}_{k/m}\| < \varepsilon \text{ and } \sum_{k=N+1}^{\infty} \|\mathbf{A}^{\tau}_{k/m}\| < \varepsilon^{\tau}.$$

Therefore, the convergent infinite term transformation in Equation (3.148) can be approximated by a finite term transformation in Equation (3.152) in the sense of ε.

Taking the derivative of Equation (3.152) with respect to time gives

$$\begin{aligned}
\dot{\mathbf{x}}^{(m)*}(t) &= \dot{\mathbf{a}}_0^{(m)} + \sum_{k=1}^{N} \left[\dot{\mathbf{b}}_{k/m} + \frac{k}{m}\Omega \mathbf{c}_{k/m}\right] \cos\left(\frac{k}{m}\theta\right) \\
&\quad + \left[\dot{\mathbf{c}}_{k/m} - \frac{k}{m}\Omega \mathbf{b}_{k/m}\right] \sin\left(\frac{k}{m}\theta\right), \\
\dot{\mathbf{x}}^{\tau(m)*}(t) &= \dot{\mathbf{a}}_0^{\tau(m)} + \sum_{k=1}^{N} \left[\dot{\mathbf{b}}^{\tau}_{k/m} + \frac{k}{m}\Omega \mathbf{c}^{\tau}_{k/m}\right] \cos\left[\frac{k}{m}(\theta - \theta^{\tau})\right] \\
&\quad + \left[\dot{\mathbf{c}}^{\tau}_{k/m} - \frac{k}{m}\Omega \mathbf{b}^{\tau}_{k/m}\right] \sin\left[\frac{k}{m}(\theta - \theta^{\tau})\right]
\end{aligned}$$

where $\mathbf{a}_0^{\tau(m)} = \mathbf{a}_0(t-\tau), \mathbf{b}^{\tau}_{k/m} = \mathbf{b}_{k/m}(t-\tau), \mathbf{c}^{\tau}_{k/m} = \mathbf{c}_{k/m}(t-\tau), \theta^{\tau} = \Omega\tau$. Substitution of the foregoing equation into the time-delayed system in Equation (3.126), and application of the virtual work principle for a basis of constant, $\cos(k\theta/m)$ and $\sin(k\theta/m)$ ($k = 1, 2, \ldots$) as a set of virtual displacements, that is,

$$\begin{aligned}
&\frac{1}{2m\pi}\int_0^{2m\pi} [\dot{\mathbf{x}}^{(m)} - \mathbf{f}(\mathbf{x}^{(m)}, \mathbf{x}^{\tau(m)}, \mathbf{p})] d\theta = 0, \\
&\frac{1}{m\pi}\int_0^{2m\pi} [\dot{\mathbf{x}}^{(m)} - \mathbf{f}(\mathbf{x}^{(m)}, \mathbf{x}^{\tau(m)}, \mathbf{p})] \cos\left(\frac{k}{m}\theta\right) d\theta = 0, \\
&\frac{1}{m\pi}\int_0^{2m\pi} [\dot{\mathbf{x}}^{(m)} - \mathbf{f}(\mathbf{x}^{(m)}, \mathbf{x}^{\tau(m)}, \mathbf{p})] \sin\left(\frac{k}{m}\theta\right) d\theta = 0
\end{aligned}$$

under $\|\mathbf{x}^{(m)} - \mathbf{x}^{(m)*}\| < \varepsilon$ and $\|\mathbf{x}^{\tau(m)} - \mathbf{x}^{\tau(m)*}\| < \varepsilon^{\tau}$ with continuity $\|\ddot{\mathbf{x}}^{(m)}\| < K$ (constant) and small $\varepsilon > 0$ and $\varepsilon^{\tau} > 0$, for $k = 1, 2, \ldots, N$,

$$\begin{aligned}
&\frac{1}{2m\pi}\int_0^{2m\pi} [\dot{\mathbf{x}}^{(m)*} - \mathbf{f}(\mathbf{x}^{(m)*}, \mathbf{x}^{\tau(m)*}, \mathbf{p})] d\theta + O(\delta) = 0, \\
&\frac{1}{m\pi}\int_0^{2m\pi} [\dot{\mathbf{x}}^{(m)*} - \mathbf{f}(\mathbf{x}^{(m)*}, \mathbf{x}^{\tau(m)*}, \mathbf{p})] \cos\left(\frac{k}{m}\theta\right) d\theta + O(\delta) = 0,
\end{aligned}$$

$$\frac{1}{m\pi}\int_0^{2m\pi}[\dot{\mathbf{x}}^{(m)*}-\mathbf{f}(\mathbf{x}^{(m)*},\mathbf{x}^{\tau(m)*},\mathbf{p})]\sin\left(\frac{k}{m}\theta\right)d\theta+O(\delta)=0$$

where $\delta=\max(\varepsilon,\varepsilon^{\tau},\varepsilon_t)$ and $\|\dot{\mathbf{x}}-\dot{\mathbf{x}}^*\|<\varepsilon_t$ with small $\varepsilon_t>0$, yields

$$\begin{aligned}
\dot{\mathbf{a}}_0^{(m)} &= \mathbf{F}_0(\mathbf{z}^{(m)},\mathbf{z}^{\tau(m)}),\\
\dot{\mathbf{b}}_{k/m}+\frac{\Omega}{m}k\mathbf{c}_{k/m} &= \mathbf{F}_{1k}^{(m)}(\mathbf{z}^{(m)},\mathbf{z}^{\tau(m)}),\\
\dot{\mathbf{c}}_{k/m}-\frac{\Omega}{m}k\mathbf{b}_{k/m} &= \mathbf{F}_{2k}^{(m)}(\mathbf{z}^{(m)},\mathbf{z}^{\tau(m)})
\end{aligned}$$

where

$$\begin{aligned}
\mathbf{F}_0^{(m)}(\mathbf{z}^{(m)},\mathbf{z}^{\tau(m)}) &= \frac{1}{2m\pi}\int_0^{2m\pi}\mathbf{f}(\mathbf{x}^{(m)*},\mathbf{x}^{\tau(m)*},\mathbf{p})d\theta;\\
\mathbf{F}_{1k}^{(m)}(\mathbf{z}^{(m)},\mathbf{z}^{\tau(m)}) &= \frac{1}{m\pi}\int_0^{2m\pi}\mathbf{f}(\mathbf{x}^{(m)*},\mathbf{x}^{\tau(m)*},\mathbf{p})\cos\left(\frac{k}{m}\theta\right)d\theta,\\
\mathbf{F}_{2k}^{(m)}(\mathbf{z}^{(m)},\mathbf{z}^{\tau(m)}) &= \frac{1}{m\pi}\int_0^{2m\pi}\mathbf{f}(\mathbf{x}^{(m)*},\mathbf{x}^{\tau(m)*},\mathbf{p})\sin\left(\frac{k}{m}\theta\right)d\theta
\end{aligned}$$

for $k=1,2,\ldots,N$;

and

$$\begin{aligned}
\mathbf{k}_1 &= diag(\mathbf{I}_{n\times n},2\mathbf{I}_{n\times n},\ldots,N\mathbf{I}_{n\times n}),\\
\mathbf{b}^{(m)} &= (\mathbf{b}_{1/m},\mathbf{b}_{2/m},\ldots,\mathbf{b}_{N/m})^{\mathrm{T}},\\
\mathbf{c}^{(m)} &= (\mathbf{c}_{1/m},\mathbf{c}_{2/m},\ldots,\mathbf{c}_{N/m})^{\mathrm{T}};\\
\mathbf{b}^{\tau(m)} &= (\mathbf{b}_{1/m}^{\tau},\mathbf{b}_{2/m}^{\tau},\ldots,\mathbf{b}_{N/m}^{\tau})^{\mathrm{T}},\\
\mathbf{c}^{\tau(m)} &= (\mathbf{c}_{1/m}^{\tau},\mathbf{c}_{2/m}^{\tau},\ldots,\mathbf{c}_{N/m}^{\tau})^{\mathrm{T}};\\
\mathbf{F}_1^{(m)} &= (\mathbf{F}_{11}^{(m)},\mathbf{F}_{12}^{(m)},\ldots,\mathbf{F}_{1N}^{(m)})^{\mathrm{T}},\\
\mathbf{F}_2^{(m)} &= (\mathbf{F}_{21}^{(m)},\mathbf{F}_{22}^{(m)},\ldots,\mathbf{F}_{2N}^{(m)})^{\mathrm{T}};\\
\mathbf{z}^{(m)} &= (\mathbf{a}_0^{(m)},\mathbf{b}^{(m)},\mathbf{c}^{(m)})^{\mathrm{T}},\\
\mathbf{z}^{\tau(m)} &= (\mathbf{a}_0^{(\tau m)},\mathbf{b}^{\tau(m)},\mathbf{c}^{\tau(m)})^{\mathrm{T}}
\end{aligned}$$

for $N=1,2,\ldots,\infty$.

Rearranging the foregoing equation gives Equation (3.153), that is,

$$\begin{aligned}
\dot{\mathbf{a}}_0^{(m)} &= \mathbf{F}_0^{(m)}(\mathbf{z}^{(m)},\mathbf{z}^{\tau(m)}),\\
\dot{\mathbf{b}}^{(m)} &= -\frac{\Omega}{m}\mathbf{k}_1\mathbf{c}^{(m)}+\mathbf{F}_1^{(m)}(\mathbf{z}^{(m)},\mathbf{z}^{\tau(m)}),\\
\dot{\mathbf{c}}^{(m)} &= \frac{\Omega}{m}\mathbf{k}_1\mathbf{b}^{(m)}+\mathbf{F}_2^{(m)}(\mathbf{z}^{(m)},\mathbf{z}^{\tau(m)}).
\end{aligned}$$

Introducing

$$\mathbf{f}^{(m)} = (\mathbf{F}_0^{(m)}, -\Omega\mathbf{k}_1\mathbf{c}^{(m)}/m + \mathbf{F}_1^{(m)}, \Omega\mathbf{k}_1\mathbf{b}^{(m)}/m + \mathbf{F}_2^{(m)})^{\mathrm{T}},$$

the standard form of equation in Equation (3.153) becomes

$$\dot{\mathbf{z}}^{(m)} = \mathbf{f}^{(m)}(\mathbf{z}^{(m)}, \mathbf{z}^{\tau(m)}).$$

Consider the equilibrium solution of the foregoing equation with $\mathbf{z}^{(m)*} = \mathbf{z}^{\tau(m)*}$ (i.e., $\mathbf{f}^{(m)}(\mathbf{z}^{(m)*}, \mathbf{z}^{\tau(m)*}) = \mathbf{0}$) by

$$\begin{aligned}
\mathbf{0} &= \mathbf{F}_0^{(m)}(\mathbf{a}_0^{(m)*}, \mathbf{b}^{(m)*}, \mathbf{c}^{(m)*}; \mathbf{a}_0^{\tau(m)*}, \mathbf{b}^{\tau(m)*}, \mathbf{c}^{\tau(m)*}),\\
\mathbf{0} &= -\frac{\Omega}{m}\mathbf{k}_1\mathbf{c}^{(m)*} + \mathbf{F}_1^{(m)}(\mathbf{a}_0^{(m)*}, \mathbf{b}^{(m)*}, \mathbf{c}^{(m)*}; \mathbf{a}_0^{\tau(m)*}, \mathbf{b}^{\tau(m)*}, \mathbf{c}^{\tau(m)*}),\\
\mathbf{0} &= \frac{\Omega}{m}\mathbf{k}_1\mathbf{b}^{(m)*} + \mathbf{F}_2^{(m)}(\mathbf{a}_0^{(m)*}, \mathbf{b}^{(m)*}, \mathbf{c}^{(m)*}; \mathbf{a}_0^{\tau(m)*}, \mathbf{b}^{\tau(m)*}, \mathbf{c}^{\tau(m)*})
\end{aligned}$$

with

$$\mathbf{a}_0^{(m)*} = \mathbf{a}_0^{\tau(m)*}, \mathbf{b}^{(m)*} = \mathbf{b}^{\tau(m)*}, \mathbf{c}^{(m)*} = \mathbf{c}^{\tau(m)*}.$$

Thus, solutions of the foregoing equation are the existence conditions of periodic solutions for time-delayed nonlinear dynamical systems. If the foregoing equation gives $\mathbf{z}^{(m)*} = \mathbf{z}^{\tau(m)*}$. In the vicinity of $(\mathbf{z}^{(m)*}, \mathbf{z}^{\tau(m)*})$, with $\mathbf{z}^{(m)} = \mathbf{z}^{(m)*} + \Delta\mathbf{z}^{(m)}$ and $\mathbf{z}^{\tau(m)} = \mathbf{z}^{\tau(m)*} + \Delta\mathbf{z}^{\tau(m)}$, the linearized equation of $\dot{\mathbf{z}}^{(m)} = \mathbf{f}^{(m)}(\mathbf{z}^{(m)}, \mathbf{z}^{\tau(m)})$ is

$$\Delta\dot{\mathbf{z}}^{(m)} = D_{\mathbf{z}^{(m)*}}\mathbf{f}^{(m)}(\mathbf{z}^{(m)}, \mathbf{z}^{\tau(m)*})\Delta\mathbf{z}^{(m)} + D_{\mathbf{z}^{\tau(m)}}\mathbf{f}^{(m)}(\mathbf{z}^{(m)*}, \mathbf{z}^{\tau(m)*})\Delta\mathbf{z}^{\tau(m)}$$

and the eigenvalue analysis of equilibrium $\mathbf{z}^{(m)*} = \mathbf{z}^{\tau(m)*}$ is completed via

$$|D_{\mathbf{z}^{(m)}}\mathbf{f}^{(m)}(\mathbf{z}^{(m)*}, \mathbf{z}^{\tau(m)*}) - \lambda\mathbf{I}_{n(2N+1)\times n(2N+1)} + D_{\mathbf{z}^{\tau(m)}}\mathbf{f}^{(m)}(\mathbf{z}^{(m)*}, \mathbf{z}^{\tau(m)*})e^{-\lambda\tau}| = 0$$

where

$$\begin{aligned}
D_{\mathbf{z}^{(m)}}\mathbf{f}^{(m)}(\mathbf{z}^{(m)*}, \mathbf{z}^{\tau(m)*}) &= \frac{\partial\mathbf{f}^{(m)}(\mathbf{z}^{(m)}, \mathbf{z}^{\tau(m)})}{\partial\mathbf{z}^{(m)}}|_{(\mathbf{z}^{(m)*}, \mathbf{z}^{\tau(m)*})},\\
D_{\mathbf{z}^{\tau(m)}}\mathbf{f}^{(m)}(\mathbf{z}^{(m)*}, \mathbf{z}^{\tau(m)*}) &= \frac{\partial\mathbf{f}^{(m)}(\mathbf{z}^{(m)}, \mathbf{z}^{\tau(m)})}{\partial\mathbf{z}^{\tau(m)}}|_{(\mathbf{z}^{(m)*}, \mathbf{z}^{\tau(m)*})}.
\end{aligned}$$

Thus, the corresponding eigenvalues are classified by

$$(n_1, n_2, n_3 \mid n_4, n_5, n_6).$$

From the stability and bifurcation theory of dynamical systems at equilibrium, the stability, and bifurcation of the periodic flow can be classified as stated in the theorem. This theorem is proved. ■

If $m \to \infty$, Equation (3.148) will give the analytical expression of chaos in time-delayed, nonlinear dynamical systems in Equation (3.126), which can be approximated by Equation (3.152) under the condition of $\|\mathbf{x}^{(m)}(t) - \mathbf{x}^{(m)*}(t)\| < \varepsilon$ and $\|\mathbf{x}^{\tau(m)}(t) - \mathbf{x}^{\tau(m)*}(t)\| < \varepsilon^{\tau}$. The route from the periodic flow to chaos for the time-delayed, nonlinear dynamical system is through the Hopf bifurcation.

3.3.2 Non-Autonomous Time-Delayed Nonlinear Systems

Periodic flows in non-autonomous, time-delayed, nonlinear dynamical systems will be discussed herein. If a time-delayed system has periodic flows with an external period $T = 2\pi/\Omega$, such periodic flows can be expressed through the Fourier series, discussed as follows.

Theorem 3.11 *Consider a non-autonomous, time-delayed, nonlinear dynamical system as*

$$\dot{\mathbf{x}} = \mathbf{F}(\mathbf{x}, \mathbf{x}^{\tau}, t, \mathbf{p}) \in \mathscr{R}^n \tag{3.162}$$

where $\mathbf{F}(\mathbf{x}, \mathbf{x}^{\tau}, t, \mathbf{p})$ is a C^r-continuous nonlinear function vector ($r \geq 1$) with an excitation period $T = 2\pi/\Omega$. If such a time-delayed, dynamical system has a periodic flow $\mathbf{x}(t)$ with finite norm $||\mathbf{x}||$ and period $T = 2\pi/\Omega$, there is a generalized coordinate transformation with $\theta = \Omega t$ for the periodic flow of Equation (3.162) in the form of

$$\begin{aligned} \mathbf{x} \equiv \mathbf{x}(t) &= \mathbf{a}_0(t) + \sum_{k=1}^{\infty} \mathbf{b}_k(t)\cos(k\theta) + \mathbf{c}_k(t)\sin(k\theta), \\ \mathbf{x}^{\tau} \equiv \mathbf{x}(t-\tau) &= \mathbf{a}_0^{\tau}(t) + \sum_{k=1}^{\infty} \mathbf{b}_k^{\tau}(t)\cos[k(\theta-\theta^{\tau})] + \mathbf{c}_k^{\tau}(t)\sin[k(\theta-\theta^{\tau})] \end{aligned} \tag{3.163}$$

with $\mathbf{a}_0^{\tau} = \mathbf{a}_0(t-\tau)$, $\mathbf{b}_k^{\tau} = \mathbf{b}_k(t-\tau)$, $\mathbf{c}_k^{\tau} = \mathbf{c}_k(t-\tau)$, $\theta^{\tau} = \Omega\tau$ and

$$\begin{aligned} \mathbf{a}_0 &= (a_{01}, a_{02}, \ldots, a_{0n})^{\mathrm{T}}, \\ \mathbf{b}_k &= (b_{k1}, b_{k2}, \ldots, b_{kn})^{\mathrm{T}}, \\ \mathbf{c}_k &= (c_{k1}, c_{k2}, \ldots, c_{kn})^{\mathrm{T}}; \\ \mathbf{a}_0^{\tau} &= (a_{01}^{\tau}, a_{02}^{\tau}, \ldots, a_{0n}^{\tau})^{\mathrm{T}}, \\ \mathbf{b}_k^{\tau} &= (b_{k1}^{\tau}, b_{k2}^{\tau}, \ldots, b_{kn}^{\tau})^{\mathrm{T}}, \\ \mathbf{c}_k^{\tau} &= (c_{k1}^{\tau}, c_{k2}^{\tau}, \ldots, c_{kn}^{\tau})^{\mathrm{T}}; \end{aligned} \tag{3.164}$$

and

$$\begin{aligned} &||\mathbf{x}|| = ||\mathbf{a}_0|| + \sum_{k=1}^{\infty} ||\mathbf{A}_k||, \text{ and } \lim_{k\to\infty} ||\mathbf{A}_k|| = 0 \text{ but not uniform} \\ &\text{with } \mathbf{A}_k = (A_{k1}, A_{k2}, \ldots, A_{kn})^{\mathrm{T}} \text{ and } A_{kj} = \sqrt{b_{kj}^2 + c_{kj}^2}\ \ (j = 1, 2, \ldots, n); \end{aligned} \tag{3.165}$$

$$\begin{aligned} &||\mathbf{x}^{\tau}|| = ||\mathbf{a}_0^{\tau}|| + \sum_{k=1}^{\infty} ||\mathbf{A}_k^{\tau}||, \text{ and } \lim_{k\to\infty} ||\mathbf{A}_k^{\tau}|| = 0 \text{ but not uniform} \\ &\text{with } \mathbf{A}_k^{\tau} = (A_{k1}^{\tau}, A_{k2}^{\tau}, \ldots, A_{kn}^{\tau})^{\mathrm{T}} \text{ and } A_{kj}^{\tau} = \sqrt{(b_{kj}^{\tau})^2 + (c_{kj}^{\tau})^2}\ \ (j = 1, 2, \ldots, n). \end{aligned} \tag{3.166}$$

For $||\mathbf{x}(t) - \mathbf{x}^*(t)|| < \varepsilon$ and $||\mathbf{x}^{\tau}(t) - \mathbf{x}^{\tau *}(t)|| < \varepsilon^{\tau}$ with prescribed small positive $\varepsilon > 0$ and $\varepsilon^{\tau} > 0$, the infinite term transformation of periodic flow $\mathbf{x}(t)$ of Equation (3.162), given by Equation (3.163), can be approximated by a finite term transformation $\mathbf{x}^*(t)$ as

$$\mathbf{x}^* = \mathbf{a}_0(t) + \sum_{k=1}^{N} \mathbf{b}_k(t)\cos(k\theta) + \mathbf{c}_k(t)\sin(k\theta)$$

$$\mathbf{x}^{\tau *} = \mathbf{a}_0^{\tau}(t) + \sum_{k=1}^{N} \mathbf{b}_k^{\tau}(t)\cos[k(\theta - \theta^{\tau})] + \mathbf{c}_k^{\tau}(t)\sin[k(\theta - \theta^{\tau})] \tag{3.167}$$

and the generalized coordinates are determined by

$$\begin{aligned} \dot{\mathbf{a}}_0 &= \mathbf{F}_0(\mathbf{z}, \mathbf{z}^{\tau}), \\ \dot{\mathbf{b}} &= -\Omega \mathbf{k}_1 \mathbf{c} + \mathbf{F}_1(\mathbf{z}, \mathbf{z}^{\tau}), \\ \dot{\mathbf{c}} &= \Omega \mathbf{k}_1 \mathbf{b} + \mathbf{F}_2(\mathbf{z}, \mathbf{z}^{\tau}); \end{aligned} \tag{3.168}$$

where

$$\begin{aligned} \mathbf{k}_1 &= diag(\mathbf{I}_{n\times n}, 2\mathbf{I}_{n\times n}, \ldots, N\mathbf{I}_{n\times n}), \\ \mathbf{b} &= (\mathbf{b}_1, \mathbf{b}_2, \ldots, \mathbf{b}_N)^{\mathrm{T}}, \\ \mathbf{c} &= (\mathbf{c}_1, \mathbf{c}_2, \ldots, \mathbf{c}_N)^{\mathrm{T}}; \\ \mathbf{b}^{\tau} &= (\mathbf{b}_1^{\tau}, \mathbf{b}_2^{\tau}, \ldots, \mathbf{b}_N^{\tau})^{\mathrm{T}}, \\ \mathbf{c}^{\tau} &= (\mathbf{c}_1^{\tau}, \mathbf{c}_2^{\tau}, \ldots, \mathbf{c}_N^{\tau})^{\mathrm{T}}; \\ \mathbf{F}_1 &= (\mathbf{F}_{11}, \mathbf{F}_{12}, \ldots, \mathbf{F}_{1N})^{\mathrm{T}}, \\ \mathbf{F}_2 &= (\mathbf{F}_{21}, \mathbf{F}_{22}, \ldots, \mathbf{F}_{2N})^{\mathrm{T}}; \\ \mathbf{z} &= (\mathbf{a}_0, \mathbf{b}, \mathbf{c})^{\mathrm{T}}, \\ \mathbf{z}^{\tau} &= (\mathbf{a}_0^{\tau}, \mathbf{b}^{\tau}, \mathbf{c}^{\tau})^{\mathrm{T}} \\ &\text{for } N = 1, 2, \ldots, \infty. \end{aligned} \tag{3.169}$$

and for $k = 1, 2, \ldots, N$

$$\begin{aligned} \mathbf{F}_0(\mathbf{z}, \mathbf{z}^{\tau}) &= \frac{1}{2\pi}\int_0^{2\pi} \mathbf{F}(\mathbf{x}, \mathbf{x}^{\tau}, t, \mathbf{p})d\theta; \\ \mathbf{F}_{1k}(\mathbf{z}, \mathbf{z}^{\tau}) &= \frac{1}{\pi}\int_0^{2\pi} \mathbf{F}(\mathbf{x}, \mathbf{x}^{\tau}, t, \mathbf{p})\cos(k\theta)dt, \\ \mathbf{F}_{2k}(\mathbf{z}, \mathbf{z}^{\tau}) &= \frac{1}{\pi}\int_0^{2\pi} \mathbf{F}(\mathbf{x}, \mathbf{x}^{\tau}, t, \mathbf{p})\sin(k\theta)dt. \end{aligned} \tag{3.170}$$

Equation (3.168) becomes

$$\dot{\mathbf{z}} = \mathbf{f}(\mathbf{z}, \mathbf{z}^{\tau}) \tag{3.171}$$

where

$$\mathbf{f} = (\mathbf{F}_0, -\Omega\mathbf{k}_1\mathbf{c} + \mathbf{F}_1, \Omega\mathbf{k}_1\mathbf{b} + \mathbf{F}_2)^{\mathrm{T}}. \tag{3.172}$$

If equilibrium $\mathbf{z}^* = \mathbf{z}^{\tau *}$ *of Equation (3.171) (i.e.,* $\mathbf{f}(\mathbf{z}^*, \mathbf{z}^{\tau *}) = \mathbf{0}$*) exists, then the approximate solution of periodic flow exists as in Equation (3.167). In the vicinity of equilibrium* $\mathbf{z}^* = \mathbf{z}^{\tau *}$*, with* $\mathbf{z} = \mathbf{z}^* + \Delta\mathbf{z}$ *and* $\mathbf{z}^{\tau} = \mathbf{z}^{*\tau} + \Delta\mathbf{z}^{\tau}$ *the linearized equation of Equation (3.171) is*

$$\Delta\dot{\mathbf{z}} = D_{\mathbf{z}}\mathbf{f}(\mathbf{z}^*, \mathbf{z}^{\tau *})\Delta\mathbf{z} + D_{\mathbf{z}^{\tau}}\mathbf{f}(\mathbf{z}^*, \mathbf{z}^{\tau *})\Delta\mathbf{z}^{\tau} \tag{3.173}$$

and the eigenvalue analysis of equilibrium $\mathbf{z}^*$ *is given by*

$$|D_{\mathbf{z}}\mathbf{f}(\mathbf{z}^*, \mathbf{z}^{\tau *}) - \lambda \mathbf{I}_{n(2N+1)\times n(2N+1)} + D_{\mathbf{z}^\tau}\mathbf{f}(\mathbf{z}^*, \mathbf{z}^{\tau *})e^{-\lambda\tau}| = 0 \tag{3.174}$$

where

$$D_{\mathbf{z}}\mathbf{f}(\mathbf{z}^*, \mathbf{z}^{\tau *}) = \frac{\partial \mathbf{f}(\mathbf{z}, \mathbf{z}^\tau)}{\partial \mathbf{z}}|_{(\mathbf{z}^*, \mathbf{z}^{\tau *})},$$

$$D_{\mathbf{z}^\tau}\mathbf{f}(\mathbf{z}^*, \mathbf{z}^{\tau *}) = \frac{\partial \mathbf{f}(\mathbf{z}, \mathbf{z}^\tau)}{\partial \mathbf{z}^\tau}|_{(\mathbf{z}^*, \mathbf{z}^{\tau *})}. \tag{3.175}$$

Thus, the stability and bifurcation of periodic flow can be classified by the eigenvalues of Equation (3.172) at equilibrium $\mathbf{z}^* = \mathbf{z}^{\tau *}$ *with*

$$(n_1, n_2, n_3 \mid n_4, n_5, n_6). \tag{3.176}$$

1. *If all eigenvalues of the equilibrium possess negative real parts, the approximate periodic solution is stable.*
2. *If at least one of the eigenvalues of the equilibrium possesses a positive real part, the approximate periodic solution is unstable.*
3. *The boundaries between stable and unstable equilibriums with higher order singularity give bifurcation and stability conditions with higher order singularity.*

Proof. The proof of this theorem is similar to Theorem 3.9. ■

Similarly, period-m flows in a non-autonomous, time-delayed, nonlinear dynamical system in Equation (3.162) can be discussed.

Theorem 3.12 *Consider a non-autonomous, time-delayed, nonlinear dynamical system in Equation (3.162) with an excitation period* $T = 2\pi/\Omega$. *If such a dynamical system has a period-*m *flow* $\mathbf{x}^{(m)}(t)$ *with finite norm* $\|\mathbf{x}^{(m)}\|$, *there is a generalized coordinate transformation for the periodic flow of Equation (3.162) in the form of*

$$\mathbf{x}^{(m)}(t) = \mathbf{a}_0^{(m)}(t) + \sum_{k=1}^{\infty} \mathbf{b}_{k/m}(t)\cos\left(\frac{k}{m}\theta\right) + \mathbf{c}_{k/m}(t)\sin\left(\frac{k}{m}\theta\right);$$

$$\mathbf{x}^{\tau(m)}(t) = \mathbf{a}_0^{\tau(m)}(t) + \sum_{k=1}^{\infty} \mathbf{b}_{k/m}^{\tau}(t)\cos\left[\frac{k}{m}(\theta - \theta^\tau)\right] + \mathbf{c}_{k/m}^{\tau}(t)\sin\left[\frac{k}{m}(\theta - \theta^\tau)\right] \tag{3.177}$$

with

$$\mathbf{a}_0^{(m)} = (a_{01}^{(m)}, a_{02}^{(m)}, \ldots, a_{0n}^{(m)})^{\mathrm{T}},$$

$$\mathbf{b}_{k/m} = (b_{k/m1}, b_{k/m2}, \ldots, b_{k/mn})^{\mathrm{T}},$$

$$\mathbf{c}_{k/m} = (c_{k/m1}, c_{k/m2}, \ldots, c_{k/mn})^{\mathrm{T}};$$

$$\mathbf{a}_0^{\tau(m)} = (a_{01}^{\tau(m)}, a_{02}^{\tau(m)}, \ldots, a_{0n}^{\tau(m)})^{\mathrm{T}},$$

$$\mathbf{b}_{k/m}^{\tau} = (b_{k/m1}^{\tau}, b_{k/m2}^{\tau}, \ldots, b_{k/mn}^{\tau})^{\mathrm{T}},$$

$$\mathbf{c}_{k/m}^{\tau} = (c_{k/m1}^{\tau}, c_{k/m2}^{\tau}, \ldots, c_{k/mn}^{\tau})^{\mathrm{T}} \tag{3.178}$$

and

$$\begin{aligned}
&\|\mathbf{x}^{(m)}\| = \|\mathbf{a}_0^{(m)}\| + \sum_{k=1}^{\infty} \|\mathbf{A}_{k/m}\|, \text{ and } \lim_{k\to\infty} \|\mathbf{A}_{k/m}\| = 0 \text{ but not uniform}\\
&\text{with } \mathbf{A}_{k/m} = (A_{k/m1}, A_{k/m2} \ldots, A_{k/mn})^T\\
&\text{and } A_{k/mj} = \sqrt{b_{k/mj}^2 + c_{k/mj}^2} \quad (j = 1, 2, \ldots, n);
\end{aligned} \tag{3.179}$$

$$\begin{aligned}
&\|\mathbf{x}^{\tau(m)}\| = \|\mathbf{a}_0^{\tau(m)}\| + \sum_{k=1}^{\infty} \|\mathbf{A}_{k/m}^{\tau}\|, \text{ and } \lim_{k\to\infty} \|\mathbf{A}_{k/m}^{\tau}\| = 0 \text{ but not uniform}\\
&\text{with } \mathbf{A}_{k/m}^{\tau} = (A_{k/m1}^{\tau}, A_{k/m2}^{\tau} \ldots, A_{k/mn}^{\tau})^T\\
&\text{and } A_{k/mj}^{\tau} = \sqrt{(b_{k/mj}^{\tau})^2 + (c_{k/mj}^{\tau})^2} \quad (j = 1, 2, \ldots, n).
\end{aligned} \tag{3.180}$$

*For $\|\mathbf{x}^{(m)}(t) - \mathbf{x}^{(m)*}(t)\| < \varepsilon$ and $\|\mathbf{x}^{\tau(m)}(t) - \mathbf{x}^{\tau(m)*}(t)\| < \varepsilon^{\tau}$ with prescribed small $\varepsilon > 0$ and $\varepsilon^{\tau} > 0$, the infinite term transformation $\mathbf{x}^{(m)}(t)$ of period-*m *flow of Equation (3.162), given by Equation (3.177), can be approximated by a finite term transformation $\mathbf{x}^{(m)*}(t)$ as*

$$\begin{aligned}
&\mathbf{x}^{(m)*}(t) = \mathbf{a}_0^{(m)}(t) + \sum_{k=1}^{N} \mathbf{b}_{k/m}(t) \cos\left(\frac{k}{m}\theta\right) + \mathbf{c}_{k/m}(t) \sin\left(\frac{k}{m}\theta\right);\\
&\mathbf{x}^{\tau(m)*}(t) = \mathbf{a}_0^{\tau(m)}(t) + \sum_{k=1}^{N} \mathbf{b}_{k/m}^{\tau}(t) \cos\left[\frac{k}{m}(\theta - \theta^{\tau})\right] + \mathbf{c}_{k/m}^{\tau}(t) \sin\left[\frac{k}{m}(\theta - \theta^{\tau})\right]
\end{aligned} \tag{3.181}$$

and the generalized coordinates are determined by

$$\begin{aligned}
\dot{\mathbf{a}}_0^{(m)} &= \mathbf{F}_0^{(m)}(\mathbf{z}^{(m)}, \mathbf{z}^{\tau(m)}),\\
\dot{\mathbf{b}}^{(m)} &= -\frac{\Omega}{m}\mathbf{k}_1\mathbf{c}^{(m)} + \mathbf{F}_1^{(m)}(\mathbf{z}^{(m)}, \mathbf{z}^{\tau(m)}),\\
\dot{\mathbf{c}}^{(m)} &= \frac{\Omega}{m}\mathbf{k}_1\mathbf{b}^{(m)} + \mathbf{F}_2^{(m)}(\mathbf{z}^{(m)}, \mathbf{z}^{\tau(m)})
\end{aligned} \tag{3.182}$$

where for $N = 1, 2, \ldots, \infty$

$$\begin{aligned}
\mathbf{k}_1 &= diag(\mathbf{I}_{n\times n}, 2\mathbf{I}_{n\times n}, \ldots, N\mathbf{I}_{n\times n}),\\
\mathbf{b}^{(m)} &= (\mathbf{b}_{1/m}, \mathbf{b}_{2/m}, \ldots, \mathbf{b}_{N/m})^{\mathrm{T}},\\
\mathbf{c}^{(m)} &= (\mathbf{c}_{1/m}, \mathbf{c}_{2/m}, \ldots, \mathbf{c}_{N/m})^{\mathrm{T}};\\
\mathbf{b}^{\tau(m)} &= (\mathbf{b}_{1/m}^{\tau}, \mathbf{b}_{2/m}^{\tau}, \ldots, \mathbf{b}_{N/m}^{\tau})^{\mathrm{T}},\\
\mathbf{c}^{\tau(m)} &= (\mathbf{c}_{1/m}^{\tau}, \mathbf{c}_{2/m}^{\tau}, \ldots, \mathbf{c}_{N/m}^{\tau})^{\mathrm{T}};\\
\mathbf{F}_1^{(m)} &= (\mathbf{F}_{11}^{(m)}, \mathbf{F}_{12}^{(m)}, \ldots, \mathbf{F}_{1N}^{(m)})^{\mathrm{T}},\\
\mathbf{F}_2^{(m)} &= (\mathbf{F}_{21}^{(m)}, \mathbf{F}_{22}^{(m)}, \ldots, \mathbf{F}_{2N}^{(m)})^{\mathrm{T}};\\
\mathbf{z}^{(m)} &= (\mathbf{a}_0^{(m)}, \mathbf{b}^{(m)}, \mathbf{c}^{(m)})^{\mathrm{T}},
\end{aligned}$$

$$\mathbf{z}^{\tau(m)} = (\mathbf{a}_0^{\tau(m)}, \mathbf{b}^{\tau(m)}, \mathbf{c}^{\tau(m)})^{\mathrm{T}}$$
$$\text{for } N = 1, 2, \ldots, \infty; \tag{3.183}$$

and

$$\mathbf{F}_0^{(m)}(\mathbf{z}^{(m)}, \mathbf{z}^{\tau(m)}) = \frac{1}{2m\pi}\int_0^{2m\pi} \mathbf{F}(\mathbf{x}^{(m)*}, \mathbf{x}^{\tau(m)*}, t, \mathbf{p})d\theta;$$
$$\mathbf{F}_{1k}^{(m)}(\mathbf{z}^{(m)}, \mathbf{z}^{\tau(m)}) = \frac{1}{m\pi}\int_0^{2m\pi} \mathbf{F}(\mathbf{x}^{(m)*}, \mathbf{x}^{\tau(m)*}, t, \mathbf{p})\cos\left(\frac{k}{m}\theta\right)d\theta,$$
$$\mathbf{F}_{2k}^{(m)}(\mathbf{z}^{(m)}, \mathbf{z}^{\tau(m)}) = \frac{1}{m\pi}\int_0^{2m\pi} \mathbf{F}(\mathbf{x}^{(m)*}, \mathbf{x}^{\tau(m)*}, t, \mathbf{p})\sin\left(\frac{k}{m}\theta\right)d\theta$$
$$\text{for } k = 1, 2, \ldots, N. \tag{3.184}$$

Equation (3.182) becomes

$$\dot{\mathbf{z}}^{(m)} = \mathbf{f}^{(m)}(\mathbf{z}^{(m)}, \mathbf{z}^{\tau(m)}) \tag{3.185}$$

where

$$\mathbf{f}^{(m)} = \left(\mathbf{F}_0^{(m)}, -\frac{\Omega}{m}\mathbf{k}_1\mathbf{c}^{(m)} + \mathbf{F}_1^{(m)}, \frac{\Omega}{m}\mathbf{k}_1\mathbf{c}^{(m)} + \mathbf{F}_2^{(m)}\right)^{\mathrm{T}}. \tag{3.186}$$

If equilibrium $\mathbf{z}^{(m)*} = \mathbf{z}^{\tau(m)*}$ *of Equation (3.185) exists (i.e.,* $\mathbf{f}^{(m)}(\mathbf{z}^{(m)*}, \mathbf{z}^{\tau(m)*}) = \mathbf{0}$*), then the approximate solution of the period-*m *flow exists as in Equation (3.161). In the vicinity of equilibrium* $\mathbf{z}^{(m)*} = \mathbf{z}^{\tau(m)*}$*, with* $\mathbf{z}^{(m)} = \mathbf{z}^{(m)*} + \Delta\mathbf{z}^{(m)}$ *and* $\mathbf{z}^{\tau(m)} = \mathbf{z}^{\tau(m)*} + \Delta\mathbf{z}^{\tau(m)}$*, the linearized equation of Equation (3.185) is*

$$\Delta\dot{\mathbf{z}}^{(m)} = D_{\mathbf{z}^{(m)}}\mathbf{f}^{(m)}(\mathbf{z}^{(m)*}, \mathbf{z}^{\tau(m)*})\Delta\mathbf{z}^{(m)} + D_{\mathbf{z}^{\tau(m)}}\mathbf{f}^{(m)}(\mathbf{z}^{(m)*}, \mathbf{z}^{\tau(m)*})\Delta\mathbf{z}^{\tau(m)} \tag{3.187}$$

and the eigenvalue analysis of equilibrium $\mathbf{z}^{(m)*} = \mathbf{z}^{\tau(m)*}$ *is given by*

$$|D_{\mathbf{z}^{(m)}}\mathbf{f}^{(m)}(\mathbf{z}^{(m)*}, \mathbf{z}^{\tau(m)*}) - \lambda\mathbf{I}_{n(2N+1)\times n(2N+1)} + D_{\mathbf{z}^{\tau(m)}}\mathbf{f}^{(m)}(\mathbf{z}^{(m)*}, \mathbf{z}^{\tau(m)*})e^{-\lambda\tau}| = 0 \tag{3.188}$$

where

$$D_{\mathbf{z}^{(m)}}\mathbf{f}^{(m)}(\mathbf{z}^{(m)*}, \mathbf{z}^{\tau(m)*}) = \frac{\partial\mathbf{f}^{(m)}(\mathbf{z}^{(m)}, \mathbf{z}^{\tau(m)})}{\partial\mathbf{z}^{(m)}}\Big|_{(\mathbf{z}^{(m)*}, \mathbf{z}^{\tau(m)*})},$$
$$D_{\mathbf{z}^{\tau(m)}}\mathbf{f}^{(m)}(\mathbf{z}^{(m)*}, \mathbf{z}^{\tau(m)*}) = \frac{\partial\mathbf{f}^{(m)}(\mathbf{z}^{(m)}, \mathbf{z}^{\tau(m)})}{\partial\mathbf{z}^{\tau(m)}}\Big|_{(\mathbf{z}^{(m)*}, \mathbf{z}^{\tau(m)*})}. \tag{3.189}$$

*The stability and bifurcation of period-*m *flow can be classified by the eigenvalues of Equation (3.187) at equilibrium* $\mathbf{z}^{(m)*} = \mathbf{z}^{\tau(m)*}$ *with*

$$(n_1, n_2, n_3 \mid n_4, n_5, n_6). \tag{3.190}$$

1. *If all eigenvalues of the equilibrium possess negative real parts, the approximate periodic solution is stable.*
2. *If at least one of the eigenvalues of the equilibrium possesses a positive real part, the approximate periodic solution is unstable.*
3. *The boundaries between stable and unstable equilibriums with higher order singularity give bifurcation and stability conditions with higher order singularity.*

Proof. The proof is similar to the proof of Theorem 3.10. ■

If $m \to \infty$, Equation (3.177) will give the analytical expression of chaos in periodically excited, time-delayed, nonlinear dynamical systems in Equation (3.162), which can be approximated by Equation (3.181) under the condition of $||\mathbf{x}^{(m)}(t) - \mathbf{x}^{(m)*}(t)|| < \varepsilon$ and $||\mathbf{x}^{\tau(m)}(t) - \mathbf{x}^{\tau(m)*}(t)|| < \varepsilon^{\tau}$.

3.4 Time-Delayed, Nonlinear Vibration Systems

In this section, the analytical solutions of periodic motion in time-delayed, nonlinear vibration systems are presented because of extensive application in vibration control, and the local stability and bifurcation theory will be applied to determine the stability and bifurcation of approximate solutions of time-delayed, nonlinear vibration systems.

3.4.1 Time-Delayed, Free Vibration Systems

Periodic flows in time-delayed, nonlinear vibration systems will be discussed herein. If such a time-delayed, vibration system has periodic flows with period $T = 2\pi/\Omega$, then such a periodic motion can be expressed by the Fourier series.

Theorem 3.13 *Consider a time-delayed, nonlinear vibration system as*

$$\ddot{\mathbf{x}} = \mathbf{f}(\mathbf{x}, \dot{\mathbf{x}}, \mathbf{x}^{\tau}, \dot{\mathbf{x}}^{\tau}, \mathbf{p}) \in \mathscr{R}^{n} \tag{3.191}$$

where $\mathbf{f}(\mathbf{x}, \dot{\mathbf{x}}, \mathbf{x}^{\tau}, \dot{\mathbf{x}}^{\tau}, \mathbf{p})$ *is a* C^{r}*-continuous nonlinear function vector* ($r \geq 1$)*. If such a time-delayed, vibration system has a periodic motion* $\mathbf{x}(t)$ *with finite norm* $||\mathbf{x}||$ *and period* $T = 2\pi/\Omega$*, there is a generalized coordinate transformation with* $\theta = \Omega t$ *for the periodic motion of Equation (3.191) in the form of*

$$\begin{aligned}
\mathbf{x} &\equiv \mathbf{x}(t) = \mathbf{a}_0(t) + \sum_{k=1}^{\infty} \mathbf{b}_k(t)\cos(k\theta) + \mathbf{c}_k(t)\sin(k\theta),\\
\mathbf{x}^{\tau} &\equiv \mathbf{x}(t-\tau) = \mathbf{a}_0^{\tau}(t) + \sum_{k=1}^{\infty} \mathbf{b}_k^{\tau}(t)\cos[k(\theta-\theta^{\tau})] + \mathbf{c}_k^{\tau}(t)\sin[k(\theta-\theta^{\tau})]
\end{aligned} \tag{3.192}$$

with $\mathbf{a}_0^{\tau} = \mathbf{a}_0(t-\tau), \mathbf{b}_k^{\tau} = \mathbf{b}_k(t-\tau), \mathbf{c}_k^{\tau} = \mathbf{c}_k(t-\tau), \theta^{\tau} = \Omega\tau$ *and*

$$\begin{aligned}
\mathbf{a}_0 &= (a_{01}, a_{02}, \ldots, a_{0n})^{\mathrm{T}},\\
\mathbf{b}_k &= (b_{k1}, b_{k2}, \ldots, b_{kn})^{\mathrm{T}},\\
\mathbf{c}_k &= (c_{k1}, c_{k2}, \ldots, c_{kn})^{\mathrm{T}};\\
\mathbf{a}_0^{\tau} &= (a_{01}^{\tau}, a_{02}^{\tau}, \ldots, a_{0n}^{\tau})^{\mathrm{T}},\\
\mathbf{b}_k^{\tau} &= (b_{k1}^{\tau}, b_{k2}^{\tau}, \ldots, b_{kn}^{\tau})^{\mathrm{T}},\\
\mathbf{c}_k^{\tau} &= (c_{k1}^{\tau}, c_{k2}^{\tau}, \ldots, c_{kn}^{\tau})^{\mathrm{T}}
\end{aligned} \tag{3.193}$$

and

$$\|\mathbf{x}\| = \|\mathbf{a}_0\| + \sum_{k=1}^{\infty} \|\mathbf{A}_k\|, \text{ and } \lim_{k\to\infty} \|\mathbf{A}_k\| = 0 \text{ but not uniform}$$

$$\text{with } \mathbf{A}_k = (A_{k1}, A_{k2}, \ldots, A_{kn})^{\mathrm{T}} \text{ and } A_{kj} = \sqrt{b_{kj}^2 + c_{kj}^2} \ (j = 1, 2, \ldots, n); \tag{3.194}$$

$$\|\mathbf{x}^{\tau}\| = \|\mathbf{a}_0^{\tau}\| + \sum_{k=1}^{\infty} \|\mathbf{A}_k^{\tau}\|, \text{ and } \lim_{k\to\infty} \|\mathbf{A}_k^{\tau}\| = 0 \text{ but not uniform}$$

$$\text{with } \mathbf{A}_k^{\tau} = (A_{k1}^{\tau}, A_{k2}^{\tau}, \ldots, A_{kn}^{\tau})^{\mathrm{T}} \text{ and } A_{kj}^{\tau} = \sqrt{(b_{kj}^{\tau})^2 + (c_{kj}^{\tau})^2} \ (j = 1, 2, \ldots, n). \tag{3.195}$$

For $\|\mathbf{x}(t) - \mathbf{x}^*(t)\| < \varepsilon$ *and* $\|\mathbf{x}^{\tau}(t) - \mathbf{x}^{\tau *}(t)\| < \varepsilon^{\tau}$ *with prescribed small positive* $\varepsilon > 0$ *and* $\varepsilon^{\tau} > 0$*, the infinite term transformation* $\mathbf{x}(t)$ *of periodic motion of Equation (3.191), given by Equation (3.192), can be approximated by a finite term transformation* $\mathbf{x}^*(t)$ *as*

$$\mathbf{x}^* = \mathbf{a}_0(t) + \sum_{k=1}^{N} \mathbf{b}_k(t)\cos(k\theta) + \mathbf{c}_k(t)\sin(k\theta)$$

$$\mathbf{x}^{\tau *} = \mathbf{a}_0^{\tau}(t) + \sum_{k=1}^{N} \mathbf{b}_k^{\tau}(t)\cos[k(\theta - \theta^{\tau})] + \mathbf{c}_k^{\tau}(t)\sin[k(\theta - \theta^{\tau})] \tag{3.196}$$

and the corresponding coefficients varying with time are determined by

$$\begin{aligned}
\ddot{\mathbf{a}}_0 &= \mathbf{F}_0(\mathbf{z}, \mathbf{z}_1; \mathbf{z}^{\tau}, \mathbf{z}_1^{\tau}),\\
\ddot{\mathbf{b}} &= -2\Omega\mathbf{k}_1\dot{\mathbf{c}} + \Omega^2\mathbf{k}_2\mathbf{b} + \mathbf{F}_1(\mathbf{z}, \mathbf{z}_1; \mathbf{z}^{\tau}, \mathbf{z}_1^{\tau}),\\
\ddot{\mathbf{c}} &= 2\Omega\mathbf{k}_1\dot{\mathbf{b}} + \Omega^2\mathbf{k}_2\mathbf{c} + \mathbf{F}_2(\mathbf{z}, \mathbf{z}_1; \mathbf{z}^{\tau}, \mathbf{z}_1^{\tau});.
\end{aligned} \tag{3.197}$$

where for $N = 1, 2, \ldots, \infty$

$$\begin{aligned}
\mathbf{k}_1 &= diag(\mathbf{I}_{n\times n}, 2\mathbf{I}_{n\times n}, \ldots, N\mathbf{I}_{n\times n}),\\
\mathbf{k}_2 &= diag(\mathbf{I}_{n\times n}, 2^2\mathbf{I}_{n\times n}, \ldots, N^2\mathbf{I}_{n\times n});\\
\mathbf{b} &= (\mathbf{b}_1, \mathbf{b}_2, \ldots, \mathbf{b}_N)^{\mathrm{T}},\\
\mathbf{c} &= (\mathbf{c}_1, \mathbf{c}_2, \ldots, \mathbf{c}_N)^{\mathrm{T}};\\
\mathbf{b}^{\tau} &= (\mathbf{b}_1^{\tau}, \mathbf{b}_2^{\tau}, \ldots, \mathbf{b}_N^{\tau})^{\mathrm{T}},\\
\mathbf{c}^{\tau} &= (\mathbf{c}_1^{\tau}, \mathbf{c}_2^{\tau}, \ldots, \mathbf{c}_N^{\tau})^{\mathrm{T}};\\
\mathbf{F}_1 &= (\mathbf{F}_{11}, \mathbf{F}_{12}, \ldots, \mathbf{F}_{1N})^{\mathrm{T}},\\
\mathbf{F}_2 &= (\mathbf{F}_{21}, \mathbf{F}_{22}, \ldots, \mathbf{F}_{2N})^{\mathrm{T}};\\
\mathbf{z} &= (\mathbf{a}_0, \mathbf{b}, \mathbf{c})^{\mathrm{T}}, \dot{\mathbf{z}} = \mathbf{z}_1;\\
\mathbf{z}^{\tau} &= (\mathbf{a}_0^{\tau}, \mathbf{b}^{\tau}, \mathbf{c}^{\tau})^{\mathrm{T}}, \dot{\mathbf{z}}^{\tau} = \mathbf{z}_1^{\tau}
\end{aligned} \tag{3.198}$$

and for $k = 1, 2, \ldots, N$

$$\mathbf{F}_0(\mathbf{z}, \mathbf{z}_1; \mathbf{z}^\tau, \mathbf{z}_1^\tau) = \frac{1}{2\pi}\int_0^{2\pi} \mathbf{f}(\mathbf{x}^*, \dot{\mathbf{x}}^*; \mathbf{x}^{\tau *}, \dot{\mathbf{x}}^{\tau *}, \mathbf{p}) d\theta;$$

$$\mathbf{F}_{1k}(\mathbf{z}, \mathbf{z}_1; \mathbf{z}^\tau, \mathbf{z}_1^\tau) = \frac{1}{\pi}\int_0^{2\pi} \mathbf{f}(\mathbf{x}^*, \dot{\mathbf{x}}^*; \mathbf{x}^{\tau *}, \dot{\mathbf{x}}^{\tau *}, \mathbf{p}) \cos(k\theta) d\theta,$$

$$\mathbf{F}_{2k}(\mathbf{z}, \mathbf{z}_1; \mathbf{z}^\tau, \mathbf{z}_1^\tau) = \frac{1}{\pi}\int_0^{2\pi} \mathbf{f}(\mathbf{x}^*, \dot{\mathbf{x}}^*; \mathbf{x}^{\tau *}, \dot{\mathbf{x}}^{\tau *}, \mathbf{p}) \sin(k\theta) d\theta. \tag{3.199}$$

The state-space form of Equation (3.197) is

$$\dot{\mathbf{z}} = \mathbf{z}_1 \text{ and } \dot{\mathbf{z}}_1 = \mathbf{g}(\mathbf{z}, \mathbf{z}_1; \mathbf{z}^\tau, \mathbf{z}_1^\tau) \tag{3.200}$$

where

$$\mathbf{g} = (\mathbf{F}_0, -2\Omega\mathbf{k}_1\dot{\mathbf{c}} + \Omega^2\mathbf{k}_2\mathbf{b} + \mathbf{F}_1, 2\Omega\mathbf{k}_1\dot{\mathbf{b}} + \Omega^2\mathbf{k}_2\mathbf{c} + \mathbf{F}_2)^{\mathrm{T}}. \tag{3.201}$$

An equivalent system of Equation (3.200) is

$$\dot{\mathbf{y}} = \mathbf{f}(\mathbf{y}, \mathbf{y}^\tau) \tag{3.202}$$

where

$$\mathbf{y} = (\mathbf{z}, \mathbf{z}_1)^{\mathrm{T}}, \mathbf{y}^\tau = (\mathbf{z}^\tau, \mathbf{z}_1^\tau)^{\mathrm{T}} \text{ and } \mathbf{f} = (\mathbf{z}_1, \mathbf{g})^{\mathrm{T}}. \tag{3.203}$$

If equilibrium $\mathbf{y}^* = \mathbf{y}^{\tau *}$ *of Equation (3.202) (i.e.,* $\mathbf{f}(\mathbf{y}^*, \mathbf{y}^{\tau *}) = \mathbf{0}$*) exists, then the approximate solution of periodic motion exists in Equation (3.196). In the vicinity of equilibrium* $\mathbf{y}^* = \mathbf{y}^{\tau *}$*, with* $\mathbf{y} = \mathbf{y}^* + \Delta\mathbf{y}$ *and* $\mathbf{y}^\tau = \mathbf{y}^{\tau *} + \Delta\mathbf{y}^\tau$*, the linearized equation of Equation (3.202) is*

$$\Delta\dot{\mathbf{y}} = D_{\mathbf{y}}\mathbf{f}(\mathbf{y}^*, \mathbf{y}^{\tau *})\Delta\mathbf{y} + D_{\mathbf{y}^\tau}\mathbf{f}(\mathbf{y}^*, \mathbf{y}^{\tau *})\Delta\mathbf{y}^\tau \tag{3.204}$$

and the eigenvalue analysis of equilibrium $\mathbf{y}^* = \mathbf{y}^{\tau *}$ *is given by*

$$|D_{\mathbf{y}}\mathbf{f}(\mathbf{y}^*, \mathbf{y}^{\tau *}) - \lambda\mathbf{I}_{2n(2N+1)\times 2n(2N+1)} + D_{\mathbf{y}^\tau}\mathbf{f}(\mathbf{y}^*, \mathbf{y}^{\tau *})e^{-\lambda\tau}| = 0 \tag{3.205}$$

where

$$D_{\mathbf{y}}\mathbf{f}(\mathbf{y}^*, \mathbf{y}^{\tau *}) = \frac{\partial\mathbf{f}(\mathbf{y}, \mathbf{y}^\tau)}{\partial\mathbf{y}}|_{(\mathbf{y}^*, \mathbf{y}^{\tau *})}, D_{\mathbf{y}^\tau}\mathbf{f}(\mathbf{y}^*, \mathbf{y}^{\tau *}) = \frac{\partial\mathbf{f}(\mathbf{y}, \mathbf{y}^\tau)}{\partial\mathbf{y}^\tau}|_{(\mathbf{y}^*, \mathbf{y}^{\tau *})}. \tag{3.206}$$

Thus, the stability and bifurcation of periodic motion can be classified by the eigenvalues of Equation (3.204) at equilibrium $\mathbf{y}^* = \mathbf{y}^{\tau *}$ *with*

$$(n_1, n_2, n_3 \mid n_4, n_5, n_6). \tag{3.207}$$

1. *If all eigenvalues of the equilibrium possess negative real parts, the approximate periodic solution is stable.*
2. *If at least one of the eigenvalues of the equilibrium possesses a positive real part, the approximate periodic solution is unstable.*
3. *The boundaries between stable and unstable equilibriums with higher order singularity give bifurcation and stability conditions with higher order singularity.*

Proof. Since $\mathbf{f}(\mathbf{x},\dot{\mathbf{x}};\mathbf{x}^\tau,\dot{\mathbf{x}}^\tau,\mathbf{p})$ is a C^r-continuous nonlinear function vector ($r \geq 1$), the velocity $\dot{\mathbf{x}}$ should be C^r-continuous ($r \geq 1$), and then the acceleration $\ddot{\mathbf{x}}$ should be bounded (i.e., $\|\ddot{\mathbf{x}}\| \leq K$). From Equation (3.192), the norms of the periodic motion are defined by

$$\|\mathbf{x}\| = \|\mathbf{a}_0\| + \sum_{k=1}^{\infty} \|\mathbf{A}_k\| \text{ and } \|\mathbf{x}^\tau\| = \|\mathbf{a}_0^\tau\| + \sum_{k=1}^{\infty} \|\mathbf{A}_k^\tau\|$$

where

$$\mathbf{A}_k = (A_{k1}, A_{k2}, \ldots, A_{kn})^{\mathrm{T}},$$
$$A_{kj} = \sqrt{b_{kj}^2 + c_{kj}^2}\ (j = 1, 2, \ldots, n),$$

$$\mathbf{A}_k^\tau = (A_{k1}^\tau, A_{k2}^\tau, \ldots, A_{kn}^\tau)^{\mathrm{T}},$$
$$A_{kj}^\tau = \sqrt{(b_{kj}^\tau)^2 + (c_{kj}^\tau)^2}\ (j = 1, 2, \ldots, n).$$

Because the periodic motion in Equation (3.191) is bounded (i.e., $\|\mathbf{x}(t)\| < C$), we have

$$\lim_{k\to\infty} \|\mathbf{A}_k\| = 0 \text{ and } \lim_{k\to\infty} \|\mathbf{A}_k^\tau\| = 0 \text{ but not uniform.}$$

Thus, the Fourier series transformation of periodic motion as in Equation (3.191) is convergent. From Equations (3.194) and (3.195), using Equations (3.192) and (3.196) gives

$$\|\mathbf{x}(t) - \mathbf{x}^*(t)\| = \left\| \sum_{k=N+1}^{\infty} \mathbf{b}_k(t)\cos(k\theta) + \mathbf{c}_k(t)\sin(k\theta) \right\| = \sum_{k=N+1}^{\infty} \|\mathbf{A}_k\|,$$

$$\|\mathbf{x}^\tau(t) - \mathbf{x}^{\tau*}(t)\| = \left\| \sum_{k=N+1}^{\infty} \mathbf{b}_k^\tau(t)\cos[k(\theta - \theta^\tau)] + \mathbf{c}_k^\tau(t)\sin[k(\theta - \theta^\tau)] \right\|$$
$$= \sum_{k=N+1}^{\infty} \|\mathbf{A}_k^\tau\|.$$

For the prescribed small positive $\varepsilon > 0$ and $\varepsilon^\tau > 0$, if $\|\mathbf{x}(t) - \mathbf{x}^*(t)\| < \varepsilon$ and $\|\mathbf{x}^\tau(t) - \mathbf{x}^{\tau*}(t)\| < \varepsilon^\tau$ exist, then we have

$$\sum_{k=N+1}^{\infty} \|\mathbf{A}_k\| < \varepsilon \text{ and } \sum_{k=N+1}^{\infty} \|\mathbf{A}_k\| < \varepsilon^\tau.$$

Therefore, the convergent infinite term transformation in Equation (3.192) can be approximated by a finite term transformation in Equation (3.196) in the sense of ε.

Taking the derivative of Equation (3.196) with respect to time gives

$$\dot{\mathbf{x}}^*(t) = \dot{\mathbf{a}}_0 + \sum_{k=1}^{N} (\dot{\mathbf{b}}_k + k\Omega\mathbf{c}_k)\cos(k\theta) + (\dot{\mathbf{c}}_k - k\Omega\mathbf{b}_k)\sin(k\theta),$$

$$\ddot{\mathbf{x}}^*(t) = \ddot{\mathbf{a}}_0 + \sum_{k=1}^{N} (\ddot{\mathbf{b}}_k + 2k\Omega\dot{\mathbf{c}}_k - k^2\Omega^2\mathbf{b}_k)\cos(k\theta)$$
$$+ (\ddot{\mathbf{c}}_k - 2k\Omega\dot{\mathbf{b}}_k - k^2\Omega^2\mathbf{c}_k)\sin(k\theta).$$

$$\dot{\mathbf{x}}^{\tau *}(t) = \dot{\mathbf{a}}_0^\tau + \sum_{k=1}^{N} (\dot{\mathbf{b}}_k^\tau + k\Omega \mathbf{c}_k^\tau) \cos[k(\theta - \theta^\tau)] + (\dot{\mathbf{c}}_k^\tau - k\Omega \mathbf{b}_k^\tau) \sin[k(\theta - \theta^\tau)]$$

$$\ddot{\mathbf{x}}^{\tau *}(t) = \ddot{\mathbf{a}}_0^\tau + \sum_{k=1}^{N} (\ddot{\mathbf{b}}_k^\tau + 2k\Omega \dot{\mathbf{c}}_k^\tau - k^2\Omega^2 \mathbf{b}_k^\tau) \cos[k(\theta - \theta^\tau)]$$
$$+ (\ddot{\mathbf{c}}_k^\tau - 2k\Omega \dot{\mathbf{b}}_k^\tau - k^2\Omega^2 \mathbf{c}_k^\tau) \sin[k(\theta - \theta^\tau)].$$

where $\mathbf{a}_0^\tau = \mathbf{a}_0(t - \tau), \mathbf{b}_k^\tau = \mathbf{b}_k(t - \tau), \mathbf{c}_k^\tau = \mathbf{c}_k(t - \tau), \theta^\tau = \Omega\tau$. Substitution of the foregoing equation into Equation (3.192), and application of the virtual work principle for constant, $\cos(k\theta)$ and $\sin(k\theta)$ $(k = 1, 2, \ldots)$ as a set of virtual displacement gives

$$\frac{1}{2\pi}\int_0^{2\pi} [\ddot{\mathbf{x}} - \mathbf{f}(\mathbf{x}, \dot{\mathbf{x}}; \mathbf{x}^\tau, \dot{\mathbf{x}}^\tau, \mathbf{p})] d\theta = 0,$$

$$\frac{1}{\pi}\int_0^{2\pi} [\ddot{\mathbf{x}} - \mathbf{f}(\mathbf{x}, \dot{\mathbf{x}}; \mathbf{x}^\tau, \dot{\mathbf{x}}^\tau, \mathbf{p})] \cos(k\theta) d\theta = 0,$$

$$\frac{1}{\pi}\int_0^{2\pi} [\ddot{\mathbf{x}} - \mathbf{f}(\mathbf{x}, \dot{\mathbf{x}}; \mathbf{x}^\tau, \dot{\mathbf{x}}^\tau, \mathbf{p})] \sin(k\theta) d\theta = 0$$

Under $||\mathbf{x}^{(m)} - \mathbf{x}^{(m)*}|| < \varepsilon$ and $||\mathbf{x}^{\tau(m)} - \mathbf{x}^{\tau(m)*}|| < \varepsilon^\tau$ with continuity $||\ddot{\mathbf{x}}^{(m)}|| < K$ (constant) and small $\varepsilon > 0$ and $\varepsilon^\tau > 0$, for $k = 1, 2, \ldots, N$, the foregoing equation gives

$$\frac{1}{2\pi}\int_0^{2\pi} [\ddot{\mathbf{x}}^* - \mathbf{f}(\mathbf{x}^*, \dot{\mathbf{x}}^*; \mathbf{x}^{\tau *}, \dot{\mathbf{x}}^{\tau *}, \mathbf{p})] d\theta + O(\delta) = 0,$$

$$\frac{1}{\pi}\int_0^{2\pi} [\ddot{\mathbf{x}}^* - \mathbf{f}(\mathbf{x}^*, \dot{\mathbf{x}}^*; \mathbf{x}^{\tau *}, \dot{\mathbf{x}}^{\tau *}, \mathbf{p})] \cos(k\theta) d\theta + O(\delta) = 0,$$

$$\frac{1}{\pi}\int_0^{2\pi} [\ddot{\mathbf{x}}^* - \mathbf{f}(\mathbf{x}^*, \dot{\mathbf{x}}^*; \mathbf{x}^{\tau *}, \dot{\mathbf{x}}^{\tau *}, \mathbf{p})] \sin(k\theta) d\theta + O(\delta) = 0$$

where $\delta = \max(\varepsilon, \varepsilon^\tau, \varepsilon_t, \varepsilon_t^\tau, \varepsilon_{tt})$ with $||\dot{\mathbf{x}} - \dot{\mathbf{x}}^*|| < \varepsilon_t$, $||\dot{\mathbf{x}}^\tau - \dot{\mathbf{x}}^{\tau *}|| < \varepsilon_t^\tau$, $||\ddot{\mathbf{x}} - \ddot{\mathbf{x}}^*|| < \varepsilon_{tt}$ for small $\{\varepsilon_t, \varepsilon_t^\tau, \varepsilon_{tt}\} > 0$. The foregoing equation produces

$$\ddot{\mathbf{a}}_0 = \mathbf{F}_0(\mathbf{z}, \mathbf{z}_1; \mathbf{z}^\tau, \mathbf{z}_1^\tau),$$
$$\ddot{\mathbf{b}}_k = -2\Omega k \dot{\mathbf{c}}_k + \Omega^2 k \mathbf{b}_k + \mathbf{F}_{1k}(\mathbf{z}, \mathbf{z}_1; \mathbf{z}^\tau, \mathbf{z}_1^\tau),$$
$$\ddot{\mathbf{c}}_k = 2\Omega k \dot{\mathbf{b}}_k + \Omega^2 k \mathbf{c}_k + \mathbf{F}_{2k}(\mathbf{z}, \mathbf{z}_1; \mathbf{z}^\tau, \mathbf{z}_1^\tau)$$

where

$$\mathbf{F}_0(\mathbf{z}, \mathbf{z}_1; \mathbf{z}^\tau, \mathbf{z}_1^\tau) = \frac{1}{2\pi}\int_0^{2\pi} \mathbf{f}(\mathbf{x}^*, \dot{\mathbf{x}}^*; \mathbf{x}^{\tau *}, \dot{\mathbf{x}}^{\tau *}, \mathbf{p}) d\theta;$$

$$\mathbf{F}_{1k}(\mathbf{z}, \mathbf{z}_1; \mathbf{z}^\tau, \mathbf{z}_1^\tau) = \frac{1}{\pi}\int_0^{2\pi} \mathbf{f}(\mathbf{x}^*, \dot{\mathbf{x}}^*; \mathbf{x}^{\tau *}, \dot{\mathbf{x}}^{\tau *}, \mathbf{p}) \cos(k\theta) d\theta,$$

$$\mathbf{F}_{2k}(\mathbf{z}, \mathbf{z}_1; \mathbf{z}^\tau, \mathbf{z}_1^\tau) = \frac{1}{\pi}\int_0^{2\pi} \mathbf{f}(\mathbf{x}^*, \dot{\mathbf{x}}^*; \mathbf{x}^{\tau *}, \dot{\mathbf{x}}^{\tau *}, \mathbf{p}) \sin(k\theta) d\theta.$$

Introduce

$$
\begin{aligned}
&\mathbf{k}_1 = diag(\mathbf{I}_{n\times n}, 2\mathbf{I}_{n\times n}, \ldots, N\mathbf{I}_{n\times n}),\\
&\mathbf{k}_2 = diag(\mathbf{I}_{n\times n}, 2^2\mathbf{I}_{n\times n}, \ldots, N^2\mathbf{I}_{n\times n})\\
&\mathbf{b} = (\mathbf{b}_1, \mathbf{b}_2, \ldots, \mathbf{b}_N)^{\mathrm{T}},\\
&\mathbf{c} = (\mathbf{c}_1, \mathbf{c}_2, \ldots, \mathbf{c}_N)^{\mathrm{T}};\\
&\mathbf{b}^{\tau} = (\mathbf{b}_1^{\tau}, \mathbf{b}_2^{\tau}, \ldots, \mathbf{b}_N^{\tau})^{\mathrm{T}},\\
&\mathbf{c}^{\tau} = (\mathbf{c}_1^{\tau}, \mathbf{c}_2^{\tau}, \ldots, \mathbf{c}_N^{\tau})^{\mathrm{T}};\\
&\mathbf{F}_1 = (\mathbf{F}_{11}, \mathbf{F}_{12}, \ldots, \mathbf{F}_{1N})^{\mathrm{T}},\\
&\mathbf{F}_2 = (\mathbf{F}_{21}, \mathbf{F}_{22}, \ldots, \mathbf{F}_{2N})^{\mathrm{T}};\\
&\mathbf{z} = (\mathbf{a}_0, \mathbf{b}, \mathbf{c})^{\mathrm{T}}, \dot{\mathbf{z}} = \mathbf{z}_1;\\
&\mathbf{z}^{\tau} = (\mathbf{a}_0^{\tau}, \mathbf{b}^{\tau}, \mathbf{c}^{\tau})^{\mathrm{T}}, \dot{\mathbf{z}}^{\tau} = \mathbf{z}_1^{\tau}.
\end{aligned}
$$

Rearranging the foregoing equation gives Equation (3.197), that is,

$$
\begin{aligned}
\ddot{\mathbf{a}}_0 &= \mathbf{F}_0(\mathbf{z}, \mathbf{z}_1; \mathbf{z}^{\tau}, \mathbf{z}_1^{\tau}),\\
\ddot{\mathbf{b}} &= -2\Omega\mathbf{k}_1\dot{\mathbf{c}} + \Omega^2\mathbf{k}_2\mathbf{b} + \mathbf{F}_1(\mathbf{z}, \mathbf{z}_1; \mathbf{z}^{\tau}, \mathbf{z}_1^{\tau}),\\
\ddot{\mathbf{c}} &= 2\Omega\mathbf{k}_1\dot{\mathbf{b}} + \Omega^2\mathbf{k}_2\mathbf{c} + \mathbf{F}_2(\mathbf{z}, \mathbf{z}_1; \mathbf{z}^{\tau}, \mathbf{z}_1^{\tau}).
\end{aligned}
$$

Let

$$
\begin{aligned}
&\mathbf{z} = (\mathbf{a}_0, \mathbf{b}, \mathbf{c})^{\mathrm{T}}, \mathbf{z}^{\tau} = (\mathbf{a}_0^{\tau}, \mathbf{b}^{\tau}, \mathbf{c}^{\tau})^{\mathrm{T}}; \dot{\mathbf{z}} = \mathbf{z}_1, \dot{\mathbf{z}}^{\tau} = \mathbf{z}_1^{\tau};\\
&\mathbf{g} = (\mathbf{F}_0, -2\Omega\mathbf{k}_1\dot{\mathbf{c}} + \Omega^2\mathbf{k}_2\mathbf{b} + \mathbf{F}_1, 2\Omega\mathbf{k}_1\dot{\mathbf{b}} + \Omega^2\mathbf{k}_2\mathbf{c} + \mathbf{F}_2)^{\mathrm{T}}.
\end{aligned}
$$

The equation in Equation (3.197) becomes

$$
\dot{\mathbf{z}} = \mathbf{z}_1 \text{ and } \dot{\mathbf{z}}_1 = \mathbf{g}(\mathbf{z}, \mathbf{z}_1; \mathbf{z}^{\tau}, \mathbf{z}_1^{\tau}).
$$

Letting $\mathbf{y} = (\mathbf{z}, \mathbf{z}_1)^{\mathrm{T}}$, $\mathbf{y}^{\tau} = (\mathbf{z}^{\tau}, \mathbf{z}_1^{\tau})^{\mathrm{T}}$, and $\mathbf{f} = (\mathbf{z}_1, \mathbf{g})^{\mathrm{T}}$, we have

$$
\dot{\mathbf{y}} = \mathbf{f}(\mathbf{y}, \mathbf{y}^{\tau}).
$$

Consider the equilibrium solution of the foregoing equation (i.e., $\mathbf{f}(\mathbf{y}^*, \mathbf{y}^{\tau *}) = \mathbf{0}$) with $\mathbf{y}^* = \mathbf{y}^{\tau *}$, that is,

$$
\begin{aligned}
\mathbf{0} &= \mathbf{F}_0(\mathbf{a}_0^*, \mathbf{b}^*, \mathbf{c}^*, \mathbf{0}, \mathbf{0}, \mathbf{0}; \mathbf{a}_0^{\tau *}, \mathbf{b}^{\tau *}, \mathbf{c}^{\tau *}, \mathbf{0}, \mathbf{0}, \mathbf{0}),\\
\mathbf{0} &= \Omega^2\mathbf{k}_2\mathbf{b}^* + \mathbf{F}_1(\mathbf{a}_0^*, \mathbf{b}^*, \mathbf{c}^*, \mathbf{0}, \mathbf{0}, \mathbf{0}; \mathbf{a}_0^{\tau *}, \mathbf{b}^{\tau *}, \mathbf{c}^{\tau *}, \mathbf{0}, \mathbf{0}, \mathbf{0}),\\
\mathbf{0} &= \Omega^2\mathbf{k}_2\mathbf{c}^* + \mathbf{F}_2(\mathbf{a}_0^*, \mathbf{b}^*, \mathbf{c}^*, \mathbf{0}, \mathbf{0}, \mathbf{0}; \mathbf{a}_0^{\tau *}, \mathbf{b}^{\tau *}, \mathbf{c}^{\tau *}, \mathbf{0}, \mathbf{0}, \mathbf{0})
\end{aligned}
$$

with

$$
\mathbf{a}_0^* = \mathbf{a}_0^{\tau *}, \mathbf{b}^* = \mathbf{b}^{\tau *}, \mathbf{c}^* = \mathbf{c}^{\tau *}.
$$

Thus, the solutions of the foregoing equation are the existence conditions of the periodic solutions for time-delayed, nonlinear vibration systems. If the foregoing equation gives the equilibrium $\mathbf{y}^* = \mathbf{y}^{\tau *}$. In vicinity of $\mathbf{y}^* = \mathbf{y}^{\tau *}$, with $\mathbf{y} = \mathbf{y}^* + \Delta\mathbf{y}$ and $\mathbf{y}^\tau = \mathbf{y}^{\tau *} + \Delta\mathbf{y}^\tau$, the linearized equation of $\dot{\mathbf{y}} = \mathbf{f}(\mathbf{y}, \mathbf{y}^\tau)$ is

$$\Delta\dot{\mathbf{y}} = D_{\mathbf{y}}\mathbf{f}(\mathbf{y}^*, \mathbf{y}^{\tau *})\Delta\mathbf{y} + D_{\mathbf{y}^\tau}\mathbf{f}(\mathbf{y}^*, \mathbf{y}^{\tau *})\Delta\mathbf{y}^\tau$$

and the eigenvalue analysis of equilibrium $\mathbf{y}^*$ is given by

$$|D_{\mathbf{y}}\mathbf{f}(\mathbf{y}^*, \mathbf{y}^{\tau *}) - \lambda\mathbf{I}_{2n(2N+1)\times 2n(2N+1)} + D_{\mathbf{y}^\tau}\mathbf{f}(\mathbf{y}^*, \mathbf{y}^{\tau *})e^{-\lambda\tau}| = 0$$

where

$$D_{\mathbf{y}}\mathbf{f}(\mathbf{y}^*, \mathbf{y}^{\tau *}) = \frac{\partial\mathbf{f}(\mathbf{y}, \mathbf{y}^\tau)}{\partial\mathbf{y}}|_{(\mathbf{y}^*, \mathbf{y}^{\tau *})}, D_{\mathbf{y}^\tau}\mathbf{f}(\mathbf{y}^*, \mathbf{y}^{\tau *}) = \frac{\partial\mathbf{f}(\mathbf{y}, \mathbf{y}^\tau)}{\partial\mathbf{y}^\tau}|_{(\mathbf{y}^*, \mathbf{y}^{\tau *})}.$$

Thus, the stability and bifurcation of the periodic solution can be classified by the eigenvalues of Equation (3.202) at equilibrium $\mathbf{y}^* = \mathbf{y}^{\tau *}$ with

$$(n_1, n_2, n_3 \,|\, n_4, n_5, n_6).$$

From the stability and bifurcation theory of dynamical systems at equilibrium, the stability, and bifurcation of the periodic solutions can be classified as stated in the theorem. This theorem is proved. ■

If the Hopf bifurcation of equilibriums of Equation (3.202) occurs, there is a periodic solution of coefficients in Equation (3.202) with a frequency ω. As discussed from Equation (3.141) to Equation (3.147), there is a period-m flow as in Equation (3.147). Herein, the period-m flow in time-delayed, nonlinear vibration systems will be discussed only.

Theorem 3.14 *Consider a time-delayed, nonlinear vibration system in Equation (3.191). If such a time-delayed, vibration system has a period-*m *motion $\mathbf{x}^{(m)}(t)$ with finite norm $\|\mathbf{x}^{(m)}\|$ and period $T = 2\pi/\Omega$, there is a generalized coordinate transformation for the period-*m *motion of Equation (3.191) in the form of*

$$\mathbf{x}^{(m)}(t) = \mathbf{a}_0^{(m)}(t) + \sum_{k=1}^{\infty} \mathbf{b}_{k/m}(t)\cos\left(\frac{k}{m}\theta\right) + \mathbf{c}_{k/m}(t)\sin\left(\frac{k}{m}\theta\right);$$

$$\mathbf{x}^{\tau(m)}(t) = \mathbf{a}_0^{\tau(m)}(t) + \sum_{k=1}^{\infty} \mathbf{b}_{k/m}^\tau(t)\cos\left[\frac{k}{m}(\theta - \theta^\tau)\right] + \mathbf{c}_{k/m}^\tau(t)\sin\left[\frac{k}{m}(\theta - \theta^\tau)\right] \tag{3.208}$$

with

$$\begin{aligned}
\mathbf{a}_0^{(m)} &= (a_{01}^{(m)}, a_{02}^{(m)}, \ldots, a_{0n}^{(m)})^{\mathrm{T}},\\
\mathbf{b}_{k/m} &= (b_{k/m1}, b_{k/m2}, \ldots, b_{k/mn})^{\mathrm{T}},\\
\mathbf{c}_{k/m} &= (c_{k/m1}, c_{k/m2}, \ldots, c_{k/mn})^{\mathrm{T}};\\
\mathbf{a}_0^{\tau(m)} &= (a_{01}^{\tau(m)}, a_{02}^{\tau(m)}, \ldots, a_{0n}^{\tau(m)})^{\mathrm{T}},\\
\mathbf{b}_{k/m}^\tau &= (b_{k/m1}^\tau, b_{k/m2}^\tau, \ldots, b_{k/mn}^\tau)^{\mathrm{T}},\\
\mathbf{c}_{k/m}^\tau &= (c_{k/m1}^\tau, c_{k/m2}^\tau, \ldots, c_{k/mn}^\tau)^{\mathrm{T}}
\end{aligned} \tag{3.209}$$

and

$$\begin{aligned}
&\|\mathbf{x}^{(m)}\| = \|\mathbf{a}_0^{(m)}\| + \sum_{k=1}^{\infty} \|\mathbf{A}_{k/m}\|, \text{ and } \lim_{k\to\infty} \|\mathbf{A}_{k/m}\| = 0 \text{ but not uniform} \\
&\text{with } \mathbf{A}_{k/m} = (A_{k/m1}, A_{k/m2} \ldots, A_{k/mn})^{\mathrm{T}} \\
&\text{and } A_{k/mj} = \sqrt{b_{k/mj}^2 + c_{k/mj}^2} \;\; (j = 1, 2, \ldots, n);
\end{aligned} \tag{3.210}$$

$$\begin{aligned}
&\|\mathbf{x}^{\tau(m)}\| = \|\mathbf{a}_0^{\tau(m)}\| + \sum_{k=1}^{\infty} \|\mathbf{A}_{k/m}^{\tau}\|, \text{ and } \lim_{k\to\infty} \|\mathbf{A}_{k/m}^{\tau}\| = 0 \text{ but not uniform} \\
&\text{with } \mathbf{A}_{k/m}^{\tau} = (A_{k/m1}^{\tau}, A_{k/m2}^{\tau} \ldots, A_{k/mn}^{\tau})^{\mathrm{T}} \\
&\text{and } A_{k/mj}^{\tau} = \sqrt{(b_{k/mj}^{\tau})^2 + (c_{k/mj}^{\tau})^2} \;\; (j = 1, 2, \ldots, n).
\end{aligned} \tag{3.211}$$

For $\|\mathbf{x}^{(m)}(t) - \mathbf{x}^{(m)*}(t)\| < \varepsilon$ *and* $\|\mathbf{x}^{\tau(m)}(t) - \mathbf{x}^{\tau(m)*}(t)\| < \varepsilon^{\tau}$ *with prescribed small* $\varepsilon > 0$ *and* $\varepsilon^{\tau} > 0$*, the infinite transformation* $\mathbf{x}^{(m)}(t)$ *of period-*m *motion of Equation (3.191), given by Equation (3.210), can be approximated by a finite term transformation* $\mathbf{x}^{(m)*}(t)$ *as*

$$\begin{aligned}
&\mathbf{x}^{(m)*}(t) = \mathbf{a}_0^{(m)}(t) + \sum_{k=1}^{N} \mathbf{b}_{k/m}(t) \cos\left(\frac{k}{m}\theta\right) + \mathbf{c}_{k/m}(t) \sin\left(\frac{k}{m}\theta\right); \\
&\mathbf{x}^{\tau(m)*}(t) = \mathbf{a}_0^{\tau(m)}(t) + \sum_{k=1}^{N} \mathbf{b}_{k/m}^{\tau}(t) \cos\left[\frac{k}{m}(\theta - \theta^{\tau})\right] + \mathbf{c}_{k/m}^{\tau}(t) \sin\left[\frac{k}{m}(\theta - \theta^{\tau})\right]
\end{aligned} \tag{3.212}$$

and the generalized coordinates are determined by

$$\begin{aligned}
\ddot{\mathbf{a}}_0^{(m)} &= \mathbf{F}_0^{(m)}(\mathbf{z}^{(m)}, \mathbf{z}_1^{(m)}; \mathbf{z}^{\tau(m)}, \mathbf{z}_1^{\tau(m)}), \\
\ddot{\mathbf{b}}^{(m)} &= -2\frac{\Omega}{m}\mathbf{k}_1\dot{\mathbf{c}}^{(m)} + \frac{\Omega^2}{m^2}\mathbf{k}_2\mathbf{b}^{(m)} + \mathbf{F}_1^{(m)}(\mathbf{z}^{(m)}, \mathbf{z}_1^{(m)}; \mathbf{z}^{\tau(m)}, \mathbf{z}_1^{\tau(m)}), \\
\ddot{\mathbf{c}}^{(m)} &= 2\frac{\Omega}{m}\mathbf{k}_1\dot{\mathbf{b}}^{(m)} + \frac{\Omega^2}{m^2}\mathbf{k}_2\mathbf{c}^{(m)} + \mathbf{F}_2^{(m)}(\mathbf{z}^{(m)}, \mathbf{z}_1^{(m)}; \mathbf{z}^{\tau(m)}, \mathbf{z}_1^{\tau(m)})
\end{aligned} \tag{3.213}$$

where

$$\begin{aligned}
\mathbf{k}_1 &= diag(\mathbf{I}_{n\times n}, 2\mathbf{I}_{n\times n}, \ldots, N\mathbf{I}_{n\times n}), \\
\mathbf{k}_2 &= diag(\mathbf{I}_{n\times n}, 2^2\mathbf{I}_{n\times n}, \ldots, N^2\mathbf{I}_{n\times n}), \\
\mathbf{b}^{(m)} &= (\mathbf{b}_{1/m}, \mathbf{b}_{2/m}, \ldots, \mathbf{b}_{N/m})^{\mathrm{T}}, \\
\mathbf{c}^{(m)} &= (\mathbf{c}_{1/m}, \mathbf{c}_{2/m}, \ldots, \mathbf{c}_{N/m})^{\mathrm{T}}, \\
\mathbf{b}^{\tau(m)} &= (\mathbf{b}_{1/m}^{\tau}, \mathbf{b}_{2/m}^{\tau}, \ldots, \mathbf{b}_{N/m}^{\tau})^{\mathrm{T}}, \\
\mathbf{c}^{\tau(m)} &= (\mathbf{c}_{1/m}^{\tau}, \mathbf{c}_{2/m}^{\tau}, \ldots, \mathbf{c}_{N/m}^{\tau})^{\mathrm{T}}; \\
\mathbf{F}_1^{(m)} &= (\mathbf{F}_{11}^{(m)}, \mathbf{F}_{12}^{(m)}, \ldots, \mathbf{F}_{1N}^{(m)})^{\mathrm{T}},
\end{aligned}$$

$$\begin{aligned}
&\mathbf{F}_2^{(m)} = (\mathbf{F}_{21}^{(m)}, \mathbf{F}_{22}^{(m)}, \ldots, \mathbf{F}_{2N}^{(m)})^{\mathrm{T}};\\
&\mathbf{z}^{(m)} = (\mathbf{a}_0^{(m)}, \mathbf{b}^{(m)}, \mathbf{c}^{(m)})^{\mathrm{T}}, \dot{\mathbf{z}}^{(m)} = \mathbf{z}_1^{(m)};\\
&\mathbf{z}^{\tau(m)} = (\mathbf{a}_0^{\tau(m)}, \mathbf{b}^{\tau(m)}, \mathbf{c}^{\tau(m)})^{\mathrm{T}}, \dot{\mathbf{z}}^{\tau(m)} = \mathbf{z}_1^{\tau(m)}\\
&\text{for } N = 1, 2, \ldots, \infty;
\end{aligned} \tag{3.214}$$

and for $k = 1, 2, \ldots, N$

$$\begin{aligned}
&\mathbf{F}_0^{(m)}(\mathbf{z}^{(m)}, \mathbf{z}_1^{(m)}; \mathbf{z}^{\tau(m)}, \mathbf{z}_1^{\tau(m)})\\
&\quad = \frac{1}{2m\pi}\int_0^{2m\pi} \mathbf{f}(\mathbf{x}^{(m)*}, \dot{\mathbf{x}}^{(m)*}; \mathbf{x}^{\tau(m)*}, \dot{\mathbf{x}}^{\tau(m)*}, \mathbf{p})d\theta;\\
&\mathbf{F}_{1k}^{(m)}(\mathbf{z}^{(m)}, \mathbf{z}_1^{(m)}; \mathbf{z}^{\tau(m)}, \mathbf{z}_1^{\tau(m)})\\
&\quad = \frac{1}{m\pi}\int_0^{2m\pi} \mathbf{f}(\mathbf{x}^{(m)*}, \dot{\mathbf{x}}^{(m)*}; \mathbf{x}^{\tau(m)*}, \dot{\mathbf{x}}^{\tau(m)*}, \mathbf{p}) \cos\left(\frac{k}{m}\theta\right)d\theta,\\
&\mathbf{F}_{2k}^{(m)}(\mathbf{z}^{(m)}, \mathbf{z}_1^{(m)}; \mathbf{z}^{\tau(m)}, \mathbf{z}_1^{\tau(m)})\\
&\quad = \frac{1}{m\pi}\int_0^{2m\pi} \mathbf{f}(\mathbf{x}^{(m)*}, \dot{\mathbf{x}}^{(m)*}; \mathbf{x}^{\tau(m)*}, \dot{\mathbf{x}}^{\tau(m)*}, \mathbf{p}) \sin\left(\frac{k}{m}\theta\right)d\theta.
\end{aligned} \tag{3.215}$$

The state-space form of Equation (3.212) is

$$\dot{\mathbf{z}}^{(m)} = \mathbf{z}_1^{(m)} \text{ and } \dot{\mathbf{z}}_1^{(m)} = \mathbf{g}^{(m)}(\mathbf{z}^{(m)}, \mathbf{z}_1^{(m)}; \mathbf{z}^{\tau(m)}, \mathbf{z}_1^{\tau(m)}) \tag{3.216}$$

where

$$\begin{aligned}
\mathbf{g}^{(m)} = (&\mathbf{F}_0^{(m)}, -2\frac{\Omega}{m}\mathbf{k}_1\dot{\mathbf{c}}^{(m)} + \frac{\Omega^2}{m^2}\mathbf{k}_2\mathbf{b}^{(m)} + \mathbf{F}_1^{(m)},\\
&2\frac{\Omega}{m}\mathbf{k}_1\dot{\mathbf{b}}^{(m)} + \frac{\Omega^2}{m^2}\mathbf{k}_2\mathbf{c}^{(m)} + \mathbf{F}_2^{(m)})^{\mathrm{T}}.
\end{aligned} \tag{3.217}$$

An equivalent system of Equation (3.215) is

$$\dot{\mathbf{y}}^{(m)} = \mathbf{f}^{(m)}(\mathbf{y}^{(m)}, \mathbf{y}^{\tau(m)}) \tag{3.218}$$

where

$$\mathbf{y}^{(m)} = (\mathbf{z}^{(m)}, \mathbf{z}_1^{(m)})^{\mathrm{T}}, \mathbf{y}^{\tau(m)} = (\mathbf{z}^{\tau(m)}, \mathbf{z}_1^{\tau(m)})^{\mathrm{T}} \text{and } \mathbf{f}^{(m)} = (\mathbf{z}_1^{(m)}, \mathbf{g}^{(m)})^{\mathrm{T}}. \tag{3.219}$$

If equilibrium $\mathbf{y}^{(m)*} = \mathbf{y}^{\tau(m)*}$ *of Equation (3.218) (i.e.,* $\mathbf{f}(\mathbf{y}^{(m)*}, \mathbf{y}^{\tau(m)*}) = \mathbf{0}$*) exists, then the approximate solution of period-*m *motion exists as in Equation (3.212). In the vicinity of equilibrium* $\mathbf{y}^{(m)*} = \mathbf{y}^{\tau(m)*}$*, with* $\mathbf{y}^{(m)} = \mathbf{y}^{(m)*} + \Delta\mathbf{y}^{(m)}$ *and* $\mathbf{y}^{\tau(m)} = \mathbf{y}^{\tau(m)*} + \Delta\mathbf{y}^{\tau(m)}$*, the linearized equation of Equation (3.218) is*

$$\Delta\dot{\mathbf{y}}^{(m)} = D_{\mathbf{y}^{(m)}}\mathbf{f}^{(m)}(\mathbf{y}^{(m)*}, \mathbf{y}^{\tau(m)*})\Delta\mathbf{y}^{(m)} + D_{\mathbf{y}^{\tau(m)}}\mathbf{f}^{(m)}(\mathbf{y}^{(m)*}, \mathbf{y}^{\tau(m)*})\Delta\mathbf{y}^{\tau(m)} \tag{3.220}$$

and the eigenvalue analysis of equilibrium $\mathbf{y}^*$ *is given by*

$$|D_{\mathbf{y}^{(m)}}\mathbf{f}^{(m)}(\mathbf{y}^{(m)*}, \mathbf{y}^{\tau(m)*}) - \lambda\mathbf{I}_{2n(2N+1)\times 2n(2N+1)} + D_{\mathbf{y}^{\tau(m)}}\mathbf{f}^{(m)}(\mathbf{y}^{(m)*}, \mathbf{y}^{\tau(m)*})e^{-\lambda\tau}| = 0 \tag{3.221}$$

where

$$D_{\mathbf{y}^{(m)}}\mathbf{f}^{(m)}(\mathbf{y}^{(m)*},\mathbf{y}^{\tau(m)*}) = \frac{\partial \mathbf{f}^{(m)}(\mathbf{y}^{(m)},\mathbf{y}^{\tau(m)})}{\partial \mathbf{y}^{(m)}}\Big|_{(\mathbf{y}^{(m)*},\mathbf{y}^{\tau(m)*})},$$

$$D_{\mathbf{y}^{\tau(m)}}\mathbf{f}^{(m)}(\mathbf{y}^{(m)*},\mathbf{y}^{\tau(m)*}) = \frac{\partial \mathbf{f}^{(m)}(\mathbf{y}^{(m)},\mathbf{y}^{\tau(m)})}{\partial \mathbf{y}^{\tau(m)}}\Big|_{(\mathbf{y}^{(m)*},\mathbf{y}^{\tau(m)*})}. \tag{3.222}$$

*Thus, the stability and bifurcation of period-*m *motion can be classified by the eigenvalues of Equation (3.220) at equilibrium* $\mathbf{y}^{(m)*} = \mathbf{y}^{\tau(m)*}$ *with*

$$(n_1, n_2, n_3 \mid n_4, n_5, n_6). \tag{3.223}$$

1. *If all eigenvalues of the equilibrium possess negative real parts, the approximate periodic solution is stable.*
2. *If at least one of the eigenvalues of the equilibrium possesses a positive real part, the approximate periodic solution is unstable.*
3. *The boundaries between stable and unstable equilibriums with higher order singularity give bifurcation and stability conditions y with higher order singularity.*

Proof. The proof is similar to the proof of Theorem 3.13. Since $\mathbf{f}(\mathbf{x},\dot{\mathbf{x}};\mathbf{x}^{\tau},\dot{\mathbf{x}}^{\tau},\mathbf{p})$ is a C^r-continuous nonlinear function vector ($r \geq 1$), the velocity $\dot{\mathbf{x}}$ should be C^r-continuous ($r \geq 1$), and then the acceleration $\ddot{\mathbf{x}}$ should be bounded (i.e., $\|\ddot{\mathbf{x}}\| < K$). From Equation (3.208), the norms of the periodic motion are defined by

$$\|\mathbf{x}^{(m)}\| = \|\mathbf{a}_0^{(m)}\| + \sum_{k=1}^{\infty}\|\mathbf{A}_{k/m}\| \text{ and } \|\mathbf{x}^{\tau(m)}\| = \|\mathbf{a}_0^{\tau(m)}\| + \sum_{k=1}^{\infty}\|\mathbf{A}_{k/m}^{\tau}\|$$

where

$$\begin{aligned}
\mathbf{A}_{k/m} &= (A_{k/m1}, A_{k/m2}, \ldots, A_{k/mn})^{\mathrm{T}},\\
A_{k/mj} &= \sqrt{b_{k/mj}^2 + c_{k/mj}^2}\ \ (j = 1, 2, \ldots, n),\\
\mathbf{A}_{k/m}^{\tau} &= (A_{k/m1}^{\tau}, A_{k/m2}^{\tau}, \ldots, A_{k/mn}^{\tau})^{\mathrm{T}},\\
A_{k/mj}^{\tau} &= \sqrt{(b_{k/mj}^{\tau})^2 + (c_{k/mj}^{\tau})^2}\ \ (j = 1, 2, \ldots, n).
\end{aligned}$$

Because the periodic motion in Equation (3.191) is bounded (i.e., $\|\mathbf{x}(t)\| < C$), we have

$$\lim_{k\to\infty}\|\mathbf{A}_{k/m}\| = 0 \text{ and } \lim_{k\to\infty}\|\mathbf{A}_{k/m}^{\tau}\| = 0 \text{ but not uniform.}$$

Thus, the Fourier series transformation of periodic motion as in Equation (3.208) is convergent. From Equations (3.210) and (3.211), using Equations (3.208) and (3.212) gives

$$\begin{aligned}
\|\mathbf{x}^{(m)}(t) - \mathbf{x}^{(m)*}(t)\| &= \left\|\sum_{k=N+1}^{\infty}\mathbf{b}_{k/m}(t)\cos\left(\frac{k}{m}\theta\right) + \mathbf{c}_{k/m}(t)\sin\left(\frac{k}{m}\theta\right)\right\|\\
&= \sum_{k=N+1}^{\infty}\|\mathbf{A}_{k/m}\|,
\end{aligned}$$

$$\begin{aligned}
&\|\mathbf{x}^{\tau(m)}(t) - \mathbf{x}^{\tau(m)*}(t)\| \\
&= \left\| \sum_{k=N+1}^{\infty} \mathbf{b}^{\tau}_{k/m}(t) \cos\left[\frac{k}{m}(\theta - \theta^{\tau})\right] + \mathbf{c}^{\tau}_{k/m}(t) \sin\left[\frac{k}{m}(\theta - \theta^{\tau})\right] \right\| \\
&= \sum_{k=N+1}^{\infty} \|\mathbf{A}^{\tau}_{k/m}\|.
\end{aligned}$$

For the prescribed small positive $\varepsilon > 0$ and $\varepsilon^{\tau} > 0$, if $\|\mathbf{x}^{(m)}(t) - \mathbf{x}^{(m)*}(t)\| < \varepsilon$ and $\|\mathbf{x}^{\tau(m)}(t) - \mathbf{x}^{\tau(m)*}(t)\| < \varepsilon^{\tau}$ exist, then we have

$$\sum_{k=N+1}^{\infty} \|\mathbf{A}_{k/m}\| < \varepsilon \text{ and } \sum_{k=N+1}^{\infty} \|\mathbf{A}_{k/m}\| < \varepsilon^{\tau}.$$

Therefore, the convergent infinite term transformation in Equation (3.208) can be approximated by a finite term transformation in Equation (3.212) in the sense of ε.

Taking the derivatives of Equation (3.212) gives

$$\begin{aligned}
\dot{\mathbf{x}}^{(m)*}(t) &= \dot{\mathbf{a}}_0^{(m)} + \sum_{k=1}^{N} \left(\dot{\mathbf{b}}_{k/m} + \frac{k}{m}\Omega\mathbf{c}_{k/m}\right)\cos\left(\frac{k}{m}\theta\right) \\
&\quad + \left(\dot{\mathbf{c}}_{k/m} - \frac{k}{m}\Omega\mathbf{b}_{k/m}\right)\sin\left(\frac{k}{m}\theta\right), \\
\ddot{\mathbf{x}}^{(m)*}(t) &= \ddot{\mathbf{a}}_0^{(m)} + \sum_{k=1}^{N} \left(\ddot{\mathbf{b}}_{k/m} + 2\frac{k}{m}\Omega\dot{\mathbf{c}}_{k/m} - \frac{k^2}{m^2}\Omega^2\mathbf{b}_{k/m}\right)\cos\left(\frac{k}{m}\theta\right) \\
&\quad + \left(\ddot{\mathbf{c}}_{k/m} - 2\frac{k}{m}\Omega\dot{\mathbf{b}}_{k/m} - \frac{k^2}{m^2}\Omega^2\mathbf{c}_{k/m}\right)\sin\left(\frac{k}{m}\theta\right);
\end{aligned}$$

$$\begin{aligned}
\dot{\mathbf{x}}^{\tau(m)*}(t) &= \dot{\mathbf{a}}_0^{\tau(m)} + \sum_{k=1}^{N} \left(\dot{\mathbf{b}}^{\tau}_{k/m} + \frac{k}{m}\Omega\mathbf{c}^{\tau}_{k/m}\right)\cos\left[\frac{k}{m}(\theta - \theta^{\tau})\right] \\
&\quad + \left(\dot{\mathbf{c}}^{\tau}_{k/m} - \frac{k}{m}\Omega\mathbf{b}^{\tau}_{k/m}\right)\sin\left[\frac{k}{m}(\theta - \theta^{\tau})\right],
\end{aligned}$$

$$\begin{aligned}
\ddot{\mathbf{x}}^{\tau(m)*}(t) &= \ddot{\mathbf{a}}_0^{\tau(m)} + \sum_{k=1}^{N} \left(\ddot{\mathbf{b}}^{\tau}_{k/m} + 2\frac{k}{m}\Omega\dot{\mathbf{c}}^{\tau}_{k/m} - \frac{k^2}{m^2}\Omega^2\mathbf{b}^{\tau}_{k/m}\right)\cos\left[\frac{k}{m}(\theta - \theta^{\tau})\right] \\
&\quad + \left(\ddot{\mathbf{c}}^{\tau}_{k/m} - 2\frac{k}{m}\Omega\dot{\mathbf{b}}^{\tau}_{k/m} - \frac{k^2}{m^2}\Omega^2\mathbf{c}^{\tau}_{k/m}\right)\sin\left[\frac{k}{m}(\theta - \theta^{\tau})\right]
\end{aligned}$$

where $\mathbf{a}_0^{\tau} = \mathbf{a}_0(t-\tau), \mathbf{b}_k^{\tau} = \mathbf{b}_k(t-\tau), \mathbf{c}_k^{\tau} = \mathbf{c}_k(t-\tau), \theta^{\tau} = \Omega\tau$. Substitution of the foregoing equation into Equation (3.192), and application of the virtual work principle for constant, $\cos(k\theta/m)$ and $\sin(k\theta/m)$ $(k = 1, 2, \ldots)$ as a set of virtual displacements gives

$$\frac{1}{2m\pi}\int_0^{2m\pi} [\ddot{\mathbf{x}}^{(m)} - \mathbf{f}(\mathbf{x}^{(m)}, \dot{\mathbf{x}}^{(m)}; \mathbf{x}^{\tau(m)}, \dot{\mathbf{x}}^{\tau(m)}, \mathbf{p})]d\theta = 0,$$

$$\frac{1}{m\pi}\int_0^{2m\pi}[\ddot{\mathbf{x}}^{(m)}-\mathbf{f}(\mathbf{x}^{(m)},\dot{\mathbf{x}}^{(m)};\mathbf{x}^{\tau(m)},\dot{\mathbf{x}}^{\tau(m)},\mathbf{p})]\cos\left(\frac{k}{m}\theta\right)d\theta=0,$$
$$\frac{1}{m\pi}\int_0^{2m\pi}[\ddot{\mathbf{x}}^{(m)}-\mathbf{f}(\mathbf{x}^{(m)},\dot{\mathbf{x}}^{(m)};\mathbf{x}^{\tau(m)},\dot{\mathbf{x}}^{\tau(m)},\mathbf{p})]\sin\left(\frac{k}{m}\theta\right)d\theta=0.$$

Under $\|\mathbf{x}^{(m)}-\mathbf{x}^{(m)*}\|<\varepsilon$ and $\|\mathbf{x}^{\tau(m)}-\mathbf{x}^{\tau(m)*}\|<\varepsilon^{\tau}$ with continuity $\|\ddot{\mathbf{x}}^{(m)}\|<K$ (constant) and small $\varepsilon>0$ and $\varepsilon^{\tau}>0$, for $k=1,2,\dots,N$, the foregoing equation gives

$$\frac{1}{2m\pi}\int_0^{2m\pi}[\ddot{\mathbf{x}}^{(m)*}-\mathbf{f}(\mathbf{x}^{(m)*},\dot{\mathbf{x}}^{(m)*};\mathbf{x}^{\tau(m)*},\dot{\mathbf{x}}^{\tau(m)*},\mathbf{p})]d\theta+O(\delta)=0,$$
$$\frac{1}{m\pi}\int_0^{2m\pi}[\ddot{\mathbf{x}}^{(m)*}-\mathbf{f}(\mathbf{x}^{(m)*},\dot{\mathbf{x}}^{(m)*};\mathbf{x}^{\tau(m)*},\dot{\mathbf{x}}^{\tau(m)*},\mathbf{p})]\cos\left(\frac{k}{m}\theta\right)d\theta+O(\delta)=0,$$
$$\frac{1}{m\pi}\int_0^{2m\pi}[\ddot{\mathbf{x}}^{(m)*}-\mathbf{f}(\mathbf{x}^{(m)*},\dot{\mathbf{x}}^{(m)*};\mathbf{x}^{\tau(m)*},\dot{\mathbf{x}}^{\tau(m)*},\mathbf{p})]\sin\left(\frac{k}{m}\theta\right)d\theta+O(\delta)=0$$

where $\delta=\max(\varepsilon,\varepsilon^{\tau},\varepsilon_t,\varepsilon_t^{\tau},\varepsilon_{tt})$ with $\|\dot{\mathbf{x}}^{(m)}-\dot{\mathbf{x}}^{(m)*}\|<\varepsilon_t$, $\|\dot{\mathbf{x}}^{\tau(m)}-\dot{\mathbf{x}}^{\tau(m)*}\|<\varepsilon_t^{\tau}$, $\|\ddot{\mathbf{x}}^{(m)}-\ddot{\mathbf{x}}^{(m)*}\|<\varepsilon_{tt}$ for small $\{\varepsilon_t,\varepsilon_t^{\tau},\varepsilon_{tt}\}>0$. The foregoing equation generates

$$\ddot{\mathbf{a}}_0^{(m)}=\mathbf{F}_0^{(m)}(\mathbf{z}^{(m)},\mathbf{z}_1^{(m)};\mathbf{z}^{\tau(m)},\mathbf{z}_1^{\tau(m)}),$$
$$\ddot{\mathbf{b}}_{k/m}=-2\frac{\Omega}{m}k\dot{\mathbf{c}}_{k/m}+\frac{\Omega^2}{m^2}k^2\mathbf{b}_{k/m}+\mathbf{F}_{1k}^{(m)}(\mathbf{z}^{(m)},\mathbf{z}_1^{(m)};\mathbf{z}^{\tau(m)},\mathbf{z}_1^{\tau(m)}),$$
$$\ddot{\mathbf{c}}_{k/m}=2\frac{\Omega}{m}k\dot{\mathbf{b}}_{k/m}+\frac{\Omega^2}{m^2}k^2\mathbf{c}_{k/m}+\mathbf{F}_{2k}^{(m)}(\mathbf{z}^{(m)},\mathbf{z}_1^{(m)};\mathbf{z}^{\tau(m)},\mathbf{z}_1^{\tau(m)})$$

where for $k=1,2,\dots,N$

$$\mathbf{F}_0^{(m)}(\mathbf{z}^{(m)},\mathbf{z}_1^{(m)};\mathbf{z}^{\tau(m)},\mathbf{z}_1^{\tau(m)})=\frac{1}{2m\pi}\int_0^{2m\pi}\mathbf{f}(\mathbf{x}^{(m)*},\dot{\mathbf{x}}^{(m)*};\mathbf{x}^{\tau(m)*},\dot{\mathbf{x}}^{\tau(m)*},\mathbf{p})d\theta;$$
$$\mathbf{F}_{1k}^{(m)}(\mathbf{z}^{(m)},\mathbf{z}_1^{(m)};\mathbf{z}^{\tau(m)},\mathbf{z}_1^{\tau(m)})=\frac{1}{m\pi}\int_0^{2m\pi}\mathbf{f}(\mathbf{x}^{(m)*},\dot{\mathbf{x}}^{(m)*};\mathbf{x}^{\tau(m)*},\dot{\mathbf{x}}^{\tau(m)*},\mathbf{p})\cos\left(\frac{k}{m}\theta\right)d\theta,$$
$$\mathbf{F}_{2k}^{(m)}(\mathbf{z}^{(m)},\mathbf{z}_1^{(m)};\mathbf{z}^{\tau(m)},\mathbf{z}_1^{\tau(m)})=\frac{1}{m\pi}\int_0^{2m\pi}\mathbf{f}(\mathbf{x}^{(m)*},\dot{\mathbf{x}}^{(m)*};\mathbf{x}^{\tau(m)*},\dot{\mathbf{x}}^{\tau(m)*},\mathbf{p})\sin\left(\frac{k}{m}\theta\right)d\theta;$$

and

$$\mathbf{k}_1=diag(\mathbf{I}_{n\times n},2\mathbf{I}_{n\times n},\dots,N\mathbf{I}_{n\times n}),$$
$$\mathbf{k}_2=diag(\mathbf{I}_{n\times n},2^2\mathbf{I}_{n\times n},\dots,N^2\mathbf{I}_{n\times n});$$

$$
\begin{aligned}
&\mathbf{b}^{(m)} = (\mathbf{b}_{1/m}, \mathbf{b}_{2/m}, \ldots, \mathbf{b}_{N/m})^{\mathrm{T}},\\
&\mathbf{c}^{(m)} = (\mathbf{c}_{1/m}, \mathbf{c}_{2/m}, \ldots, \mathbf{c}_{N/m})^{\mathrm{T}};\\
&\mathbf{b}^{\tau(m)} = (\mathbf{b}^{\tau}_{1/m}, \mathbf{b}^{\tau}_{2/m}, \ldots, \mathbf{b}^{\tau}_{N/m})^{\mathrm{T}},\\
&\mathbf{c}^{\tau(m)} = (\mathbf{c}^{\tau}_{1/m}, \mathbf{c}^{\tau}_{2/m}, \ldots, \mathbf{c}^{\tau}_{N/m})^{\mathrm{T}};\\
&\mathbf{F}_1^{(m)} = (\mathbf{F}_{11}^{(m)}, \mathbf{F}_{12}^{(m)}, \ldots, \mathbf{F}_{1N}^{(m)})^{\mathrm{T}},\\
&\mathbf{F}_2^{(m)} = (\mathbf{F}_{21}^{(m)}, \mathbf{F}_{22}^{(m)}, \ldots, \mathbf{F}_{2N}^{(m)})^{\mathrm{T}};\\
&\mathbf{z}^{(m)} = (\mathbf{a}_0^{(m)}, \mathbf{b}^{(m)}, \mathbf{c}^{(m)})^{\mathrm{T}}, \dot{\mathbf{z}}^{(m)} = \mathbf{z}_1^{(m)};\\
&\mathbf{z}^{\tau(m)} = (\mathbf{a}_0^{\tau(m)}, \mathbf{b}^{\tau(m)}, \mathbf{c}^{\tau(m)})^{\mathrm{T}}, \dot{\mathbf{z}}^{\tau(m)} = \mathbf{z}_1^{\tau(m)}\\
&\text{for } N = 1, 2, \ldots, \infty.
\end{aligned}
$$

Rearranging the foregoing equation gives Equation (3.213), that is,

$$
\begin{aligned}
\ddot{\mathbf{a}}_0^{(m)} &= \mathbf{F}_0^{(m)}(\mathbf{z}^{(m)}, \mathbf{z}_1^{(m)}; \mathbf{z}^{\tau(m)}, \mathbf{z}_1^{\tau(m)}),\\
\ddot{\mathbf{b}}^{(m)} &= -2\frac{\Omega}{m}\mathbf{k}_1\dot{\mathbf{c}}^{(m)} + \frac{\Omega^2}{m^2}\mathbf{k}_2\mathbf{b}^{(m)} + \mathbf{F}_1^{(m)}(\mathbf{z}^{(m)}, \mathbf{z}_1^{(m)}; \mathbf{z}^{\tau(m)}, \mathbf{z}_1^{\tau(m)}),\\
\ddot{\mathbf{c}}^{(m)} &= 2\frac{\Omega}{m}\mathbf{k}_1\dot{\mathbf{b}}^{(m)} + \frac{\Omega^2}{m^2}\mathbf{k}_2\mathbf{c}^{(m)} + \mathbf{F}_2^{(m)}(\mathbf{z}^{(m)}, \mathbf{z}_1^{(m)}; \mathbf{z}^{\tau(m)}, \mathbf{z}_1^{\tau(m)})
\end{aligned}
$$

Introduce

$$
\begin{aligned}
\mathbf{g}^{(m)} = (\mathbf{F}_0^{(m)}, &-2\frac{\Omega}{m}\mathbf{k}_1\dot{\mathbf{c}}^{(m)} + \frac{\Omega^2}{m^2}\mathbf{k}_2\mathbf{b}^{(m)} + \mathbf{F}_1^{(m)},\\
&2\frac{\Omega}{m}\mathbf{k}_1\dot{\mathbf{b}}^{(m)} + \frac{\Omega^2}{m^2}\mathbf{k}_2\mathbf{c}^{(m)} + \mathbf{F}_2^{(m)})^{\mathrm{T}}.
\end{aligned}
$$

The equation in Equation (3.213) becomes

$$
\dot{\mathbf{z}}^{(m)} = \mathbf{z}_1^{(m)} \text{ and } \dot{\mathbf{z}}_1^{(m)} = \mathbf{g}^{(m)}(\mathbf{z}^{(m)}, \mathbf{z}_1^{(m)}; \mathbf{z}^{\tau(m)}, \mathbf{z}_1^{\tau(m)}).
$$

Letting

$$
\mathbf{y}^{(m)} = (\mathbf{z}^{(m)}, \mathbf{z}_1^{(m)})^{\mathrm{T}}, \mathbf{y}^{\tau(m)} = (\mathbf{z}^{\tau(m)}, \mathbf{z}_1^{\tau(m)})^{\mathrm{T}}, \mathbf{f}^{(m)} = (\mathbf{z}_1^{(m)}, \mathbf{g}^{(m)})^{\mathrm{T}}
$$

we have

$$
\dot{\mathbf{y}}^{(m)} = \mathbf{f}^{(m)}(\mathbf{y}^{(m)}, \mathbf{y}^{\tau(m)}).
$$

Consider equilibriums of the foregoing equation (i.e., $\mathbf{f}^{(m)}(\mathbf{y}^{(m)*}, \mathbf{y}^{\tau(m)*}) = \mathbf{0}$), that is,

$$
\begin{aligned}
&\mathbf{0} = \mathbf{F}_0(\mathbf{a}_0^{(m)*}, \mathbf{b}^{(m)*}, \mathbf{c}^{(m)*}, \mathbf{0}, \mathbf{0}, \mathbf{0}; \mathbf{a}_0^{\tau(m)*}, \mathbf{b}^{\tau(m)*}, \mathbf{c}^{\tau(m)*}, \mathbf{0}, \mathbf{0}, \mathbf{0}),\\
&\mathbf{0} = \Omega^2\mathbf{k}_2\mathbf{b}^* + \mathbf{F}_1(\mathbf{a}_0^{(m)*}, \mathbf{b}^{(m)*}, \mathbf{c}^{(m)*}, \mathbf{0}, \mathbf{0}, \mathbf{0}; \mathbf{a}_0^{\tau(m)*}, \mathbf{b}^{\tau(m)*}, \mathbf{c}^{\tau(m)*}, \mathbf{0}, \mathbf{0}, \mathbf{0}),\\
&\mathbf{0} = \Omega^2\mathbf{k}_2\mathbf{c}^* + \mathbf{F}_2(\mathbf{a}_0^{(m)*}, \mathbf{b}^{(m)*}, \mathbf{c}^{(m)*}, \mathbf{0}, \mathbf{0}, \mathbf{0}; \mathbf{a}_0^{\tau(m)*}, \mathbf{b}^{\tau(m)*}, \mathbf{c}^{\tau(m)*}, \mathbf{0}, \mathbf{0}, \mathbf{0})
\end{aligned}
$$

with

$$\mathbf{a}_0^{(m)*} = \mathbf{a}_0^{\tau(m)*}, \mathbf{b}^{(m)*} = \mathbf{b}^{\tau(m)*}, \mathbf{c}^{(m)*} = \mathbf{c}^{\tau(m)*}.$$

Thus, the solutions of the foregoing equation are the existence conditions of the periodic solutions for time-delayed, nonlinear vibration systems. If the foregoing equation gives equilibrium $\mathbf{y}^{(m)*} = \mathbf{y}^{\tau(m)*}$, in vicinity of $\mathbf{y}^{(m)*} = \mathbf{y}^{\tau(m)*}$, with $\mathbf{y}^{(m)} = \mathbf{y}^{(m)*} + \Delta\mathbf{y}^{(m)}$ and $\mathbf{y}^{\tau(m)} = \mathbf{y}^{\tau(m)*} + \Delta\mathbf{y}^{\tau(m)}$, the linearized equation of $\dot{\mathbf{y}}^{(m)} = \mathbf{f}^{(m)}(\mathbf{y}^{(m)}, \mathbf{y}^{\tau(m)})$ is

$$\Delta\dot{\mathbf{y}}^{(m)} = D_{\mathbf{y}^{(m)}}\mathbf{f}^{(m)}(\mathbf{y}^{(m)*}, \mathbf{y}^{\tau(m)*})\Delta\mathbf{y}^{(m)} + D_{\mathbf{y}^{\tau(m)}}\mathbf{f}^{(m)}(\mathbf{y}^{(m)*}, \mathbf{y}^{\tau(m)*})\Delta\mathbf{y}^{\tau(m)}$$

and the eigenvalue analysis of equilibrium $\mathbf{y}^*$ is given by

$$|D_{\mathbf{y}^{(m)}}\mathbf{f}^{(m)}(\mathbf{y}^{(m)*}, \mathbf{y}^{\tau(m)*}) - \lambda\mathbf{I}_{2n(2N+1)\times 2n(2N+1)} + D_{\mathbf{y}^{\tau(m)}}\mathbf{f}^{(m)}(\mathbf{y}^{(m)*}, \mathbf{y}^{\tau(m)*})e^{-\lambda\tau}| = 0$$

where

$$D_{\mathbf{y}^{(m)}}\mathbf{f}^{(m)}(\mathbf{y}^{(m)*}, \mathbf{y}^{\tau(m)*}) = \frac{\partial\mathbf{f}^{(m)}(\mathbf{y}^{(m)}, \mathbf{y}^{\tau(m)})}{\partial\mathbf{y}^{(m)}}\Big|_{(\mathbf{y}^{(m)*}, \mathbf{y}^{\tau(m)*})},$$

$$D_{\mathbf{y}^{\tau(m)}}\mathbf{f}^{(m)}(\mathbf{y}^{(m)*}, \mathbf{y}^{\tau(m)*}) = \frac{\partial\mathbf{f}^{(m)}(\mathbf{y}^{(m)}, \mathbf{y}^{\tau(m)})}{\partial\mathbf{y}^{\tau(m)}}\Big|_{(\mathbf{y}^{(m)*}, \mathbf{y}^{\tau(m)*})}.$$

The periodic solution stability and bifurcation can be classified by the eigenvalues of Equation (3.220) at equilibrium $\mathbf{y}^{(m)*} = \mathbf{y}^{\tau(m)*}$ with

$$(n_1, n_2, n_3 \mid n_4, n_5, n_6).$$

From the stability and bifurcation theory of dynamical systems at such equilibrium, the stability, and bifurcation of the periodic solutions can be classified as stated in the theorem. This theorem is proved. ■

Similarly, if the Hopf bifurcation of period-m motion occurs, the period-doubling solution of period-m motion can be expressed by

$$\mathbf{x}^{(2m)}(t) = \mathbf{a}_0^{(2m)}(t) + \sum_{k=1}^{\infty} \mathbf{b}_{k/2m}(t)\cos\left(\frac{k}{2m}\theta\right) + \mathbf{c}_{k/2m}(t)\sin\left(\frac{k}{2m}\theta\right),$$

$$\mathbf{x}^{\tau(2m)}(t) = \mathbf{a}_0^{\tau(2m)}(t) + \sum_{k=1}^{\infty} \mathbf{b}_{k/2m}^{\tau}(t)\cos\left[\frac{k}{2m}(\theta - \theta^{\tau})\right]$$
$$+ \mathbf{c}_{k/2m}^{\tau}(t)\sin\left[\frac{k}{2m}(\theta - \theta^{\tau})\right]. \quad (3.224)$$

Under the conditions of $||\mathbf{x}^{(2m)}(t) - \mathbf{x}^{(2m)*}(t)|| < \varepsilon$ and $||\mathbf{x}^{\tau(2m)}(t) - \mathbf{x}^{\tau(2m)*}(t)|| < \varepsilon^{\tau}$ with prescribed small $\varepsilon > 0$ and $\varepsilon^{\tau} > 0$, the period-$2m$ motion in time-delayed, nonlinear vibration systems can be approximated by

$$\mathbf{x}^{(2m)*}(t) = \mathbf{a}_0^{(2m)}(t) + \sum_{k=1}^{N} \mathbf{b}_{k/2m}(t)\cos\left(\frac{k}{2m}\theta\right) + \mathbf{c}_{k/2m}(t)\sin\left(\frac{k}{2m}\theta\right),$$

$$\mathbf{x}^{\tau(2m)*}(t) = \mathbf{a}_0^{\tau(2m)}(t) + \sum_{k=1}^{N} \mathbf{b}^{\tau}_{k/2m}(t)\cos\left[\frac{k}{2m}(\theta-\theta^{\tau})\right] + \mathbf{c}^{\tau}_{k/2m}(t)\sin\left[\frac{k}{2m}(\theta-\theta^{\tau})\right] \tag{3.225}$$

where for $k = 1, 2, \ldots, N$

$$\begin{aligned}
&\ddot{\mathbf{a}}_0^{(2m)} = \mathbf{F}_0^{(2m)}(\mathbf{z}^{(2m)}, \mathbf{z}_1^{(2m)}; \mathbf{z}^{\tau(2m)}, \mathbf{z}_1^{\tau(2m)}),\\
&\ddot{\mathbf{b}}^{(2m)} + 2\frac{\Omega}{(2m)}\mathbf{k}_1\dot{\mathbf{c}}^{(2m)} - \frac{\Omega^2}{(2m)^2}\mathbf{k}_2\mathbf{b}^{(2m)}\\
&\quad = \mathbf{F}_{1k}^{(m)}(\mathbf{z}^{(2m)}, \mathbf{z}_1^{(2m)}; \mathbf{z}^{\tau(2m)}, \mathbf{z}_1^{\tau(2m)}),\\
&\ddot{\mathbf{c}}^{(2m)} - 2\frac{\Omega}{(2m)}\mathbf{k}_1\dot{\mathbf{b}}^{(2m)} - \frac{\Omega^2}{(2m)^2}\mathbf{k}_2\mathbf{c}^{(2m)}\\
&\quad = \mathbf{F}_{2k}^{(m)}(\mathbf{z}^{(2m)}, \mathbf{z}_1^{(2m)}; \mathbf{z}^{\tau(2m)}, \mathbf{z}_1^{\tau(2m)})
\end{aligned} \tag{3.226}$$

where for $N = 1, 2, \ldots, \infty$

$$\begin{aligned}
&\mathbf{k}_1 = diag(\mathbf{I}_{n\times n}, 2\mathbf{I}_{n\times n}, \ldots, N\mathbf{I}_{n\times n}),\\
&\mathbf{k}_2 = diag(\mathbf{I}_{n\times n}, 2^2\mathbf{I}_{n\times n}, \ldots, N^2\mathbf{I}_{n\times n});\\
&\mathbf{b}^{(2m)} = (\mathbf{b}_{1/2m}, \mathbf{b}_{2/2m}, \ldots, \mathbf{b}_{N/2m})^{\mathrm{T}},\\
&\mathbf{c}^{(2m)} = (\mathbf{c}_{1/2m}, \mathbf{c}_{2/2m}, \ldots, \mathbf{c}_{N/2m})^{\mathrm{T}};\\
&\mathbf{b}^{\tau(2m)} = (\mathbf{b}^{\tau}_{1/2m}, \mathbf{b}^{\tau}_{2/2m}, \ldots, \mathbf{b}^{\tau}_{N/2m})^{\mathrm{T}},\\
&\mathbf{c}^{\tau(2m)} = (\mathbf{c}^{\tau}_{1/2m}, \mathbf{c}^{\tau}_{2/2m}, \ldots, \mathbf{c}^{\tau}_{N/2m})^{\mathrm{T}};\\
&\mathbf{F}_1^{(2m)} = (\mathbf{F}_{11}^{(2m)}, \mathbf{F}_{12}^{(2m)}, \ldots, \mathbf{F}_{1N}^{(2m)})^{\mathrm{T}},\\
&\mathbf{F}_2^{(2m)} = (\mathbf{F}_{21}^{(2m)}, \mathbf{F}_{22}^{(2m)}, \ldots, \mathbf{F}_{2N}^{(2m)})^{\mathrm{T}};\\
&\mathbf{z}^{(2m)} = (\mathbf{a}_0^{(2m)}, \mathbf{b}^{(2m)}, \mathbf{c}^{(2m)})^{\mathrm{T}}, \dot{\mathbf{z}}^{(2m)} = \mathbf{z}_1^{(2m)};\\
&\mathbf{z}^{\tau(2m)} = (\mathbf{a}_0^{\tau(2m)}, \mathbf{b}^{\tau(2m)}, \mathbf{c}^{\tau(2m)})^{\mathrm{T}}, \dot{\mathbf{z}}^{\tau(2m)} = \mathbf{z}_1^{\tau(2m)}
\end{aligned} \tag{3.227}$$

and for $k = 1, 2, \ldots, N$

$$\begin{aligned}
&\mathbf{F}_0^{(2m)}(\mathbf{z}^{(2m)}, \mathbf{z}_1^{(2m)}; \mathbf{z}^{\tau(2m)}, \mathbf{z}_1^{\tau(2m)})\\
&\quad = \frac{1}{2(2m\pi)}\int_0^{2(2m\pi)} \mathbf{f}(\mathbf{x}^{(2m)*}, \dot{\mathbf{x}}^{(2m)*}; \mathbf{x}^{\tau(2m)*}, \dot{\mathbf{x}}^{\tau(2m)*}, \mathbf{p})d\theta;
\end{aligned}$$

$$
\begin{aligned}
&\mathbf{F}^{(2m)}_{1k}(\mathbf{z}^{(2m)},\mathbf{z}^{(2m)}_1;\mathbf{z}^{\tau(2m)},\mathbf{z}^{\tau(2m)}_1)\\
&=\frac{1}{2m\pi}\int_0^{2(2m\pi)}\mathbf{f}(\mathbf{x}^{(2m)*},\dot{\mathbf{x}}^{(2m)*};\mathbf{x}^{\tau(2m)*},\dot{\mathbf{x}}^{\tau(2m)*},\mathbf{p})\cos\left(\frac{k}{2m}\theta\right)d\theta,\\
&\mathbf{F}^{(2m)}_{2k}(\mathbf{z}^{(2m)},\mathbf{z}^{(2m)}_1;\mathbf{z}^{\tau(2m)},\mathbf{z}^{\tau(2m)}_1)\\
&=\frac{1}{2m\pi}\int_0^{2(2m\pi)}\mathbf{f}(\mathbf{x}^{(2m)*},\dot{\mathbf{x}}^{(2m)*};\mathbf{x}^{\tau(2m)*},\dot{\mathbf{x}}^{\tau(2m)*},\mathbf{p})\sin\left(\frac{k}{2m}\theta\right)dt.
\end{aligned}
\tag{3.228}
$$

If the Hopf bifurcation of period-$2m$ motion occurs again and again, the analytical expression for period-$2^l m$ solutions can be expressed by

$$
\begin{aligned}
\mathbf{x}^{(2^l m)}(t)&=\mathbf{a}_0^{(2^l m)}(t)+\sum_{k=1}^{\infty}\mathbf{b}_{k/2^l m}(t)\cos\left(\frac{k}{2^l m}\theta\right)+\mathbf{c}_{k/2^l m}(t)\sin\left(\frac{k}{2^l m}\theta\right),\\
\mathbf{x}^{\tau(2^l m)}(t)&=\mathbf{a}_0^{\tau(2^l m)}(t)+\sum_{k=1}^{\infty}\mathbf{b}^{\tau}_{k/2^l m}(t)\cos\left[\frac{k}{2^l m}(\theta-\theta^{\tau})\right]\\
&\quad+\mathbf{c}^{\tau}_{k/2^l m}(t)\sin\left[\frac{k}{2^l m}(\theta-\theta^{\tau})\right],
\end{aligned}
\tag{3.229}
$$

Under conditions of $\|\mathbf{x}^{(2^l m)}(t)-\mathbf{x}^{(2^l m)*}(t)\|<\varepsilon$ and $\|\mathbf{x}^{\tau(2^l m)}(t)-\mathbf{x}^{\tau(2^l m)*}(t)\|<\varepsilon^{\tau}$ with prescribed small $\varepsilon>0$ and $\varepsilon^{\tau}>0$, the solution of period-$2^l m$ motion in the time-delayed, nonlinear vibration systems can be approximated by

$$
\begin{aligned}
\mathbf{x}^{(2^l m)*}(t)&=\mathbf{a}_0^{(2^l m)}(t)+\sum_{k=1}^{N}\mathbf{b}_{k/2^l m}(t)\cos\left(\frac{k}{2^l m}\theta\right)+\mathbf{c}_{k/2^l m}(t)\sin\left(\frac{k}{2^l m}\theta\right),\\
\mathbf{x}^{\tau(2^l m)*}(t)&=\mathbf{a}_0^{\tau(2^l m)}(t)+\sum_{k=1}^{N}\mathbf{b}^{\tau}_{k/2^l m}(t)\cos\left[\frac{k}{2^l m}(\theta-\theta^{\tau})\right]\\
&\quad+\mathbf{c}^{\tau}_{k/2^l m}(t)\sin\left[\frac{k}{2^l m}(\theta-\theta^{\tau})\right],
\end{aligned}
\tag{3.230}
$$

where for $k=1,2,\ldots,N$

$$
\begin{aligned}
&\ddot{\mathbf{a}}_0^{(2^l m)}=\mathbf{F}_0^{(2^l m)}(\mathbf{z}^{(2^l m)},\mathbf{z}_1^{(2^l m)};\mathbf{z}^{\tau(2^l m)},\mathbf{z}_1^{\tau(2^l m)}),\\
&\ddot{\mathbf{b}}^{(2^l m)}+2\frac{\Omega}{2^l m}\mathbf{k}_1\dot{\mathbf{c}}^{(2^l m)}-\frac{\Omega^2}{(2^l m)^2}\mathbf{k}_2\mathbf{b}^{(2^l m)}\\
&\quad=\mathbf{F}_{1k}^{(2^l m)}(\mathbf{z}^{(2^l m)},\mathbf{z}_1^{(2^l m)};\mathbf{z}^{\tau(2^l m)},\mathbf{z}_1^{\tau(2^l m)}),\\
&\ddot{\mathbf{c}}^{(2^l m)}-2\frac{\Omega}{2^l m}\mathbf{k}_1\dot{\mathbf{b}}^{(2^l m)}-\frac{\Omega^2}{(2^l m)^2}\mathbf{k}_2\mathbf{c}^{(2^l m)}\\
&\quad=\mathbf{F}_{2k}^{(2^l m)}(\mathbf{z}^{(2^l m)},\mathbf{z}_1^{(2^l m)};\mathbf{z}^{\tau(2^l m)},\mathbf{z}_1^{\tau(2^l m)});
\end{aligned}
\tag{3.231}
$$

and for $N=1,2,\ldots,\infty$

$$
\mathbf{k}_1=diag(\mathbf{I}_{n\times n},2\mathbf{I}_{n\times n},\ldots,N\mathbf{I}_{n\times n}),
$$

$$
\begin{aligned}
&\mathbf{k}_2 = diag(\mathbf{I}_{n\times n}, 2^2\mathbf{I}_{n\times n}, \ldots, N^2\mathbf{I}_{n\times n}),\\
&\mathbf{b}^{(2^lm)} = (\mathbf{b}_{1/2^lm}, \mathbf{b}_{2/2^lm}, \ldots, \mathbf{b}_{N/2^lm})^{\mathrm{T}},\\
&\mathbf{c}^{(2^lm)} = (\mathbf{c}_{1/2^lm}, \mathbf{c}_{2/2^lm}, \ldots, \mathbf{c}_{N/2^lm})^{\mathrm{T}};\\
&\mathbf{b}^{\tau(2^lm)} = (\mathbf{b}^{\tau}_{1/2^lm}, \mathbf{b}^{\tau}_{2/2^lm}, \ldots, \mathbf{b}^{\tau}_{N/2^lm})^{\mathrm{T}},\\
&\mathbf{c}^{\tau(2^lm)} = (\mathbf{c}^{\tau}_{1/2^lm}, \mathbf{c}^{\tau}_{2/2^lm}, \ldots, \mathbf{c}^{\tau}_{N/2^lm})^{\mathrm{T}};\\
&\mathbf{F}_1^{(2^lm)} = (\mathbf{F}_{11}^{(2^lm)}, \mathbf{F}_{12}^{(2^lm)}, \ldots, \mathbf{F}_{1N}^{(2^lm)})^{\mathrm{T}},\\
&\mathbf{F}_2^{(2^lm)} = (\mathbf{F}_{21}^{(2^lm)}, \mathbf{F}_{22}^{(2^lm)}, \ldots, \mathbf{F}_{2N}^{(2^lm)})^{\mathrm{T}};\\
&\mathbf{z}^{(2^lm)} = (\mathbf{a}_0^{(2^lm)}, \mathbf{b}^{(2^lm)}, \mathbf{c}^{(2^lm)})^{\mathrm{T}}, \dot{\mathbf{z}}^{(2^lm)} = \mathbf{z}_1^{(2^lm)};\\
&\mathbf{z}^{\tau(2^lm)} = (\mathbf{a}_0^{\tau(2^lm)}, \mathbf{b}^{\tau(2^lm)}, \mathbf{c}^{\tau(2^lm)})^{\mathrm{T}}, \dot{\mathbf{z}}^{\tau(2^lm)} = \mathbf{z}_1^{\tau(2^lm)}.
\end{aligned}
\tag{3.232}
$$

$$
\begin{aligned}
&\mathbf{F}_0^{(2^lm)}(\mathbf{z}^{(2^lm)}, \mathbf{z}_1^{(2^lm)}; \mathbf{z}^{\tau(2^lm)}, \mathbf{z}_1^{\tau(2^lm)})\\
&\quad = \frac{1}{2^l(2m\pi)}\int_0^{2^l(2m\pi)} \mathbf{f}(\mathbf{x}^{(2^lm)*}, \dot{\mathbf{x}}^{(2^lm)*}; \mathbf{x}^{\tau(2^lm)*}, \dot{\mathbf{x}}^{\tau(2^lm)*}, \mathbf{p})d\theta;\\
&\mathbf{F}_{1k}^{(2^lm)}(\mathbf{z}^{(2^lm)}, \mathbf{z}_1^{(2^lm)}; \mathbf{z}^{\tau(2^lm)}, \mathbf{z}_1^{\tau(2^lm)})\\
&\quad = \frac{1}{2^{l-1}(2m\pi)}\int_0^{2^l(2m\pi)} \mathbf{f}(\mathbf{x}^{(2^lm)*}, \dot{\mathbf{x}}^{(2^lm)*}; \mathbf{x}^{\tau(2^lm)*}, \dot{\mathbf{x}}^{\tau(2^lm)*}, \mathbf{p})\cos\left(\frac{k}{2^lm}\theta\right)d\theta,\\
&\mathbf{F}_{2k}^{(2^lm)}(\mathbf{z}^{(2^lm)}, \mathbf{z}_1^{(2^lm)}; \mathbf{z}^{\tau(2^lm)}, \mathbf{z}_1^{\tau(2^lm)})\\
&\quad = \frac{1}{2^{l-1}(2m\pi)}\int_0^{2^l(2m\pi)} \mathbf{f}(\mathbf{x}^{(2^lm)*}, \dot{\mathbf{x}}^{(2^lm)*}; \mathbf{x}^{\tau(2^lm)*}, \dot{\mathbf{x}}^{\tau(2^lm)*}, \mathbf{p})\sin\left(\frac{k}{2^lm}\theta\right)d\theta.
\end{aligned}
\tag{3.233}
$$

The solution of period-2^lm motion can be determined by the equilibrium of coefficient dynamical system in the time-delayed, nonlinear vibration system, and the corresponding stability and bifurcation can be done. As $l \to \infty$, the stable and unstable chaos with $(n_1, n_2, n_3 \mid n_4, n_5, n_6)$ in Equation (3.223) can be obtained where

$$
\sum_{i=1}^{3} n_i + 2\sum_{i=4}^{6} n_i = 2n(1 + 2N). \tag{3.234}
$$

With increasing l, the Fourier truncated number $N >> 2^lm$ will dramatically increase. If period-1 motion possesses at least N_1 harmonic vector terms, then the total harmonic vector terms for period-2^lm motion should be $N \geq 2^lmN_1$. The chaotic motion classifications with specific cases for the time-delayed, nonlinear systems are given as follows.

1. For the chaotic motion of $(n_1, 0, 0 \mid 0, 0, 0)$ with $n_1 = 2n(1 + 2N)$, the chaotic motion is called the hyperbolic stable chaos.

2. For the chaotic motion of $(0, 0, 0 \mid n_4, 0, 0)$ with $n_4 = n(1 + 2N)$, the chaotic motion is called the spiral stable chaos.
3. For the chaotic motion of $(n_1, 0, 0 \mid n_4, 0, 0)$ with $n_1 + 2n_4 = 2n(1 + 2N)$, the chaotic motion is called the hyperbolic-spiral stable chaos.
4. For the chaotic motion of $(0, n_2, 0 \mid 0, 0, 0)$ with $n_2 = 2n(1 + 2N)$, the chaotic motion is called the hyperbolic unstable chaos.
5. For the chaotic motion of $(0, 0, 0 \mid 0, n_5, 0)$ with $n_5 = n(1 + 2N)$, the chaotic motion is called the spiral unstable chaos.
6. For the chaotic motion of $(0, n_2, 0 \mid 0, n_5, 0)$ with $n_2 + 2n_5 = 2n(1 + 2N)$, the chaotic motion is called the hyperbolic-spiral unstable chaos.
7. For the chaotic motion of $(n_1, n_2, 0 \mid 0, 0, 0)$ with $n_1 + n_2 = 2n(1 + 2N)$, the chaotic motion is called the saddle unstable chaos.
8. For the chaotic motion of $(n_1, n_2, 0 \mid n_4, n_5, 0)$ with $n_1 + n_2 + 2n_4 + 2n_5 = 2n(1 + 2N)$, the chaotic motion is called the spiral saddle unstable chaos.

Since m is an arbitrary positive integer number, it includes $(2^l m_1)$ for period-$2^l m_1$ motion. Thus, the expression for period-m motion can be employed for any periodic motions. The expression in Equation (3.208) can be used to express the solution for chaotic motion as $m \to \infty$, which can be approximated by Equation (3.212) under the condition of $\|\mathbf{x}^{(m)}(t) - \mathbf{x}^{(m)*}(t)\| < \varepsilon$ and $\|\mathbf{x}^{\tau(m)}(t) - \mathbf{x}^{\tau(m)*}(t)\| < \varepsilon^\tau$. The chaotic solutions can be classified as discussed for period-$2^l m_1$ motion.

3.4.2 Periodically Excited Vibration Systems with Time-Delay

Periodic motions in periodically excited, time-delayed, vibration systems will be presented herein. If such an excited, time-delayed, vibration system possesses periodic motions with period $T = 2\pi/\Omega$, then such periodic motions can be expressed by the Fourier series, discussed as follows.

Theorem 3.15 *Consider a periodically excited, time-delayed, nonlinear vibration system as*

$$\ddot{\mathbf{x}} = \mathbf{F}(\mathbf{x}, \dot{\mathbf{x}}; \mathbf{x}^\tau, \dot{\mathbf{x}}^\tau, t, \mathbf{p}) \in \mathscr{R}^n \tag{3.235}$$

where $\mathbf{F}(\mathbf{x}, \dot{\mathbf{x}}; \mathbf{x}^\tau, \dot{\mathbf{x}}^\tau, t, \mathbf{p})$ *is a* C^r*-continuous nonlinear function vector* $(r \geq 1)$ *with an excitation period* $T = 2\pi/\Omega$. *If such a vibration system has a periodic motion with finite norm* $\|\mathbf{x}\|$, *there is a generalized coordinate transformation with* $\theta = \Omega t$ *for the periodic motion of Equation (3.235) in the form of*

$$\mathbf{x} \equiv \mathbf{x}(t) = \mathbf{a}_0(t) + \sum_{k=1}^{\infty} \mathbf{b}_k(t)\cos(k\theta) + \mathbf{c}_k(t)\sin(k\theta);$$

$$\mathbf{x}^\tau \equiv \mathbf{x}(t - \tau) = \mathbf{a}_0^\tau(t) + \sum_{k=1}^{\infty} \mathbf{b}_k^\tau(t)\cos[k(\theta - \theta^\tau)] + \mathbf{c}_k^\tau(t)\sin[k(\theta - \theta^\tau)] \tag{3.236}$$

with $\mathbf{a}_0^\tau = \mathbf{a}_0(t - \tau), \mathbf{b}_k^\tau = \mathbf{b}_k(t - \tau), \mathbf{c}_k^\tau = \mathbf{c}_k(t - \tau), \theta^\tau = \Omega\tau$ *and*

$$\mathbf{a}_0 = (a_{01}, a_{02}, \ldots, a_{0n})^{\mathrm{T}},$$

$$\begin{aligned}
\mathbf{b}_k &= (b_{k1}, b_{k2}, \ldots, b_{kn})^{\mathrm{T}},\\
\mathbf{c}_k &= (c_{k1}, c_{k2}, \ldots, c_{kn})^{\mathrm{T}};\\
\mathbf{a}_0^\tau &= (a_{01}^\tau, a_{02}^\tau, \ldots, a_{0n}^\tau)^{\mathrm{T}},\\
\mathbf{b}_k^\tau &= (b_{k1}^\tau, b_{k2}^\tau, \ldots, b_{kn}^\tau)^{\mathrm{T}},\\
\mathbf{c}_k^\tau &= (c_{k1}^\tau, c_{k2}^\tau, \ldots, c_{kn}^\tau)^{\mathrm{T}};
\end{aligned} \tag{3.237}$$

and

$$\begin{aligned}
&\|\mathbf{x}\| = \|\mathbf{a}_0\| + \sum_{k=1}^{\infty} \|\mathbf{A}_k\|, \text{ and } \lim_{k\to\infty} \|\mathbf{A}_k\| = 0 \text{ but not uniform}\\
&\text{with } \mathbf{A}_k = (A_{k1}, A_{k2}, \ldots, A_{kn})^{\mathrm{T}} \text{ and } A_{kj} = \sqrt{b_{kj}^2 + c_{kj}^2}\ \ (j = 1, 2, \ldots, n);
\end{aligned} \tag{3.238}$$

$$\begin{aligned}
&\|\mathbf{x}^\tau\| = \|\mathbf{a}_0^\tau\| + \sum_{k=1}^{\infty} \|\mathbf{A}_k^\tau\|, \text{ and } \lim_{k\to\infty} \|\mathbf{A}_k^\tau\| = 0 \text{ but not uniform}\\
&\text{with } \mathbf{A}_k^\tau = (A_{k1}^\tau, A_{k2}^\tau, \ldots, A_{kn}^\tau)^{\mathrm{T}} \text{ and } A_{kj}^\tau = \sqrt{(b_{kj}^\tau)^2 + (c_{kj}^\tau)^2}\ \ (j = 1, 2, \ldots, n).
\end{aligned} \tag{3.239}$$

For $\|\mathbf{x}(t) - \mathbf{x}^*(t)\| < \varepsilon$ *and* $\|\mathbf{x}^\tau(t) - \mathbf{x}^{\tau*}(t)\| < \varepsilon^\tau$ *with prescribed small positive* $\varepsilon > 0$ *and* $\varepsilon^\tau > 0$*, the infinite term transformation of periodic motion* $\mathbf{x}(t)$ *of Equation (3.235), given by Equation (3.236), can be approximated by a finite term transformation* $\mathbf{x}^*(t)$ *as*

$$\begin{aligned}
\mathbf{x}^* &= \mathbf{a}_0(t) + \sum_{k=1}^{N} \mathbf{b}_k(t)\cos(k\theta) + \mathbf{c}_k(t)\sin(k\theta),\\
\mathbf{x}^{\tau*} &= \mathbf{a}_0^\tau(t) + \sum_{k=1}^{N} \mathbf{b}_k^\tau(t)\cos[k(\theta - \theta^\tau)] + \mathbf{c}_k^\tau(t)\sin[k(\theta - \theta^\tau)]
\end{aligned} \tag{3.240}$$

and the generalized coordinates are determined by

$$\begin{aligned}
\ddot{\mathbf{a}}_0 &= \mathbf{F}_0(\mathbf{z}, \mathbf{z}_1; \mathbf{z}^\tau, \mathbf{z}_1^\tau),\\
\ddot{\mathbf{b}} &= -2\Omega\mathbf{k}_1\dot{\mathbf{c}} + \Omega^2\mathbf{k}_2\mathbf{b} + \mathbf{F}_1(\mathbf{z}, \mathbf{z}_1; \mathbf{z}^\tau, \mathbf{z}_1^\tau),\\
\ddot{\mathbf{c}} &= 2\Omega\mathbf{k}_1\dot{\mathbf{b}} + \Omega^2\mathbf{k}_2\mathbf{c} + \mathbf{F}_2(\mathbf{z}, \mathbf{z}_1; \mathbf{z}^\tau, \mathbf{z}_1^\tau);
\end{aligned} \tag{3.241}$$

where for $N = 1, 2, \ldots, \infty$

$$\begin{aligned}
\mathbf{k}_1 &= diag(\mathbf{I}_{n\times n}, 2\mathbf{I}_{n\times n}, \ldots, N\mathbf{I}_{n\times n}),\\
\mathbf{k}_2 &= diag(\mathbf{I}_{n\times n}, 2^2\mathbf{I}_{n\times n}, \ldots, N^2\mathbf{I}_{n\times n}),\\
\mathbf{b} &= (\mathbf{b}_1, \mathbf{b}_2, \ldots, \mathbf{b}_N)^{\mathrm{T}},\\
\mathbf{c} &= (\mathbf{c}_1, \mathbf{c}_2, \ldots, \mathbf{c}_N)^{\mathrm{T}};
\end{aligned}$$

$$\begin{aligned}
\mathbf{b}^{\tau} &= (\mathbf{b}_1^{\tau}, \mathbf{b}_2^{\tau}, \ldots, \mathbf{b}_N^{\tau})^{\mathrm{T}},\\
\mathbf{c}^{\tau} &= (\mathbf{c}_1^{\tau}, \mathbf{c}_2^{\tau}, \ldots, \mathbf{c}_N^{\tau})^{\mathrm{T}};\\
\mathbf{F}_1 &= (\mathbf{F}_{11}, \mathbf{F}_{12}, \ldots, \mathbf{F}_{1N})^{\mathrm{T}},\\
\mathbf{F}_2 &= (\mathbf{F}_{21}, \mathbf{F}_{22}, \ldots, \mathbf{F}_{2N})^{\mathrm{T}};\\
\mathbf{z} &= (\mathbf{a}_0, \mathbf{b}, \mathbf{c})^{\mathrm{T}}, \dot{\mathbf{z}} = \mathbf{z}_1;\\
\mathbf{z}^{\tau} &= (\mathbf{a}_0^{\tau}, \mathbf{b}^{\tau}, \mathbf{c}^{\tau})^{\mathrm{T}}, \dot{\mathbf{z}}^{\tau} = \mathbf{z}_1^{\tau}
\end{aligned} \tag{3.242}$$

and for $k = 1, 2, \ldots, N$

$$\begin{aligned}
\mathbf{F}_0(\mathbf{z}, \mathbf{z}_1; \mathbf{z}^{\tau}, \mathbf{z}_1^{\tau}) &= \frac{1}{2\pi}\int_0^{2\pi} \mathbf{F}(\mathbf{x}^*, \dot{\mathbf{x}}^*; \mathbf{x}^{\tau *}, \dot{\mathbf{x}}^{\tau *}, t, \mathbf{p})d\theta;\\
\mathbf{F}_{1k}(\mathbf{z}, \mathbf{z}_1; \mathbf{z}^{\tau}, \mathbf{z}_1^{\tau}) &= \frac{1}{\pi}\int_0^{2\pi} \mathbf{F}(\mathbf{x}^*, \dot{\mathbf{x}}^*; \mathbf{x}^{\tau *}, \dot{\mathbf{x}}^{\tau *}, t, \mathbf{p})\cos(k\theta)d\theta,\\
\mathbf{F}_{2k}(\mathbf{z}, \mathbf{z}_1; \mathbf{z}^{\tau}, \mathbf{z}_1^{\tau}) &= \frac{1}{\pi}\int_0^{2\pi} \mathbf{F}(\mathbf{x}^*, \dot{\mathbf{x}}^*; \mathbf{x}^{\tau *}, \dot{\mathbf{x}}^{\tau *}, t, \mathbf{p})\sin(k\theta)d\theta.
\end{aligned} \tag{3.243}$$

The state-space form of Equation (3.241) is

$$\dot{\mathbf{z}} = \mathbf{z}_1 \text{ and } \dot{\mathbf{z}}_1 = \mathbf{g}(\mathbf{z}, \mathbf{z}_1; \mathbf{z}^{\tau}, \mathbf{z}_1^{\tau}) \tag{3.244}$$

where

$$\mathbf{g} = (\mathbf{F}_0, -2\Omega\mathbf{k}_1\dot{\mathbf{c}} + \Omega^2\mathbf{k}_2\mathbf{b} + \mathbf{F}_1, 2\Omega\mathbf{k}_1\dot{\mathbf{b}} + \Omega^2\mathbf{k}_2\mathbf{c} + \mathbf{F}_2)^{\mathrm{T}}. \tag{3.245}$$

An equivalent system of Equation (3.244) is

$$\dot{\mathbf{y}} = \mathbf{f}(\mathbf{y}, \mathbf{y}^{\tau}) \tag{3.246}$$

where

$$\mathbf{y} = (\mathbf{z}, \mathbf{z}_1)^{\mathrm{T}}, \mathbf{y}^{\tau} = (\mathbf{z}^{\tau}, \mathbf{z}_1^{\tau})^{\mathrm{T}} \text{ and } \mathbf{f} = (\mathbf{z}_1, \mathbf{g})^{\mathrm{T}} \tag{3.247}$$

If equilibrium $\mathbf{y}^* = \mathbf{y}^{\tau *}$ *of Equation (3.246) (i.e.,* $\mathbf{f}(\mathbf{y}^*, \mathbf{y}^{\tau *}) = \mathbf{0}$*) exists, then the analytical solution of periodic motion exists as in Equation (3.240). In the vicinity of equilibrium* $\mathbf{y}^* = \mathbf{y}^{\tau *}$*, with* $\mathbf{y} = \mathbf{y}^* + \Delta\mathbf{y}$ *and* $\mathbf{y}^{\tau} = \mathbf{y}^{\tau *} + \Delta\mathbf{y}^{\tau}$*, the linearized equation of Equation (3.246) is*

$$\Delta\dot{\mathbf{y}} = D_{\mathbf{y}}\mathbf{f}(\mathbf{y}^*, \mathbf{y}^{\tau *})\Delta\mathbf{y} + D_{\mathbf{y}^{\tau}}\mathbf{f}(\mathbf{y}^*, \mathbf{y}^{\tau *})\Delta\mathbf{y}^{\tau} \tag{3.248}$$

and the eigenvalue analysis of equilibrium $\mathbf{y}^* = \mathbf{y}^{\tau *}$ *is given by*

$$|D_{\mathbf{y}}\mathbf{f}(\mathbf{y}^*, \mathbf{y}^{\tau *}) - \lambda\mathbf{I}_{2n(2N+1)\times 2n(2N+1)} + D_{\mathbf{y}^{\tau}}\mathbf{f}(\mathbf{y}^*, \mathbf{y}^{\tau *})e^{-\lambda\tau}| = 0 \tag{3.249}$$

where

$$D_{\mathbf{y}}\mathbf{f}(\mathbf{y}^*, \mathbf{y}^{\tau *}) = \frac{\partial\mathbf{f}(\mathbf{y}, \mathbf{y}^{\tau})}{\partial\mathbf{y}}|_{(\mathbf{y}^*, \mathbf{y}^{\tau *})}, D_{\mathbf{y}^{\tau}}\mathbf{f}(\mathbf{y}^*, \mathbf{y}^{\tau *}) = \frac{\partial\mathbf{f}(\mathbf{y}, \mathbf{y}^{\tau})}{\partial\mathbf{y}^{\tau}}|_{(\mathbf{y}^*, \mathbf{y}^{\tau *})}. \tag{3.250}$$

Thus, the stability and bifurcation of periodic motion can be classified by the eigenvalues of Equation (3.248) at equilibrium $\mathbf{y}^* = \mathbf{y}^{\tau *}$ *with*

$$(n_1, n_2, n_3 \mid n_4, n_5, n_6). \tag{3.251}$$

1. *If all eigenvalues of the equilibrium possess negative real parts, the approximate periodic solution is stable.*
2. *If at least one of the eigenvalues of the equilibrium possesses a positive real part, the approximate periodic solution is unstable.*
3. *The boundary between the stable and unstable equilibriums with higher order singularity gives the bifurcation conditions and stability with higher order singularity.*

Proof. The proof of this theorem is similar to Theorem 3.13. ∎

Similarly, the periodic-m motion in periodically excited, nonlinear vibration system will be discussed.

Theorem 3.16 *Consider a periodically excited, time-delayed nonlinear vibration system in Equation (3.235) with an excitation period $T = 2\pi/\Omega$. If such a time-delayed vibration system has a period-m motion $\mathbf{x}^{(m)}(t)$ with finite norm $\|\mathbf{x}^{(m)}\|$, there is a generalized coordinate transformation with $\theta = \Omega t$ for the period-m motion of Equation (3.235) in the form of*

$$\mathbf{x}^{(m)}(t) = \mathbf{a}_0^{(m)}(t) + \sum_{k=1}^{\infty} \mathbf{b}_{k/m}(t)\cos\left(\frac{k}{m}\theta\right) + \mathbf{c}_{k/m}(t)\sin\left(\frac{k}{m}\theta\right);$$

$$\mathbf{x}^{\tau(m)}(t) = \mathbf{a}_0^{\tau(m)}(t) + \sum_{k=1}^{\infty} \mathbf{b}_{k/m}^{\tau}(t)\cos\left[\frac{k}{m}(\theta-\theta^{\tau})\right] + \mathbf{c}_{k/m}^{\tau}(t)\sin\left[\frac{k}{m}(\theta-\theta^{\tau})\right] \tag{3.252}$$

with

$$\begin{aligned}
\mathbf{a}_0^{(m)} &= (a_{01}^{(m)}, a_{02}^{(m)}, \ldots, a_{0n}^{(m)})^{\mathrm{T}},\\
\mathbf{b}_{k/m} &= (b_{k/m1}, b_{k/m2}, \ldots, b_{k/mn})^{\mathrm{T}},\\
\mathbf{c}_{k/m} &= (c_{k/m1}, c_{k/m2}, \ldots, c_{k/mn})^{\mathrm{T}};\\
\mathbf{a}_0^{\tau(m)} &= (a_{01}^{\tau(m)}, a_{02}^{\tau(m)}, \ldots, a_{0n}^{\tau(m)})^{\mathrm{T}},\\
\mathbf{b}_{k/m}^{\tau} &= (b_{k/m1}^{\tau}, b_{k/m2}^{\tau}, \ldots, b_{k/mn}^{\tau})^{\mathrm{T}},\\
\mathbf{c}_{k/m}^{\tau} &= (c_{k/m1}^{\tau}, c_{k/m2}^{\tau}, \ldots, c_{k/mn}^{\tau})^{\mathrm{T}}
\end{aligned} \tag{3.253}$$

and

$$\begin{aligned}
&\|\mathbf{x}^{(m)}\| = \|\mathbf{a}_0^{(m)}\| + \sum_{k=1}^{\infty}\|\mathbf{A}_{k/m}\|, \text{ and } \lim_{k\to\infty}\|\mathbf{A}_{k/m}\| = 0 \text{ but not uniform}\\
&\text{with } \mathbf{A}_{k/m} = (A_{k/m1}, A_{k/m2} \ldots, A_{k/mn})^{T}\\
&\text{and } A_{k/mj} = \sqrt{b_{k/mj}^2 + c_{k/mj}^2}\ (j = 1, 2, \ldots, n);
\end{aligned} \tag{3.254}$$

$$\begin{aligned}
&\|\mathbf{x}^{\tau(m)}\| = \|\mathbf{a}_0^{\tau(m)}\| + \sum_{k=1}^{\infty}\|\mathbf{A}_{k/m}^{\tau}\|, \text{ and } \lim_{k\to\infty}\|\mathbf{A}_{k/m}^{\tau}\| = 0 \text{ but not uniform}\\
&\text{with } \mathbf{A}_{k/m}^{\tau} = (A_{k/m1}^{\tau}, A_{k/m2}^{\tau} \ldots, A_{k/mn}^{\tau})^{T}\\
&\text{and } A_{k/mj}^{\tau} = \sqrt{(b_{k/mj}^{\tau})^2 + (c_{k/mj}^{\tau})^2}\ (j = 1, 2, \ldots, n).
\end{aligned} \tag{3.255}$$

For $\|\mathbf{x}^{(m)}(t) - \mathbf{x}^{(m)*}(t)\| < \varepsilon$ *and* $\|\mathbf{x}^{\tau(m)}(t) - \mathbf{x}^{\tau(m)*}(t)\| < \varepsilon^{\tau}$ *with prescribed small* $\varepsilon > 0$ *and* $\varepsilon^{\tau} > 0$*, the infinite term transformation* $\mathbf{x}^{(m)}(t)$ *of period-*m *motion of Equation (3.235), given by Equation (3.252), can be approximated by a finite term transformation* $\mathbf{x}^{(m)*}(t)$ *as*

$$\mathbf{x}^{(m)*}(t) = \mathbf{a}_0^{(m)}(t) + \sum_{k=1}^{N} \mathbf{b}_{k/m}(t)\cos\left(\frac{k}{m}\theta\right) + \mathbf{c}_{k/m}(t)\sin\left(\frac{k}{m}\theta\right);$$

$$\mathbf{x}^{(m)\tau *}(t) = \mathbf{a}_0^{\tau(m)}(t) + \sum_{k=1}^{N} \mathbf{b}^{\tau}_{k/m}(t)\cos\left[\frac{k}{m}(\theta - \theta^{\tau})\right] + \mathbf{c}^{\tau}_{k/m}(t)\sin\left[\frac{k}{m}(\theta - \theta^{\tau})\right] \tag{3.256}$$

and the generalized coordinates are determined by

$$\begin{aligned}
\ddot{\mathbf{a}}_0^{(m)} &= \mathbf{F}_0^{(m)}(\mathbf{z}^{(m)}, \mathbf{z}_1^{(m)}; \mathbf{z}^{\tau(m)}, \mathbf{z}_1^{\tau(m)}),\\
\ddot{\mathbf{b}}^{(m)} &= -2\frac{\Omega}{m}\mathbf{k}_1\dot{\mathbf{c}}^{(m)} + \frac{\Omega^2}{m^2}\mathbf{k}_2\mathbf{b}^{(m)} + \mathbf{F}_1^{(m)}(\mathbf{z}^{(m)}, \mathbf{z}_1^{(m)}; \mathbf{z}^{\tau(m)}, \mathbf{z}_1^{\tau(m)}),\\
\ddot{\mathbf{c}}^{(m)} &= 2\frac{\Omega}{m}\mathbf{k}_1\dot{\mathbf{b}}^{(m)} + \frac{\Omega^2}{m^2}\mathbf{k}_2\mathbf{c}^{(m)} + \mathbf{F}_2^{(m)}(\mathbf{z}^{(m)}, \mathbf{z}_1^{(m)}; \mathbf{z}^{\tau(m)}, \mathbf{z}_1^{\tau(m)})
\end{aligned} \tag{3.257}$$

where for $N = 1, 2, \ldots, \infty$

$$\begin{aligned}
\mathbf{k}_1 &= diag(\mathbf{I}_{n\times n}, 2\mathbf{I}_{n\times n}, \ldots, N\mathbf{I}_{n\times n}),\\
\mathbf{k}_2 &= diag(\mathbf{I}_{n\times n}, 2^2\mathbf{I}_{n\times n}, \ldots, N^2\mathbf{I}_{n\times n});\\
\mathbf{b}^{(m)} &= (\mathbf{b}_{1/m}, \mathbf{b}_{2/m}, \ldots, \mathbf{b}_{N/m})^{\mathrm{T}},\\
\mathbf{c}^{(m)} &= (\mathbf{c}_{1/m}, \mathbf{c}_{2/m}, \ldots, \mathbf{c}_{N/m})^{\mathrm{T}};\\
\mathbf{b}^{\tau(m)} &= (\mathbf{b}^{\tau}_{1/m}, \mathbf{b}^{\tau}_{2/m}, \ldots, \mathbf{b}^{\tau}_{N/m})^{\mathrm{T}},\\
\mathbf{c}^{\tau(m)} &= (\mathbf{c}^{\tau}_{1/m}, \mathbf{c}^{\tau}_{2/m}, \ldots, \mathbf{c}^{\tau}_{N/m})^{\mathrm{T}};\\
\mathbf{F}_1^{(m)} &= (\mathbf{F}_{11}^{(m)}, \mathbf{F}_{12}^{(m)}, \ldots, \mathbf{F}_{1N}^{(m)})^{\mathrm{T}},\\
\mathbf{F}_2^{(m)} &= (\mathbf{F}_{21}^{(m)}, \mathbf{F}_{22}^{(m)}, \ldots, \mathbf{F}_{2N}^{(m)})^{\mathrm{T}};\\
\mathbf{z}^{(m)} &= (\mathbf{a}_0^{(m)}, \mathbf{b}^{(m)}, \mathbf{c}^{(m)})^{\mathrm{T}}, \dot{\mathbf{z}}^{(m)} = \mathbf{z}_1^{(m)};\\
\mathbf{z}^{\tau(m)} &= (\mathbf{a}_0^{\tau(m)}, \mathbf{b}^{\tau(m)}, \mathbf{c}^{\tau(m)})^{\mathrm{T}}, \dot{\mathbf{z}}^{\tau(m)} = \mathbf{z}_1^{\tau(m)}
\end{aligned} \tag{3.258}$$

and for $k = 1, 2, \ldots, N$

$$\begin{aligned}
&\mathbf{F}_0^{(m)}(\mathbf{z}^{(m)}, \mathbf{z}_1^{(m)}; \mathbf{z}^{\tau(m)}, \mathbf{z}_1^{\tau(m)})\\
&\quad = \frac{1}{2m\pi}\int_0^{2m\pi} \mathbf{F}(\mathbf{x}^{(m)*}, \dot{\mathbf{x}}^{(m)*}; \mathbf{x}^{\tau(m)*}, \dot{\mathbf{x}}^{\tau(m)*}, t, \mathbf{p})d\theta;
\end{aligned}$$

$$\mathbf{F}_{1k}^{(m)}(\mathbf{z}^{(m)}, \mathbf{z}_1^{(m)}; \mathbf{z}^{\tau(m)}, \mathbf{z}_1^{\tau(m)})$$
$$= \frac{1}{m\pi}\int_0^{2m\pi} \mathbf{F}(\mathbf{x}^{(m)*}, \dot{\mathbf{x}}^{(m)*}; \mathbf{x}^{\tau(m)*}, \dot{\mathbf{x}}^{\tau(m)*}, t, \mathbf{p}) \cos\left(\frac{k}{m}\theta\right) d\theta,$$
$$\mathbf{F}_{2k}^{(m)}(\mathbf{z}^{(m)}, \mathbf{z}_1^{(m)}; \mathbf{z}^{\tau(m)}, \mathbf{z}_1^{\tau(m)})$$
$$= \frac{1}{m\pi}\int_0^{2m\pi} \mathbf{F}(\mathbf{x}^{(m)*}, \dot{\mathbf{x}}^{(m)*}; \mathbf{x}^{\tau(m)*}, \dot{\mathbf{x}}^{\tau(m)*}, t, \mathbf{p}) \sin\left(\frac{k}{m}\theta\right) d\theta. \tag{3.259}$$

The state-space form of Equation (3.257) is

$$\dot{\mathbf{z}}^{(m)} = \mathbf{z}_1^{(m)} \text{ and } \dot{\mathbf{z}}_1^{(m)} = \mathbf{g}^{(m)}(\mathbf{z}^{(m)}, \mathbf{z}_1^{(m)}; \mathbf{z}^{\tau(m)}, \mathbf{z}_1^{\tau(m)}) \tag{3.260}$$

where

$$\mathbf{g}^{(m)} = (\mathbf{F}_0^{(m)}, -2\frac{\Omega}{m}\mathbf{k}_1\dot{\mathbf{c}}^{(m)} + \frac{\Omega^2}{m^2}\mathbf{k}_2\mathbf{b}^{(m)} + \mathbf{F}_1^{(m)},$$
$$2\frac{\Omega}{m}\mathbf{k}_1\dot{\mathbf{b}}^{(m)} + \frac{\Omega^2}{m^2}\mathbf{k}_2\mathbf{c}^{(m)} + \mathbf{F}_2^{(m)})^{\mathrm{T}}. \tag{3.261}$$

An equivalent system of Equation (3.257) is

$$\dot{\mathbf{y}}^{(m)} = \mathbf{f}^{(m)}(\mathbf{y}^{(m)}, \mathbf{y}^{\tau(m)}) \tag{3.262}$$

where

$$\mathbf{y}^{(m)} = (\mathbf{z}^{(m)}, \mathbf{z}_1^{(m)})^{\mathrm{T}}, \mathbf{y}^{\tau(m)} = (\mathbf{z}^{\tau(m)}, \mathbf{z}_1^{\tau(m)})^{\mathrm{T}} \text{ and } \mathbf{f}^{(m)} = (\mathbf{z}_1^{(m)}, \mathbf{g}^{(m)})^{\mathrm{T}}. \tag{3.263}$$

*If equilibrium $\mathbf{y}^{(m)*} = \mathbf{y}^{\tau(m)*}$ of Equation (3.262) (i.e., $\mathbf{f}^{(m)}(\mathbf{y}^{(m)*}, \mathbf{y}^{\tau(m)*}) = \mathbf{0}$) exists, then the analytical solution of period-*m *motion exists as in Equation (3.256). In the vicinity of equilibrium $\mathbf{y}^{(m)*} = \mathbf{y}^{\tau(m)*}$, with $\mathbf{y}^{(m)} = \mathbf{y}^{(m)*} + \Delta\mathbf{y}^{(m)}$ and $\mathbf{y}^{\tau(m)} = \mathbf{y}^{\tau(m)*} + \Delta\mathbf{y}^{\tau(m)}$, the linearized equation of Equation (3.262) is*

$$\Delta\dot{\mathbf{y}}^{(m)} = D_{\mathbf{y}^{(m)}}\mathbf{f}^{(m)}(\mathbf{y}^{(m)*}, \mathbf{y}^{\tau(m)*})\Delta\mathbf{y}^{(m)} + D_{\mathbf{y}^{\tau(m)}}\mathbf{f}^{(m)}(\mathbf{y}^{(m)*}, \mathbf{y}^{\tau(m)*})\Delta\mathbf{y}^{\tau(m)} \tag{3.264}$$

and the eigenvalue analysis of equilibrium $\mathbf{y}^{(m)*} = \mathbf{y}^{\tau(m)*}$ is given by

$$|D_{\mathbf{y}^{(m)}}\mathbf{f}^{(m)}(\mathbf{y}^{(m)*}, \mathbf{y}^{\tau(m)*}) - \lambda\mathbf{I}_{2n(2N+1)\times 2n(2N+1)} + D_{\mathbf{y}^{\tau(m)}}\mathbf{f}^{(m)}(\mathbf{y}^{(m)*}, \mathbf{y}^{\tau(m)*})e^{-\lambda\tau}| = 0 \tag{3.265}$$

where

$$D_{\mathbf{y}^{(m)}}\mathbf{f}^{(m)}(\mathbf{y}^{(m)*}, \mathbf{y}^{\tau(m)*}) = \frac{\partial\mathbf{f}^{(m)}(\mathbf{y}^{(m)}, \mathbf{y}^{\tau(m)})}{\partial\mathbf{y}^{(m)}}\bigg|_{(\mathbf{y}^{(m)*}, \mathbf{y}^{\tau(m)*})},$$
$$D_{\mathbf{y}^{\tau(m)}}\mathbf{f}^{(m)}(\mathbf{y}^{(m)*}, \mathbf{y}^{\tau(m)*}) = \frac{\partial\mathbf{f}^{(m)}(\mathbf{y}^{(m)}, \mathbf{y}^{\tau(m)})}{\partial\mathbf{y}^{\tau(m)}}\bigg|_{(\mathbf{y}^{(m)*}, \mathbf{y}^{\tau(m)*})}. \tag{3.266}$$

*Thus, the stability and bifurcation of period-*m *motion can be classified by the eigenvalues of Equation (3.264) at equilibrium* $\mathbf{y}^{(m)*} = \mathbf{y}^{\tau(m)*}$ *with*

$$(n_1, n_2, n_3 \mid n_4, n_5, n_6). \tag{3.267}$$

1. *If all eigenvalues of the equilibrium possess negative real parts, the approximate periodic solution is stable.*
2. *If at least one of the eigenvalues of the equilibrium possesses a positive real part, the approximate periodic solution is unstable.*
3. *The boundaries between stable and unstable equilibriums with higher order singularity give bifurcation and stability conditions with higher order singularity.*

Proof. The proof is similar to the proof of Theorem 3.14. ■

4

Analytical Periodic to Quasi-Periodic Flows

In this chapter, from the idea of Luo (2012, 2013), period-m flows to quasi-periodic flows in nonlinear dynamical systems will be presented. The analytical solutions of quasi-periodic flows in autonomous systems will be discussed, and the analytical solutions of quasi-periodic flows in periodically forced nonlinear dynamical systems will be presented. The analytical solutions of quasi-periodic motions in free and periodically forced vibration systems will be presented. The analytical solutions of quasi-periodic flows for time-delayed nonlinear systems will be presented with/without periodic excitations, and time-delayed nonlinear vibration systems will be discussed as well.

4.1 Nonlinear Dynamical Systems

In this section, analytical quasi-periodic flows in autonomous and periodically forced, nonlinear dynamical systems will be discussed. Consider analytical quasi-periodic flows in autonomous nonlinear systems first.

Theorem 4.1 *Consider a nonlinear dynamical system as*

$$\dot{\mathbf{x}} = \mathbf{f}(\mathbf{x}, \mathbf{p}) \in \mathscr{R}^n \tag{4.1}$$

where $\mathbf{f}(\mathbf{x}, \mathbf{p})$ *is a* C^r*-continuous nonlinear function vector* $(r \geq 1)$.

1. *If such a dynamical system has a period-*m *flow* $\mathbf{x}^{(m)}(t)$ *with finite norm* $\|\mathbf{x}^{(m)}\|$ *and period* $T = 2\pi/\Omega$*, there is a generalized coordinate transformation with* $\theta = \Omega t$ *for the periodic flow of Equation (4.1) in a form of*

$$\mathbf{x}^{(m)}(t) = \mathbf{a}_0^{(m)}(t) + \sum_{k=1}^{\infty} \mathbf{b}_{k/m}(t) \cos\left(\frac{k}{m}\theta\right) + \mathbf{c}_{k/m}(t) \sin\left(\frac{k}{m}\theta\right) \tag{4.2}$$

with

$$\begin{aligned}
\mathbf{a}_1^{(0)} &\equiv \mathbf{a}_0^{(m)} = (a_{01}^{(m)}, a_{02}^{(m)}, \ldots, a_{0n}^{(m)})^{\mathrm{T}}, \\
\mathbf{a}_2^{(k)} &\equiv \mathbf{b}_{k/m} = (b_{k/m1}, b_{k/m2}, \ldots, b_{k/mn})^{\mathrm{T}}, \\
\mathbf{a}_3^{(k)} &\equiv \mathbf{c}_{k/m} = (c_{k/m1}, c_{k/m2}, \ldots, c_{k/mn})^{\mathrm{T}}
\end{aligned} \tag{4.3}$$

Toward Analytical Chaos in Nonlinear Systems, First Edition. Albert C. J. Luo.
© 2014 John Wiley & Sons, Ltd. Published 2014 by John Wiley & Sons, Ltd.

which, under $\|\mathbf{x}^{(m)}(t) - \mathbf{x}^{(m)*}(t)\| < \varepsilon$ *with a prescribed small* $\varepsilon > 0$*, can be approximated by a finite term transformation* $\mathbf{x}^{(m)*}(t)$

$$\mathbf{x}^{(m)*}(t) = \mathbf{a}_0^{(m)}(t) + \sum_{k=1}^{N_0} \mathbf{b}_{k/m}(t)\cos\left(\frac{k}{m}\theta\right) + \mathbf{c}_{k/m}(t)\sin\left(\frac{k}{m}\theta\right) \tag{4.4}$$

and the generalized coordinates are determined by

$$\dot{\mathbf{a}}_{s_0} = \mathbf{f}_{s_0}(\mathbf{a}_{s_0}, \mathbf{p}) \tag{4.5}$$

where

$$\begin{aligned}
\mathbf{k}_0 &= diag(\mathbf{I}_{n\times n}, 2\mathbf{I}_{n\times n}, \ldots, N_0\mathbf{I}_{n\times n}),\\
\mathbf{a}_1^{(0)} &\equiv \mathbf{a}_0^{(m)}, \mathbf{a}_2^{(k)} \equiv \mathbf{b}_{k/m}, \mathbf{a}_3^{(k)} \equiv \mathbf{c}_{k/m};\\
\mathbf{a}_1 &= \mathbf{a}_1^{(0)},\\
\mathbf{a}_2 &= (\mathbf{a}_2^{(1)}, \mathbf{a}_2^{(2)}, \ldots, \mathbf{a}_2^{(N_0)})^{\mathrm{T}} \equiv \mathbf{b}^{(m)},\\
\mathbf{a}_3 &= (\mathbf{a}_3^{(1)}, \mathbf{a}_3^{(2)}, \ldots, \mathbf{a}_3^{(N_0)})^{\mathrm{T}} \equiv \mathbf{c}^{(m)},\\
\mathbf{F}_1 &= \mathbf{F}_0^{(m)},\\
\mathbf{F}_2 &= (\mathbf{F}_{11}^{(m)}, \mathbf{F}_{12}^{(m)}, \ldots, \mathbf{F}_{1N_0}^{(m)})^{\mathrm{T}},\\
\mathbf{F}_3 &= (\mathbf{F}_{21}^{(m)}, \mathbf{F}_{22}^{(m)}, \ldots, \mathbf{F}_{2N_0}^{(m)})^{\mathrm{T}};\\
\mathbf{a}_{s_0} &= (\mathbf{a}_1, \mathbf{a}_2, \mathbf{a}_3)^{\mathrm{T}},\\
\mathbf{f}_{s_0} &= \left(\mathbf{F}_1, -\frac{\Omega}{m}\mathbf{k}_0\mathbf{a}_3 + \mathbf{F}_2, \frac{\Omega}{m}\mathbf{k}_0\mathbf{a}_2 + \mathbf{F}_3\right)^{\mathrm{T}}\\
&\text{for } N_0 = 1, 2, \ldots, \infty;
\end{aligned} \tag{4.6}$$

and

$$\begin{aligned}
\mathbf{F}_0^{(m)}(\mathbf{a}_0^{(m)}, \mathbf{b}^{(m)}, \mathbf{c}^{(m)}) &= \frac{1}{2m\pi}\int_0^{2m\pi} \mathbf{f}(\mathbf{x}^{(m)*}, \mathbf{p})d\theta;\\
\mathbf{F}_{1k}^{(m)}(\mathbf{a}_0^{(m)}, \mathbf{b}^{(m)}, \mathbf{c}^{(m)}) &= \frac{1}{m\pi}\int_0^{2m\pi} \mathbf{f}(\mathbf{x}^{(m)*}, \mathbf{p})\cos\left(\frac{k}{m}\theta\right)d\theta,\\
\mathbf{F}_{2k}^{(m)}(\mathbf{a}_0^{(m)}, \mathbf{b}^{(m)}, \mathbf{c}^{(m)}) &= \frac{1}{m\pi}\int_0^{2m\pi} \mathbf{f}(\mathbf{x}^{(m)*}, \mathbf{p})\sin\left(\frac{k}{m}\theta\right)d\theta\\
&\text{for } k = 1, 2, \ldots, N_0.
\end{aligned} \tag{4.7}$$

2. *After the* kth *Hopf bifurcation with* $p_k\omega_k = \omega_{k-1}(k = 1, 2, \ldots)$ *and* $\omega_0 = \Omega/m$*, there is a dynamical system of coefficients as*

$$\dot{\mathbf{a}}_{s_0s_1\ldots s_k} = \mathbf{f}_{s_0s_1\ldots s_k}(\mathbf{a}_{s_0s_1\ldots s_k}, \mathbf{p}) \tag{4.8}$$

where

$$\begin{aligned}
\mathbf{a}_{s_0s_1\ldots s_k} &= (\mathbf{a}_{s_0s_1\ldots s_{k-1}1}, \mathbf{a}_{s_0s_1\ldots s_{k-1}2}, \mathbf{a}_{s_0s_1\ldots s_{k-1}3})^{\mathrm{T}},\\
\mathbf{f}_{s_0s_1\ldots s_k} &= (\mathbf{F}_{s_0s_1\ldots s_{k-1}1}, -\omega_k\mathbf{k}_k\mathbf{a}_{s_0s_1\ldots s_{k-1}3} + \mathbf{F}_{s_0s_1\ldots s_{k-1}2},\\
&\qquad \omega_k\mathbf{k}_k\mathbf{a}_{s_0s_1\ldots s_{k-1}2} + \mathbf{F}_{s_0s_1\ldots s_{k-1}3})^{\mathrm{T}},
\end{aligned}$$

$$\mathbf{k}_k = diag(\mathbf{I}_{n_{k-1}\times n_{k-1}}, 2\mathbf{I}_{n_{k-1}\times n_{k-1}}, \dots, N_k\mathbf{I}_{n_{k-1}\times n_{k-1}}),$$
$$n_{k-1} = n(2N_0+1)(2N_1+1)\dots(2N_{k-1}+1) \tag{4.9}$$

with a periodic solution as

$$\mathbf{a}_{s_0s_1\dots s_k} = \mathbf{a}^{(0)}_{s_0s_1\dots s_k1}(t) + \sum_{l_{k+1}=1}^{\infty} \mathbf{a}^{(l_{k+1})}_{s_0s_1\dots s_k2}(t)\cos(l_{k+1}\theta_{k+1}) + \mathbf{a}^{(l_{k+1})}_{s_0s_1\dots s_k3}(t)\sin(l_{k+1}\theta_{k+1}) \tag{4.10}$$

with

$$\begin{aligned}
&s_i = 1,2,3\ (i=0,1,2,\dots,k),\\
&\mathbf{a}_{s_0s_1\dots s_k1} = \mathbf{a}^{(0)}_{s_0s_1\dots s_k1}\\
&\mathbf{a}_{s_0s_1\dots s_k2} = (\mathbf{a}^{(1)}_{s_0s_1\dots s_k2}, \mathbf{a}^{(2)}_{s_0s_1\dots s_k2}, \dots, \mathbf{a}^{(N_{k+1})}_{s_0s_1\dots s_k2})^{\mathrm{T}},\\
&\mathbf{a}_{s_0s_1\dots s_k3} = (\mathbf{a}^{(1)}_{s_0s_1\dots s_k3}, \mathbf{a}^{(2)}_{s_0s_1\dots s_k3}, \dots, \mathbf{a}^{(N_{k+1})}_{s_0s_1\dots s_k3})^{\mathrm{T}};\\
&\mathbf{a}_{s_0s_1\dots s_{k-1}1} = \mathbf{a}^{(0)}_{s_0s_1\dots s_{k-1}1}\\
&\mathbf{a}_{s_0s_1\dots s_{k-1}2} = (\mathbf{a}^{(1)}_{s_0s_1\dots s_{k-1}2}, \mathbf{a}^{(2)}_{s_0s_1\dots s_{k-1}2}, \dots, \mathbf{a}^{(N_k)}_{s_0s_1\dots s_{k-1}2})^{\mathrm{T}},\\
&\mathbf{a}_{s_0s_1\dots s_{k-1}3} = (\mathbf{a}^{(1)}_{s_0s_1\dots s_{k-1}3}, \mathbf{a}^{(2)}_{s_0s_1\dots s_{k-1}3}, \dots, \mathbf{a}^{(N_k)}_{s_0s_1\dots s_{k-1}3})^{\mathrm{T}};\\
&\vdots\\
&\mathbf{a}_1 = \mathbf{a}^{(0)}_1\\
&\mathbf{a}_2 = (\mathbf{a}^{(1)}_2, \mathbf{a}^{(2)}_2, \dots, \mathbf{a}^{(N_0)}_2)^{\mathrm{T}},\\
&\mathbf{a}_3 = (\mathbf{a}^{(1)}_3, \mathbf{a}^{(2)}_3, \dots, \mathbf{a}^{(N_0)}_3)^{\mathrm{T}};
\end{aligned} \tag{4.11}$$

which, under $\|\mathbf{a}_{s_0s_1\dots s_k}(t) - \mathbf{a}^*_{s_0s_1\dots s_k}(t)\| < \varepsilon$ *with a prescribed small* $\varepsilon > 0$, *can be approximated by a finite term transformation* $\mathbf{a}^*_{s_0s_1\dots s_k}(t)$

$$\mathbf{a}^*_{s_0s_1\dots s_k} = \mathbf{a}^{(0)}_{s_0s_1\dots s_k1}(t) + \sum_{l_{k+1}=1}^{N_{k+1}} \mathbf{a}^{(l_{k+1})}_{s_0s_1\dots s_k2}(t)\cos(l_{k+1}\theta_{k+1}) + \mathbf{a}^{(l_{k+1})}_{s_0s_1\dots s_k3}(t)\sin(l_{k+1}\theta_{k+1}) \tag{4.12}$$

and the generalized coordinates are determined by

$$\dot{\mathbf{a}}_{s_0s_1\dots s_{k+1}} = \mathbf{f}_{s_0s_1\dots s_{k+1}}(\mathbf{a}_{s_0s_1\dots s_{k+1}}, \mathbf{p}) \tag{4.13}$$

where

$$\begin{aligned}
\mathbf{a}_{s_0s_1\dots s_{k+1}} &= (\mathbf{a}_{s_0s_1\dots s_k1}, \mathbf{a}_{s_0s_1\dots s_k2}, \mathbf{a}_{s_0s_1\dots s_k3})^{\mathrm{T}},\\
\mathbf{f}_{s_0s_1\dots s_{k+1}} &= (\mathbf{F}_{s_0s_1\dots s_k1}, -\omega_{k+1}\mathbf{k}_{k+1}\mathbf{a}_{s_0s_1\dots s_k3} + \mathbf{F}_{s_0s_1\dots s_k2},\\
&\qquad \omega_{k+1}\mathbf{k}_{k+1}\mathbf{a}_{s_0s_1\dots s_k2} + \mathbf{F}_{s_0s_1\dots s_k3})^{\mathrm{T}}
\end{aligned} \tag{4.14}$$

and

$$
\begin{aligned}
&\mathbf{k}_{k+1} = diag(\mathbf{I}_{n_k \times n_k}, 2\mathbf{I}_{n_k \times n_k}, \ldots, N_{k+1}\mathbf{I}_{n_k \times n_k}),\\
&n_k = n(2N_0+1)(2N_1+1)\ldots(2N_k+1);\\
&\mathbf{a}_{s_0 s_1 \ldots s_k 1} = \mathbf{a}^{(0)}_{s_0 s_1 \ldots s_k 1},\\
&\mathbf{a}_{s_0 s_1 \ldots s_k 2} = (\mathbf{a}^{(1)}_{s_0 s_1 \ldots s_k 2}, \mathbf{a}^{(2)}_{s_0 s_1 \ldots s_k 2}, \ldots, \mathbf{a}^{(N_{k+1})}_{s_0 s_1 \ldots s_k 2})^{\mathrm{T}},\\
&\mathbf{a}_{s_0 s_1 \ldots s_k 3} = (\mathbf{a}^{(1)}_{s_0 s_1 \ldots s_k 3}, \mathbf{a}^{(2)}_{s_0 s_1 \ldots s_k 3}, \ldots, \mathbf{a}^{(N_{k+1})}_{s_0 s_1 \ldots s_k 3})^{\mathrm{T}};\\
&\mathbf{F}_{s_0 s_1 \ldots s_k 1} = \mathbf{F}^{(0)}_{s_0 s_1 \ldots s_k 1},\\
&\mathbf{F}_{s_0 s_1 \ldots s_k 2} = (\mathbf{F}^{(1)}_{s_0 s_1 \ldots s_k 2}, \mathbf{F}^{(2)}_{s_0 s_1 \ldots s_k 2}, \ldots, \mathbf{F}^{(N_{k+1})}_{s_0 s_1 \ldots s_k 2})^{\mathrm{T}},\\
&\mathbf{F}_{s_0 s_1 \ldots s_k 3} = (\mathbf{F}^{(1)}_{s_0 s_1 \ldots s_k 3}, \mathbf{F}^{(2)}_{s_0 s_1 \ldots s_k 3}, \ldots, \mathbf{F}^{(N_{k+1})}_{s_0 s_1 \ldots s_k 3})^{\mathrm{T}}\\
&\text{for } N_{k+1} = 1, 2, \ldots, \infty;
\end{aligned}
\tag{4.15}
$$

and

$$
\begin{aligned}
&\mathbf{F}_{s_0 s_1 \ldots s_k 1}(\mathbf{a}_{s_0 s_1 \ldots s_{k+1}}, \mathbf{p}) = \frac{1}{2\pi}\int_0^{2\pi} \mathbf{f}_{s_0 s_1 \ldots s_k}(\mathbf{a}^{*}_{s_0 s_1 \ldots s_k}, \mathbf{p}) d\theta_{k+1};\\
&\mathbf{F}^{(l_{k+1})}_{s_0 s_1 \ldots s_k 2}(\mathbf{a}_{s_0 s_1 \ldots s_{k+1}}, \mathbf{p}) = \frac{1}{\pi}\int_0^{2\pi} \mathbf{f}_{s_0 s_1 \ldots s_k}(\mathbf{a}^{*}_{s_0 s_1 \ldots s_k}, \mathbf{p})) \cos(l_{k+1}\theta_{k+1}) d\theta_{k+1},\\
&\mathbf{F}^{(l_{k+1})}_{s_0 s_1 \ldots s_k 3}(\mathbf{a}_{s_0 s_1 \ldots s_{k+1}}, \mathbf{p}) = \frac{1}{\pi}\int_0^{2\pi} \mathbf{f}_{s_0 s_1 \ldots s_k}(\mathbf{a}^{*}_{s_0 s_1 \ldots s_k}, \mathbf{p}) \sin(l_{k+1}\theta_{k+1}) d\theta_{k+1}\\
&\text{for } l_{k+1} = 1, 2, \ldots, N_{k+1}.
\end{aligned}
\tag{4.16}
$$

3. *Equation (4.13) becomes*

$$\dot{\mathbf{z}}_{s_0 s_1 \ldots s_{k+1}} = \mathbf{f}_{s_0 s_1 \ldots s_{k+1}}(\mathbf{z}_{s_0 s_1 \ldots s_{k+1}}) \tag{4.17}$$

where

$$
\begin{aligned}
\mathbf{z}_{s_0 s_1 \ldots s_{k+1}} &= (\mathbf{a}_{s_0 s_1 \ldots s_k 1}, \mathbf{a}_{s_0 s_1 \ldots s_k 2}, \mathbf{a}_{s_0 s_1 \ldots s_k 3})^{\mathrm{T}},\\
\mathbf{f}_{s_0 s_1 \ldots s_{k+1}} &= (\mathbf{F}_{s_0 s_1 \ldots s_k 1}, -\omega_{k+1}\mathbf{k}_{k+1}\mathbf{a}_{s_0 s_1 \ldots s_k 3} + \mathbf{F}_{s_0 s_1 \ldots s_k 2},\\
&\qquad \omega_{k+1}\mathbf{k}_{k+1}\mathbf{a}_{s_0 s_1 \ldots s_k 2} + \mathbf{F}_{s_0 s_1 \ldots s_k 3})^{\mathrm{T}}.
\end{aligned}
\tag{4.18}
$$

If equilibrium $\mathbf{z}^{*}_{s_0 s_1 \ldots s_{k+1}}$ of Equation (4.17) (i.e., $\mathbf{f}_{s_0 s_1 \ldots s_{k+1}}(\mathbf{z}^{*}_{s_0 s_1 \ldots s_{k+1}}) = \mathbf{0}$) exists, then the approximate solution of the periodic flow of kth *generalized coordinates for the period-*m *flow exists as in Equation (4.12). In the vicinity of equilibrium $\mathbf{z}^{*}_{s_0 s_1 \ldots s_{k+1}}$, with*

$$\mathbf{z}_{s_0 s_1 \ldots s_{k+1}} = \mathbf{z}^{*}_{s_0 s_1 \ldots s_{k+1}} + \Delta\mathbf{z}_{s_0 s_1 \ldots s_{k+1}}, \tag{4.19}$$

the linearized equation of Equation (4.17) is

$$\Delta\dot{\mathbf{z}}_{s_0s_1\ldots s_{k+1}} = D\mathbf{f}_{s_0s_1\ldots s_{k+1}}(\mathbf{z}^*_{s_0s_1\ldots s_{k+1}})\Delta\mathbf{z}_{s_0s_1\ldots s_{k+1}} \tag{4.20}$$

and the eigenvalue analysis of equilibrium $\mathbf{z}^*$ *is given by*

$$|D\mathbf{f}_{s_0s_1\ldots s_{k+1}}(\mathbf{z}^*_{s_0s_1\ldots s_{k+1}}) - \lambda\mathbf{I}_{n_k(2N_{k+1}+1)\times n_k(2N_{k+1}+1)}| = 0 \tag{4.21}$$

where

$$D\mathbf{f}_{s_0s_1\ldots s_{k+1}}(\mathbf{z}^*_{s_0s_1\ldots s_{k+1}}) = \left.\frac{\partial\mathbf{f}_{s_0s_1\ldots s_{k+1}}\left(\mathbf{z}_{s_0s_1\ldots s_{k+1}}\right)}{\partial\mathbf{z}_{s_0s_1\ldots s_{k+1}}}\right|_{\mathbf{z}^*_{s_0s_1\ldots s_{k+1}}}. \tag{4.22}$$

The stability and bifurcation of such a periodic flow of the kth *generalized coordinates can be classified by the eigenvalues of* $D\mathbf{f}_{s_0s_1\ldots s_{k+1}}(\mathbf{z}^*_{s_0s_1\ldots s_{k+1}})$ *with*

$$(n_1, n_2, n_3|n_4, n_5, n_6). \tag{4.23}$$

a. *If all eigenvalues of the equilibrium possess negative real parts, the approximate quasi-periodic solution is stable.*
b. *If at least one of the eigenvalues of the equilibrium possesses a positive real part, the approximate quasi-periodic solution is unstable.*
c. *The boundaries between stable and unstable equilibriums with higher order singularity give bifurcation and stability conditions with higher order singularity.*

4. *For the* kth *order Hopf bifurcation of period-*m *flow, a relation exists as*

$$p_k\omega_k = \omega_{k-1}. \tag{4.24}$$

a. *If* p_k *is an irrational number, the* kth-*order Hopf bifurcation of the period-*m *flow is called the quasi-period-*p_k *Hopf bifurcation, and the corresponding solution of the* kth *generalized coordinates is* p_k*-quasi-periodic to the system of the* $(k-1)$th *generalized coordinates.*
b. *If* $p_k = 2$*, the* kth-*order Hopf bifurcation of the period-*m *flow is called a period-doubling Hopf bifurcation (or a period-2 Hopf bifurcation), and the corresponding solution of the* kth *generalized coordinates is period-doubling to the system of the* $(k-1)$th *generalized coordinates.*
c. *If* $p_k = q$ *with an integer* q*, the* kth-*order Hopf bifurcation of the period-*m *flow is called a period-*q *Hopf bifurcation, and the corresponding solution of the* kth *generalized coordinates is of* q*-times period to the system of the* $(k-1)$th *generalized coordinates.*
d. *If* $p_k = p/q$ $(p, q$ *are irreducible integer), the* kth-*order Hopf bifurcation of the period-*m *flow is called a period-*p/q *Hopf bifurcation, and the corresponding solution of the* kth *generalized coordinates is* p/q*-periodic to the system of the* $(k-1)$th *generalized coordinates.*

Proof. The proof of this theorem is similar to the proof of Theorem 3.2. ■

From the theorem, the mathematical structure of the analytical solutions of quasi-periodic flow relative to the period-*m* flow in dynamical systems in Equation (4.1) is discussed as follows.

For the zero-order quasi-periodic flow relative to the period-m flow, it is the period-m flow as in Equation (4.2), that is,

$$\mathbf{x}(t) \equiv \mathbf{x}_0(t) = \mathbf{a}_1^{(0)}(t) + \sum_{l_0=1}^{\infty} \mathbf{a}_2^{(l_0)}(t)\cos(l_0\theta_0) + \mathbf{a}_3^{(l_0)}(t)\sin(l_0\theta_0) \tag{4.25}$$

where

$$\begin{aligned}
\theta_0 &= \frac{1}{m}\Omega t,\\
\mathbf{F}_1^{(0)}(\mathbf{a}_1,\mathbf{a}_2,\mathbf{a}_3) &= \frac{1}{2\pi}\int_0^{2\pi} \mathbf{f}(\mathbf{x}^{(m)*},\mathbf{p})d\theta_0\\
&= \frac{1}{2m\pi}\int_0^{2m\pi} \mathbf{f}(\mathbf{x}^{(m)*},\mathbf{p})d\theta\\
&= \frac{1}{mT}\int_0^{mT} \mathbf{f}(\mathbf{x}^{(m)*},\mathbf{p})dt;\\
\mathbf{F}_2^{(l_0)}(\mathbf{a}_1,\mathbf{a}_2,\mathbf{a}_3) &= \frac{1}{\pi}\int_0^{2\pi} \mathbf{f}(\mathbf{x}^{(m)*},\mathbf{p})\cos(l_0\theta_0)d\theta_0\\
&= \frac{1}{m\pi}\int_0^{2m\pi} \mathbf{f}(\mathbf{x}^{(m)*},\mathbf{p})\cos\left(\frac{l_0}{m}\theta\right)d\theta\\
&= \frac{2}{mT}\int_0^{mT} \mathbf{f}(\mathbf{x}^{(m)*},\mathbf{p})\cos\left(\frac{l_0}{m}\Omega t\right)dt;\\
\mathbf{F}_3^{(l_0)}(\mathbf{a}_1,\mathbf{a}_2,\mathbf{a}_3) &= \frac{1}{\pi}\int_0^{2\pi} \mathbf{f}(\mathbf{x}^{(m)*},\mathbf{p})\cos(l_0\theta_0)d\theta_0\\
&= \frac{1}{m\pi}\int_0^{2m\pi} \mathbf{f}(\mathbf{x}^{(m)*},\mathbf{p})\sin\left(\frac{l_0}{m}\theta\right)d\theta\\
&= \frac{2}{mT}\int_0^{mT} \mathbf{f}(\mathbf{x}^{(m)*},\mathbf{p})\sin\left(\frac{l_0}{m}\Omega t\right)dt.
\end{aligned} \tag{4.26}$$

The approximate solution of the zero-order quasi-periodic flows can be expressed as in Equation (4.4), that is,

$$\mathbf{x}^*(t) \equiv \mathbf{x}_0^*(t) = \mathbf{a}_1^{(0)}(t) + \sum_{l_0=1}^{N_0} \mathbf{a}_2^{(l_0)}(t)\cos(l_0\theta_0) + \mathbf{a}_3^{(l_0)}(t)\sin(l_0\theta_0). \tag{4.27}$$

For periodic flows, $\mathbf{a}_1^{(0)}(t)$, $\mathbf{a}_2^{(l_0)}(t)$, and $\mathbf{a}_3^{(l_0)}(t)$ for $l_0 = 1, 2, \ldots$ are constant, independent of time t. However, for stability and bifurcation, such coefficients will change with time t.

For the first-order quasi-periodic flow relative to the period-m flow, the coefficients (or generalized coordinates) with time are periodic with oscillation frequency $\omega_1 = p_1\omega_0 = p_1\Omega/m$ (p_1 is an irrational number). Thus, with $\theta_1 = \omega_1 t$,

$$\mathbf{a}_{s_0} = \mathbf{a}_{s_0 1}^{(0)}(t) + \sum_{l_1=1}^{\infty} \mathbf{a}_{s_0 2}^{(l_1)}(t)\cos(l_1\theta_1) + \mathbf{a}_{s_0 3}^{(l_1)}(t)\sin(l_1\theta_1) \tag{4.28}$$

from which

$$\mathbf{a}_1 = \mathbf{a}_{11}^{(0)}(t) + \sum_{l_1=1}^{\infty} \mathbf{a}_{12}^{(l_1)}(t)\cos(l_1\theta_1) + \mathbf{a}_{13}^{(l_1)}(t)\sin(l_1\theta_1),$$

$$\mathbf{a}_2 = \mathbf{a}_{21}^{(0)}(t) + \sum_{l_1=1}^{\infty} \mathbf{a}_{22}^{(l_1)}(t)\cos(l_1\theta_1) + \mathbf{a}_{23}^{(l_1)}(t)\sin(l_1\theta_1),$$

$$\mathbf{a}_3 = \mathbf{a}_{31}^{(0)}(t) + \sum_{l_1=1}^{\infty} \mathbf{a}_{32}^{(l_1)}(t)\cos(l_1\theta_1) + \mathbf{a}_{33}^{(l_1)}(t)\sin(l_1\theta_1) \tag{4.29}$$

and for $N_0 = 1, 2, \ldots, \infty$

$$\mathbf{a}_1 = \mathbf{a}_1^{(0)}, \mathbf{a}_2 = (\mathbf{a}_2^{(1)}, \mathbf{a}_2^{(2)}, \ldots, \mathbf{a}_2^{(N_0)})^{\mathrm{T}}, \mathbf{a}_3 = (\mathbf{a}_3^{(1)}, \mathbf{a}_3^{(2)}, \ldots, \mathbf{a}_3^{(N_0)})^{\mathrm{T}} \tag{4.30}$$

with

$$\mathbf{a}_1^{(0)} = \mathbf{a}_{11}^{(00)}(t) + \sum_{l_1=1}^{\infty} \mathbf{a}_{12}^{(0l_1)}(t)\cos(l_1\theta_1) + \mathbf{a}_{13}^{(0l_1)}(t)\sin(l_1\theta_1),$$

$$\mathbf{a}_2^{(l_0)} = \mathbf{a}_{21}^{(l_00)}(t) + \sum_{l_1=1}^{\infty} \mathbf{a}_{22}^{(l_0l_1)}(t)\cos(l_1\theta_1) + \mathbf{a}_{23}^{(l_0l_1)}(t)\sin(l_1\theta_1),$$

$$\mathbf{a}_2^{(l_0)} = \mathbf{a}_{31}^{(l_00)}(t) + \sum_{l_1=1}^{\infty} \mathbf{a}_{32}^{(l_0l_1)}(t)\cos(l_1\theta_1) + \mathbf{a}_{33}^{(l_0l_1)}(t)\sin(l_1\theta_1) \tag{4.31}$$

where

$$\begin{aligned}
&\mathbf{a}_{s_0s_1} = (\mathbf{a}_{s_01}, \mathbf{a}_{s_02}, \mathbf{a}_{s_03})^{\mathrm{T}};\\
&\mathbf{a}_{s_01} = \mathbf{a}_{s_01}^{(0)}, \mathbf{a}_{s_02} = (\mathbf{a}_{s_02}^{(1)}, \mathbf{a}_{s_02}^{(2)}, \ldots, \mathbf{a}_{s_02}^{(N_1)})^{\mathrm{T}}, \mathbf{a}_{s_03} = (\mathbf{a}_{s_03}^{(1)}, \mathbf{a}_{s_03}^{(2)}, \ldots, \mathbf{a}_{s_03}^{(N_1)})^{\mathrm{T}};\\
&\mathbf{a}_{s_01}^{(0)} = (\mathbf{a}_{11}^{(00)}, \mathbf{a}_{21}^{(l_00)}, \mathbf{a}_{31}^{(l_00)})^{\mathrm{T}}, \mathbf{a}_{s_02}^{(l_1)} = (\mathbf{a}_{12}^{(0l_1)}, \mathbf{a}_{22}^{(l_0l_1)}, \mathbf{a}_{32}^{(l_0l_1)})^{\mathrm{T}},\\
&\mathbf{a}_{s_03}^{(l_1)} = (\mathbf{a}_{13}^{(0l_1)}, \mathbf{a}_{23}^{(l_0l_1)}, \mathbf{a}_{33}^{(l_0l_1)})^{\mathrm{T}}\\
&\text{for } l_0 = 1, 2, \ldots, l_1 = 1, 2, 3, \ldots.
\end{aligned} \tag{4.32}$$

Thus

$$\begin{aligned}
\mathbf{x}_{s_0s_1}(t) &= \mathbf{a}_{11}^{(00)}(t) + \sum_{l_1=1}^{\infty} \mathbf{a}_{12}^{(0l_1)}(t)\cos(l_1\theta_1) + \mathbf{a}_{13}^{(0l_1)}(t)\sin(l_1\theta_1)\\
&+ \sum_{l_0=1}^{\infty}\left[\mathbf{a}_{21}^{(l_00)} + \sum_{l_1=1}^{\infty}[\mathbf{a}_{22}^{(l_0l_1)}(t)\cos(l_1\theta_1) + \mathbf{a}_{23}^{(l_0l_1)}(t)\sin(l_1\theta_1)\right]\cos(l_0\theta_0)\\
&+ \sum_{l_0=1}^{\infty}\left[\mathbf{a}_{31}^{(l_00)}(t) + \sum_{l_1=1}^{\infty}\mathbf{a}_{32}^{(l_0l_1)}(t)\cos(l_1\theta_1) + \mathbf{a}_{33}^{(l_0l_1)}(t)\sin(l_1\theta_1)\right]\sin(l_0\theta_0).
\end{aligned} \tag{4.33}$$

If an approximate solution for the periodic flow of the coefficient system is

$$\mathbf{a}_{s_0}^* = \mathbf{a}_{s_0 1}^{(0)}(t) + \sum_{l_1=1}^{N_1} \mathbf{a}_{s_0 2}^{(l_1)}(t)\cos(l_1\theta_1) + \mathbf{a}_{s_0 3}^{(l_1)}(t)\sin(l_1\theta_1) \tag{4.34}$$

the approximate solution relative to the period-m flow in such a dynamical system is

$$\begin{aligned}
\mathbf{x}_{s_0 s_1}^*(t) &= \mathbf{a}_{11}^{(00)}(t) + \sum_{l_1=1}^{N_1} \mathbf{a}_{12}^{(0l_1)}(t)\cos(l_1\theta_1) + \mathbf{a}_{13}^{(0l_1)}(t)\sin(l_1\theta_1) \\
&\quad + \sum_{l_0=1}^{N_0} \left[\mathbf{a}_{21}^{(l_0 0)} + \sum_{l_1=1}^{N_1} [\mathbf{a}_{22}^{(l_0 l_1)}(t)\cos(l_1\theta_1) + \mathbf{a}_{23}^{(l_0 l_1)}(t)\sin(l_1\theta_1) \right] \cos(l_0\theta_0) \\
&\quad + \sum_{l_0=1}^{N_0} [\mathbf{a}_{31}^{(l_0 0)}(t) + \sum_{l_1=1}^{N_1} \mathbf{a}_{32}^{(l_0 l_1)}(t)\cos(l_1\theta_1) + \mathbf{a}_{33}^{(l_0 l_1)}(t)\sin(l_1\theta_1)]\sin(l_0\theta_0).
\end{aligned} \tag{4.35}$$

For the Hopf bifurcation of period-m motion, we assume

$$p_1\omega_1 = \omega_0. \tag{4.36}$$

1. If p_1 is an irrational number, the first-order Hopf bifurcation of the period-m flow is called a quasi-period-p_1 Hopf bifurcation, the corresponding solution in Equation (4.33) is p_1-quasi-periodic.
2. If $p_k = 2$, the first-order Hopf bifurcation of the period-m flow is called a period-doubling Hopf bifurcation (or a period-2 Hopf bifurcation), and the corresponding solution in Equation (4.33) is period-doubling as

$$\mathbf{x}^{(2)}(t) = \mathbf{a}_0^{(2)}(t) + \sum_{k=1}^{\infty} \mathbf{b}_{k/2}(t)\cos\left(\frac{k}{2}\theta_0\right) + \mathbf{c}_{k/2}(t)\sin\left(\frac{k}{2}\theta_0\right) \tag{4.37}$$

   where k is formed from l_0, l_1 with 2. $\mathbf{a}_0^{(2)}(t), \mathbf{b}_{k/2}(t)$ and $\mathbf{c}_{k/2}(t)$ are linear combinations of $\mathbf{a}_{11}^{(00)}(t), \dots, \mathbf{a}_{33}^{(l_0 l_1)}$, but they can be determined by Equation (4.7) with m replaced by $2m$.
3. If $p_1 = q$ with an integer q, the first order Hopf bifurcation of the period-m flow is called a period-q Hopf bifurcation, and the corresponding solution in Equation (4.33) is q-times periodic as

$$\mathbf{x}^{(q)}(t) = \mathbf{a}_0^{(q)}(t) + \sum_{k=1}^{\infty} \mathbf{b}_{k/q}(t)\cos\left(\frac{k}{q}\theta_0\right) + \mathbf{c}_{k/q}(t)\sin\left(\frac{k}{q}\theta_0\right) \tag{4.38}$$

   where k is formed from l_0, l_1 with q. $\mathbf{a}_0^{(q)}(t), \mathbf{b}_{k/q}(t)$ and $\mathbf{c}_{k/q}(t)$ are linear combinations of $\mathbf{a}_{11}^{(00)}(t), \dots, \mathbf{a}_{33}^{(l_0 l_1)}$, but they can be determined by Equation (4.7) with m replaced by qm.
4. If $p_1 = p/q$ where p, q are irreducible, the first order Hopf bifurcation of the period-m flow is called a period-p/q Hopf bifurcation, and the corresponding solution in Equation (4.33) is p/q-periodic as

$$\mathbf{x}^{(p/q)}(t) = \mathbf{a}_0^{(p/q)}(t) + \sum_{k=1}^{\infty} \mathbf{b}_{k/p_1}(t)\cos\left(\frac{k}{p_1}\theta_0\right) + \mathbf{c}_{k/p_1}(t)\sin\left(\frac{k}{p_1}\theta_0\right) \tag{4.39}$$

where k is formed from l_0, l_1 with p_1. $\mathbf{a}_0^{(p/q)}(t), \mathbf{b}_{k/p_1}(t)$ and $\mathbf{c}_{k/p_1}(t)$ are linear combinations of $\mathbf{a}_{11}^{(00)}(t), \ldots, \mathbf{a}_{33}^{(l_0l_1)}$, but they can be determined by Equation (4.7) with m replaced by $p_1 m = pm/q$.

In addition, for the p/q-periodic flow of the first generalized coordinates, the Fourier coefficients are computed by

$$\begin{aligned}
\mathbf{F}_{s_01}^{(l_00)}(\mathbf{a}_{s_01}, \mathbf{a}_{s_02}, \mathbf{a}_{s_03}) &= \frac{1}{2\pi}\int_0^{2\pi} \mathbf{f}_{s_0}(\mathbf{a}_{s_0}^*, \mathbf{p})d\theta_1 \\
&= \frac{q}{2p\pi}\int_0^{2p\pi/q} \mathbf{f}_{s_0}(\mathbf{a}_{s_0}^*, \mathbf{p})d\theta_0 \\
&= \frac{q}{pT_0}\int_0^{pT_0/q} \mathbf{f}_{s_0}(\mathbf{a}_{s_0}^*, \mathbf{p})dt; \\
\mathbf{F}_{s_02}^{(l_0l_1)}(\mathbf{a}_{s_01}, \mathbf{a}_{s_02}, \mathbf{a}_{s_03}) &= \frac{1}{\pi}\int_0^{2\pi} \mathbf{f}_{s_0}(\mathbf{a}_{s_0}^*, \mathbf{p})\cos(l_1\theta_1)d\theta_1 \\
&= \frac{q}{p\pi}\int_0^{2p\pi/q} \mathbf{f}_{s_0}(\mathbf{a}_{s_0}^*, \mathbf{p})\cos\left(\frac{ql_1}{p}\theta_0\right)d\theta_0 \\
&= \frac{2q}{pT_0}\int_0^{pT_0/q} \mathbf{f}_{s_0}(\mathbf{a}_{s_0}^*, \mathbf{p})\cos\left(\frac{ql_1}{p}\omega_0 t\right)dt; \\
\mathbf{F}_3^{(l_1)}(\mathbf{a}_{s_01}, \mathbf{a}_{s_02}, \mathbf{a}_{s_03}) &= \frac{1}{\pi}\int_0^{2\pi} \mathbf{f}_{s_0}(\mathbf{a}_{s_0}^*, \mathbf{p})\cos(l_1\theta_1)d\theta_1 \\
&= \frac{q}{p\pi}\int_0^{2p\pi/q} \mathbf{f}_{s_0}(\mathbf{a}_{s_0}^*, \mathbf{p})\sin\left(\frac{ql_1}{p}\theta_0\right)d\theta_0 \\
&= \frac{2q}{pT_0}\int_0^{pT_0/q} \mathbf{f}_{s_0}(\mathbf{a}_{s_0}^*, \mathbf{p})\sin\left(\frac{ql_1}{p}\omega_0 t\right)dt.
\end{aligned} \tag{4.40}$$

Similarly, for the second Hopf bifurcation of the period-m flow, the approximate solution of periodic flow is expressed as

$$\begin{aligned}
\mathbf{x}_{s_0s_1s_2}^*(t) &= \mathbf{a}_{111}^{(000)}(t) + \sum_{l_2=1}^{N_2} \mathbf{a}_{112}^{(00l_2)}(t)\cos(l_2\theta_2) + \mathbf{a}_{113}^{(00l_2)}(t)\sin(l_2\theta_2) \\
&+ \sum_{l_1=1}^{N_1}\left[\mathbf{a}_{121}^{(0l_10)} + \sum_{l_2=1}^{N_2} \mathbf{a}_{122}^{(0l_1l_2)}(t)\cos(l_2\theta_2) + \mathbf{a}_{123}^{(0l_1l_2)}(t)\sin(l_2\theta_2)\right]\cos(l_1\theta_1) \\
&+ \sum_{l_1=1}^{N_1}\left[\mathbf{a}_{131}^{(0l_10)} + \sum_{l_2=1}^{N_2} \mathbf{a}_{132}^{(0l_1l_2)}(t)\cos(l_2\theta_2) + \mathbf{a}_{133}^{(0l_1l_2)}(t)\sin(l_2\theta_2)\right]\sin(l_1\theta_1) \\
&+ \sum_{l_0=1}^{N_0}\left[\mathbf{a}_{211}^{(l_000)} + \sum_{l_2=1}^{N_2} \mathbf{a}_{212}^{(l_00l_2)}(t)\cos(l_2\theta_2) + \mathbf{a}_{213}^{(l_00l_2)}(t)\sin(l_2\theta_2)\right]\cos(l_0\theta_0)
\end{aligned}$$

$$
\begin{aligned}
&+\sum_{l=1}^{N_0}\sum_{l_1=1}^{N_1}\left[\mathbf{a}_{221}^{(l_0 l_1 0)}(t)+\sum_{l_2=1}^{N_2}\mathbf{a}_{222}^{(l_0 l_1 l_2)}(t)\cos(l_2\theta_2)+\mathbf{a}_{223}^{(l_0 l_1 l_2)}(t)\sin(l_2\theta_2)\right]\cos(l_1\theta_1)\cos(l_0\theta_0)\\
&+\sum_{l=1}^{N_0}\sum_{l_1=1}^{N_1}\left[\mathbf{a}_{231}^{(l_0 l_1 0)}(t)+\sum_{l_2=1}^{N_2}\mathbf{a}_{232}^{(l_0 l_1 l_2)}(t)\cos(l_2\theta_2)+\mathbf{a}_{233}^{(l_0 l_1 l_2)}(t)\sin(l_2\theta_2)\right]\sin(l_1\theta_1)\cos(l_0\theta_0)\\
&+\sum_{l_0=1}^{N_0}\left[\mathbf{a}_{311}^{(l_0 0 0)}+\sum_{l_2=1}^{N_2}\mathbf{a}_{312}^{(l_0 0 l_2)}(t)\cos(l_2\theta_2)+\mathbf{a}_{313}^{(l_0 0 l_2)}(t)\sin(l_2\theta_2)\right]\sin(l_0\theta_0)\\
&+\sum_{l_0=1}^{N_0}\sum_{l_1=1}^{N_1}\left[\mathbf{a}_{321}^{(l_0 l_1 0)}(t)+\sum_{l_2=1}^{N_2}\mathbf{a}_{322}^{(l_0 l_1 l_2)}(t)\cos(l_2\theta_2)+\mathbf{a}_{323}^{(l_0 l_1 l_2)}(t)\sin(l_2\theta_2)\right]\cos(l_1\theta_1)\sin(l_0\theta_0)\\
&+\sum_{l_0=1}^{N_0}\sum_{l_1=1}^{N_1}\left[\mathbf{a}_{331}^{(l_0 l_1 0)}(t)+\sum_{l_2=1}^{N_2}\mathbf{a}_{332}^{(l_0 l_1 l_2)}(t)\cos(l_2\theta_2)+\mathbf{a}_{333}^{(l_0 l_1 l_2)}(t)\sin(l_2\theta_2)\right]\sin(l_1\theta_1)\sin(l_0\theta_0).
\end{aligned}
\tag{4.41}
$$

When $N_i \to \infty$ $(i = 0, 1, 2)$, the foregoing expression gives the exact solution for the periodic motion after the second order Hopf bifurcation of the period-m flow in such a nonlinear dynamical system. After the kth order Hopf bifurcation of period-m motion, the approximate solution of periodic flow is expressed by

$$
\begin{aligned}
&\mathbf{x}^*_{s_0 s_1 \ldots s_k}(t)=\mathbf{a}_{1\ldots 11}^{(0\ldots 00)}(t)+\sum_{l_k=1}^{N_k}\mathbf{a}_{1\ldots 12}^{(0\ldots 0 l_k)}(t)\cos(l_k\theta_k)+\mathbf{a}_{1\ldots 13}^{(0\ldots 0 l_k)}(t)\sin(l_k\theta_k)\\
&+\sum_{l_{k-1}=1}^{N_{k-1}}\left[\mathbf{a}_{1\ldots 21}^{(0\ldots l_{k-1}0)}+\sum_{l_k=1}^{N_k}\mathbf{a}_{1\ldots 122}^{(0\ldots l_{k-1}l_k)}(t)\cos(l_k\theta_k)+\mathbf{a}_{1\ldots 123}^{(0\ldots l_{k-1}l_k)}(t)\sin(l_k\theta_k)\right]\cos(l_{k-1}\theta_{k-1})\\
&+\sum_{l_{k-1}=1}^{N_{k-1}}\left[\mathbf{a}_{1\ldots 131}^{(0\ldots l_{k-1}0)}+\sum_{l_k=1}^{N_k}\mathbf{a}_{1\ldots 132}^{(0\ldots l_{k-1}l_k)}(t)\cos(l_1\theta_1)+\mathbf{a}_{1\ldots 133}^{(0\ldots l_{k-1}l_{k-2})}(t)\sin(l_k\theta_k)\right]\sin(l_{k-1}\theta_{k-1})\\
&\vdots\\
&+\sum_{l_0=1}^{N_0}\sum_{l_1=1}^{N_1}\cdots\sum_{l_{k-2}=1}^{N_{k-2}}\left[\mathbf{a}_{22\ldots 211}^{(l_0 l_1\ldots l_{k-2}00)}+\sum_{l_k=1}^{N_k}\mathbf{a}_{22\ldots 212}^{(l_0 l_1\ldots l_{k-2}0 l_k)}(t)\cos(l_k\theta_k)+\mathbf{a}_{22\ldots 213}^{(l_0 l_1\ldots l_{k-2}0 l_k)}(t)\sin(l_k\theta_k)\right]\\
&\times\cos(l_{k-2}\theta_{k-2})\ldots\cos(l_1\theta_1)\cos(l_0\theta_0)\\
&+\sum_{l_0=1}^{N_0}\sum_{l_1=1}^{N_1}\cdots\sum_{l_{k-1}=1}^{N_{k-1}}\left[\mathbf{a}_{22\ldots 221}^{(l_0 l_1\ldots l_{k-1}0)}+\sum_{l_k=1}^{N_k}\mathbf{a}_{22\ldots 222}^{(l_0\ldots l_{k-1}l_k)}(t)\cos(l_k\theta_k)+\mathbf{a}_{22\ldots 223}^{(l_0\ldots l_{k-1}l_k)}(t)\sin(l_k\theta_k)\right]\\
&\times\cos(l_{k-1}\theta_{k-1})\cos(l_{k-2}\theta_{k-2})\ldots\cos(l_1\theta_1)\cos(l_0\theta_0)\\
&+\sum_{l_0=1}^{N_0}\sum_{l_1=1}^{N_1}\cdots\sum_{l_{k-2}=1}^{N_{k-2}}\sum_{l_{k-1}=1}^{N_{k-1}}\left[\mathbf{a}_{22\ldots 231}^{(l_0 l_1\ldots l_{k-1}0)}+\sum_{l_k=1}^{N_k}\mathbf{a}_{22\ldots 232}^{(l_0\ldots l_{k-1}l_k)}(t)\cos(l_k\theta_k)+\mathbf{a}_{22\ldots 233}^{(l_0\ldots l_{k-1}l_k)}(t)\sin(l_k\theta_k)\right]
\end{aligned}
$$

$$\times \sin(l_{k-1}\theta_{k-1})\cos(l_{k-2}\theta_{k-2})\ldots\cos(l_1\theta_1)\cos(l_0\theta_0)$$

$$\vdots$$

$$+\sum_{l_0=1}^{N_0}\sum_{l_1=1}^{N_1}\cdots\sum_{l_{k-2}=1}^{N_{k-2}}\left[\mathbf{a}_{33\ldots311}^{(l_0l_1\ldots l_{k-2}00)}+\sum_{l_k=1}^{N_k}\mathbf{a}_{33\ldots312}^{(l_0l_1\ldots l_{k-2}0l_k)}(t)\cos(l_k\theta_k)+\mathbf{a}_{33\ldots313}^{(l_0l_1\ldots l_{k-2}0l_k)}(t)\sin(l_k\theta_k)\right]$$

$$\times \sin(l_{k-2}\theta_{k-2})\ldots\sin(l_1\theta_1)\sin(l_0\theta_0)$$

$$+\sum_{l_0=1}^{N_0}\sum_{l_1=1}^{N_1}\cdots\sum_{l_{k-1}=1}^{N_{k-1}}\left[\mathbf{a}_{33\ldots321}^{(l_0l_1\ldots l_{k-1}0)}+\sum_{l_k=1}^{N_k}\mathbf{a}_{33\ldots322}^{(l_0\ldots l_{k-1}l_k)}(t)\cos(l_k\theta_k)+\mathbf{a}_{33\ldots323}^{(l_0\ldots l_{k-1}l_k)}(t)\sin(l_k\theta_k)\right]$$

$$\times \cos(l_{k-1}\theta_{k-1})\sin(l_{k-2}\theta_{k-2})\ldots\sin(l_1\theta_1)\sin(l_0\theta_0)$$

$$+\sum_{l_0=1}^{N_0}\sum_{l_1=1}^{N_1}\cdots\sum_{l_{k-2}=1}^{N_{k-2}}\sum_{l_{k-1}=1}^{N_{k-1}}\left[\mathbf{a}_{33\ldots331}^{(l_0l_1\ldots l_{k-1}0)}+\sum_{l_k=1}^{N_k}\mathbf{a}_{33\ldots332}^{(l_0\ldots l_{k-1}l_k)}(t)\cos(l_k\theta_k)+\mathbf{a}_{33\ldots333}^{(l_0\ldots l_{k-1}l_k)}(t)\sin(l_k\theta_k)\right]$$

$$\times \sin(l_{k-1}\theta_{k-1})\sin(l_{k-2}\theta_{k-2})\ldots\sin(l_1\theta_1)\sin(l_0\theta_0). \tag{4.42}$$

When $N_i \to \infty$ $(i = 0, 1, 2, \ldots, k)$, the foregoing solution gives the exact solution for quasi-periodic flow after the kth-order Hopf bifurcation of the period-m flows in such a nonlinear dynamical system. If $k \to \infty$, the chaotic flow is generated via the periodic flows. If one of all p_k $(k = 1, 2, \ldots)$ is an irrational number, the periodic flow is quasi-periodic. If all p_k $(k = 1, 2, \ldots)$ are irrational numbers, the periodic flow are formed by quasi-periodic.

For the kth order Hopf bifurcation, we assume

$$p_k\omega_k = \omega_{k-1} \tag{4.43}$$

1. If p_k is an irrational number, the kth-order Hopf bifurcation of the period-m flow is called the quasi-period-p_k Hopf bifurcation, and the corresponding solution is p_k-quasi-periodic.
2. If $p_k = 2$ is two, the kth-order Hopf bifurcation of the period-m flow is called a period-doubling Hopf bifurcation (or a period-2 Hopf bifurcation), and the corresponding solution is period-doubling.
3. If $p_k = q$ is an integer q, the kth-order Hopf bifurcation of the period-m flow is called a period-q Hopf bifurcation, and the corresponding solution is q times-periodic.
4. If $p_k = p/q$ is a fractional number (p, q are irreducible), the kth-order Hopf bifurcation of the period-m flow is called a period-p/q Hopf bifurcation, and the corresponding solution is p/q-periodic.

Consider quasi-periodic flows in a periodically forced, nonlinear system.

Theorem 4.2 *Consider a periodically forced, nonlinear dynamical system as*

$$\dot{\mathbf{x}} = \mathbf{F}(\mathbf{x}, t, \mathbf{p}) \in \mathscr{R}^n \tag{4.44}$$

where $\mathbf{F}(\mathbf{x}, t, \mathbf{p})$ *is a* C^r*-continuous nonlinear function vector* $(r \geq 1)$ *with a forcing period* $T = 2\pi/\Omega$.

1. *If such a dynamical system has a period-*m *flow* $\mathbf{x}^{(m)}(t)$ *with finite norm* $\|\mathbf{x}^{(m)}\|$ *and period* $T = 2\pi/\Omega$, *there is a generalized coordinate transformation with* $\theta = \Omega t$ *for the periodic flow of Equation (4.44) in the form of*

$$\mathbf{x}^{(m)}(t) = \mathbf{a}_0^{(m)}(t) + \sum_{k=1}^{\infty} \mathbf{b}_{k/m}(t)\cos\left(\frac{k}{m}\theta\right) + \mathbf{c}_{k/m}(t)\sin\left(\frac{k}{m}\theta\right) \tag{4.45}$$

with

$$\begin{aligned}
\mathbf{a}_1^{(0)} &\equiv \mathbf{a}_0^{(m)} = (a_{01}^{(m)}, a_{02}^{(m)}, \ldots, a_{0n}^{(m)})^{\mathrm{T}}, \\
\mathbf{a}_2^{(k)} &\equiv \mathbf{b}_{k/m} = (b_{k/m1}, b_{k/m2}, \ldots, b_{k/mn})^{\mathrm{T}}, \\
\mathbf{a}_3^{(k)} &\equiv \mathbf{c}_{k/m} = (c_{k/m1}, c_{k/m2}, \ldots, c_{k/mn})^{\mathrm{T}}
\end{aligned} \tag{4.46}$$

which, under $\|\mathbf{x}^{(m)}(t) - \mathbf{x}^{(m)*}(t)\| < \varepsilon$ *with a prescribed small* $\varepsilon > 0$, *can be approximated by a finite term transformation* $\mathbf{x}^{(m)*}(t)$

$$\mathbf{x}^{(m)*}(t) = \mathbf{a}_0^{(m)}(t) + \sum_{k=1}^{N_0} \mathbf{b}_{k/m}(t)\cos\left(\frac{k}{m}\theta\right) + \mathbf{c}_{k/m}(t)\sin\left(\frac{k}{m}\theta\right) \tag{4.47}$$

and the generalized coordinates are determined by

$$\dot{\mathbf{a}}_{s_0} = \mathbf{f}_{s_0}(\mathbf{a}_{s_0}, \mathbf{p}) \tag{4.48}$$

where

$$\begin{aligned}
&\mathbf{k}_0 = diag(\mathbf{I}_{n\times n}, 2\mathbf{I}_{n\times n}, \ldots, N_0\mathbf{I}_{n\times n}), \\
&\mathbf{a}_1^{(0)} \equiv \mathbf{a}_0^{(m)}, \mathbf{a}_2^{(k)} \equiv \mathbf{b}_{k/m}, \mathbf{a}_3^{(k)} \equiv \mathbf{c}_{k/m}; \\
&\mathbf{a}_1 = \mathbf{a}_1^{(0)}, \\
&\mathbf{a}_2 = (\mathbf{a}_2^{(1)}, \mathbf{a}_2^{(2)}, \ldots, \mathbf{a}_2^{(N_0)})^{\mathrm{T}} \equiv \mathbf{b}^{(m)}, \\
&\mathbf{a}_3 = (\mathbf{a}_3^{(1)}, \mathbf{a}_3^{(2)}, \ldots, \mathbf{a}_3^{(N_0)})^{\mathrm{T}} \equiv \mathbf{c}^{(m)}, \\
&\mathbf{F}_1 = \mathbf{F}_0^{(m)}, \\
&\mathbf{F}_2 = (\mathbf{F}_{11}^{(m)}, \mathbf{F}_{12}^{(m)}, \ldots, \mathbf{F}_{1N_0}^{(m)})^{\mathrm{T}}, \\
&\mathbf{F}_3 = (\mathbf{F}_{21}^{(m)}, \mathbf{F}_{22}^{(m)}, \ldots, \mathbf{F}_{2N_0}^{(m)})^{\mathrm{T}}; \\
&\mathbf{a}_{s_0} = (\mathbf{a}_1, \mathbf{a}_2, \mathbf{a}_3)^{\mathrm{T}}, \\
&\mathbf{f}_{s_0} = \left(\mathbf{F}_1, -\frac{\Omega}{m}\mathbf{k}_0\mathbf{a}_3 + \mathbf{F}_2, \frac{\Omega}{m}\mathbf{k}_0\mathbf{a}_2 + \mathbf{F}_3\right)^{\mathrm{T}} \\
&\text{for } N_0 = 1, 2, \ldots, \infty;
\end{aligned} \tag{4.49}$$

and

$$\mathbf{F}_0^{(m)}(\mathbf{a}_0^{(m)},\mathbf{b}^{(m)},\mathbf{c}^{(m)}) = \frac{1}{2m\pi}\int_0^{2m\pi}\mathbf{F}(\mathbf{x}^{(m)*},t,\mathbf{p})d\theta;$$
$$\mathbf{F}_{1k}^{(m)}(\mathbf{a}_0^{(m)},\mathbf{b}^{(m)},\mathbf{c}^{(m)}) = \frac{1}{m\pi}\int_0^{2m\pi}\mathbf{F}(\mathbf{x}^{(m)*},t,\mathbf{p})\cos\left(\frac{k}{m}\theta\right)d\theta,$$
$$\mathbf{F}_{2k}^{(m)}(\mathbf{a}_0^{(m)},\mathbf{b}^{(m)},\mathbf{c}^{(m)}) = \frac{1}{m\pi}\int_0^{2m\pi}\mathbf{F}(\mathbf{x}^{(m)*},t,\mathbf{p})\sin\left(\frac{k}{m}\theta\right)d\theta$$
$$\text{for } k = 1,2,\ldots,N_0. \tag{4.50}$$

2. *After the* kth *Hopf bifurcation with* $p_k\omega_k = \omega_{k-1}(k = 1,2,\ldots)$ *and* $\omega_0 = \Omega/m$, *there is a dynamical system of coefficients as*

$$\dot{\mathbf{a}}_{s_0s_1\ldots s_k} = \mathbf{f}_{s_0s_1\ldots s_k}(\mathbf{a}_{s_0s_1\ldots s_k},\mathbf{p}) \tag{4.51}$$

where

$$\begin{aligned}
\mathbf{a}_{s_0s_1\ldots s_k} &= (\mathbf{a}_{s_0s_1\ldots s_{k-1}1},\mathbf{a}_{s_0s_1\ldots s_{k-1}2},\mathbf{a}_{s_0s_1\ldots s_{k-1}3})^{\mathrm{T}},\\
\mathbf{f}_{s_0s_1\ldots s_k} &= (\mathbf{F}_{s_0s_1\ldots s_{k-1}1},-\omega_k\mathbf{k}_k\mathbf{a}_{s_0s_1\ldots s_{k-1}3}+\mathbf{F}_{s_0s_1\ldots s_{k-1}2},\\
&\qquad \omega_k\mathbf{k}_k\mathbf{a}_{s_0s_1\ldots s_{k-1}2}+\mathbf{F}_{s_0s_1\ldots s_{k-1}3})^{\mathrm{T}},\\
\mathbf{k}_k &= diag(\mathbf{I}_{n_{k-1}\times n_{k-1}},2\mathbf{I}_{n_{k-1}\times n_{k-1}},\ldots,N_k\mathbf{I}_{n_{k-1}\times n_{k-1}}),\\
n_{k-1} &= n(2N_0+1)(2N_1+1)\ldots(2N_{k-1}+1)
\end{aligned} \tag{4.52}$$

with a periodic solution as a new transformation

$$\begin{aligned}
\mathbf{a}_{s_0s_1\ldots s_k} &= \mathbf{a}^{(0)}_{s_0s_1\ldots s_k1}(t) + \sum_{l_{k+1}=1}^{\infty}\mathbf{a}^{(l_{k+1})}_{s_0s_1\ldots s_k2}(t)\cos(l_{k+1}\theta_{k+1})\\
&\quad + \mathbf{a}^{(l_{k+1})}_{s_0s_1\ldots s_k3}(t)\sin(l_{k+1}\theta_{k+1})
\end{aligned} \tag{4.53}$$

with

$$\begin{aligned}
& s_i = 1,2,3\ (i = 0,1,2,\ldots,k),\\
& \mathbf{a}_{s_0s_1\ldots s_k1} = \mathbf{a}^{(0)}_{s_0s_1\ldots s_k1}\\
& \mathbf{a}_{s_0s_1\ldots s_k2} = (\mathbf{a}^{(1)}_{s_0s_1\ldots s_k2},\mathbf{a}^{(2)}_{s_0s_1\ldots s_k2},\ldots,\mathbf{a}^{(N_{k+1})}_{s_0s_1\ldots s_k2})^{\mathrm{T}},\\
& \mathbf{a}_{s_0s_1\ldots s_k3} = (\mathbf{a}^{(1)}_{s_0s_1\ldots s_k3},\mathbf{a}^{(2)}_{s_0s_1\ldots s_k3},\ldots,\mathbf{a}^{(N_{k+1})}_{s_0s_1\ldots s_k3})^{\mathrm{T}};\\
& \mathbf{a}_{s_0s_1\ldots s_{k-1}1} = \mathbf{a}^{(0)}_{s_0s_1\ldots s_{k-1}1}\\
& \mathbf{a}_{s_0s_1\ldots s_{k-1}2} = (\mathbf{a}^{(1)}_{s_0s_1\ldots s_{k-1}2},\mathbf{a}^{(2)}_{s_0s_1\ldots s_{k-1}2},\ldots,\mathbf{a}^{(N_k)}_{s_0s_1\ldots s_{k-1}2})^{\mathrm{T}},\\
& \mathbf{a}_{s_0s_1\ldots s_{k-1}3} = (\mathbf{a}^{(1)}_{s_0s_1\ldots s_{k-1}3},\mathbf{a}^{(2)}_{s_0s_1\ldots s_{k-1}3},\ldots,\mathbf{a}^{(N_k)}_{s_0s_1\ldots s_{k-1}3})^{\mathrm{T}};
\end{aligned}$$

$$\begin{aligned}
&\vdots \\
&\mathbf{a}_1 = \mathbf{a}_1^{(0)} \\
&\mathbf{a}_2 = (\mathbf{a}_2^{(1)}, \mathbf{a}_2^{(2)}, \ldots, \mathbf{a}_2^{(N_0)})^{\mathrm{T}}, \\
&\mathbf{a}_3 = (\mathbf{a}_3^{(1)}, \mathbf{a}_3^{(2)}, \ldots, \mathbf{a}_3^{(N_0)})^{\mathrm{T}};
\end{aligned} \tag{4.54}$$

which, under $\|\mathbf{a}_{s_0 s_1 \ldots s_k}(t) - \mathbf{a}^*_{s_0 s_1 \ldots s_k}(t)\| < \varepsilon$ *with a prescribed small* $\varepsilon > 0$*, can be approximated by a finite term transformation* $\mathbf{a}^*_{s_0 s_1 \ldots s_k}(t)$

$$\begin{aligned}
\mathbf{a}^*_{s_0 s_1 \ldots s_k} &= \mathbf{a}^{(0)}_{s_0 s_1 \ldots s_k 1}(t) + \sum_{l_{k+1}=1}^{N_{k+1}} \mathbf{a}^{(l_{k+1})}_{s_0 s_1 \ldots s_k 2}(t) \cos(l_{k+1}\theta_{k+1}) \\
&\quad + \mathbf{a}^{(l_{k+1})}_{s_0 s_1 \ldots s_k 3}(t) \sin(l_{k+1}\theta_{k+1})
\end{aligned} \tag{4.55}$$

and the generalized coordinates are determined by

$$\dot{\mathbf{a}}_{s_0 s_1 \ldots s_{k+1}} = \mathbf{f}_{s_0 s_1 \ldots s_{k+1}}(\mathbf{a}_{s_0 s_1 \ldots s_{k+1}}, \mathbf{p}) \tag{4.56}$$

where

$$\begin{aligned}
\mathbf{a}_{s_0 s_1 \ldots s_{k+1}} &= (\mathbf{a}_{s_0 s_1 \ldots s_k 1}, \mathbf{a}_{s_0 s_1 \ldots s_k 2}, \mathbf{a}_{s_0 s_1 \ldots s_k 3})^{\mathrm{T}}, \\
\mathbf{f}_{s_0 s_1 \ldots s_{k+1}} &= (\mathbf{F}_{s_0 s_1 \ldots s_k 1}, -\omega_{k+1}\mathbf{k}_{k+1}\mathbf{a}_{s_0 s_1 \ldots s_k 3} + \mathbf{F}_{s_0 s_1 \ldots s_k 2}, \\
&\qquad \omega_{k+1}\mathbf{k}_{k+1}\mathbf{a}_{s_0 s_1 \ldots s_k 2} + \mathbf{F}_{s_0 s_1 \ldots s_k 3})^{\mathrm{T}}
\end{aligned} \tag{4.57}$$

and

$$\begin{aligned}
&\mathbf{k}_{k+1} = diag(\mathbf{I}_{n_k \times n_k}, 2\mathbf{I}_{n_k \times n_k}, \ldots, N_{k+1}\mathbf{I}_{n_k \times n_k}), \\
&n_k = n(2N_0 + 1)(2N_1 + 1)\ldots(2N_k + 1); \\
&\mathbf{a}_{s_0 s_1 \ldots s_k 1} = \mathbf{a}^{(0)}_{s_0 s_1 \ldots s_k 1} \\
&\mathbf{a}_{s_0 s_1 \ldots s_k 2} = (\mathbf{a}^{(1)}_{s_0 s_1 \ldots s_k 2}, \mathbf{a}^{(2)}_{s_0 s_1 \ldots s_k 2}, \ldots, \mathbf{a}^{(N_{k+1})}_{s_0 s_1 \ldots s_k 2})^{\mathrm{T}}, \\
&\mathbf{a}_{s_0 s_1 \ldots s_k 3} = (\mathbf{a}^{(1)}_{s_0 s_1 \ldots s_k 3}, \mathbf{a}^{(2)}_{s_0 s_1 \ldots s_k 3}, \ldots, \mathbf{a}^{(N_{k+1})}_{s_0 s_1 \ldots s_k 3})^{\mathrm{T}}; \\
&\mathbf{a}_{s_0 s_1 \ldots s_{k+1}} = (\mathbf{a}_{s_0 s_1 \ldots s_k 1}, \mathbf{a}_{s_0 s_1 \ldots s_k 2}, \mathbf{a}_{s_0 s_1 \ldots s_k 3})^{\mathrm{T}}; \\
&\mathbf{F}_{s_0 s_1 \ldots s_k 1} = \mathbf{F}^{(0)}_{s_0 s_1 \ldots s_k 1}, \\
&\mathbf{F}_{s_0 s_1 \ldots s_k 2} = (\mathbf{F}^{(1)}_{s_0 s_1 \ldots s_k 2}, \mathbf{F}^{(2)}_{s_0 s_1 \ldots s_k 2}, \ldots, \mathbf{F}^{(N_{k+1})}_{s_0 s_1 \ldots s_k 2})^{\mathrm{T}}, \\
&\mathbf{F}_{s_0 s_1 \ldots s_k 3} = (\mathbf{F}^{(1)}_{s_0 s_1 \ldots s_k 3}, \mathbf{F}^{(2)}_{s_0 s_1 \ldots s_k 3}, \ldots, \mathbf{F}^{(N_{k+1})}_{s_0 s_1 \ldots s_k 3})^{\mathrm{T}} \\
&\text{for } N_{k+1} = 1, 2, \ldots, \infty;
\end{aligned} \tag{4.58}$$

and

$$
\begin{aligned}
\mathbf{F}_{s_0s_1\ldots s_k1}(\mathbf{a}_{s_0s_1\ldots s_{k+1}},\mathbf{p}) &= \frac{1}{2\pi}\int_0^{2\pi}\mathbf{f}_{s_0s_1\ldots s_k}(\mathbf{a}^*_{s_0s_1\ldots s_k},\mathbf{p})d\theta_{k+1};\\
\mathbf{F}^{(l_{k+1})}_{s_0s_1\ldots s_k2}(\mathbf{a}_{s_0s_1\ldots s_{k+1}},\mathbf{p}) &= \frac{1}{\pi}\int_0^{2\pi}\mathbf{f}_{s_0s_1\ldots s_k}(\mathbf{a}^*_{s_0s_1\ldots s_k},\mathbf{p})\cos(l_{k+1}\theta_{k+1})d\theta_{k+1},\\
\mathbf{F}^{(l_{k+1})}_{s_0s_1\ldots s_k3}(\mathbf{a}_{s_0s_1\ldots s_{k+1}},\mathbf{p}) &= \frac{1}{\pi}\int_0^{2\pi}\mathbf{f}_{s_0s_1\ldots s_k}(\mathbf{a}^*_{s_0s_1\ldots s_k},\mathbf{p})\sin(l_{k+1}\theta_{k+1})d\theta_{k+1}\\
&\text{for } l_{k+1}=1,2,\ldots,N_{k+1}.
\end{aligned}
\tag{4.59}
$$

3. *Equation (4.51) becomes*

$$
\dot{\mathbf{z}}_{s_0s_1\ldots s_{k+1}} = \mathbf{f}_{s_0s_1\ldots s_{k+1}}(\mathbf{z}_{s_0s_2\ldots s_{k+1}})
\tag{4.60}
$$

where

$$
\begin{aligned}
\mathbf{z}_{s_0s_1\ldots s_{k+1}} &= (\mathbf{a}_{s_0s_1\ldots s_k1},\mathbf{a}_{s_0s_1\ldots s_k2},\mathbf{a}_{s_0s_1\ldots s_k3})^{\mathrm{T}},\\
\mathbf{f}_{s_0s_1\ldots s_{k+1}} &= (\mathbf{F}_{s_0s_1\ldots s_k1},-\omega_{k+1}\mathbf{k}_{k+1}\mathbf{a}_{s_0s_1\ldots s_k3}+\mathbf{F}_{s_0s_1\ldots s_k2},\\
&\qquad \omega_{k+1}\mathbf{k}_{k+1}\mathbf{b}_{s_0s_1\ldots s_k2}+\mathbf{F}_{s_0s_1\ldots s_k3})^{\mathrm{T}}.
\end{aligned}
\tag{4.61}
$$

If equilibrium $\mathbf{z}^*_{s_0s_1\ldots s_{k+1}}$ of Equation (4.51) (i.e., $\mathbf{f}_{s_0s_1\ldots s_{k+1}}(\mathbf{z}^*_{s_0s_1\ldots s_{k+1}})=\mathbf{0}$) exists, then the approximate solution of the periodic flow of kth *generalized coordinates for the period-*m *flow exists as in Equation (4.55). In the vicinity of equilibrium $\mathbf{z}^*_{s_0s_1\ldots s_{k+1}}$, with*

$$
\mathbf{z}_{s_0s_1\ldots s_{k+1}} = \mathbf{z}^*_{s_0s_1\ldots s_{k+1}} + \Delta\mathbf{z}_{s_0s_1\ldots s_{k+1}},
\tag{4.62}
$$

the linearized equation of Equation (4.60) is

$$
\Delta\dot{\mathbf{z}}_{s_0s_1\ldots s_{k+1}} = D\mathbf{f}_{s_0s_1\ldots s_{k+1}}(\mathbf{z}^*_{s_0s_1\ldots s_{k+1}})\Delta\mathbf{z}_{s_0s_1\ldots s_{k+1}}
\tag{4.63}
$$

and the eigenvalue analysis of equilibrium $\mathbf{z}^*$ is given by

$$
|D\mathbf{f}_{s_0s_1\ldots s_{k+1}}(\mathbf{z}^*_{s_0s_1\ldots s_{k+1}}) - \lambda\mathbf{I}_{n_k(2N_{k+1}+1)\times n_k(2N_{k+1}+1)}| = 0
\tag{4.64}
$$

where

$$
D\mathbf{f}_{s_0s_1\ldots s_{k+1}}(\mathbf{z}^*_{s_0s_1\ldots s_{k+1}}) = \left.\frac{\partial\mathbf{f}_{s_0s_1\ldots s_{k+1}}\left(\mathbf{z}_{s_0s_1\ldots s_{k+1}}\right)}{\partial\mathbf{z}_{s_0s_1\ldots s_{k+1}}}\right|_{\mathbf{z}^*_{s_0s_1\ldots s_{k+1}}}.
\tag{4.65}
$$

The stability and bifurcation of such a periodic flow of the kth *generalized coordinates can be classified by the eigenvalues of $D\mathbf{f}_{s_0s_1\ldots s_{k+1}}(\mathbf{z}^*_{s_0s_1\ldots s_{k+1}})$ with*

$$
(n_1,n_2,n_3|n_4,n_5,n_6).
\tag{4.66}
$$

a. *If all eigenvalues of the equilibrium possess negative real parts, the approximate quasi-periodic solution is stable.*
b. *If at least one of the eigenvalues of the equilibrium possesses a positive real part, the approximate quasi-periodic solution is unstable.*

   c. *The boundaries between stable and unstable equilibriums with higher order singularity give bifurcation and stability conditions with higher order singularity.*
4. *For the* kth *order Hopf bifurcation of period-*m *motion, a relation exists as*

$$p_k \omega_k = \omega_{k-1}. \tag{4.67}$$

   a. *If p_k is an irrational number, the* kth*-order Hopf bifurcation of the period-*m *flow is called the quasi-period-p_k Hopf bifurcation, and the corresponding solution of the* kth *generalized coordinates is p_k-quasi-periodic to the system of the* $(k-1)$th *generalized coordinates.*
   b. *If $p_k = 2$, the* kth*-order Hopf bifurcation of the period-*m *flow is called a period-doubling Hopf bifurcation (or a period-2 Hopf bifurcation), and the corresponding solution of the* kth *generalized coordinates is period-doubling to the system of the* $(k-1)$th *generalized coordinates.*
   c. *If $p_k = q$ with an integer q, the* kth*-order Hopf bifurcation of the period-*m *flow is called a period-q Hopf bifurcation, and the corresponding solution of the* kth *generalized coordinates is of q-times period to the system of the* $(k-1)$th *generalized coordinates.*
   d. *If $p_k = p/q$ (p, q are irreducible integer), the* kth*-order Hopf bifurcation of the period-*m *flow is called a period-p/q Hopf bifurcation, and the corresponding solution of the* kth *generalized coordinates is p/q-periodic to the system of the* $(k-1)$th *generalized coordinates.*

■

Proof. The proof of this theorem is similar to the proof of Theorem 3.4.

4.2 Nonlinear Vibration Systems

In this section, the analytical solutions of quasi-periodic motion for nonlinear vibration systems are presented, and the stability and bifurcation of approximate solutions of quasi-periodic motions in nonlinear vibration systems will be discussed.

Theorem 4.3 *Consider a nonlinear vibration system as*

$$\ddot{\mathbf{x}} = \mathbf{f}(\mathbf{x}, \dot{\mathbf{x}}, \mathbf{p}) \in \mathscr{R}^n \tag{4.68}$$

where $\mathbf{f}(\mathbf{x}, \dot{\mathbf{x}}, \mathbf{p})$ *is a C^r-continuous nonlinear function vector ($r \geq 1$).*

1. *If such a dynamical system has a period-*m *motion $\mathbf{x}^{(m)}(t)$ with finite norm $\|\mathbf{x}^{(m)}\|$ and period $T = 2\pi/\Omega$, there is a generalized coordinate transformation with $\theta = \Omega t$ for the periodic motion of Equation (4.68) in a form of*

$$\mathbf{x}^{(m)}(t) = \mathbf{a}_0^{(m)}(t) + \sum_{k=1}^{\infty} \mathbf{b}_{k/m}(t) \cos\left(\frac{k}{m}\theta\right) + \mathbf{c}_{k/m}(t) \sin\left(\frac{k}{m}\theta\right) \tag{4.69}$$

with

$$\begin{aligned} \mathbf{a}_1^{(0)} &\equiv \mathbf{a}_0^{(m)} = (a_{01}^{(m)}, a_{02}^{(m)}, \dots, a_{0n}^{(m)})^{\mathrm{T}}, \\ \mathbf{a}_2^{(k)} &\equiv \mathbf{b}_{k/m} = (b_{k/m1}, b_{k/m2}, \dots, b_{k/mn})^{\mathrm{T}}, \\ \mathbf{a}_3^{(k)} &\equiv \mathbf{c}_{k/m} = (c_{k/m1}, c_{k/m2}, \dots, c_{k/mn})^{\mathrm{T}} \end{aligned} \tag{4.70}$$

which, under $\|\mathbf{x}^{(m)}(t) - \mathbf{x}^{(m)*}(t)\| < \varepsilon$ *with a prescribed small* $\varepsilon > 0$, *can be approximated by a finite term transformation* $\mathbf{x}^{(m)*}(t)$

$$\mathbf{x}^{(m)*}(t) = \mathbf{a}_0^{(m)}(t) + \sum_{k=1}^{N_0} \mathbf{b}_{k/m}(t)\cos\left(\frac{k}{m}\theta\right) + \mathbf{c}_{k/m}(t)\sin\left(\frac{k}{m}\theta\right) \tag{4.71}$$

and the generalized coordinates are determined by

$$\ddot{\mathbf{a}}_{s_0} = \mathbf{g}_{s_0}(\mathbf{a}_{s_0}, \dot{\mathbf{a}}_{s_0}, \mathbf{p}) \tag{4.72}$$

where

$$\begin{aligned}
\mathbf{k}_0^{(1)} &= diag(\mathbf{I}_{n\times n}, 2\mathbf{I}_{n\times n}, \ldots, N_0\mathbf{I}_{n\times n}),\\
\mathbf{k}_0^{(2)} &= diag(\mathbf{I}_{n\times n}, 2^2\mathbf{I}_{n\times n}, \ldots, N_0^2\mathbf{I}_{n\times n}),\\
\mathbf{a}_1^{(0)} &\equiv \mathbf{a}_0^{(m)}, \mathbf{a}_2^{(k)} \equiv \mathbf{b}_{k/m}, \mathbf{a}_3^{(k)} \equiv \mathbf{c}_{k/m};\\
\mathbf{a}_1 &= \mathbf{a}_1^{(0)},\\
\mathbf{a}_2 &= (\mathbf{a}_2^{(1)}, \mathbf{a}_2^{(2)}, \ldots, \mathbf{a}_2^{(N_0)})^{\mathrm{T}} \equiv \mathbf{b}^{(m)},\\
\mathbf{a}_3 &= (\mathbf{a}_3^{(1)}, \mathbf{a}_3^{(2)}, \ldots, \mathbf{a}_3^{(N_0)})^{\mathrm{T}} \equiv \mathbf{c}^{(m)},\\
\mathbf{F}_1 &= \mathbf{F}_0^{(m)},\\
\mathbf{F}_2 &= (\mathbf{F}_{11}^{(m)}, \mathbf{F}_{12}^{(m)}, \ldots, \mathbf{F}_{1N_0}^{(m)})^{\mathrm{T}},\\
\mathbf{F}_3 &= (\mathbf{F}_{21}^{(m)}, \mathbf{F}_{22}^{(m)}, \ldots, \mathbf{F}_{2N_0}^{(m)})^{\mathrm{T}};\\
\mathbf{a}_{s_0} &= (\mathbf{a}_1, \mathbf{a}_2, \mathbf{a}_3)^{\mathrm{T}},\\
\mathbf{g}_{s_0} &= \left(\mathbf{F}_1^{(m)}, -2\frac{\Omega}{m}\mathbf{k}_0^{(1)}\dot{\mathbf{a}}_3 + \frac{\Omega^2}{m^2}\mathbf{k}_0^{(2)}\mathbf{a}_2 + \mathbf{F}_2,\right.\\
&\qquad \left. 2\frac{\Omega}{m}\mathbf{k}_0^{(1)}\dot{\mathbf{a}}_2 + \frac{\Omega^2}{m^2}\mathbf{k}_0^{(2)}\mathbf{a}_3 + \mathbf{F}_3\right)^{\mathrm{T}}\\
&\text{for } N_0 = 1, 2, \ldots, \infty;
\end{aligned} \tag{4.73}$$

and

$$\begin{aligned}
\mathbf{F}_0^{(m)}(\mathbf{a}_0^{(m)}, \mathbf{b}^{(m)}, \mathbf{c}^{(m)}) &= \frac{1}{2m\pi}\int_0^{2m\pi} \mathbf{f}(\mathbf{x}^{(m)*}, \dot{\mathbf{x}}^{(m)*}, \mathbf{p})d\theta;\\
\mathbf{F}_{1k}^{(m)}(\mathbf{a}_0^{(m)}, \mathbf{b}^{(m)}, \mathbf{c}^{(m)}) &= \frac{1}{m\pi}\int_0^{2m\pi} \mathbf{f}(\mathbf{x}^{(m)*}, \dot{\mathbf{x}}^{(m)*}, \mathbf{p})\cos\left(\frac{k}{m}\theta\right)d\theta,\\
\mathbf{F}_{2k}^{(m)}(\mathbf{a}_0^{(m)}, \mathbf{b}^{(m)}, \mathbf{c}^{(m)}) &= \frac{1}{m\pi}\int_0^{2m\pi} \mathbf{f}(\mathbf{x}^{(m)*}, \dot{\mathbf{x}}^{(m)*}, \mathbf{p})\sin\left(\frac{k}{m}\theta\right)d\theta\\
&\text{for } k = 1, 2, \ldots, N_0.
\end{aligned} \tag{4.74}$$

2. *If after the* kth *Hopf bifurcation with* $p_k\omega_k = \omega_{k-1}$ $(k = 1, 2, \ldots)$ *and* $\omega_0 = \Omega/m$, *there is a dynamical system of coefficients as*

$$\ddot{\mathbf{a}}_{s_0s_1\ldots s_k} = \mathbf{g}_{s_0s_1\ldots s_k}(\mathbf{a}_{s_0s_1\ldots s_k}, \dot{\mathbf{a}}_{s_0s_1\ldots s_k}, \mathbf{p}) \tag{4.75}$$

where

$$\begin{aligned}
\mathbf{a}_{s_0s_1\ldots s_k} &= (\mathbf{a}_{s_0s_1\ldots s_{k-1}1}, \mathbf{a}_{s_0s_1\ldots s_{k-1}2}, \mathbf{a}_{s_0s_1\ldots s_{k-1}3})^{\mathrm{T}},\\
\mathbf{g}_{s_0s_1\ldots s_k} &= (\mathbf{F}_{s_0s_1\ldots s_{k-1}1}, -2\omega_k\mathbf{k}_k^{(1)}\dot{\mathbf{a}}_{s_0s_1\ldots s_{k-1}3} + \omega_k^2\mathbf{k}_k^{(2)}\mathbf{a}_{s_0s_1\ldots s_{k-1}2} + \mathbf{F}_{s_0s_1\ldots s_{k-1}2},\\
&\quad 2\omega_k\mathbf{k}_k^{(1)}\dot{\mathbf{a}}_{s_0s_1\ldots s_{k-1}2} + \omega_k^2\mathbf{k}_k^{(2)}\mathbf{a}_{s_0s_1\ldots s_{k-1}3} + \mathbf{F}_{s_0s_1\ldots s_{k-1}3})^{\mathrm{T}},\\
\mathbf{k}_k^{(1)} &= diag(\mathbf{I}_{n_{k-1}\times n_{k-1}}, 2\mathbf{I}_{n_{k-1}\times n_{k-1}}, \ldots, N_k\mathbf{I}_{n_{k-1}\times n_{k-1}}),\\
\mathbf{k}_k^{(2)} &= diag(\mathbf{I}_{n_{k-1}\times n_{k-1}}, 2^2\mathbf{I}_{n_{k-1}\times n_{k-1}}, \ldots, N_k^2\mathbf{I}_{n_{k-1}\times n_{k-1}})\\
n_{k-1} &= n(2N_0+1)(2N_1+1)\ldots(2N_{k-1}+1)
\end{aligned} \tag{4.76}$$

with a periodic solution as

$$\begin{aligned}
\mathbf{a}_{s_0s_1\ldots s_k} &= \mathbf{a}^{(0)}_{s_0s_1\ldots s_k1}(t) + \sum_{l_{k+1}=1}^{\infty} \mathbf{a}^{(l_{k+1})}_{s_0s_1\ldots s_k2}(t)\cos(l_{k+1}\theta_{k+1})\\
&\quad + \mathbf{a}^{(l_{k+1})}_{s_0s_1\ldots s_k3}(t)\sin(l_{k+1}\theta_{k+1})
\end{aligned} \tag{4.77}$$

with

$$\begin{aligned}
& s_i = 1, 2, 3\ (i = 0, 1, 2, \ldots, k),\\
& \mathbf{a}_{s_0s_1\ldots s_k1} = \mathbf{a}^{(0)}_{s_0s_1\ldots s_k1}\\
& \mathbf{a}_{s_0s_1\ldots s_k2} = (\mathbf{a}^{(1)}_{s_0s_1\ldots s_k2}, \mathbf{a}^{(2)}_{s_0s_1\ldots s_k2}, \ldots, \mathbf{a}^{(N_{k+1})}_{s_0s_1\ldots s_k2})^{\mathrm{T}},\\
& \mathbf{a}_{s_0s_1\ldots s_k3} = (\mathbf{a}^{(1)}_{s_0s_1\ldots s_k3}, \mathbf{a}^{(2)}_{s_0s_1\ldots s_k3}, \ldots, \mathbf{a}^{(N_{k+1})}_{s_0s_1\ldots s_k3})^{\mathrm{T}};\\
& \mathbf{a}_{s_0s_1\ldots s_{k-1}1} = \mathbf{a}^{(0)}_{s_0s_1\ldots s_{k-1}1}\\
& \mathbf{a}_{s_0s_1\ldots s_{k-1}2} = (\mathbf{a}^{(1)}_{s_0s_1\ldots s_{k-1}2}, \mathbf{a}^{(2)}_{s_0s_1\ldots s_{k-1}2}, \ldots, \mathbf{a}^{(N_k)}_{s_0s_1\ldots s_{k-1}2})^{\mathrm{T}},\\
& \mathbf{a}_{s_0s_1\ldots s_{k-1}3} = (\mathbf{a}^{(1)}_{s_0s_1\ldots s_{k-1}3}, \mathbf{a}^{(2)}_{s_0s_1\ldots s_{k-1}3}, \ldots, \mathbf{a}^{(N_k)}_{s_0s_1\ldots s_{k-1}3})^{\mathrm{T}};\\
& \vdots\\
& \mathbf{a}_1 = \mathbf{a}_1^{(0)}\\
& \mathbf{a}_2 = (\mathbf{a}_2^{(1)}, \mathbf{a}_2^{(2)}, \ldots, \mathbf{a}_2^{(N_0)})^{\mathrm{T}},\\
& \mathbf{a}_3 = (\mathbf{a}_3^{(1)}, \mathbf{a}_3^{(2)}, \ldots, \mathbf{a}_3^{(N_0)})^{\mathrm{T}};
\end{aligned} \tag{4.78}$$

which, under $\|\mathbf{a}_{s_0s_1\ldots s_k}(t) - \mathbf{a}^*_{s_0s_1\ldots s_k}(t)\| < \varepsilon$ *with a prescribed small* $\varepsilon > 0$, *can be approximated by a finite term transformation* $\mathbf{a}^*_{s_0s_1\ldots s_k}(t)$

$$\mathbf{a}^*_{s_0s_1\ldots s_k} = \mathbf{a}^{(0)}_{s_0s_1\ldots s_k1}(t) + \sum_{l_{k+1}=1}^{N_{k+1}} \mathbf{a}^{(l_{k+1})}_{s_0s_1\ldots s_k2}(t)\cos(l_{k+1}\theta_{k+1}) + \mathbf{a}^{(l_{k+1})}_{s_0s_1\ldots s_k3}(t)\sin(l_{k+1}\theta_{k+1}) \tag{4.79}$$

and the generalized coordinates are determined by

$$\ddot{\mathbf{a}}_{s_0s_1\ldots s_{k+1}} = \mathbf{g}_{s_0s_1\ldots s_{k+1}}(\mathbf{a}_{s_0s_1\ldots s_{k+1}}, \dot{\mathbf{a}}_{s_0s_1\ldots s_{k+1}}, \mathbf{p}) \tag{4.80}$$

where

$$\begin{aligned}
\mathbf{a}_{s_0s_1\ldots s_{k+1}} &= (\mathbf{a}_{s_0s_1\ldots s_k1}, \mathbf{a}_{s_0s_1\ldots s_k2}, \mathbf{a}_{s_0s_1\ldots s_k3})^{\mathrm{T}},\\
\mathbf{g}_{s_0s_1\ldots s_{k+1}} &= (\mathbf{F}_{s_0s_1\ldots s_k1}, -2\omega_{k+1}\mathbf{k}^{(1)}_{k+1}\dot{\mathbf{a}}_{s_0s_1\ldots s_k3} + \omega^2_{k+1}\mathbf{k}^{(2)}_{k+1}\mathbf{a}_{s_0s_1\ldots s_k2} + \mathbf{F}_{s_0s_1\ldots s_k2},\\
&\quad 2\omega_{k+1}\mathbf{k}^{(1)}_{k+1}\dot{\mathbf{a}}_{s_0s_1\ldots s_k2} + \omega^2_{k+1}\mathbf{k}^{(2)}_{k+1}\mathbf{a}_{s_0s_1\ldots s_k3} + \mathbf{F}_{s_0s_1\ldots s_k3})^{\mathrm{T}};
\end{aligned} \tag{4.81}$$

and

$$\begin{aligned}
\mathbf{k}^{(1)}_{k+1} &= diag(\mathbf{I}_{n_k\times n_k}, 2\mathbf{I}_{n_k\times n_k}, \ldots, N_{k+1}\mathbf{I}_{n_k\times n_k}),\\
\mathbf{k}^{(2)}_{k+1} &= diag(\mathbf{I}_{n_k\times n_k}, 2^2\mathbf{I}_{n_k\times n_k}, \ldots, N^2_{k+1}\mathbf{I}_{n_k\times n_k})\\
n_k &= n(2N_0+1)(2N_1+1)\ldots(2N_k+1);\\
\mathbf{a}_{s_0s_1\ldots s_k1} &= \mathbf{a}^{(0)}_{s_0s_1\ldots s_k1},\\
\mathbf{a}_{s_0s_1\ldots s_k2} &= (\mathbf{a}^{(1)}_{s_0s_1\ldots s_k2}, \mathbf{a}^{(2)}_{s_0s_1\ldots s_k2}, \ldots, \mathbf{a}^{(N_{k+1})}_{s_0s_1\ldots s_k2})^{\mathrm{T}},\\
\mathbf{a}_{s_0s_1\ldots s_k3} &= (\mathbf{a}^{(1)}_{s_0s_1\ldots s_k3}, \mathbf{a}^{(2)}_{s_0s_1\ldots s_k3}, \ldots, \mathbf{a}^{(N_{k+1})}_{s_0s_1\ldots s_k3})^{\mathrm{T}};\\
\mathbf{F}_{s_0s_1\ldots s_k1} &= \mathbf{F}^{(0)}_{s_0s_1\ldots s_k1},\\
\mathbf{F}_{s_0s_1\ldots s_k2} &= (\mathbf{F}^{(1)}_{s_0s_1\ldots s_k2}, \mathbf{F}^{(2)}_{s_0s_1\ldots s_k2}, \ldots, \mathbf{F}^{(N_{k+1})}_{s_0s_1\ldots s_k2})^{\mathrm{T}},\\
\mathbf{F}_{s_0s_1\ldots s_k3} &= (\mathbf{F}^{(1)}_{s_0s_1\ldots s_k3}, \mathbf{F}^{(2)}_{s_0s_1\ldots s_k3}, \ldots, \mathbf{F}^{(N_{k+1})}_{s_0s_1\ldots s_k3})^{\mathrm{T}}\\
\text{for } N_{k+1} &= 1, 2, \ldots, \infty;
\end{aligned} \tag{4.82}$$

and

$$\begin{aligned}
&\mathbf{F}_{s_0s_1\ldots s_k1}(\mathbf{a}_{s_0s_1\ldots s_{k+1}}, \dot{\mathbf{a}}_{s_0s_1\ldots s_{k+1}}, \mathbf{p})\\
&= \frac{1}{2\pi}\int_0^{2\pi} \mathbf{g}_{s_0s_1\ldots s_k}(\mathbf{a}^*_{s_0s_1\ldots s_k}, \dot{\mathbf{a}}^*_{s_0s_1\ldots s_k}, \mathbf{p})d\theta_{k+1};
\end{aligned}$$

$$\mathbf{F}^{(l_{k+1})}_{s_0 s_1 \ldots s_k 2}(\mathbf{a}_{s_0 s_1 \ldots s_{k+1}}, \dot{\mathbf{a}}_{s_0 s_1 \ldots s_{k+1}}, \mathbf{p})$$
$$= \frac{1}{\pi}\int_0^{2\pi} \mathbf{g}_{s_0 s_1 \ldots s_k}(\mathbf{a}^*_{s_0 s_1 \ldots s_k}, \dot{\mathbf{a}}^*_{s_0 s_1 \ldots s_k}, \mathbf{p}) \cos(l_{k+1}\theta_{k+1}) d\theta_{k+1},$$
$$\mathbf{F}^{(l_{k+1})}_{s_0 s_1 \ldots s_k 3}(\mathbf{a}_{s_0 s_1 \ldots s_{k+1}}, \dot{\mathbf{a}}_{s_0 s_1 \ldots s_{k+1}}, \mathbf{p})$$
$$= \frac{1}{\pi}\int_0^{2\pi} \mathbf{g}_{s_0 s_1 \ldots s_k}(\mathbf{a}^*_{s_0 s_1 \ldots s_k}, \dot{\mathbf{a}}^*_{s_0 s_1 \ldots s_k}, \mathbf{p}) \sin(l_{k+1}\theta_{k+1}) d\theta_{k+1}$$
$$\text{for } l_{k+1} = 1, 2, \ldots, N_{k+1}. \tag{4.83}$$

3. *Equation (4.80) becomes*

$$\dot{\mathbf{z}}_{s_0 s_1 \ldots s_{k+1}} = \mathbf{f}_{s_0 s_1 \ldots s_{k+1}}(\mathbf{z}_{s_0 s_1 \ldots s_{k+1}}) \tag{4.84}$$

where

$$\begin{aligned} \mathbf{z}_{s_0 s_1 \ldots s_{k+1}} &= (\mathbf{a}_{s_0 s_1 \ldots s_{k+1}}, \dot{\mathbf{a}}_{s_0 s_1 \ldots s_{k+1}})^{\mathrm{T}}, \\ \mathbf{f}_{s_0 s_1 \ldots s_{k+1}} &= (\dot{\mathbf{a}}_{s_0 s_1 \ldots s_{k+1}}, \mathbf{g}_{s_0 s_1 \ldots s_{k+1}})^{\mathrm{T}}. \end{aligned} \tag{4.85}$$

If equilibrium $\mathbf{z}^*_{s_0 s_1 \ldots s_{k+1}}$ *of Equation (4.84) (i.e.,* $\mathbf{f}_{s_0 s_1 \ldots s_{k+1}}(\mathbf{z}^*_{s_0 s_1 \ldots s_{k+1}}) = \mathbf{0}$*) exists, then the approximate solution of the periodic motion of* kth *generalized coordinates for the period-*m *flow exists as in Equation (4.79). In the vicinity of equilibrium* $\mathbf{z}^*_{s_0 s_1 \ldots s_{k+1}}$, *with*

$$\mathbf{z}_{s_0 s_1 \ldots s_{k+1}} = \mathbf{z}^*_{s_0 s_1 \ldots s_{k+1}} + \Delta\mathbf{z}_{s_0 s_1 \ldots s_{k+1}}, \tag{4.86}$$

the linearized equation of Equation (4.84) is

$$\Delta\dot{\mathbf{z}}_{s_0 s_1 \ldots s_{k+1}} = D\mathbf{f}_{s_0 s_1 \ldots s_{k+1}}(\mathbf{z}^*_{s_0 s_1 \ldots s_{k+1}})\Delta\mathbf{z}_{s_0 s_1 \ldots s_{k+1}} \tag{4.87}$$

and the eigenvalue analysis of equilibrium $\mathbf{z}^*$ *is given by*

$$|D\mathbf{f}_{s_0 s_1 \ldots s_{k+1}}(\mathbf{z}^*_{s_0 s_1 \ldots s_{k+1}}) - \lambda\mathbf{I}_{2n_k(2N_{k+1}+1)\times 2n_k(2N_{k+1}+1)}| = 0 \tag{4.88}$$

where

$$D\mathbf{f}_{s_0 s_1 \ldots s_{k+1}}(\mathbf{z}^*_{s_0 s_1 \ldots s_{k+1}}) = \left.\frac{\partial \mathbf{f}_{s_0 s_1 \ldots s_{k+1}}\left(\mathbf{z}_{s_0 s_1 \ldots s_{k+1}}\right)}{\partial \mathbf{z}_{s_0 s_1 \ldots s_{k+1}}}\right|_{\mathbf{z}^*_{s_0 s_1 \ldots s_{k+1}}}. \tag{4.89}$$

The stability and bifurcation of such a periodic motion of the kth *generalized coordinates can be classified by the eigenvalues of* $\mathbf{Df}_{s_0 s_1 \ldots s_{k+1}}(\mathbf{z}^*_{s_0 s_1 \ldots s_{k+1}})$ *with*

$$(n_1, n_2, n_3 | n_4, n_5, n_6). \tag{4.90}$$

a. *If all eigenvalues of the equilibrium possess negative real parts, the approximate quasi-periodic solution is stable.*
b. *If at least one of the eigenvalues of the equilibrium possesses a positive real part, the approximate quasi-periodic solution is unstable.*

   c. *The boundaries between stable and unstable equilibriums with higher order singularity give bifurcation and stability conditions with higher order singularity.*
4. *For the* kth *order Hopf bifurcation of period-*m *motion, a relation exists as*

$$p_k \omega_k = \omega_{k-1}. \tag{4.91}$$

   a. *If p_k is an irrational number, the* kth*-order Hopf bifurcation of the period-*m *motion is called the quasi-period-p_k Hopf bifurcation, and the corresponding solution of the* kth *generalized coordinates is p_k-quasi-periodic to the system of the* $(k-1)$th *generalized coordinates.*
   b. *If $p_k = 2$, the* kth*-order Hopf bifurcation of the period-*m *motion is called a period-doubling Hopf bifurcation (or a period-2 Hopf bifurcation), and the corresponding solution of the* kth *generalized coordinates is period-doubling to the system of the* $(k-1)$th *generalized coordinates.*
   c. *If $p_k = q$ with an integer q, the* kth*-order Hopf bifurcation of the period-*m *motion is called a period-q Hopf bifurcation, and the corresponding solution of the* kth *generalized coordinates is of q-times period to the system of the* $(k-1)$th *generalized coordinates.*
   d. *If $p_k = p/q$ (p, q are irreducible integer), the* kth*-order Hopf bifurcation of the period-*m *motion is called a period-p/q Hopf bifurcation, and the corresponding solution of the* kth *generalized coordinates is p/q-periodic to the system of the* $(k-1)$th *generalized coordinates.*

Proof. The proof of this theorem is similar to the proof of Theorem 3.6. ■

Similarly, for periodically forced vibration systems, the analytical solution of quasi-periodic motions can be presented as follows.

Theorem 4.4 *Consider a periodically forced, nonlinear vibration system as*

$$\ddot{\mathbf{x}} = \mathbf{F}(\mathbf{x}, \dot{\mathbf{x}}, t, \mathbf{p}) \in \mathscr{R}^n \tag{4.92}$$

where $\mathbf{F}(\mathbf{x}, \dot{\mathbf{x}}, t, \mathbf{p})$ is a C^r-continuous nonlinear function vector ($r \geq 1$) with forcing period $T = 2\pi/\Omega$.

1. *If such a vibration system has a period-*m *motion $\mathbf{x}^{(m)}(t)$ with finite norm $\|\mathbf{x}^{(m)}\|$, there is a generalized coordinate transformation with $\theta = \Omega t$ for the periodic motion of Equation (4.92) in a form of*

$$\mathbf{x}^{(m)}(t) = \mathbf{a}_0^{(m)}(t) + \sum_{k=1}^{\infty} \mathbf{b}_{k/m}(t) \cos\left(\frac{k}{m}\theta\right) + \mathbf{c}_{k/m}(t) \sin\left(\frac{k}{m}\theta\right) \tag{4.93}$$

   with

$$\begin{aligned}
\mathbf{a}_1^{(0)} &\equiv \mathbf{a}_0^{(m)} = (a_{01}^{(m)}, a_{02}^{(m)}, \ldots, a_{0n}^{(m)})^{\mathrm{T}}, \\
\mathbf{a}_2^{(k)} &\equiv \mathbf{b}_{k/m} = (b_{k/m1}, b_{k/m2}, \ldots, b_{k/mn})^{\mathrm{T}}, \\
\mathbf{a}_3^{(k)} &\equiv \mathbf{c}_{k/m} = (c_{k/m1}, c_{k/m2}, \ldots, c_{k/mn})^{\mathrm{T}}
\end{aligned} \tag{4.94}$$

which, under $\|\mathbf{x}^{(m)}(t) - \mathbf{x}^{(m)*}(t)\| < \varepsilon$ *with a prescribed small* $\varepsilon > 0$, *can be approximated by a finite term transformation* $\mathbf{x}^{(m)*}(t)$

$$\mathbf{x}^{(m)*}(t) = \mathbf{a}_0^{(m)}(t) + \sum_{k=1}^{N_0} \mathbf{b}_{k/m}(t)\cos\left(\frac{k}{m}\theta\right) + \mathbf{c}_{k/m}(t)\sin\left(\frac{k}{m}\theta\right) \tag{4.95}$$

and the generalized coordinates are determined by

$$\ddot{\mathbf{a}}_{s_0} = \mathbf{g}_{s_0}(\mathbf{a}_{s_0}, \dot{\mathbf{a}}_{s_0}, \mathbf{p}) \tag{4.96}$$

where

$$\begin{aligned}
\mathbf{k}_0^{(1)} &= diag(\mathbf{I}_{n\times n}, 2\mathbf{I}_{n\times n}, \ldots, N_0\mathbf{I}_{n\times n}),\\
\mathbf{k}_0^{(2)} &= diag(\mathbf{I}_{n\times n}, 2^2\mathbf{I}_{n\times n}, \ldots, N_0^2\mathbf{I}_{n\times n}),\\
\mathbf{a}_1^{(0)} &\equiv \mathbf{a}_0^{(m)}, \mathbf{a}_2^{(k)} \equiv \mathbf{b}_{k/m}, \mathbf{a}_3^{(k)} \equiv \mathbf{c}_{k/m};\\
\mathbf{a}_1 &= \mathbf{a}_1^{(0)},\\
\mathbf{a}_2 &= (\mathbf{a}_2^{(1)}, \mathbf{a}_2^{(2)}, \ldots, \mathbf{a}_2^{(N_0)})^{\mathrm{T}} \equiv \mathbf{b}^{(m)},\\
\mathbf{a}_3 &= (\mathbf{a}_3^{(1)}, \mathbf{a}_3^{(2)}, \ldots, \mathbf{a}_3^{(N_0)})^{\mathrm{T}} \equiv \mathbf{c}^{(m)},\\
\mathbf{F}_2 &= (\mathbf{F}_{11}^{(m)}, \mathbf{F}_{12}^{(m)}, \ldots, \mathbf{F}_{1N_0}^{(m)})^{\mathrm{T}},\\
\mathbf{F}_3 &= (\mathbf{F}_{21}^{(m)}, \mathbf{F}_{22}^{(m)}, \ldots, \mathbf{F}_{2N_0}^{(m)})^{\mathrm{T}};\\
\mathbf{a}_{s_0} &= (\mathbf{a}_1, \mathbf{a}_2, \mathbf{a}_3)^{\mathrm{T}},\\
\mathbf{g}_{s_0} &= \left(\mathbf{F}_1^{(m)}, -2\frac{\Omega}{m}\mathbf{k}_0^{(1)}\dot{\mathbf{a}}_3 + \frac{\Omega^2}{m^2}\mathbf{k}_0^{(2)}\mathbf{a}_2 + \mathbf{F}_2\,,\right.\\
&\qquad \left. 2\frac{\Omega}{m}\mathbf{k}_0^{(1)}\dot{\mathbf{a}}_2 + \frac{\Omega^2}{m^2}\mathbf{k}_0^{(2)}\mathbf{a}_3 + \mathbf{F}_3\right)^{\mathrm{T}}\\
&\text{for } N_0 = 1, 2, \ldots, \infty;
\end{aligned} \tag{4.97}$$

and

$$\begin{aligned}
&\mathbf{F}_0^{(m)}(\mathbf{a}_0^{(m)}, \mathbf{b}^{(m)}, \mathbf{c}^{(m)}, \dot{\mathbf{a}}_0^{(m)}, \dot{\mathbf{b}}^{(m)}, \dot{\mathbf{c}}^{(m)})\\
&\quad = \frac{1}{2m\pi}\int_0^{2m\pi} \mathbf{F}(\mathbf{x}^{(m)*}, \dot{\mathbf{x}}^{(m)*}, t, \mathbf{p})d\theta;\\
&\mathbf{F}_{1k}^{(m)}(\mathbf{a}_0^{(m)}, \mathbf{b}^{(m)}, \mathbf{c}^{(m)}, \dot{\mathbf{a}}_0^{(m)}, \dot{\mathbf{b}}^{(m)}, \dot{\mathbf{c}}^{(m)})\\
&\quad = \frac{1}{m\pi}\int_0^{2m\pi} \mathbf{F}(\mathbf{x}^{(m)*}, \dot{\mathbf{x}}^{(m)*}, t, \mathbf{p})\cos\left(\frac{k}{m}\theta\right)d\theta,\\
&\mathbf{F}_{2k}^{(m)}(\mathbf{a}_0^{(m)}, \mathbf{b}^{(m)}, \mathbf{c}^{(m)}, \dot{\mathbf{a}}_0^{(m)}, \dot{\mathbf{b}}^{(m)}, \dot{\mathbf{c}}^{(m)})\\
&\quad = \frac{1}{m\pi}\int_0^{2m\pi} \mathbf{F}(\mathbf{x}^{(m)*}, \dot{\mathbf{x}}^{(m)*}, t, \mathbf{p})\sin\left(\frac{k}{m}\theta\right)d\theta.
\end{aligned} \tag{4.98}$$

2. *For the* kth *Hopf bifurcation with* $p_k\omega_k = \omega_{k-1}$ $(k = 1, 2, \ldots)$ *and* $\omega_0 = \Omega/m$, *there is a dynamical system of coefficients as*

$$\ddot{\mathbf{a}}_{s_0 s_1 \ldots s_k} = \mathbf{g}_{s_0 s_1 \ldots s_k}(\mathbf{a}_{s_0 s_1 \ldots s_k}, \dot{\mathbf{a}}_{s_0 s_1 \ldots s_k}, \mathbf{p}) \tag{4.99}$$

where

$$\begin{aligned}
\mathbf{a}_{s_0 s_1 \ldots s_k} &= (\mathbf{a}_{s_0 s_1 \ldots s_{k-1} 1}, \mathbf{a}_{s_0 s_1 \ldots s_{k-1} 2}, \mathbf{a}_{s_0 s_1 \ldots s_{k-1} 3})^{\mathrm{T}}, \\
\mathbf{g}_{s_0 s_1 \ldots s_k} &= (\mathbf{F}_{s_0 s_1 \ldots s_{k-1} 1}, -2\omega_k \mathbf{k}_k^{(1)} \dot{\mathbf{a}}_{s_0 s_1 \ldots s_{k-1} 3} + \omega_k^2 \mathbf{k}_k^{(2)} \mathbf{a}_{s_0 s_1 \ldots s_{k-1} 2} + \mathbf{F}_{s_0 s_1 \ldots s_{k-1} 2}, \\
&\quad 2\omega_k \mathbf{k}_k^{(1)} \dot{\mathbf{a}}_{s_0 s_1 \ldots s_{k-1} 2} + \omega_k^2 \mathbf{k}_k^{(2)} \mathbf{a}_{s_0 s_1 \ldots s_{k-1} 3} + \mathbf{F}_{s_0 s_1 \ldots s_{k-1} 3})^{\mathrm{T}}, \\
\mathbf{k}_k^{(1)} &= diag(\mathbf{I}_{n_{k-1} \times n_{k-1}}, 2\mathbf{I}_{n_{k-1} \times n_{k-1}}, \ldots, N_k \mathbf{I}_{n_{k-1} \times n_{k-1}}), \\
\mathbf{k}_k^{(2)} &= diag(\mathbf{I}_{n_{k-1} \times n_{k-1}}, 2^2 \mathbf{I}_{n_{k-1} \times n_{k-1}}, \ldots, N_k^2 \mathbf{I}_{n_{k-1} \times n_{k-1}}) \\
n_{k-1} &= n(2N_0 + 1)(2N_1 + 1) \ldots (2N_{k-1} + 1)
\end{aligned} \tag{4.100}$$

with a periodic solution as

$$\begin{aligned}
\mathbf{a}_{s_0 s_1 \ldots s_k} &= \mathbf{a}^{(0)}_{s_0 s_1 \ldots s_k 1}(t) + \sum_{l_{k+1}=1}^{\infty} \mathbf{a}^{(l_{k+1})}_{s_0 s_1 \ldots s_k 2}(t) \cos(l_{k+1}\theta_{k+1}) \\
&\quad + \mathbf{a}^{(l_{k+1})}_{s_0 s_1 \ldots s_k 3}(t) \sin(l_{k+1}\theta_{k+1})
\end{aligned} \tag{4.101}$$

with

$$\begin{aligned}
&s_i = 1, 2, 3 \; (i = 0, 1, 2, \ldots, k), \\
&\mathbf{a}_{s_0 s_1 \ldots s_k 1} = \mathbf{a}^{(0)}_{s_0 s_1 \ldots s_k 1} \\
&\mathbf{a}_{s_0 s_1 \ldots s_k 2} = (\mathbf{a}^{(1)}_{s_0 s_1 \ldots s_k 2}, \mathbf{a}^{(2)}_{s_0 s_1 \ldots s_k 2}, \ldots, \mathbf{a}^{(N_{k+1})}_{s_0 s_1 \ldots s_k 2})^{\mathrm{T}}, \\
&\mathbf{a}_{s_0 s_1 \ldots s_k 3} = (\mathbf{a}^{(1)}_{s_0 s_1 \ldots s_k 3}, \mathbf{a}^{(2)}_{s_0 s_1 \ldots s_k 3}, \ldots, \mathbf{a}^{(N_{k+1})}_{s_0 s_1 \ldots s_k 3})^{\mathrm{T}}; \\
&\mathbf{a}_{s_0 s_1 \ldots s_{k-1} 1} = \mathbf{a}^{(0)}_{s_0 s_1 \ldots s_{k-1} 1} \\
&\mathbf{a}_{s_0 s_1 \ldots s_{k-1} 2} = (\mathbf{a}^{(1)}_{s_0 s_1 \ldots s_{k-1} 2}, \mathbf{a}^{(2)}_{s_0 s_1 \ldots s_{k-1} 2}, \ldots, \mathbf{a}^{(N_k)}_{s_0 s_1 \ldots s_{k-1} 2})^{\mathrm{T}}, \\
&\mathbf{a}_{s_0 s_1 \ldots s_{k-1} 3} = (\mathbf{a}^{(1)}_{s_0 s_1 \ldots s_{k-1} 3}, \mathbf{a}^{(2)}_{s_0 s_1 \ldots s_{k-1} 3}, \ldots, \mathbf{a}^{(N_k)}_{s_0 s_1 \ldots s_{k-1} 3})^{\mathrm{T}}; \\
&\vdots \\
&\mathbf{a}_1 = \mathbf{a}_1^{(0)} \\
&\mathbf{a}_2 = (\mathbf{a}_2^{(1)}, \mathbf{a}_2^{(2)}, \ldots, \mathbf{a}_2^{(N_0)})^{\mathrm{T}}, \\
&\mathbf{a}_3 = (\mathbf{a}_3^{(1)}, \mathbf{a}_3^{(2)}, \ldots, \mathbf{a}_3^{(N_0)})^{\mathrm{T}};
\end{aligned} \tag{4.102}$$

which, under $\|\mathbf{a}_{s_0 s_1 \ldots s_k}(t) - \mathbf{a}^*_{s_0 s_1 \ldots s_k}(t)\| < \varepsilon$ *with a prescribed small* $\varepsilon > 0$, *can be approximated by a finite term transformation* $\mathbf{a}^*_{s_0 s_1 \ldots s_k}(t)$

$$\mathbf{a}^*_{s_0 s_1 \ldots s_k} = \mathbf{a}^{(0)}_{s_0 s_1 \ldots s_k 1}(t) + \sum_{l_{k+1}=1}^{N_{k+1}} \mathbf{a}^{(l_{k+1})}_{s_0 s_1 \ldots s_k 2}(t) \cos(l_{k+1}\theta_{k+1}) + \mathbf{a}^{(l_{k+1})}_{s_0 s_1 \ldots s_k 3}(t) \sin(l_{k+1}\theta_{k+1}) \tag{4.103}$$

and the generalized coordinates are determined by

$$\ddot{\mathbf{a}}_{s_0 s_1 \ldots s_{k+1}} = \mathbf{g}_{s_0 s_1 \ldots s_{k+1}}(\mathbf{a}_{s_0 s_1 \ldots s_{k+1}}, \dot{\mathbf{a}}_{s_0 s_1 \ldots s_{k+1}}, \mathbf{p}) \tag{4.104}$$

where

$$\begin{aligned}
\mathbf{a}_{s_0 s_1 \ldots s_{k+1}} &= (\mathbf{a}_{s_0 s_1 \ldots s_k 1}, \mathbf{a}_{s_0 s_1 \ldots s_k 2}, \mathbf{a}_{s_0 s_1 \ldots s_k 3})^{\mathrm{T}},\\
\mathbf{g}_{s_0 s_1 \ldots s_{k+1}} &= (\mathbf{F}_{s_0 s_1 \ldots s_k 1}, -2\omega_{k+1}\mathbf{k}^{(1)}_{k+1}\dot{\mathbf{a}}_{s_0 s_1 \ldots s_k 3} + \omega^2_{k+1}\mathbf{k}^{(2)}_{k+1}\mathbf{a}_{s_0 s_1 \ldots s_k 2} + \mathbf{F}_{s_0 s_1 \ldots s_k 2},\\
&\quad 2\omega_{k+1}\mathbf{k}^{(1)}_{k+1}\dot{\mathbf{a}}_{s_0 s_1 \ldots s_k 2} + \omega^2_{k+1}\mathbf{k}^{(2)}_{k+1}\mathbf{a}_{s_0 s_1 \ldots s_k 3} + \mathbf{F}_{s_0 s_1 \ldots s_k 3})^{\mathrm{T}};
\end{aligned} \tag{4.105}$$

and

$$\begin{aligned}
\mathbf{k}^{(1)}_{k+1} &= diag(\mathbf{I}_{n_k \times n_k}, 2\mathbf{I}_{n_k \times n_k}, \ldots, N_{k+1}\mathbf{I}_{n_k \times n_k}),\\
\mathbf{k}^{(2)}_{k+1} &= diag(\mathbf{I}_{n_k \times n_k}, 2^2\mathbf{I}_{n_k \times n_k}, \ldots, N^2_{k+1}\mathbf{I}_{n_k \times n_k})\\
n_k &= n(2N_0 + 1)(2N_1 + 1) \ldots (2N_k + 1);\\
\mathbf{a}_{s_0 s_1 \ldots s_k 1} &= \mathbf{a}^{(0)}_{s_0 s_1 \ldots s_k 1},\\
\mathbf{a}_{s_0 s_1 \ldots s_k 2} &= (\mathbf{a}^{(1)}_{s_0 s_1 \ldots s_k 2}, \mathbf{a}^{(2)}_{s_0 s_1 \ldots s_k 2}, \ldots, \mathbf{a}^{(N_{k+1})}_{s_0 s_1 \ldots s_k 2})^{\mathrm{T}},\\
\mathbf{a}_{s_0 s_1 \ldots s_k 3} &= (\mathbf{a}^{(1)}_{s_0 s_1 \ldots s_k 3}, \mathbf{a}^{(2)}_{s_0 s_1 \ldots s_k 3}, \ldots, \mathbf{a}^{(N_{k+1})}_{s_1 s_1 \ldots s_k 3})^{\mathrm{T}};\\
\mathbf{a}_{s_0 s_1 \ldots s_{k+1}} &= (\mathbf{a}_{s_0 s_1 \ldots s_k 1}, \mathbf{a}_{s_0 s_1 \ldots s_k 2}, \mathbf{a}_{s_0 s_1 \ldots s_k 3})^{\mathrm{T}};\\
\mathbf{F}_{s_0 s_1 \ldots s_k 1} &= \mathbf{F}^{(0)}_{s_0 s_1 \ldots s_k 1},\\
\mathbf{F}_{s_0 s_1 \ldots s_k 2} &= (\mathbf{F}^{(1)}_{s_0 s_1 \ldots s_k 2}, \mathbf{F}^{(2)}_{s_0 s_1 \ldots s_k 2}, \ldots, \mathbf{F}^{(N_{k+1})}_{s_0 s_1 \ldots s_k 2})^{\mathrm{T}},\\
\mathbf{F}_{s_0 s_1 \ldots s_k 3} &= (\mathbf{F}^{(1)}_{s_0 s_1 \ldots s_k 3}, \mathbf{F}^{(2)}_{s_0 s_1 \ldots s_k 3}, \ldots, \mathbf{F}^{(N_{k+1})}_{s_0 s_1 \ldots s_k 3})^{\mathrm{T}}\\
&\text{for } N_{k+1} = 1, 2, \ldots, \infty;
\end{aligned} \tag{4.106}$$

and

$$\begin{aligned}
&\mathbf{F}_{s_1 s_2 \ldots s_k 1}(\mathbf{a}_{s_0 s_1 \ldots s_{k+1}}, \dot{\mathbf{a}}_{s_0 s_1 \ldots s_{k+1}}, \mathbf{p})\\
&= \frac{1}{2\pi}\int_0^{2\pi} \mathbf{g}_{s_0 s_1 \ldots s_k}(\mathbf{a}^*_{s_0 s_1 \ldots s_k}, \dot{\mathbf{a}}^*_{s_0 s_1 \ldots s_k}, \mathbf{p}) d\theta_{k+1};
\end{aligned}$$

$$\begin{aligned}
&\mathbf{F}^{(l_{k+1})}_{s_1 s_2 \ldots s_k 2}(\mathbf{a}_{s_0 s_1 \ldots s_{k+1}}, \dot{\mathbf{a}}_{s_0 s_1 \ldots s_{k+1}}, \mathbf{p}) \\
&\quad = \frac{1}{\pi}\int_0^{2\pi} \mathbf{g}_{s_0 s_1 \ldots s_k}(\mathbf{a}^*_{s_0 s_1 \ldots s_k}, \dot{\mathbf{a}}^*_{s_0 s_1 \ldots s_k}, \mathbf{p}) \cos(l_{k+1}\theta_{k+1}) d\theta_{k+1}, \\
&\mathbf{F}^{(l_{k+1})}_{s_1 s_2 \ldots s_k 3}(\mathbf{a}_{s_0 s_1 \ldots s_{k+1}}, \dot{\mathbf{a}}_{s_0 s_1 \ldots s_{k+1}}, \mathbf{p}) \\
&\quad = \frac{1}{\pi}\int_0^{2\pi} \mathbf{g}_{s_0 s_1 \ldots s_k}(\mathbf{a}^*_{s_0 s_1 \ldots s_k}, \dot{\mathbf{a}}^*_{s_0 s_1 \ldots s_k}, \mathbf{p}) \sin(l_{k+1}\theta_{k+1}) d\theta_{k+1} \\
&\text{for } l_{k+1} = 1, 2, \ldots, N_{k+1}.
\end{aligned} \tag{4.107}$$

3. *Equation (4.104) becomes*

$$\dot{\mathbf{z}}_{s_0 s_1 \ldots s_{k+1}} = \mathbf{f}_{s_0 s_1 \ldots s_{k+1}}(\mathbf{z}_{s_0 s_1 \ldots s_{k+1}}) \tag{4.108}$$

where

$$\begin{aligned}
\mathbf{z}_{s_0 s_1 \ldots s_{k+1}} &= (\mathbf{a}_{s_0 s_1 \ldots s_{k+1}}, \dot{\mathbf{a}}_{s_0 s_1 \ldots s_{k+1}})^{\mathrm{T}}, \\
\mathbf{f}_{s_0 s_1 \ldots s_{k+1}} &= (\dot{\mathbf{a}}_{s_0 s_1 \ldots s_{k+1}}, \mathbf{g}_{s_0 s_1 \ldots s_{k+1}})^{\mathrm{T}}.
\end{aligned} \tag{4.109}$$

If equilibrium $\mathbf{z}^*_{s_1 s_2 \ldots s_{k+1}}$ of Equation (4.108) (i.e., $\mathbf{f}_{s_1 s_2 \ldots s_{k+1}}(\mathbf{z}^*_{s_1 s_2 \ldots s_{k+1}}) = \mathbf{0}$) exists, then the approximate solution of the periodic motion of kth *generalized coordinates for the period*-m *motion exists as in Equation (4.103). In vicinity of equilibrium $\mathbf{z}^*_{s_1 s_2 \ldots s_{k+1}}$, with*

$$\mathbf{z}_{s_0 s_1 \ldots s_{k+1}} = \mathbf{z}^*_{s_0 s_1 \ldots s_{k+1}} + \Delta\mathbf{z}_{s_0 s_1 \ldots s_{k+1}}, \tag{4.110}$$

the linearized equation of Equation (4.108) is

$$\Delta\dot{\mathbf{z}}_{s_0 s_1 \ldots s_{k+1}} = D\mathbf{f}_{s_0 s_1 \ldots s_{k+1}}(\mathbf{z}^*_{s_0 s_1 \ldots s_{k+1}})\Delta\mathbf{z}_{s_0 s_1 \ldots s_{k+1}} \tag{4.111}$$

and the eigenvalue analysis of equilibrium $\mathbf{z}^*$ is given by

$$|D\mathbf{f}_{s_0 s_1 \ldots s_{k+1}}(\mathbf{z}^*_{s_0 s_1 \ldots s_{k+1}}) - \lambda \mathbf{I}_{2n_k(2N_{k+1}+1)\times 2n_k(2N_{k+1}+1)}| = 0 \tag{4.112}$$

where

$$D\mathbf{f}_{s_0 s_1 \ldots s_{k+1}}(\mathbf{z}^*_{s_0 s_1 \ldots s_{k+1}}) = \left.\frac{\partial \mathbf{f}_{s_0 s_1 \ldots s_{k+1}}\left(\mathbf{z}_{s_0 s_1 \ldots s_{k+1}}\right)}{\partial \mathbf{z}_{s_0 s_1 \ldots s_{k+1}}}\right|_{\mathbf{z}^*_{s_0 s_1 \ldots s_{k+1}}}. \tag{4.113}$$

The stability and bifurcation of such a periodic motion of the kth *generalized coordinates can be classified by the eigenvalues of $D\mathbf{f}_{s_0 s_1 \ldots s_{k+1}}(\mathbf{z}^*_{s_0 s_1 \ldots s_{k+1}})$ with*

$$(n_1, n_2, n_3 | n_4, n_5, n_6). \tag{4.114}$$

a. *If all eigenvalues of the equilibrium possess negative real parts, the approximate quasi-periodic solution is stable.*

   b. *If at least one of the eigenvalues of the equilibrium possesses a positive real part, the approximate quasi-periodic solution is unstable.*
   c. *The boundaries between stable and unstable equilibriums with higher order singularity give bifurcation and stability conditions with higher order singularity.*
4. *For the* kth *order Hopf bifurcation of period*-m *motion, a relation exists as*

$$p_k \omega_k = \omega_{k-1}. \tag{4.115}$$

   a. *If p_k is an irrational number, the* kth-*order Hopf bifurcation of the period*-m *motion is called the quasi-period-p_k Hopf bifurcation, and the corresponding solution of the* kth *generalized coordinates is p_k-quasi-periodic to the system of the* $(k-1)$th *generalized coordinates.*
   b. *If $p_k = 2$, the* kth-*order Hopf bifurcation of the period*-m *motion is called a period-doubling Hopf bifurcation (or a period-2 Hopf bifurcation), and the corresponding solution of the* kth *generalized coordinates is period-doubling to the system of the* $(k-1)$th *generalized coordinates.*
   c. *If $p_k = q$ with an integer q, the* kth-*order Hopf bifurcation of the period*-m *motion is called a period-q Hopf bifurcation, and the corresponding solution of the* kth *generalized coordinates is of q-times period to the system of the* $(k-1)$th *generalized coordinates.*
   d. *If $p_k = p/q$ (p, q are irreducible integer), the* kth-*order Hopf bifurcation of the period*-m *motion is called a period-p/q Hopf bifurcation, and the corresponding solution of the* kth *generalized coordinates is p/q-periodic to the system of the* $(k-1)$th *generalized coordinates.*

Proof. The proof of this theorem is similar to the proof of Theorem 3.8. ■

4.3 Time-Delayed Nonlinear Systems

In this section, analytical quasi-periodic flows in time-delayed, nonlinear dynamical systems will be discussed. Consider quasi-periodic flows in autonomous, time-delayed nonlinear systems first, and the analytical solution of quasi-periodic motion relative to period-m flow is given as follows.

Theorem 4.5 *Consider a time-delayed, nonlinear system as*

$$\dot{\mathbf{x}} = \mathbf{f}(\mathbf{x}, \mathbf{x}^{\tau}, \mathbf{p}) \in \mathscr{R}^n \tag{4.116}$$

where $\mathbf{f}(\mathbf{x}, \mathbf{x}^{\tau}, \mathbf{p})$ is a C^r-continuous nonlinear function vector ($r \geq 1$).

1. *If such a time-delayed dynamical system has a period*-m *flow $\mathbf{x}^{(m)}(t)$ with finite norm $\|\mathbf{x}^{(m)}\|$ and period $T = 2\pi/\Omega$, there is a generalized coordinate transformation for the period*-m *flow of Equation (4.116) in the form of*

$$\mathbf{x}^{(m)}(t) = \mathbf{a}_0^{(m)}(t) + \sum_{k=1}^{\infty} \mathbf{b}_{k/m}(t) \cos\left(\frac{k}{m}\theta\right) + \mathbf{c}_{k/m}(t) \sin\left(\frac{k}{m}\theta\right);$$

$$\mathbf{x}^{\tau(m)}(t) = \mathbf{a}_0^{\tau(m)}(t) + \sum_{k=1}^{\infty} \mathbf{b}_{k/m}^{\tau}(t) \cos\left[\frac{k}{m}(\theta - \theta^{\tau})\right] + \mathbf{c}_{k/m}^{\tau}(t) \sin\left[\frac{k}{m}(\theta - \theta^{\tau})\right] \tag{4.117}$$

with $\mathbf{a}_0^{\tau(m)} = \mathbf{a}_0^{(m)}(t-\tau), \mathbf{b}_k^{\tau(m)} = \mathbf{b}_k^{(m)}(t-\tau), \mathbf{c}_k^{\tau(m)} = \mathbf{c}_k^{(m)}(t-\tau), \theta^\tau = \Omega\tau$ *and*

$$\begin{aligned}
\mathbf{a}_1^{(0)} &\equiv \mathbf{a}_0^{(m)} = (a_{01}^{(m)}, a_{02}^{(m)}, \ldots, a_{0n}^{(m)})^{\mathrm{T}},\\
\mathbf{a}_2^{(k)} &\equiv \mathbf{b}_{k/m} = (b_{k/m1}, b_{k/m2}, \ldots, b_{k/mn})^{\mathrm{T}},\\
\mathbf{a}_3^{(k)} &\equiv \mathbf{c}_{k/m} = (c_{k/m1}, c_{k/m2}, \ldots, c_{k/mn})^{\mathrm{T}}\\
\mathbf{a}_1^{\tau(0)} &\equiv \mathbf{a}_0^{\tau(m)} = (a_{01}^{\tau(m)}, a_{02}^{\tau(m)}, \ldots, a_{0n}^{\tau(m)})^{\mathrm{T}},\\
\mathbf{a}_2^{\tau(k)} &\equiv \mathbf{b}_{k/m}^{\tau} = (b_{k/m1}^{\tau}, b_{k/m2}^{\tau}, \ldots, b_{k/mn}^{\tau})^{\mathrm{T}},\\
\mathbf{a}_3^{\tau(k)} &\equiv \mathbf{c}_{k/m}^{\tau} = (c_{k/m1}^{\tau}, c_{k/m2}^{\tau}, \ldots, c_{k/mn}^{\tau})^{\mathrm{T}}
\end{aligned} \tag{4.118}$$

which, under $\|\mathbf{x}^{(m)}(t) - \mathbf{x}^{(m)*}(t)\| < \varepsilon$ *and* $\|\mathbf{x}^{\tau(m)}(t) - \mathbf{x}^{\tau(m)*}(t)\| < \varepsilon^\tau$ *with prescribed small* $\varepsilon > 0$ *and* $\varepsilon^\tau > 0$*, can be approximated by a finite term transformation* $\mathbf{x}^{(m)*}(t)$ *as*

$$\begin{aligned}
\mathbf{x}^{(m)*}(t) &= \mathbf{a}_0^{(m)}(t) + \sum_{k=1}^{N_0} \mathbf{b}_{k/m}(t)\cos\left(\frac{k}{m}\theta\right) + \mathbf{c}_{k/m}(t)\sin\left(\frac{k}{m}\theta\right);\\
\mathbf{x}^{\tau(m)*}(t) &= \mathbf{a}_0^{\tau(m)}(t) + \sum_{k=1}^{N_0} \mathbf{b}_{k/m}^{\tau}(t)\cos\left[\frac{k}{m}(\theta-\theta^\tau)\right] + \mathbf{c}_{k/m}^{\tau}(t)\sin\left[\frac{k}{m}(\theta-\theta^\tau)\right]
\end{aligned} \tag{4.119}$$

and the generalized coordinates are determined by

$$\dot{\mathbf{a}}_{s_0} = \mathbf{f}_{s_0}(\mathbf{a}_{s_0}, \mathbf{a}_{s_0}^{\tau}, \mathbf{p}) \tag{4.120}$$

where

$$\begin{aligned}
\mathbf{k}_0 &= diag(\mathbf{I}_{n\times n}, 2\mathbf{I}_{n\times n}, \ldots, N_0\mathbf{I}_{n\times n}),\\
\mathbf{a}_1^{(0)} &\equiv \mathbf{a}_0^{(m)}, \mathbf{a}_2^{(k)} \equiv \mathbf{b}_{k/m}, \mathbf{a}_3^{(k)} \equiv \mathbf{c}_{k/m};\\
\mathbf{a}_1^{\tau(0)} &\equiv \mathbf{a}_0^{\tau(m)}, \mathbf{a}_2^{\tau(k)} \equiv \mathbf{b}_{k/m}^{\tau}, \mathbf{a}_3^{(k)} \equiv \mathbf{c}_{k/m}^{\tau}\\
\mathbf{a}_1 &= \mathbf{a}_1^{(0)},\\
\mathbf{a}_2 &= (\mathbf{a}_2^{(1)}, \mathbf{a}_2^{(2)}, \ldots, \mathbf{a}_2^{(N_0)})^{\mathrm{T}} \equiv \mathbf{b}^{(m)},\\
\mathbf{a}_3 &= (\mathbf{a}_3^{(1)}, \mathbf{a}_3^{(2)}, \ldots, \mathbf{a}_3^{(N_0)})^{\mathrm{T}} \equiv \mathbf{c}^{(m)},\\
\mathbf{a}_1^{\tau} &= \mathbf{a}_1^{\tau(0)},\\
\mathbf{a}_2^{\tau} &= (\mathbf{a}_2^{\tau(1)}, \mathbf{a}_2^{\tau(2)}, \ldots, \mathbf{a}_2^{\tau(N_0)})^{\mathrm{T}} \equiv \mathbf{b}^{\tau(m)},\\
\mathbf{a}_3^{\tau} &= (\mathbf{a}_3^{\tau(1)}, \mathbf{a}_3^{\tau(2)}, \ldots, \mathbf{a}_3^{\tau(N_0)})^{\mathrm{T}} \equiv \mathbf{c}^{\tau(m)},\\
\mathbf{F}_1 &= \mathbf{F}_0^{(m)}\\
\mathbf{F}_2 &= (\mathbf{F}_{11}^{(m)}, \mathbf{F}_{12}^{(m)}, \ldots, \mathbf{F}_{1N_0}^{(m)})^{\mathrm{T}},
\end{aligned}$$

$$\begin{aligned}
&\mathbf{F}_3 = (\mathbf{F}_{21}^{(m)}, \mathbf{F}_{22}^{(m)}, \ldots, \mathbf{F}_{2N_0}^{(m)})^{\mathrm{T}};\\
&\mathbf{a}_{s_0} = (\mathbf{a}_1, \mathbf{a}_2, \mathbf{a}_3)^{\mathrm{T}},\\
&\mathbf{a}_{s_0}^{\tau} = (\mathbf{a}_1^{\tau}, \mathbf{a}_2^{\tau}, \mathbf{a}_3^{\tau})^{\mathrm{T}}\\
&\mathbf{f}_{s_0} = \left(\mathbf{F}_1, -\frac{\Omega}{m}\mathbf{k}_0\mathbf{a}_3 + \mathbf{F}_2, \frac{\Omega}{m}\mathbf{k}_0\mathbf{a}_2 + \mathbf{F}_3\right)^{\mathrm{T}}\\
&\text{for } N = 1, 2, \ldots, \infty;
\end{aligned} \tag{4.121}$$

and

$$\begin{aligned}
&\mathbf{F}_0^{(m)}(\mathbf{a}_{s_0}, \mathbf{a}_{s_0}^{\tau}) = \frac{1}{2m\pi}\int_0^{2m\pi} \mathbf{f}(\mathbf{x}^{(m)*}, \mathbf{x}^{\tau(m)*}, \mathbf{p})d\theta;\\
&\mathbf{F}_{1k}^{(m)}(\mathbf{a}_{s_0}, \mathbf{a}_{s_0}^{\tau}) = \frac{1}{m\pi}\int_0^{2m\pi} \mathbf{f}(\mathbf{x}^{(m)*}, \mathbf{x}^{\tau(m)*}, \mathbf{p})\cos\left(\frac{k}{m}\theta\right)d\theta,\\
&\mathbf{F}_{2k}^{(m)}(\mathbf{a}_{s_0}, \mathbf{a}_{s_0}^{\tau}) = \frac{1}{m\pi}\int_0^{2m\pi} \mathbf{f}(\mathbf{x}^{(m)*}, \mathbf{x}^{\tau(m)*}, \mathbf{p})\sin\left(\frac{k}{m}\theta\right)d\theta\\
&\text{for } k = 1, 2, \ldots, N_0.
\end{aligned} \tag{4.122}$$

2. *After the* k*th Hopf bifurcation with* $p_k\omega_k = \omega_{k-1}((k = 1, 2, \ldots))$ *and* $\omega_0 = \Omega/m$, *there is a dynamical system of coefficients as*

$$\dot{\mathbf{a}}_{s_0s_1\ldots s_k} = \mathbf{f}_{s_0s_1\ldots s_k}(\mathbf{a}_{s_0s_1\ldots s_k}, \mathbf{a}_{s_0s_1\ldots s_k}^{\tau}, \mathbf{p}) \tag{4.123}$$

where

$$\begin{aligned}
&\mathbf{a}_{s_0s_1\ldots s_k} = (\mathbf{a}_{s_0s_1\ldots s_{k-1}1}, \mathbf{a}_{s_0s_1\ldots s_{k-1}2}, \mathbf{a}_{s_0s_1\ldots s_{k-1}3})^{\mathrm{T}},\\
&\mathbf{a}_{s_0s_1\ldots s_k}^{\tau} = (\mathbf{a}_{s_0s_1\ldots s_{k-1}1}^{\tau}, \mathbf{a}_{s_0s_1\ldots s_{k-1}2}^{\tau}, \mathbf{a}_{s_0s_1\ldots s_{k-1}3}^{\tau})^{\mathrm{T}}\\
&\mathbf{f}_{s_0s_1\ldots s_k} = (\mathbf{F}_{s_0s_1\ldots s_{k-1}1}, -\omega_k\mathbf{k}_k\mathbf{a}_{s_0s_1\ldots s_{k-1}3} + \mathbf{F}_{s_0s_1\ldots s_{k-1}2},\\
&\qquad \omega_k\mathbf{k}_k\mathbf{a}_{s_0s_1\ldots s_{k-1}2} + \mathbf{F}_{s_0s_1\ldots s_{k-1}3})^{\mathrm{T}},\\
&\mathbf{k}_k = diag(\mathbf{I}_{n_{k-1}\times n_{k-1}}, 2\mathbf{I}_{n_{k-1}\times n_{k-1}}, \ldots, N_k\mathbf{I}_{n_{k-1}\times n_{k-1}}),\\
&n_k = n(2N_0+1)(2N_1+1)\ldots(2N_{k-1}+1)
\end{aligned} \tag{4.124}$$

with a periodic solution as

$$\begin{aligned}
\mathbf{a}_{s_0s_1\ldots s_k} &= \mathbf{a}_{s_0s_1\ldots s_k1}^{(0)}(t) + \sum_{l_{k+1}=1}^{\infty} \mathbf{a}_{s_0s_1\ldots s_k2}^{(l_{k+1})}(t)\cos(l_{k+1}\theta_{k+1})\\
&\quad + \mathbf{a}_{s_0s_1\ldots s_k3}^{(l_{k+1})}(t)\sin(l_{k+1}\theta_{k+1}),\\
\mathbf{a}_{s_0s_1\ldots s_k}^{\tau} &= \mathbf{a}_{s_0s_1\ldots s_k1}^{\tau(0)}(t) + \sum_{l_{k+1}=1}^{\infty} \mathbf{a}_{s_0s_1\ldots s_k2}^{\tau(l_{k+1})}(t)\cos[l_{k+1}(\theta_{k+1} - \theta_{k+1}^{\tau})]\\
&\quad + \mathbf{a}_{s_0s_1\ldots s_k3}^{\tau(l_{k+1})}(t)[l_{k+1}(\theta_{k+1} - \theta_{k+1}^{\tau})]
\end{aligned} \tag{4.125}$$

with

$$\begin{aligned}
&\mathbf{a}^{\tau(0)}_{s_0s_1\ldots s_k1}(t)=\mathbf{a}^{(0)}_{s_0s_1\ldots s_k1}(t-\tau),\\
&\mathbf{a}^{\tau(l_{k+1})}_{s_0s_1\ldots s_k2}(t)=\mathbf{a}^{(l_{k+1})}_{s_0s_1\ldots s_k2}(t-\tau),\\
&\mathbf{a}^{\tau(l_{k+1})}_{s_0s_1\ldots s_k3}(t)=\mathbf{a}^{(l_{k+1})}_{s_0s_1\ldots s_k3}(t-\tau),\\
&\theta^{\tau}_{k+1}=\omega_{k+1}\tau
\end{aligned}\tag{4.126}$$

and

$$\begin{aligned}
&s_i=1,2,3\ (i=0,1,2,\ldots,k),\\
&\mathbf{a}_{s_0s_1\ldots s_k1}=\mathbf{a}^{(0)}_{s_0s_1\ldots s_k1},\\
&\mathbf{a}_{s_0s_1\ldots s_k2}=(\mathbf{a}^{(1)}_{s_0s_1\ldots s_k2},\mathbf{a}^{(2)}_{s_0s_1\ldots s_k2},\ldots,\mathbf{a}^{(N_{k+1})}_{s_0s_1\ldots s_k2})^{\mathrm{T}},\\
&\mathbf{a}_{s_0s_1\ldots s_k3}=(\mathbf{a}^{(1)}_{s_0s_1\ldots s_k3},\mathbf{a}^{(2)}_{s_0s_1\ldots s_k3},\ldots,\mathbf{a}^{(N_{k+1})}_{s_0s_1\ldots s_k3})^{\mathrm{T}};\\
&\mathbf{a}_{s_0s_1\ldots s_{k-1}1}=\mathbf{a}^{(0)}_{s_0s_1\ldots s_{k-1}1},\\
&\mathbf{a}_{s_0s_1\ldots s_{k-1}2}=(\mathbf{a}^{(1)}_{s_0s_1\ldots s_{k-1}2},\mathbf{a}^{(2)}_{s_0s_1\ldots s_{k-1}2},\ldots,\mathbf{a}^{(N_k)}_{s_0s_1\ldots s_{k-1}2})^{\mathrm{T}},\\
&\mathbf{a}_{s_0s_1\ldots s_{k-1}3}=(\mathbf{a}^{(1)}_{s_0s_1\ldots s_{k-1}3},\mathbf{a}^{(2)}_{s_0s_1\ldots s_{k-1}3},\ldots,\mathbf{a}^{(N_k)}_{s_0s_1\ldots s_{k-1}3})^{\mathrm{T}};\\
&\vdots\\
&\mathbf{a}_1=\mathbf{a}^{(0)}_1,\\
&\mathbf{a}_2=(\mathbf{a}^{(1)}_2,\mathbf{a}^{(2)}_2,\ldots,\mathbf{a}^{(N_0)}_2)^{\mathrm{T}},\\
&\mathbf{a}_3=(\mathbf{a}^{(1)}_3,\mathbf{a}^{(2)}_3,\ldots,\mathbf{a}^{(N_0)}_3)^{\mathrm{T}};
\end{aligned}\tag{4.127}$$

and

$$\begin{aligned}
&s_i=1,2,3\ (i=0,1,2,\ldots,k),\\
&\mathbf{a}^{\tau}_{s_0s_1\ldots s_k1}=\mathbf{a}^{\tau(0)}_{s_0s_1\ldots s_k1},\\
&\mathbf{a}^{\tau}_{s_0s_1\ldots s_k2}=(\mathbf{a}^{\tau(1)}_{s_0s_1\ldots s_k2},\mathbf{a}^{\tau(2)}_{s_0s_1\ldots s_k2},\ldots,\mathbf{a}^{\tau(N_{k+1})}_{s_0s_1\ldots s_k2})^{\mathrm{T}},\\
&\mathbf{a}^{\tau}_{s_0s_1\ldots s_k2}=(\mathbf{a}^{\tau(1)}_{s_0s_1\ldots s_k3},\mathbf{a}^{\tau(2)}_{s_0s_1\ldots s_k3},\ldots,\mathbf{a}^{\tau(N_{k+1})}_{s_0s_1\ldots s_k3})^{\mathrm{T}};\\
&\mathbf{a}^{\tau}_{s_0s_1\ldots s_{k-1}1}=\mathbf{a}^{\tau(0)}_{s_0s_1\ldots s_{k-1}1},\\
&\mathbf{a}^{\tau}_{s_0s_1\ldots s_{k-1}2}=(\mathbf{a}^{\tau(1)}_{s_0s_1\ldots s_{k-1}2},\mathbf{a}^{\tau(2)}_{s_0s_1\ldots s_{k-1}2},\ldots,\mathbf{a}^{\tau(N_k)}_{s_0s_1\ldots s_{k-1}2})^{\mathrm{T}},\\
&\mathbf{a}^{\tau}_{s_0s_1\ldots s_{k-1}3}=(\mathbf{a}^{\tau(1)}_{s_0s_1\ldots s_{k-1}3},\mathbf{a}^{\tau(2)}_{s_0s_1\ldots s_{k-1}3},\ldots,\mathbf{a}^{\tau(N_k)}_{s_0s_1\ldots s_{k-1}3})^{\mathrm{T}};\\
&\vdots\\
&\mathbf{a}^{\tau}_1=\mathbf{a}^{\tau(0)}_1,\\
&\mathbf{a}^{\tau}_2=(\mathbf{a}^{\tau(1)}_2,\mathbf{a}^{\tau(2)}_2,\ldots,\mathbf{a}^{\tau(N_0)}_2)^{\mathrm{T}},\\
&\mathbf{a}^{\tau}_3=(\mathbf{a}^{\tau(1)}_3,\mathbf{a}^{\tau(2)}_3,\ldots,\mathbf{a}^{\tau(N_0)}_3)^{\mathrm{T}};
\end{aligned}\tag{4.128}$$

which, under $\|\mathbf{a}_{s_0 s_1 \ldots s_k}(t) - \mathbf{a}^*_{s_0 s_1 \ldots s_k}(t)\| < \varepsilon$ *and* $\|\mathbf{a}^\tau_{s_0 s_1 \ldots s_k}(t) - \mathbf{a}^{\tau *}_{s_0 s_1 \ldots s_k}(t)\| < \varepsilon_\tau$ *with a prescribed small* $\varepsilon > 0$ *and* $\varepsilon_\tau > 0$, *can be approximated by a finite term transformation, that is,*

$$\begin{aligned}
\mathbf{a}^*_{s_0 s_1 \ldots s_k} &= \mathbf{a}^{(0)}_{s_0 s_1 \ldots s_k 1}(t) + \sum_{l_{k+1}=1}^{N_{k+1}} \mathbf{a}^{(l_{k+1})}_{s_0 s_1 \ldots s_k 2}(t) \cos(l_{k+1}\theta_{k+1}) \\
&\quad + \mathbf{a}^{(l_{k+1})}_{s_0 s_1 \ldots s_k 3}(t) \sin(l_{k+1}\theta_{k+1}), \\
\mathbf{a}^{\tau *}_{s_0 s_1 \ldots s_k} &= \mathbf{a}^{\tau(0)}_{s_0 s_1 \ldots s_k 1}(t) + \sum_{l_{k+1}=1}^{N_{k+1}} \mathbf{a}^{\tau(l_{k+1})}_{s_0 s_1 \ldots s_k 2}(t) \cos[l_{k+1}(\theta_{k+1} - \theta^\tau_{k+1})] \\
&\quad + \mathbf{a}^{\tau(l_{k+1})}_{s_0 s_1 \ldots s_k 3}(t) \sin[l_{k+1}(\theta_{k+1} - \theta^\tau_{k+1})]
\end{aligned} \tag{4.129}$$

and the generalized coordinates are determined by

$$\dot{\mathbf{a}}_{s_0 s_1 \ldots s_{k+1}} = \mathbf{f}_{s_0 s_1 \ldots s_{k+1}}(\mathbf{a}_{s_0 s_1 \ldots s_{k+1}}, \mathbf{a}^\tau_{s_0 s_1 \ldots s_{k+1}}, \mathbf{p}) \tag{4.130}$$

where

$$\begin{aligned}
\mathbf{a}_{s_0 s_1 \ldots s_{k+1}} &= (\mathbf{a}_{s_0 s_1 \ldots s_k 1}, \mathbf{a}_{s_0 s_1 \ldots s_k 2}, \mathbf{a}_{s_0 s_1 \ldots s_k 3})^{\mathrm{T}}, \\
\mathbf{a}^\tau_{s_0 s_1 \ldots s_{k+1}} &= (\mathbf{a}^\tau_{s_0 s_1 \ldots s_k 1}, \mathbf{a}^\tau_{s_0 s_1 \ldots s_k 2}, \mathbf{a}^\tau_{s_0 s_1 \ldots s_k 3})^{\mathrm{T}}, \\
\mathbf{f}_{s_0 s_1 \ldots s_{k+1}} &= (\mathbf{F}_{s_0 s_1 \ldots s_k 1}, -\omega_{k+1}\mathbf{k}_{k+1}\mathbf{a}_{s_0 s_1 \ldots s_k 3} + \mathbf{F}_{s_0 s_1 \ldots s_k 2}, \\
&\quad \omega_{k+1}\mathbf{k}_{k+1}\mathbf{a}_{s_0 s_1 \ldots s_k 2} + \mathbf{F}_{s_0 s_1 \ldots s_k 3})^{\mathrm{T}}
\end{aligned} \tag{4.131}$$

and

$$\begin{aligned}
\mathbf{k}_{k+1} &= diag(\mathbf{I}_{n_k \times n_k}, 2\mathbf{I}_{n_k \times n_k}, \ldots, N_{k+1}\mathbf{I}_{n_k \times n_k}), \\
n_k &= n(2N_0 + 1)(2N_1 + 1)\ldots(2N_k + 1); \\
\mathbf{a}_{s_0 s_1 \ldots s_k 1} &= \mathbf{a}^{(0)}_{s_0 s_1 \ldots s_k 1}, \\
\mathbf{a}_{s_0 s_1 \ldots s_k 2} &= (\mathbf{a}^{(1)}_{s_0 s_1 \ldots s_k 2}, \mathbf{a}^{(2)}_{s_0 s_1 \ldots s_k 2}, \ldots, \mathbf{a}^{(N_{k+1})}_{s_0 s_1 \ldots s_k 2})^{\mathrm{T}}, \\
\mathbf{a}_{s_0 s_1 \ldots s_k 3} &= (\mathbf{a}^{(1)}_{s_0 s_1 \ldots s_k 3}, \mathbf{a}^{(2)}_{s_0 s_1 \ldots s_k 3}, \ldots, \mathbf{a}^{(N_{k+1})}_{s_0 s_1 \ldots s_k 3})^{\mathrm{T}}; \\
\mathbf{a}^\tau_{s_0 s_1 \ldots s_k 1} &= \mathbf{a}^{\tau(0)}_{s_0 s_1 \ldots s_k 1}, \\
\mathbf{a}^\tau_{s_0 s_1 \ldots s_k 2} &= (\mathbf{a}^{\tau(1)}_{s_0 s_1 \ldots s_k 2}, \mathbf{a}^{\tau(2)}_{s_0 s_1 \ldots s_k 2}, \ldots, \mathbf{a}^{\tau(N_{k+1})}_{s_0 s_1 \ldots s_k 2})^{\mathrm{T}}, \\
\mathbf{a}^\tau_{s_0 s_1 \ldots s_k 3} &= (\mathbf{a}^{\tau(1)}_{s_0 s_1 \ldots s_k 3}, \mathbf{a}^{\tau(2)}_{s_0 s_1 \ldots s_k 2}, \ldots, \mathbf{a}^{\tau(N_{k+1})}_{s_0 s_1 \ldots s_k 3})^{\mathrm{T}}; \\
\mathbf{F}_{s_0 s_1 \ldots s_k 1} &= \mathbf{F}^{(0)}_{s_1 s_2 \ldots s_k 1},
\end{aligned}$$

$$\begin{aligned}
&\mathbf{F}_{s_0 s_1 \ldots s_k 2}=(\mathbf{F}_{s_0 s_1 \ldots s_k 2}^{(1)}, \mathbf{F}_{s_0 s_1 \ldots s_k 2}^{(2)}, \ldots, \mathbf{F}_{s_0 s_1 \ldots s_k 2}^{(N_{k+1})})^{\mathrm{T}}, \\
&\mathbf{F}_{s_0 s_1 \ldots s_k 3}=(\mathbf{F}_{s_0 s_1 \ldots s_k 3}^{(1)}, \mathbf{F}_{s_0 s_1 \ldots s_k 3}^{(2)}, \ldots, \mathbf{F}_{s_0 s_1 \ldots s_k 3}^{(N_{k+1})})^{\mathrm{T}} \\
&\text{for } N_{k+1}=1,2, \ldots, \infty;
\end{aligned} \tag{4.132}$$

and

$$\begin{aligned}
&\mathbf{F}_{s_0 s_1 \ldots s_k 1}(\mathbf{a}_{s_0 s_1 \ldots s_{k+1}}, \mathbf{a}_{s_0 s_1 \ldots s_{k+1}}^{\tau}, \mathbf{p}) \\
&\quad=\frac{1}{2 \pi} \int_0^{2 \pi} \mathbf{f}_{s_0 s_1 \ldots s_k}(\mathbf{a}_{s_0 s_1 \ldots s_k}^*, \mathbf{a}_{s_0 s_1 \ldots s_k}^{\tau *}, \mathbf{p}) d \theta_{k+1}; \\
&\mathbf{F}_{s_0 s_1 \ldots s_k 2}^{(l_{k+1})}(\mathbf{a}_{s_0 s_1 \ldots s_{k+1}}, \mathbf{a}_{s_0 s_1 \ldots s_{k+1}}^{\tau}, \mathbf{p}) \\
&\quad=\frac{1}{\pi} \int_0^{2 \pi} \mathbf{f}_{s_0 s_1 \ldots s_k}(\mathbf{a}_{s_0 s_1 \ldots s_k}^*, \mathbf{a}_{s_0 s_1 \ldots s_k}^{\tau *}, \mathbf{p}) \cos(l_{k+1} \theta_{k+1}) d \theta_{k+1}, \\
&\mathbf{F}_{s_0 s_1 \ldots s_k 3}^{(l_{k+1})}(\mathbf{a}_{s_0 s_1 \ldots s_{k+1}}, \mathbf{a}_{s_0 s_1 \ldots s_{k+1}}^{\tau}, \mathbf{p}) \\
&\quad=\frac{1}{\pi} \int_0^{2 \pi} \mathbf{f}_{s_0 s_1 \ldots s_k}(\mathbf{a}_{s_0 s_1 \ldots s_k}^*, \mathbf{a}_{s_0 s_1 \ldots s_k}^{\tau *}, \mathbf{p}) \sin(l_{k+1} \theta_{k+1}) d \theta_{k+1} \\
&\text{for } l_{k+1}=1,2, \ldots, N_{k+1}.
\end{aligned} \tag{4.133}$$

3. *Equation (4.130) becomes*

$$\dot{\mathbf{z}}_{s_0 s_1 \ldots s_{k+1}}=\mathbf{f}_{s_0 s_1 \ldots s_{k+1}}(\mathbf{z}_{s_0 s_1 \ldots s_{k+1}}, \mathbf{z}_{s_0 s_1 \ldots s_{k+1}}^{\tau}) \tag{4.134}$$

where

$$\begin{aligned}
\mathbf{z}_{s_0 s_1 \ldots s_{k+1}} &=(\mathbf{a}_{s_0 s_1 \ldots s_k 1}, \mathbf{a}_{s_0 s_1 \ldots s_k 2}, \mathbf{a}_{s_0 s_1 \ldots s_k 3})^{\mathrm{T}}, \\
\mathbf{z}_{s_0 s_1 \ldots s_{k+1}}^{\tau} &=(\mathbf{a}_{s_0 s_1 \ldots s_k 1}^{\tau}, \mathbf{a}_{s_0 s_1 \ldots s_k 2}^{\tau}, \mathbf{a}_{s_0 s_1 \ldots s_k 3}^{\tau})^{\mathrm{T}}, \\
\mathbf{f}_{s_0 s_1 \ldots s_{k+1}} &=(\mathbf{F}_{s_0 s_1 \ldots s_k 1},-\omega_{k+1} \mathbf{k}_{k+1} \mathbf{a}_{s_0 s_1 \ldots s_k 3}+\mathbf{F}_{s_0 s_1 \ldots s_k 2}, \\
&\qquad \omega_{k+1} \mathbf{k}_{k+1} \mathbf{b}_{s_0 s_1 \ldots s_k 2}+\mathbf{F}_{s_0 s_1 \ldots s_k 3})^{\mathrm{T}}.
\end{aligned} \tag{4.135}$$

If equilibrium $\mathbf{z}_{s_0 s_1 \ldots s_{k+1}}^*=\mathbf{z}_{s_0 s_1 \ldots s_{k+1}}^{\tau *}$ *of Equation (4.134) (i.e.,* $\mathbf{f}_{s_0 s_1 \ldots s_{k+1}}(\mathbf{z}_{s_0 s_1 \ldots s_{k+1}}^*, \mathbf{z}_{s_0 s_1 \ldots s_{k+1}}^{\tau *})=\mathbf{0}$*) exists, then the approximate solution of the periodic flow of* kth *generalized coordinates for the period-*m *flow exists as in Equation (4.125). In the vicinity of equilibrium* $\mathbf{z}_{s_0 s_1 \ldots s_{k+1}}^*=\mathbf{z}_{s_0 s_1 \ldots s_{k+1}}^{\tau *}$*, with*

$$\begin{aligned}
\mathbf{z}_{s_0 s_1 \ldots s_{k+1}} &=\mathbf{z}_{s_0 s_1 \ldots s_{k+1}}^*+\Delta \mathbf{z}_{s_0 s_1 \ldots s_{k+1}}, \\
\mathbf{z}_{s_0 s_1 \ldots s_{k+1}}^{\tau} &=\mathbf{z}_{s_0 s_1 \ldots s_{k+1}}^{\tau *}+\Delta \mathbf{z}_{s_0 s_1 \ldots s_{k+1}}^{\tau},
\end{aligned} \tag{4.136}$$

the linearized equation of Equation (4.134) is

$$\Delta \dot{\mathbf{z}}_{s_0 s_1 \ldots s_{k+1}}=D_{\mathbf{z}_{s_0 s_1 \ldots s_{k+1}}} \mathbf{f}_{s_0 s_1 \ldots s_{k+1}} \Delta \mathbf{z}_{s_0 s_1 \ldots s_{k+1}}+D_{\mathbf{z}_{s_0 s_1 \ldots s_{k+1}}^{\tau}} \mathbf{f}_{s_0 s_1 \ldots s_{k+1}} \Delta \mathbf{z}_{s_0 s_1 \ldots s_{k+1}}^{\tau} \tag{4.137}$$

and the eigenvalue analysis of equilibrium $\mathbf{z}^*$ *is given by*

$$|D_{\mathbf{z}_{s_0 s_1 \ldots s_{k+1}}} \mathbf{f}_{s_0 s_1 \ldots s_{k+1}} + e^{\lambda \tau} D_{\mathbf{z}^{\tau}_{s_0 s_1 \ldots s_{k+1}}} \mathbf{f}_{s_0 s_1 \ldots s_{k+1}} - \lambda \mathbf{I}_{n_k(2N_{k+1}+1)\times n_k(2N_{k+1}+1)}| = 0 \quad (4.138)$$

where

$$D_{\mathbf{z}_{s_0 s_1 \ldots s_{k+1}}} \mathbf{f}_{s_0 s_1 \ldots s_{k+1}} = \left. \frac{\partial \mathbf{f}_{s_0 s_1 \ldots s_{k+1}} \left(\mathbf{z}_{s_0 s_1 \ldots s_{k+1}}, \mathbf{z}^{\tau}_{s_0 s_1 \ldots s_{k+1}} \right)}{\partial \mathbf{z}_{s_0 s_1 \ldots s_{k+1}}} \right|_{(\mathbf{z}^*_{s_0 s_1 \ldots s_{k+1}}, \mathbf{z}^{\tau *}_{s_0 s_1 \ldots s_{k+1}})},$$

$$D_{\mathbf{z}^{\tau}_{s_0 s_1 \ldots s_{k+1}}} \mathbf{f}_{s_0 s_1 \ldots s_{k+1}} = \left. \frac{\partial \mathbf{f}_{s_0 s_1 \ldots s_{k+1}} \left(\mathbf{z}_{s_0 s_1 \ldots s_{k+1}}, \mathbf{z}^{\tau}_{s_0 s_1 \ldots s_{k+1}} \right)}{\partial \mathbf{z}^{\tau}_{s_0 s_1 \ldots s_{k+1}}} \right|_{(\mathbf{z}^*_{s_0 s_1 \ldots s_{k+1}}, \mathbf{z}^{\tau *}_{s_0 s_1 \ldots s_{k+1}})}. \quad (4.139)$$

The stability and bifurcation of such a periodic flow of the kth *generalized coordinates can be classified by the eigenvalues of Equation (4.134) at equilibrium* $\mathbf{z}^*_{s_1 s_2 \ldots s_{k+1}} = \mathbf{z}^{\tau *}_{s_1 s_2 \ldots s_{k+1}}$ *with*

$$(n_1, n_2, n_3 | n_4, n_5, n_6). \quad (4.140)$$

a. *If all eigenvalues of the equilibrium possess negative real parts, the approximate quasi-periodic solution is stable.*
b. *If at least one of the eigenvalues of the equilibrium possesses a positive real part, the approximate quasi-periodic solution is unstable.*
c. *The boundaries between stable and unstable equilibriums with higher order singularity give bifurcation and stability conditions with higher order singularity.*

4. *For the* kth *order Hopf bifurcation of period-*m *flow, a relation exists as*

$$p_k \omega_k = \omega_{k-1}. \quad (4.141)$$

a. *If* p_k *is an irrational number, the* kth*-order Hopf bifurcation of the period-*m *flow is called the quasi-period-*p_k *Hopf bifurcation, and the corresponding solution of the* kth *generalized coordinates is* p_k*-quasi-periodic to the system of the* $(k-1)$th *generalized coordinates.*
b. *If* $p_k = 2$, *the* kth*-order Hopf bifurcation of the period-*m *flow is called a period-doubling Hopf bifurcation (or a period-2 Hopf bifurcation), and the corresponding solution of the* kth *generalized coordinates is period-doubling to the system of the* $(k-1)$th *generalized coordinates.*
c. *If* $p_k = q$ *with an integer q, the* kth*-order Hopf bifurcation of the period-*m *flow is called a period-q Hopf bifurcation, and the corresponding solution of the* kth *generalized coordinates is of q-times period to the system of the* $(k-1)$th *generalized coordinates.*
d. *If* $p_k = p/q$ *(p, q are irreducible integer), the* kth*-order Hopf bifurcation of the period-*m *flow is called a period-p/q Hopf bifurcation, and the corresponding solution of the* kth *generalized coordinates is p/q-periodic to the system of the* $(k-1)$th *generalized coordinates.*

Proof. The proof of this theorem is similar to the proof of Theorem 3.10. ■

Consider quasi-periodic flows in periodically forced, time-delayed nonlinear systems, and the analytical solution of quasi-periodic motion relative to period-m flow is stated from the following theorem.

Theorem 4.6 *Consider a periodically forced, time-delayed, nonlinear system as*

$$\dot{\mathbf{x}} = \mathbf{F}(\mathbf{x}, \mathbf{x}^{\tau}, t, \mathbf{p}) \in \mathscr{R}^{n} \tag{4.142}$$

where $\mathbf{F}(\mathbf{x}, \mathbf{x}^{\tau}, t, \mathbf{p})$ *is a* C^{r}*-continuous nonlinear function vector* ($r \geq 1$) *with forcing period* $T = 2\pi/\Omega$.

1. *If such a time-delayed dynamical system has a period-*m *flow* $\mathbf{x}^{(m)}(t)$ *with finite norm* $\|\mathbf{x}^{(m)}\|$ *and period* $T = 2\pi/\Omega$, *there is a generalized coordinate transformation with* $\theta = \Omega t$ *for the period-*m *flow of Equation (4.142) in the form of*

$$\mathbf{x}^{(m)}(t) = \mathbf{a}_0^{(m)}(t) + \sum_{k=1}^{\infty} \mathbf{b}_{k/m}(t) \cos\left(\frac{k}{m}\theta\right) + \mathbf{c}_{k/m}(t) \sin\left(\frac{k}{m}\theta\right);$$
$$\mathbf{x}^{\tau(m)}(t) = \mathbf{a}_0^{\tau(m)}(t) + \sum_{k=1}^{\infty} \mathbf{b}_{k/m}^{\tau}(t) \cos\left[\frac{k}{m}(\theta - \theta^{\tau})\right] + \mathbf{c}_{k/m}^{\tau}(t) \sin\left[\frac{k}{m}(\theta - \theta^{\tau})\right] \tag{4.143}$$

with $\mathbf{a}_0^{\tau(m)} = \mathbf{a}_0^{(m)}(t-\tau), \mathbf{b}_k^{\tau(m)} = \mathbf{b}_k^{(m)}(t-\tau), \mathbf{c}_k^{\tau(m)} = \mathbf{c}_k^{(m)}(t-\tau), \theta^{\tau} = \Omega\tau$ *and*

$$\begin{aligned}
\mathbf{a}_1^{(0)} &\equiv \mathbf{a}_0^{(m)} = (a_{01}^{(m)}, a_{02}^{(m)}, \ldots, a_{0n}^{(m)})^{\mathrm{T}}, \\
\mathbf{a}_2^{(k)} &\equiv \mathbf{b}_{k/m} = (b_{k/m1}, b_{k/m2}, \ldots, b_{k/mn})^{\mathrm{T}}, \\
\mathbf{a}_3^{(k)} &\equiv \mathbf{c}_{k/m} = (c_{k/m1}, c_{k/m2}, \ldots, c_{k/mn})^{\mathrm{T}}; \\
\mathbf{a}_1^{\tau(0)} &\equiv \mathbf{a}_0^{\tau(m)} = (a_{01}^{\tau(m)}, a_{02}^{\tau(m)}, \ldots, a_{0n}^{\tau(m)})^{\mathrm{T}}, \\
\mathbf{a}_2^{\tau(k)} &\equiv \mathbf{b}_{k/m}^{\tau} = (b_{k/m1}^{\tau}, b_{k/m2}^{\tau}, \ldots, b_{k/mn}^{\tau})^{\mathrm{T}}, \\
\mathbf{a}_3^{\tau(k)} &\equiv \mathbf{c}_{k/m}^{\tau} = (c_{k/m1}^{\tau}, c_{k/m2}^{\tau}, \ldots, c_{k/mn}^{\tau})^{\mathrm{T}}
\end{aligned} \tag{4.144}$$

which, under $\|\mathbf{x}^{(m)}(t) - \mathbf{x}^{(m)*}(t)\| < \varepsilon$ *and* $\|\mathbf{x}^{\tau(m)}(t) - \mathbf{x}^{\tau(m)*}(t)\| < \varepsilon^{\tau}$ *with prescribed small* $\varepsilon > 0$ *and* $\varepsilon^{\tau} > 0$, *can be approximated by a finite term transformation* $\mathbf{x}^{(m)*}(t)$ *as*

$$\mathbf{x}^{(m)*}(t) = \mathbf{a}_0^{(m)}(t) + \sum_{k=1}^{N_0} \mathbf{b}_{k/m}(t) \cos\left(\frac{k}{m}\theta\right) + \mathbf{c}_{k/m}(t) \sin\left(\frac{k}{m}\theta\right);$$
$$\mathbf{x}^{\tau(m)*}(t) = \mathbf{a}_0^{\tau(m)}(t) + \sum_{k=1}^{N_0} \mathbf{b}_{k/m}^{\tau}(t) \cos\left[\frac{k}{m}(\theta - \theta^{\tau})\right] + \mathbf{c}_{k/m}^{\tau}(t) \sin\left[\frac{k}{m}(\theta - \theta^{\tau})\right] \tag{4.145}$$

and the generalized coordinates are determined by

$$\dot{\mathbf{a}}_{s_0} = \mathbf{f}_{s_0}(\mathbf{a}_{s_0}, \mathbf{a}_{s_0}^{\tau}, \mathbf{p}) \tag{4.146}$$

where

$$
\begin{aligned}
&\mathbf{k}_0 = diag(\mathbf{I}_{n\times n}, 2\mathbf{I}_{n\times n}, \ldots, N_0\mathbf{I}_{n\times n}),\\
&\mathbf{a}_1^{(0)} \equiv \mathbf{a}_0^{(m)}, \mathbf{a}_2^{(k)} \equiv \mathbf{b}_{k/m}, \mathbf{a}_3^{(k)} \equiv \mathbf{c}_{k/m};\\
&\mathbf{a}_1^{\tau(0)} \equiv \mathbf{a}_0^{\tau(m)}, \mathbf{a}_2^{\tau(k)} \equiv \mathbf{b}_{k/m}^{\tau}, \mathbf{a}_3^{(k)} \equiv \mathbf{c}_{k/m}^{\tau}\\
&\mathbf{a}_1 = \mathbf{a}_1^{(0)},\\
&\mathbf{a}_2 = (\mathbf{a}_2^{(1)}, \mathbf{a}_2^{(2)}, \ldots, \mathbf{a}_2^{(N_0)})^{\mathrm{T}} \equiv \mathbf{b}^{(m)},\\
&\mathbf{a}_3 = (\mathbf{a}_3^{(1)}, \mathbf{a}_3^{(2)}, \ldots, \mathbf{a}_3^{(N_0)})^{\mathrm{T}} \equiv \mathbf{c}^{(m)},\\
&\mathbf{a}_1^{\tau} = \mathbf{a}_1^{\tau(0)},\\
&\mathbf{a}_2^{\tau} = (\mathbf{a}_2^{\tau(1)}, \mathbf{a}_2^{\tau(2)}, \ldots, \mathbf{a}_2^{\tau(N_0)})^{\mathrm{T}} \equiv \mathbf{b}^{\tau(m)},\\
&\mathbf{a}_3^{\tau} = (\mathbf{a}_3^{\tau(1)}, \mathbf{a}_3^{\tau(2)}, \ldots, \mathbf{a}_3^{\tau(N_0)})^{\mathrm{T}} \equiv \mathbf{c}^{\tau(m)},\\
&\mathbf{F}_1 = \mathbf{F}_0^{(m)},\\
&\mathbf{F}_2 = (\mathbf{F}_{11}^{(m)}, \mathbf{F}_{12}^{(m)}, \ldots, \mathbf{F}_{1N_0}^{(m)})^{\mathrm{T}},\\
&\mathbf{F}_3 = (\mathbf{F}_{21}^{(m)}, \mathbf{F}_{22}^{(m)}, \ldots, \mathbf{F}_{2N_0}^{(m)})^{\mathrm{T}};\\
&\mathbf{a}_{s_0} = (\mathbf{a}_1, \mathbf{a}_2, \mathbf{a}_3)^{\mathrm{T}},\\
&\mathbf{a}_{s_0}^{\tau} = (\mathbf{a}_1^{\tau}, \mathbf{a}_2^{\tau}, \mathbf{a}_3^{\tau})^{\mathrm{T}},\\
&\mathbf{f}_{s_0} = \left(\mathbf{F}_1, -\frac{\Omega}{m}\mathbf{k}_0\mathbf{a}_3 + \mathbf{F}_2, \frac{\Omega}{m}\mathbf{k}_0\mathbf{a}_2 + \mathbf{F}_3\right)^{\mathrm{T}}\\
&\text{for } N_0 = 1, 2, \ldots, \infty;
\end{aligned}
\tag{4.147}
$$

and

$$
\begin{aligned}
&\mathbf{F}_0^{(m)}(\mathbf{a}_{s_0}, \mathbf{a}_{s_0}^{\tau}) = \frac{1}{2m\pi}\int_0^{2m\pi} \mathbf{F}(\mathbf{x}^{(m)*}, \mathbf{x}^{\tau(m)*}, t, \mathbf{p})d\theta;\\
&\mathbf{F}_{1k}^{(m)}(\mathbf{a}_{s_0}, \mathbf{a}_{s_0}^{\tau}) = \frac{1}{m\pi}\int_0^{2m\pi} \mathbf{F}(\mathbf{x}^{(m)*}, \mathbf{x}^{\tau(m)*}, t, \mathbf{p})\cos\left(\frac{k}{m}\theta\right)d\theta,\\
&\mathbf{F}_{2k}^{(m)}(\mathbf{a}_{s_0}, \mathbf{a}_{s_0}^{\tau}) = \frac{1}{m\pi}\int_0^{2m\pi} \mathbf{F}(\mathbf{x}^{(m)*}, \mathbf{x}^{\tau(m)*}, t, \mathbf{p})\sin\left(\frac{k}{m}\theta\right)d\theta\\
&\text{for } k = 1, 2, \ldots, N_0.
\end{aligned}
\tag{4.148}
$$

2. *After the* k*th Hopf bifurcation with* $p_k\omega_k = \omega_{k-1}(k = 1, 2, \ldots)$ *and* $\omega_0 = \Omega/m$, *there is a dynamical system of coefficients as*

$$
\dot{\mathbf{a}}_{s_0s_1\ldots s_k} = \mathbf{f}_{s_0s_1\ldots s_k}(\mathbf{a}_{s_0s_1\ldots s_k}, \mathbf{a}_{s_0s_1\ldots s_k}^{\tau}, \mathbf{p}) \tag{4.149}
$$

where

$$
\begin{aligned}
\mathbf{a}_{s_0 s_1 \ldots s_k} &= (\mathbf{a}_{s_0 s_1 \ldots s_{k-1} 1}, \mathbf{a}_{s_0 s_1 \ldots s_{k-1} 2}, \mathbf{a}_{s_0 s_1 \ldots s_{k-1} 3})^{\mathrm{T}}, \\
\mathbf{a}^{\tau}_{s_0 s_1 \ldots s_k} &= (\mathbf{a}^{\tau}_{s_0 s_1 \ldots s_{k-1} 1}, \mathbf{a}^{\tau}_{s_0 s_1 \ldots s_{k-1} 2}, \mathbf{a}^{\tau}_{s_0 s_1 \ldots s_{k-1} 3})^{\mathrm{T}} \\
\mathbf{f}_{s_0 s_1 \ldots s_k} &= (\mathbf{F}_{s_0 s_1 \ldots s_{k-1} 1}, -\omega_k \mathbf{k}_k \mathbf{a}_{s_0 s_1 \ldots s_{k-1} 3} + \mathbf{F}_{s_0 s_1 \ldots s_{k-1} 2}, \\
&\quad \omega_k \mathbf{k}_k \mathbf{a}_{s_0 s_1 \ldots s_{k-1} 2} + \mathbf{F}_{s_0 s_1 \ldots s_{k-1} 3})^{\mathrm{T}}, \\
\mathbf{k}_k &= diag(\mathbf{I}_{n_{k-1} \times n_{k-1}}, 2\mathbf{I}_{n_{k-1} \times n_{k-1}}, \ldots, N_k \mathbf{I}_{n_{k-1} \times n_{k-1}}), \\
n_k &= n(2N_0 + 1)(2N_1 + 1) \ldots (2N_{k-1} + 1)
\end{aligned} \tag{4.150}
$$

with a periodic solution as

$$
\begin{aligned}
\mathbf{a}_{s_0 s_1 \ldots s_k} &= \mathbf{a}^{(0)}_{s_0 s_1 \ldots s_k 1}(t) + \sum_{l_{k+1}=1}^{\infty} \mathbf{a}^{(l_{k+1})}_{s_0 s_1 \ldots s_k 2}(t) \cos(l_{k+1} \theta_{k+1}) \\
&\quad + \mathbf{a}^{(l_{k+1})}_{s_0 s_1 \ldots s_k 3}(t) \sin(l_{k+1} \theta_{k+1}), \\
\mathbf{a}^{\tau}_{s_0 s_1 \ldots s_k} &= \mathbf{a}^{\tau(0)}_{s_0 s_1 \ldots s_k 1}(t) + \sum_{l_{k+1}=1}^{\infty} \mathbf{a}^{\tau(l_{k+1})}_{s_0 s_1 \ldots s_k 2}(t) \cos[l_{k+1}(\theta_{k+1} - \theta^{\tau}_{k+1})] \\
&\quad + \mathbf{a}^{\tau(l_{k+1})}_{s_0 s_1 \ldots s_k 3}(t)[l_{k+1}(\theta_{k+1} - \theta^{\tau}_{k+1})]
\end{aligned} \tag{4.151}
$$

with

$$
\begin{aligned}
&\mathbf{a}^{\tau(0)}_{s_0 s_1 \ldots s_k 1}(t) = \mathbf{a}^{(0)}_{s_0 s_1 \ldots s_k 1}(t - \tau), \\
&\mathbf{a}^{\tau(l_{k+1})}_{s_0 s_1 \ldots s_k 2}(t) = \mathbf{a}^{(l_{k+1})}_{s_0 s_1 \ldots s_k 2}(t - \tau), \\
&\mathbf{a}^{\tau(l_{k+1})}_{s_0 s_1 \ldots s_k 3}(t) = \mathbf{a}^{(l_{k+1})}_{s_0 s_1 \ldots s_k 3}(t - \tau), \\
&\theta^{\tau}_{k+1} = \omega_{k+1} \tau
\end{aligned} \tag{4.152}
$$

and

$$
\begin{aligned}
&s_i = 1, 2, 3 \ (i = 0, 1, 2, \ldots, k), \\
&\mathbf{a}_{s_0 s_1 \ldots s_k 1} = \mathbf{a}^{(0)}_{s_0 s_1 \ldots s_k 1}, \\
&\mathbf{a}_{s_0 s_1 \ldots s_k 2} = (\mathbf{a}^{(1)}_{s_0 s_1 \ldots s_k 2}, \mathbf{a}^{(2)}_{s_0 s_1 \ldots s_k 2}, \ldots, \mathbf{a}^{(N_{k+1})}_{s_0 s_1 \ldots s_k 2})^{\mathrm{T}}, \\
&\mathbf{a}_{s_0 s_1 \ldots s_k 3} = (\mathbf{a}^{(1)}_{s_0 s_1 \ldots s_k 3}, \mathbf{a}^{(2)}_{s_0 s_1 \ldots s_k 3}, \ldots, \mathbf{a}^{(N_{k+1})}_{s_0 s_1 \ldots s_k 3})^{\mathrm{T}}; \\
&\mathbf{a}_{s_0 s_1 \ldots s_{k-1} 1} = \mathbf{a}^{(0)}_{s_0 s_1 \ldots s_{k-1} 1}, \\
&\mathbf{a}_{s_0 s_1 \ldots s_{k-1} 2} = (\mathbf{a}^{(1)}_{s_0 s_1 \ldots s_{k-1} 2}, \mathbf{a}^{(2)}_{s_0 s_1 \ldots s_{k-1} 2}, \ldots, \mathbf{a}^{(N_k)}_{s_0 s_1 \ldots s_{k-1} 2})^{\mathrm{T}},
\end{aligned}
$$

$$
\begin{aligned}
&\mathbf{a}_{s_0 s_1 \ldots s_{k-1} 3} = (\mathbf{a}^{(1)}_{s_0 s_1 \ldots s_{k-1} 3}, \mathbf{a}^{(2)}_{s_0 s_1 \ldots s_{k-1} 3}, \ldots, \mathbf{a}^{(N_k)}_{s_0 s_1 \ldots s_{k-1} 3})^{\mathrm{T}};\\
&\vdots\\
&\mathbf{a}_1 = \mathbf{a}^{(0)}_1,\\
&\mathbf{a}_2 = (\mathbf{a}^{(1)}_2, \mathbf{a}^{(2)}_2, \ldots, \mathbf{a}^{(N)}_2)^{\mathrm{T}},\\
&\mathbf{a}_3 = (\mathbf{a}^{(1)}_3, \mathbf{a}^{(2)}_3, \ldots, \mathbf{a}^{(N)}_3)^{\mathrm{T}};
\end{aligned} \tag{4.153}
$$

and

$$
\begin{aligned}
&s_i = 1, 2, 3 \; (i = 0, 1, 2, \ldots, k),\\
&\mathbf{a}^{\tau}_{s_0 s_1 \ldots s_k 1} = \mathbf{a}^{\tau(0)}_{s_0 s_1 \ldots s_k 1},\\
&\mathbf{a}^{\tau}_{s_0 s_1 \ldots s_k 2} = (\mathbf{a}^{\tau(1)}_{s_0 s_1 \ldots s_k 2}, \mathbf{a}^{\tau(2)}_{s_0 s_1 \ldots s_k 2}, \ldots, \mathbf{a}^{\tau(N_{k+1})}_{s_0 s_1 \ldots s_k 2})^{\mathrm{T}},\\
&\mathbf{a}^{\tau}_{s_0 s_1 \ldots s_k 2} = (\mathbf{a}^{\tau(1)}_{s_0 s_1 \ldots s_k 3}, \mathbf{a}^{\tau(2)}_{s_0 s_1 \ldots s_k 3}, \ldots, \mathbf{a}^{\tau(N_{k+1})}_{s_0 s_1 \ldots s_k 3})^{\mathrm{T}};\\
&\mathbf{a}^{\tau}_{s_0 s_1 \ldots s_{k-1} 1} = \mathbf{a}^{\tau(0)}_{s_0 s_1 \ldots s_{k-1} 1},\\
&\mathbf{a}^{\tau}_{s_0 s_1 \ldots s_{k-1} 2} = (\mathbf{a}^{\tau(1)}_{s_0 s_1 \ldots s_{k-1} 2}, \mathbf{a}^{\tau(2)}_{s_0 s_1 \ldots s_{k-1} 2}, \ldots, \mathbf{a}^{\tau(N_k)}_{s_0 s_1 \ldots s_{k-1} 2})^{\mathrm{T}},\\
&\mathbf{a}^{\tau}_{s_0 s_1 \ldots s_{k-1} 3} = (\mathbf{a}^{\tau(1)}_{s_0 s_1 \ldots s_{k-1} 3}, \mathbf{a}^{\tau(2)}_{s_0 s_1 \ldots s_{k-1} 3}, \ldots, \mathbf{a}^{\tau(N_k)}_{s_0 s_1 \ldots s_{k-1} 3})^{\mathrm{T}};\\
&\vdots\\
&\mathbf{a}^{\tau}_1 = \mathbf{a}^{\tau(0)}_1,\\
&\mathbf{a}^{\tau}_2 = (\mathbf{a}^{\tau(1)}_2, \mathbf{a}^{\tau(2)}_2, \ldots, \mathbf{a}^{\tau(N_0)}_2)^{\mathrm{T}},\\
&\mathbf{a}^{\tau}_3 = (\mathbf{a}^{\tau(1)}_3, \mathbf{a}^{\tau(2)}_3, \ldots, \mathbf{a}^{\tau(N_0)}_3)^{\mathrm{T}};
\end{aligned} \tag{4.154}
$$

which, under $\|\mathbf{a}_{s_0 s_1 \ldots s_k}(t) - \mathbf{a}^*_{s_0 s_1 \ldots s_k}(t)\| < \varepsilon$ *and* $\|\mathbf{a}^{\tau}_{s_0 s_1 \ldots s_k}(t) - \mathbf{a}^{\tau *}_{s_0 s_1 \ldots s_k}(t)\| < \varepsilon_\tau$ *with a prescribed small* $\varepsilon > 0$ *and* $\varepsilon_\tau > 0$, *can be approximated by a finite term transformation, that is,*

$$
\begin{aligned}
\mathbf{a}^*_{s_0 s_1 \ldots s_k} &= \mathbf{a}^{(0)}_{s_0 s_1 \ldots s_k 1}(t) + \sum_{l_{k+1}=1}^{N_{k+1}} \mathbf{a}^{(l_{k+1})}_{s_0 s_1 \ldots s_k 2}(t) \cos(l_{k+1}\theta_{k+1})\\
&\quad + \mathbf{a}^{(l_{k+1})}_{s_0 s_1 \ldots s_k 3}(t) \sin(l_{k+1}\theta_{k+1}),\\
\mathbf{a}^{\tau *}_{s_0 s_1 \ldots s_k} &= \mathbf{a}^{\tau(0)}_{s_0 s_1 \ldots s_k 1}(t) + \sum_{l_{k+1}=1}^{N_{k+1}} \mathbf{a}^{\tau(l_{k+1})}_{s_0 s_1 \ldots s_k 2}(t) \cos[l_{k+1}(\theta_{k+1} - \theta^{\tau}_{k+1})]\\
&\quad + \mathbf{a}^{\tau(l_{k+1})}_{s_0 s_1 \ldots s_k 3}(t) \sin[l_{k+1}(\theta_{k+1} - \theta^{\tau}_{k+1})]
\end{aligned} \tag{4.155}
$$

and the generalized coordinates are determined by

$$\dot{\mathbf{a}}_{s_0 s_1 \ldots s_{k+1}} = \mathbf{f}_{s_0 s_1 \ldots s_{k+1}}(\mathbf{a}_{s_0 s_1 \ldots s_{k+1}}, \mathbf{a}^{\tau}_{s_0 s_1 \ldots s_{k+1}}, \mathbf{p}) \tag{4.156}$$

where

$$\begin{aligned}
\mathbf{a}_{s_0 s_1 \ldots s_{k+1}} &= (\mathbf{a}_{s_0 s_1 \ldots s_k 1}, \mathbf{a}_{s_0 s_1 \ldots s_k 2}, \mathbf{a}_{s_0 s_1 \ldots s_k 3})^{\mathrm{T}}, \\
\mathbf{a}^{\tau}_{s_0 s_1 \ldots s_{k+1}} &= (\mathbf{a}^{\tau}_{s_0 s_1 \ldots s_k 1}, \mathbf{a}^{\tau}_{s_0 s_1 \ldots s_k 2}, \mathbf{a}^{\tau}_{s_0 s_1 \ldots s_k 3})^{\mathrm{T}}, \\
\mathbf{f}_{s_0 s_1 \ldots s_{k+1}} &= (\mathbf{F}_{s_0 s_1 \ldots s_k 1}, -\omega_{k+1}\mathbf{k}_{k+1}\mathbf{a}_{s_0 s_1 \ldots s_k 3} + \mathbf{F}_{s_0 s_1 \ldots s_k 2}, \\
&\qquad \omega_{k+1}\mathbf{k}_{k+1}\mathbf{a}_{s_0 s_1 \ldots s_k 2} + \mathbf{F}_{s_0 s_1 \ldots s_k 3})^{\mathrm{T}}
\end{aligned} \tag{4.157}$$

and

$$\begin{aligned}
&\mathbf{k}_{k+1} = diag(\mathbf{I}_{n_k \times n_k}, 2\mathbf{I}_{n_k \times n_k}, \ldots, N_{k+1}\mathbf{I}_{n_k \times n_k}), \\
&n_k = n(2N_0 + 1)(2N_1 + 1)\ldots(2N_k + 1); \\
&\mathbf{a}_{s_0 s_1 \ldots s_k 1} = \mathbf{a}^{(0)}_{s_0 s_1 \ldots s_k 1}, \\
&\mathbf{a}_{s_0 s_1 \ldots s_k 2} = (\mathbf{a}^{(1)}_{s_0 s_1 \ldots s_k 2}, \mathbf{a}^{(2)}_{s_0 s_1 \ldots s_k 2}, \ldots, \mathbf{a}^{(N_{k+1})}_{s_0 s_1 \ldots s_k 2})^{\mathrm{T}}, \\
&\mathbf{a}_{s_0 s_1 \ldots s_k 3} = (\mathbf{a}^{(1)}_{s_0 s_1 \ldots s_k 3}, \mathbf{a}^{(2)}_{s_0 s_1 \ldots s_k 3}, \ldots, \mathbf{a}^{(N_{k+1})}_{s_0 s_1 \ldots s_k 3})^{\mathrm{T}}; \\
&\mathbf{a}^{\tau}_{s_0 s_1 \ldots s_k 1} = \mathbf{a}^{\tau(0)}_{s_0 s_1 \ldots s_k 1}, \\
&\mathbf{a}^{\tau}_{s_0 s_1 \ldots s_k 2} = (\mathbf{a}^{\tau(1)}_{s_0 s_1 \ldots s_k 2}, \mathbf{a}^{\tau(2)}_{s_0 s_1 \ldots s_k 2}, \ldots, \mathbf{a}^{\tau(N_{k+1})}_{s_0 s_1 \ldots s_k 2})^{\mathrm{T}}, \\
&\mathbf{a}^{\tau}_{s_0 s_1 \ldots s_k 3} = (\mathbf{a}^{\tau(1)}_{s_0 s_1 \ldots s_k 3}, \mathbf{a}^{\tau(2)}_{s_0 s_1 \ldots s_k 3}, \ldots, \mathbf{a}^{\tau(N_{k+1})}_{s_0 s_1 \ldots s_k 3})^{\mathrm{T}}; \\
&\mathbf{F}_{s_0 s_1 \ldots s_k 1} = \mathbf{F}^{(0)}_{s_0 s_1 \ldots s_k 1}, \\
&\mathbf{F}_{s_0 s_1 \ldots s_k 2} = (\mathbf{F}^{(1)}_{s_0 s_1 \ldots s_k 2}, \mathbf{F}^{(2)}_{s_0 s_1 \ldots s_k 2}, \ldots, \mathbf{F}^{(N_{k+1})}_{s_0 s_1 \ldots s_k 2})^{\mathrm{T}}, \\
&\mathbf{F}_{s_0 s_1 \ldots s_k 3} = (\mathbf{F}^{(1)}_{s_0 s_1 \ldots s_k 3}, \mathbf{F}^{(2)}_{s_0 s_1 \ldots s_k 3}, \ldots, \mathbf{F}^{(N_{k+1})}_{s_0 s_1 \ldots s_k 3})^{\mathrm{T}} \\
&\text{for } N_{k+1} = 1, 2, \ldots, \infty;
\end{aligned} \tag{4.158}$$

and

$$\begin{aligned}
&\mathbf{F}_{s_0 s_1 \ldots s_k 1}(\mathbf{a}_{s_0 s_1 \ldots s_{k+1}}, \mathbf{a}^{\tau}_{s_0 s_1 \ldots s_{k+1}}, \mathbf{p}) \\
&\quad = \frac{1}{2\pi}\int_0^{2\pi} \mathbf{f}_{s_0 s_1 \ldots s_k}(\mathbf{a}^{*}_{s_0 s_1 \ldots s_k}, \mathbf{a}^{\tau *}_{s_0 s_1 \ldots s_k}, \mathbf{p})d\theta_{k+1}; \\
&\mathbf{F}^{(l_{k+1})}_{s_0 s_1 \ldots s_k 2}(\mathbf{a}_{s_0 s_1 \ldots s_{k+1}}, \mathbf{a}^{\tau}_{s_0 s_1 \ldots s_{k+1}}, \mathbf{p}) \\
&\quad = \frac{1}{\pi}\int_0^{2\pi} \mathbf{f}_{s_0 s_1 \ldots s_k}(\mathbf{a}^{*}_{s_0 s_1 \ldots s_k}, \mathbf{a}^{\tau *}_{s_0 s_1 \ldots s_k}, \mathbf{p})\cos(l_{k+1}\theta_{k+1})d\theta_{k+1},
\end{aligned}$$

$$
\begin{aligned}
&\mathbf{F}^{(l_{k+1})}_{s_0 s_1 \ldots s_k 3}(\mathbf{a}_{s_0 s_1 \ldots s_{k+1}}, \mathbf{a}^{\tau}_{s_0 s_1 \ldots s_{k+1}}, \mathbf{p}) \\
&= \frac{1}{\pi}\int_0^{2\pi} \mathbf{f}_{s_0 s_1 \ldots s_k}(\mathbf{a}^{*}_{s_0 s_1 \ldots s_k}, \mathbf{a}^{\tau *}_{s_0 s_1 \ldots s_k}, \mathbf{p}) \sin(l_{k+1}\theta_{k+1}) d\theta_{k+1} \\
&\text{for } l_{k+1} = 1, 2, \ldots, N_{k+1}.
\end{aligned}
\tag{4.159}
$$

3. *Equation (4.156) becomes*

$$
\dot{\mathbf{z}}_{s_0 s_1 \ldots s_{k+1}} = \mathbf{f}_{s_0 s_1 \ldots s_{k+1}}(\mathbf{z}_{s_0 s_1 \ldots s_{k+1}}, \mathbf{z}^{\tau}_{s_1 s_0 \ldots s_{k+1}}) \tag{4.160}
$$

where

$$
\begin{aligned}
\mathbf{z}_{s_0 s_1 \ldots s_{k+1}} =& (\mathbf{a}_{s_0 s_1 \ldots s_k 1}, \mathbf{a}_{s_0 s_1 \ldots s_k 2}, \mathbf{a}_{s_0 s_1 \ldots s_k 3})^{\mathrm{T}}, \\
\mathbf{z}^{\tau}_{s_0 s_1 \ldots s_{k+1}} =& (\mathbf{a}^{\tau}_{s_0 s_1 \ldots s_k 1}, \mathbf{a}^{\tau}_{s_0 s_1 \ldots s_k 2}, \mathbf{a}^{\tau}_{s_0 s_1 \ldots s_k 3})^{\mathrm{T}}, \\
\mathbf{f}_{s_0 s_1 \ldots s_{k+1}} =& (\mathbf{F}_{s_0 s_1 \ldots s_k 1}, -\omega_{k+1}\mathbf{k}_{k+1}\mathbf{a}_{s_0 s_1 \ldots s_k 3} + \mathbf{F}_{s_0 s_1 \ldots s_k 2}, \\
& \omega_{k+1}\mathbf{k}_{k+1}\mathbf{a}_{s_0 s_1 \ldots s_k 2} + \mathbf{F}_{s_0 s_1 \ldots s_k 3})^{\mathrm{T}}.
\end{aligned}
\tag{4.161}
$$

If equilibrium $\mathbf{z}^{*}_{s_0 s_1 \ldots s_{k+1}} = \mathbf{z}^{\tau *}_{s_0 s_1 \ldots s_{k+1}}$ *of Equation (4.160) (i.e.,* $\mathbf{f}_{s_0 s_1 \ldots s_{k+1}}(\mathbf{z}^{*}_{s_0 s_1 \ldots s_{k+1}}, \mathbf{z}^{\tau *}_{s_0 s_1 \ldots s_{k+1}}) = \mathbf{0}$*) exists, then the approximate solution of the periodic flow of* kth *generalized coordinates for the period-*m *flow exists as in Equation (4.155). In the vicinity of equilibrium* $\mathbf{z}^{*}_{s_0 s_1 \ldots s_{k+1}} = \mathbf{z}^{\tau *}_{s_0 s_1 \ldots s_{k+1}}$*, with*

$$
\begin{aligned}
\mathbf{z}_{s_0 s_1 \ldots s_{k+1}} &= \mathbf{z}^{*}_{s_0 s_1 \ldots s_{k+1}} + \Delta\mathbf{z}_{s_0 s_1 \ldots s_{k+1}}, \\
\mathbf{z}^{\tau}_{s_0 s_1 \ldots s_{k+1}} &= \mathbf{z}^{\tau *}_{s_0 s_1 \ldots s_{k+1}} + \Delta\mathbf{z}^{\tau}_{s_0 s_1 \ldots s_{k+1}}
\end{aligned}
\tag{4.162}
$$

the linearized equation of Equation (4.160) is

$$
\Delta\dot{\mathbf{z}}_{s_0 s_1 \ldots s_{k+1}} = D_{\mathbf{z}_{s_0 s_1 \ldots s_{k+1}}} \mathbf{f}_{s_0 s_1 \ldots s_{k+1}} \Delta\mathbf{z}_{s_0 s_1 \ldots s_{k+1}} + D_{\mathbf{z}^{\tau}_{s_0 s_1 \ldots s_{k+1}}} \mathbf{f}_{s_0 s_1 \ldots s_{k+1}} \Delta\mathbf{z}^{\tau}_{s_0 s_1 \ldots s_{k+1}} \tag{4.163}
$$

and the eigenvalue analysis of equilibrium $\mathbf{z}^{*}$ *is given by*

$$
|D_{\mathbf{z}_{s_0 s_1 \ldots s_{k+1}}} \mathbf{f}_{s_0 s_1 \ldots s_{k+1}} + e^{\lambda\tau} D_{\mathbf{z}^{\tau}_{s_0 s_1 \ldots s_{k+1}}} \mathbf{f}_{s_0 s_1 \ldots s_{k+1}} - \lambda \mathbf{I}_{n_k(2N_{k+1}+1)\times n_k(2N_{k+1}+1)}| = 0 \tag{4.164}
$$

where

$$
\begin{aligned}
D_{\mathbf{z}_{s_0 s_1 \ldots s_{k+1}}} \mathbf{f}_{s_0 s_1 \ldots s_{k+1}} &= \left.\frac{\partial \mathbf{f}_{s_0 s_1 \ldots s_{k+1}}\left(\mathbf{z}_{s_0 s_1 \ldots s_{k+1}}, \mathbf{z}^{\tau}_{s_0 s_1 \ldots s_{k+1}}\right)}{\partial \mathbf{z}_{s_1 s_2 \ldots s_{k+1}}}\right|_{(\mathbf{z}^{*}_{s_0 s_1 \ldots s_{k+1}}, \mathbf{z}^{\tau *}_{s_0 s_1 \ldots s_{k+1}})}, \\
D_{\mathbf{z}^{\tau}_{s_0 s_1 \ldots s_{k+1}}} \mathbf{f}_{s_0 s_1 \ldots s_{k+1}} &= \left.\frac{\partial \mathbf{f}_{s_0 s_1 \ldots s_{k+1}}\left(\mathbf{z}_{s_0 s_1 \ldots s_{k+1}}, \mathbf{z}^{\tau}_{s_0 s_1 \ldots s_{k+1}}\right)}{\partial \mathbf{z}^{\tau}_{s_0 s_1 \ldots s_{k+1}}}\right|_{(\mathbf{z}^{*}_{s_0 s_1 \ldots s_{k+1}}, \mathbf{z}^{\tau *}_{s_0 s_1 \ldots s_{k+1}})}.
\end{aligned}
\tag{4.165}
$$

The stability and bifurcation of such a periodic flow of the kth *generalized coordinates can be classified by the eigenvalues of Equation (4.163) at equilibrium* $\mathbf{z}^{*}_{s_0 s_1 \ldots s_{k+1}} = \mathbf{z}^{\tau *}_{s_0 s_1 \ldots s_{k+1}}$ *with*

$$
(n_1, n_2, n_3 \mid n_4, n_5, n_6). \tag{4.166}
$$

a. *If all eigenvalues of the equilibrium possess negative real parts, the approximate quasi-periodic solution is stable.*
b. *If at least one of the eigenvalues of the equilibrium possesses a positive real part, the approximate quasi-periodic solution is unstable.*
c. *The boundaries between stable and unstable equilibriums with higher order singularity give bifurcation and stability conditions with higher order singularity.*

4. *For the* kth *order Hopf bifurcation of period-*m *motion, a relation exists as*

$$p_k \omega_k = \omega_{k-1}. \tag{4.167}$$

a. *If p_k is an irrational number, the* kth-*order Hopf bifurcation of the period-*m *flow is called the quasi-period-p_k Hopf bifurcation, and the corresponding solution of the* kth *generalized coordinates is p_k-quasi-periodic to the system of the* $(k-1)$th *generalized coordinates.*
b. *If $p_k = 2$, the* kth-*order Hopf bifurcation of the period-*m *flow is called a period-doubling Hopf bifurcation (or a period-2 Hopf bifurcation), and the corresponding solution of the* kth *generalized coordinates is period-doubling to the system of the* $(k-1)$th *generalized coordinates.*
c. *If $p_k = q$ with an integer q, the* kth-*order Hopf bifurcation of the period-*m *flow is called a period-q Hopf bifurcation, and the corresponding solution of the* kth *generalized coordinates is of q-times period to the system of the* $(k-1)$th *generalized coordinates.*
d. *If $p_k = p/q$ (p, q are irreducible integer), the* kth-*order Hopf bifurcation of the period-*m *flow is called a period-p/q Hopf bifurcation, and the corresponding solution of the* kth *generalized coordinates is p/q-periodic to the system of the* $(k-1)$th *generalized coordinates.*

■

Proof. The proof of this theorem is similar to the proof of Theorem 3.12.

4.4 Time-Delayed, Nonlinear Vibration Systems

In this section, analytical quasi-periodic flows in time-delayed, nonlinear vibration systems will be discussed. Consider time-delayed, nonlinear vibration systems, and the analytical solution of quasi-periodic motion relative to period-m motion in such a time-delayed, free vibration system is given as follows.

Theorem 4.7 *Consider a time-delayed, free vibration system as*

$$\ddot{\mathbf{x}} = \mathbf{f}(\mathbf{x}, \dot{\mathbf{x}}, \mathbf{x}^{\tau}, \dot{\mathbf{x}}^{\tau}, \mathbf{p}) \in \mathscr{R}^n \tag{4.168}$$

where $\mathbf{f}(\mathbf{x}, \dot{\mathbf{x}}, \mathbf{x}^{\tau}, \dot{\mathbf{x}}^{\tau}, \mathbf{p})$ is a C^r-continuous nonlinear function vector ($r \geq 1$).

1. *If such a time-delayed, vibration system has a period-*m *motion $\mathbf{x}^{(m)}(t)$ with finite norm $||\mathbf{x}^{(m)}||$ and period $T = 2\pi/\Omega$, there is a generalized coordinate transformation with $\theta = \Omega t$ for the period-*m *motion of Equation (4.168) in the form of*

$$\mathbf{x}^{(m)}(t) = \mathbf{a}_0^{(m)}(t) + \sum_{k=1}^{\infty} \mathbf{b}_{k/m}(t) \cos\left(\frac{k}{m}\theta\right) + \mathbf{c}_{k/m}(t) \sin\left(\frac{k}{m}\theta\right);$$

$$\mathbf{x}^{\tau(m)}(t) = \mathbf{a}_0^{\tau(m)}(t) + \sum_{k=1}^{\infty} \mathbf{b}_{k/m}^{\tau}(t) \cos\left[\frac{k}{m}(\theta - \theta^{\tau})\right] + \mathbf{c}_{k/m}^{\tau}(t) \sin\left[\frac{k}{m}(\theta - \theta^{\tau})\right] \tag{4.169}$$

with $\mathbf{a}_0^{\tau(m)} = \mathbf{a}_0^{(m)}(t-\tau), \mathbf{b}_k^{\tau(m)} = \mathbf{b}_k^{(m)}(t-\tau), \mathbf{c}_k^{\tau(m)} = \mathbf{c}_k^{(m)}(t-\tau), \theta^\tau = \Omega\tau$ *and*

$$
\begin{aligned}
\mathbf{a}_1^{(0)} &\equiv \mathbf{a}_0^{(m)} = (a_{01}^{(m)}, a_{02}^{(m)}, \ldots, a_{0n}^{(m)})^{\mathrm{T}},\\
\mathbf{a}_2^{(k)} &\equiv \mathbf{b}_{k/m} = (b_{k/m1}, b_{k/m2}, \ldots, b_{k/mn})^{\mathrm{T}},\\
\mathbf{a}_3^{(k)} &\equiv \mathbf{c}_{k/m} = (c_{k/m1}, c_{k/m2}, \ldots, c_{k/mn})^{\mathrm{T}}\\
\mathbf{a}_1^{\tau(0)} &\equiv \mathbf{a}_0^{\tau(m)} = (a_{01}^{\tau(m)}, a_{02}^{\tau(m)}, \ldots, a_{0n}^{\tau(m)})^{\mathrm{T}},\\
\mathbf{a}_2^{\tau(k)} &\equiv \mathbf{b}_{k/m}^{\tau} = (b_{k/m1}^{\tau}, b_{k/m2}^{\tau}, \ldots, b_{k/mn}^{\tau})^{\mathrm{T}},\\
\mathbf{a}_3^{\tau(k)} &\equiv \mathbf{c}_{k/m}^{\tau} = (c_{k/m1}^{\tau}, c_{k/m2}^{\tau}, \ldots, c_{k/mn}^{\tau})^{\mathrm{T}}
\end{aligned}
\tag{4.170}
$$

which, under $\|\mathbf{x}^{(m)}(t) - \mathbf{x}^{(m)*}(t)\| < \varepsilon$ *and* $\|\mathbf{x}^{\tau(m)}(t) - \mathbf{x}^{\tau(m)*}(t)\| < \varepsilon^\tau$ *with prescribed small* $\varepsilon > 0$ *and* $\varepsilon^\tau > 0$*, can be approximated by a finite term transformation* $\mathbf{x}^{(m)*}(t)$ *as*

$$
\begin{aligned}
\mathbf{x}^{(m)*}(t) &= \mathbf{a}_0^{(m)}(t) + \sum_{k=1}^{N_0} \mathbf{b}_{k/m}(t)\cos\left(\frac{k}{m}\theta\right) + \mathbf{c}_{k/m}(t)\sin\left(\frac{k}{m}\theta\right);\\
\mathbf{x}^{\tau(m)*}(t) &= \mathbf{a}_0^{\tau(m)}(t) + \sum_{k=1}^{N_0} \mathbf{b}_{k/m}^{\tau}(t)\cos\left[\frac{k}{m}(\theta-\theta^\tau)\right] + \mathbf{c}_{k/m}^{\tau}(t)\sin\left[\frac{k}{m}(\theta-\theta^\tau)\right]
\end{aligned}
\tag{4.171}
$$

and the generalized coordinates are determined by

$$
\ddot{\mathbf{a}}_{s_0} = \mathbf{g}_{s_0}(\mathbf{a}_{s_0}, \mathbf{a}_{s_0}^{\tau}, \dot{\mathbf{a}}_{s_0}, \dot{\mathbf{a}}_{s_0}^{\tau}, \mathbf{p}) \tag{4.172}
$$

where

$$
\begin{aligned}
\mathbf{k}_0^{(1)} &= diag(\mathbf{I}_{n\times n}, 2\mathbf{I}_{n\times n}, \ldots, N_0\mathbf{I}_{n\times n}),\\
\mathbf{k}_0^{(2)} &= diag(\mathbf{I}_{n\times n}, 2^2\mathbf{I}_{n\times n}, \ldots, N_0^2\mathbf{I}_{n\times n});\\
\mathbf{a}_1^{(0)} &\equiv \mathbf{a}_0^{(m)}, \mathbf{a}_2^{(k)} \equiv \mathbf{b}_{k/m}, \mathbf{a}_3^{(k)} \equiv \mathbf{c}_{k/m};\\
\mathbf{a}_1^{\tau(0)} &\equiv \mathbf{a}_0^{\tau(m)}, \mathbf{a}_2^{\tau(k)} \equiv \mathbf{b}_{k/m}^{\tau}, \mathbf{a}_3^{(k)} \equiv \mathbf{c}_{k/m}^{\tau};\\
\mathbf{a}_1 &= \mathbf{a}_1^{(0)},\\
\mathbf{a}_2 &= (\mathbf{a}_2^{(1)}, \mathbf{a}_2^{(2)}, \ldots, \mathbf{a}_2^{(N_0)})^{\mathrm{T}} \equiv \mathbf{b}^{(m)},\\
\mathbf{a}_3 &= (\mathbf{a}_3^{(1)}, \mathbf{a}_3^{(2)}, \ldots, \mathbf{a}_3^{(N_0)})^{\mathrm{T}} \equiv \mathbf{c}^{(m)},\\
\mathbf{a}_1^{\tau} &= \mathbf{a}_1^{\tau(0)},\\
\mathbf{a}_2^{\tau} &= (\mathbf{a}_2^{\tau(1)}, \mathbf{a}_2^{\tau(2)}, \ldots, \mathbf{a}_2^{\tau(N_0)})^{\mathrm{T}} \equiv \mathbf{b}^{\tau(m)},\\
\mathbf{a}_3^{\tau} &= (\mathbf{a}_3^{\tau(1)}, \mathbf{a}_3^{\tau(2)}, \ldots, \mathbf{a}_3^{\tau(N_0)})^{\mathrm{T}} \equiv \mathbf{c}^{\tau(m)},\\
\mathbf{F}_1 &= \mathbf{F}_0^{(m)},\\
\mathbf{F}_2 &= (\mathbf{F}_{11}^{(m)}, \mathbf{F}_{12}^{(m)}, \ldots, \mathbf{F}_{1N_0}^{(m)})^{\mathrm{T}},\\
\mathbf{F}_3 &= (\mathbf{F}_{21}^{(m)}, \mathbf{F}_{22}^{(m)}, \ldots, \mathbf{F}_{2N_0}^{(m)})^{\mathrm{T}};
\end{aligned}
$$

$$
\begin{aligned}
\mathbf{a}_{s_0} &= (\mathbf{a}_1, \mathbf{a}_2, \mathbf{a}_3)^{\mathrm{T}}, \\
\mathbf{a}^{\tau}_{s_0} &= (\mathbf{a}^{\tau}_1, \mathbf{a}^{\tau}_2, \mathbf{a}^{\tau}_3)^{\mathrm{T}} \\
\mathbf{g}_{s_0} &= \left(\mathbf{F}_1, -2\frac{\Omega}{m}\mathbf{k}^{(1)}_0\dot{\mathbf{a}}_3 + \frac{\Omega^2}{m^2}\mathbf{k}^{(2)}_0\mathbf{a}_2 + \mathbf{F}_2, 2\frac{\Omega}{m}\mathbf{k}^{(1)}_0\dot{\mathbf{a}}_2 + \frac{\Omega^2}{m^2}\mathbf{k}^{(2)}_0\mathbf{a}_3 + \mathbf{F}_3\right)^T \\
&\text{for } N_0 = 1, 2, \ldots, \infty;
\end{aligned} \tag{4.173}
$$

and

$$
\begin{aligned}
\mathbf{F}^{(m)}_0(\mathbf{a}_{s_0}, \mathbf{a}^{\tau}_{s_0}, \dot{\mathbf{a}}_{s_0}, \dot{\mathbf{a}}^{\tau}_{s_0}) &= \frac{1}{2m\pi}\int_0^{2m\pi} \mathbf{f}(\mathbf{x}^{(m)*}, \mathbf{x}^{\tau(m)*}, \dot{\mathbf{x}}^{(m)*}, \dot{\mathbf{x}}^{\tau(m)*}, \mathbf{p})d\theta; \\
\mathbf{F}^{(m)}_{1k}(\mathbf{a}_{s_0}, \mathbf{a}^{\tau}_{s_0}, \dot{\mathbf{a}}_{s_0}, \dot{\mathbf{a}}^{\tau}_{s_0}) &= \frac{1}{m\pi}\int_0^{2m\pi} \mathbf{f}(\mathbf{x}^{(m)*}, \mathbf{x}^{\tau(m)*}, \dot{\mathbf{x}}^{(m)*}, \dot{\mathbf{x}}^{\tau(m)*}, \mathbf{p})\cos\left(\frac{k}{m}\theta\right)d\theta, \\
\mathbf{F}^{(m)}_{2k}(\mathbf{a}_{s_0}, \mathbf{a}^{\tau}_{s_0}, \dot{\mathbf{a}}_{s_0}, \dot{\mathbf{a}}^{\tau}_{s_0}) &= \frac{1}{m\pi}\int_0^{2m\pi} \mathbf{f}(\mathbf{x}^{(m)*}, \mathbf{x}^{\tau(m)*}, \dot{\mathbf{x}}^{(m)*}, \dot{\mathbf{x}}^{\tau(m)*}, \mathbf{p})\sin\left(\frac{k}{m}\theta\right)d\theta \\
&\text{for } k = 1, 2, \ldots, N_0.
\end{aligned} \tag{4.174}
$$

2. *After the* kth *Hopf bifurcation with* $p_k\omega_k = \omega_{k-1}(k = 1, 2, \ldots)$ *and* $\omega_0 = \Omega/m$, *there is a dynamical system of coefficients as*

$$
\ddot{\mathbf{a}}_{s_0s_1\ldots s_k} = \mathbf{g}_{s_0s_1\ldots s_k}(\mathbf{a}_{s_0s_1\ldots s_k}, \mathbf{a}^{\tau}_{s_0s_1\ldots s_k}, \dot{\mathbf{a}}_{s_0s_1\ldots s_k}, \dot{\mathbf{a}}^{\tau}_{s_0s_1\ldots s_k}, \mathbf{p}) \tag{4.175}
$$

where

$$
\begin{aligned}
\mathbf{a}_{s_0s_1\ldots s_k} &= (\mathbf{a}_{s_0s_1\ldots s_{k-1}1}, \mathbf{a}_{s_0s_1\ldots s_{k-1}2}, \mathbf{a}_{s_0s_1\ldots s_{k-1}3})^{\mathrm{T}}, \\
\mathbf{a}^{\tau}_{s_0s_1\ldots s_k} &= (\mathbf{a}^{\tau}_{s_0s_1\ldots s_{k-1}1}, \mathbf{a}^{\tau}_{s_0s_1\ldots s_{k-1}2}, \mathbf{a}^{\tau}_{s_0s_1\ldots s_{k-1}3})^{\mathrm{T}}; \\
\mathbf{f}_{s_0s_1\ldots s_k} &= (\mathbf{F}_{s_0s_1\ldots s_{k-1}1}, -2\omega_k\mathbf{k}^{(1)}_k\dot{\mathbf{a}}_{s_0s_1\ldots s_{k-1}3} + \omega_k^2\mathbf{k}^{(2)}_k\mathbf{a}_{s_0s_1\ldots s_{k-1}2} + \mathbf{F}_{s_0s_1\ldots s_{k-1}2}, \\
&\quad -2\omega_k\mathbf{k}^{(1)}_k\dot{\mathbf{a}}_{s_0s_1\ldots s_{k-1}2} + \omega_k^2\mathbf{k}^{(2)}_k\mathbf{a}_{s_0s_1\ldots s_{k-1}3} + \mathbf{F}_{s_0s_1\ldots s_{k-1}3})^{\mathrm{T}}, \\
\mathbf{k}^{(1)}_k &= diag(\mathbf{I}_{n_{k-1}\times n_{k-1}}, 2\mathbf{I}_{n_{k-1}\times n_{k-1}}, \ldots, N_k\mathbf{I}_{n_{k-1}\times n_{k-1}}), \\
\mathbf{k}^{(2)}_k &= diag(\mathbf{I}_{n_{k-1}\times n_{k-1}}, 2^2\mathbf{I}_{n_{k-1}\times n_{k-1}}, \ldots, N_k^2\mathbf{I}_{n_{k-1}\times n_{k-1}}), \\
n_k &= n(2N_0 + 1)(2N_1 + 1)\ldots(2N_{k-1} + 1)
\end{aligned} \tag{4.176}
$$

with a periodic solution as

$$
\begin{aligned}
\mathbf{a}_{s_0s_1\ldots s_k} &= \mathbf{a}^{(0)}_{s_0s_1\ldots s_k1}(t) + \sum_{l_{k+1}=1}^{\infty} \mathbf{a}^{(l_{k+1})}_{s_0s_1\ldots s_k2}(t)\cos(l_{k+1}\theta_{k+1}) \\
&\quad + \mathbf{a}^{(l_{k+1})}_{s_0s_1\ldots s_k3}(t)\sin(l_{k+1}\theta_{k+1}), \\
\mathbf{a}^{\tau}_{s_0s_1\ldots s_k} &= \mathbf{a}^{\tau(0)}_{s_0s_1\ldots s_k1}(t) + \sum_{l_{k+1}=1}^{\infty} \mathbf{a}^{\tau(l_{k+1})}_{s_0s_1\ldots s_k2}(t)\cos[l_{k+1}(\theta_{k+1} - \theta^{\tau}_{k+1})] \\
&\quad + \mathbf{a}^{\tau(l_{k+1})}_{s_0s_1\ldots s_k3}(t)[l_{k+1}(\theta_{k+1} - \theta^{\tau}_{k+1})]
\end{aligned} \tag{4.177}
$$

with

$$\begin{aligned}
&\mathbf{a}^{\tau(0)}_{s_0s_1\ldots s_k1}(t)=\mathbf{a}^{(0)}_{s_0s_1\ldots s_k1}(t-\tau),\\
&\mathbf{a}^{\tau(l_{k+1})}_{s_0s_1\ldots s_k2}(t)=\mathbf{a}^{\tau(l_{k+1})}_{s_0s_1\ldots s_k2}(t-\tau),\\
&\mathbf{a}^{\tau(l_{k+1})}_{s_0s_1\ldots s_k3}(t)=\mathbf{a}^{\tau(l_{k+1})}_{s_0s_1\ldots s_k3}(t-\tau),\\
&\theta^{\tau}_{k+1}=\omega_{k+1}\tau
\end{aligned}\tag{4.178}$$

and

$$\begin{aligned}
&s_i=1,2,3\ (i=0,1,2,\ldots,k),\\
&\mathbf{a}_{s_0s_1\ldots s_k1}=\mathbf{a}^{(0)}_{s_0s_1\ldots s_k1},\\
&\mathbf{a}_{s_0s_1\ldots s_k2}=(\mathbf{a}^{(1)}_{s_0s_1\ldots s_k2},\mathbf{a}^{(2)}_{s_0s_1\ldots s_k2},\ldots,\mathbf{a}^{(N_{k+1})}_{s_0s_1\ldots s_k2})^{\mathrm{T}},\\
&\mathbf{a}_{s_0s_1\ldots s_k3}=(\mathbf{a}^{(1)}_{s_0s_1\ldots s_k3},\mathbf{a}^{(2)}_{s_0s_1\ldots s_k3},\ldots,\mathbf{a}^{(N_{k+1})}_{s_0s_1\ldots s_k3})^{\mathrm{T}};\\
&\mathbf{a}_{s_0s_1\ldots s_{k-1}1}=\mathbf{a}^{(0)}_{s_0s_1\ldots s_{k-1}1},\\
&\mathbf{a}_{s_0s_1\ldots s_{k-1}2}=(\mathbf{a}^{(1)}_{s_0s_1\ldots s_{k-1}2},\mathbf{a}^{(2)}_{s_0s_1\ldots s_{k-1}2},\ldots,\mathbf{a}^{(N_k)}_{s_0s_1\ldots s_{k-1}2})^{\mathrm{T}},\\
&\mathbf{a}_{s_0s_1\ldots s_{k-1}3}=(\mathbf{a}^{(1)}_{s_0s_1\ldots s_{k-1}3},\mathbf{a}^{(2)}_{s_0s_1\ldots s_{k-1}3},\ldots,\mathbf{a}^{(N_k)}_{s_0s_1\ldots s_{k-1}3})^{\mathrm{T}};\\
&\vdots\\
&\mathbf{a}_1=\mathbf{a}^{(0)}_1,\\
&\mathbf{a}_2=(\mathbf{a}^{(1)}_2,\mathbf{a}^{(2)}_2,\ldots,\mathbf{a}^{(N)}_2)^{\mathrm{T}},\\
&\mathbf{a}_3=(\mathbf{a}^{(1)}_3,\mathbf{a}^{(2)}_3,\ldots,\mathbf{a}^{(N)}_3)^{\mathrm{T}};
\end{aligned}\tag{4.179}$$

and

$$\begin{aligned}
&s_i=1,2,3\ (i=0,1,2,\ldots,k),\\
&\mathbf{a}^{\tau}_{s_0s_1\ldots s_k1}=\mathbf{a}^{\tau(0)}_{s_0s_1\ldots s_k1},\\
&\mathbf{a}^{\tau}_{s_0s_1\ldots s_k2}=(\mathbf{a}^{\tau(1)}_{s_0s_1\ldots s_k2},\mathbf{a}^{\tau(2)}_{s_0s_1\ldots s_k2},\ldots,\mathbf{a}^{\tau(N_{k+1})}_{s_0s_1\ldots s_k2})^{\mathrm{T}},\\
&\mathbf{a}^{\tau}_{s_0s_1\ldots s_k2}=(\mathbf{a}^{\tau(1)}_{s_0s_1\ldots s_k3},\mathbf{a}^{\tau(2)}_{s_0s_1\ldots s_k3},\ldots,\mathbf{a}^{\tau(N_{k+1})}_{s_0s_1\ldots s_k3})^{\mathrm{T}};\\
&\mathbf{a}^{\tau}_{s_0s_1\ldots s_{k-1}1}=\mathbf{a}^{\tau(0)}_{s_0s_1\ldots s_{k-1}1},\\
&\mathbf{a}^{\tau}_{s_0s_1\ldots s_{k-1}2}=(\mathbf{a}^{\tau(1)}_{s_0s_1\ldots s_{k-1}2},\mathbf{a}^{\tau(2)}_{s_0s_1\ldots s_{k-1}2},\ldots,\mathbf{a}^{\tau(N_k)}_{s_0s_1\ldots s_{k-1}2})^{\mathrm{T}},\\
&\mathbf{a}^{\tau}_{s_0s_1\ldots s_{k-1}3}=(\mathbf{a}^{\tau(1)}_{s_0s_1\ldots s_{k-1}3},\mathbf{a}^{\tau(2)}_{s_0s_1\ldots s_{k-1}3},\ldots,\mathbf{a}^{\tau(N_k)}_{s_0s_1\ldots s_{k-1}3})^{\mathrm{T}};\\
&\vdots\\
&\mathbf{a}^{\tau}_1=\mathbf{a}^{\tau(0)}_1,\\
&\mathbf{a}^{\tau}_2=(\mathbf{a}^{\tau(1)}_2,\mathbf{a}^{\tau(2)}_2,\ldots,\mathbf{a}^{\tau(N)}_2)^{\mathrm{T}},\\
&\mathbf{a}^{\tau}_3=(\mathbf{a}^{\tau(1)}_3,\mathbf{a}^{\tau(2)}_3,\ldots,\mathbf{a}^{\tau(N)}_3)^{\mathrm{T}};
\end{aligned}\tag{4.180}$$

which, under $\|\mathbf{a}_{s_0s_1\ldots s_k}(t)-\mathbf{a}^{*}_{s_0s_1\ldots s_k}(t)\|<\varepsilon$ *and* $\|\mathbf{a}^{\tau}_{s_0s_1\ldots s_k}(t)-\mathbf{a}^{\tau *}_{s_0s_1\ldots s_k}(t)\|<\varepsilon_\tau$ *with a prescribed small* $\varepsilon>0$ *and* $\varepsilon_\tau>0$, *can be approximated by a finite term transformation, that is,*

$$
\begin{aligned}
\mathbf{a}^{*}_{s_0s_1\ldots s_k} &= \mathbf{a}^{(0)}_{s_0s_1\ldots s_k1}(t)+\sum_{l_{k+1}=1}^{N_{k+1}}\mathbf{a}^{(l_{k+1})}_{s_0s_1\ldots s_k2}(t)\cos(l_{k+1}\theta_{k+1})\\
&\quad+\mathbf{a}^{(l_{k+1})}_{s_0s_1\ldots s_k3}(t)\sin(l_{k+1}\theta_{k+1}),\\
\mathbf{a}^{\tau *}_{s_0s_1\ldots s_k} &= \mathbf{a}^{\tau(0)}_{s_0s_1\ldots s_k1}(t)+\sum_{l_{k+1}=1}^{N_{k+1}}\mathbf{a}^{\tau(l_{k+1})}_{s_0s_1\ldots s_k2}(t)\cos[l_{k+1}(\theta_{k+1}-\theta^{\tau}_{k+1})]\\
&\quad+\mathbf{a}^{\tau(l_{k+1})}_{s_0s_1\ldots s_k3}(t)\sin[l_{k+1}(\theta_{k+1}-\theta^{\tau}_{k+1})]
\end{aligned}
\tag{4.181}
$$

and the generalized coordinates are determined by

$$
\ddot{\mathbf{a}}_{s_0s_1\ldots s_{k+1}}=\mathbf{g}_{s_0s_1\ldots s_{k+1}}(\mathbf{a}_{s_0s_1\ldots s_{k+1}},\mathbf{a}^{\tau}_{s_0s_1\ldots s_{k+1}},\dot{\mathbf{a}}_{s_0s_1\ldots s_{k+1}},\dot{\mathbf{a}}^{\tau}_{s_0s_1\ldots s_{k+1}},\mathbf{p})
\tag{4.182}
$$

where

$$
\begin{aligned}
\mathbf{a}_{s_0s_1\ldots s_{k+1}} &= (\mathbf{a}_{s_0s_1\ldots s_k1},\mathbf{a}_{s_0s_1\ldots s_k2},\mathbf{a}_{s_0s_1\ldots s_k3})^{\mathrm{T}},\\
\mathbf{a}^{\tau}_{s_0s_1\ldots s_{k+1}} &= (\mathbf{a}^{\tau}_{s_0s_1\ldots s_k1},\mathbf{a}^{\tau}_{s_0s_1\ldots s_k2},\mathbf{a}^{\tau}_{s_0s_1\ldots s_k3})^{\mathrm{T}},\\
\mathbf{g}_{s_0s_1\ldots s_{k+1}} &= (\mathbf{F}_{s_0s_1\ldots s_k1},-2\omega_{k+1}\mathbf{k}^{(1)}_{k+1}\dot{\mathbf{a}}_{s_0s_1\ldots s_k3}+\omega^2_{k+1}\mathbf{k}^{(2)}_{k+1}\mathbf{a}_{s_0s_1\ldots s_k2}+\mathbf{F}_{s_0s_1\ldots s_k2},\\
&\quad 2\omega_{k+1}\mathbf{k}^{(1)}_{k+1}\dot{\mathbf{a}}_{s_0s_1\ldots s_k2}+\omega^2_{k+1}\mathbf{k}^{(2)}_{k+1}\mathbf{a}_{s_0s_1\ldots s_k3}+\mathbf{F}_{s_0s_1\ldots s_k3})^{\mathrm{T}}
\end{aligned}
\tag{4.183}
$$

and

$$
\begin{aligned}
\mathbf{k}^{(1)}_{k+1} &= diag(\mathbf{I}_{n_k\times n_k},2\mathbf{I}_{n_k\times n_k},\ldots,N_{k+1}\mathbf{I}_{n_k\times n_k}),\\
\mathbf{k}^{(2)}_{k+1} &= diag(\mathbf{I}_{n_k\times n_k},2^2\mathbf{I}_{n_k\times n_k},\ldots,N^2_{k+1}\mathbf{I}_{n_k\times n_k})\\
n_k &= n(2N_0+1)(2N_1+1)\ldots(2N_k+1);\\
\mathbf{a}_{s_0s_1\ldots s_k1} &= \mathbf{a}^{(0)}_{s_0s_1\ldots s_k1},\\
\mathbf{a}_{s_0s_1\ldots s_k2} &= (\mathbf{a}^{(1)}_{s_0s_1\ldots s_k2},\mathbf{a}^{(2)}_{s_0s_1\ldots s_k2},\ldots,\mathbf{a}^{(N_{k+1})}_{s_0s_1\ldots s_k2})^{\mathrm{T}},\\
\mathbf{a}_{s_0s_1\ldots s_k3} &= (\mathbf{a}^{(1)}_{s_0s_1\ldots s_k3},\mathbf{a}^{(2)}_{s_0s_1\ldots s_k3},\ldots,\mathbf{a}^{(N_{k+1})}_{s_0s_1\ldots s_k3})^{\mathrm{T}};\\
\mathbf{a}^{\tau}_{s_0s_1\ldots s_k1} &= \mathbf{a}^{\tau(0)}_{s_0s_1\ldots s_k1},\\
\mathbf{a}^{\tau}_{s_0s_1\ldots s_k2} &= (\mathbf{a}^{\tau(1)}_{s_0s_1\ldots s_k2},\mathbf{a}^{\tau(2)}_{s_0s_1\ldots s_k2},\ldots,\mathbf{a}^{\tau(N_{k+1})}_{s_0s_1\ldots s_k2})^{\mathrm{T}},\\
\mathbf{a}^{\tau}_{s_0s_1\ldots s_k3} &= (\mathbf{a}^{\tau(1)}_{s_0s_1\ldots s_k3},\mathbf{a}^{\tau(2)}_{s_0s_1\ldots s_k3},\ldots,\mathbf{a}^{\tau(N_{k+1})}_{s_0s_1\ldots s_k3})^{\mathrm{T}};
\end{aligned}
$$

$$
\begin{aligned}
&\mathbf{F}_{s_0 s_1 \ldots s_k 1} = \mathbf{F}^{(0)}_{s_0 s_1 \ldots s_k 1},\\
&\mathbf{F}_{s_0 s_1 \ldots s_k 2} = (\mathbf{F}^{(1)}_{s_0 s_1 \ldots s_k 2}, \mathbf{F}^{(2)}_{s_0 s_1 \ldots s_k 2}, \ldots, \mathbf{F}^{(N_{k+1})}_{s_0 s_1 \ldots s_k 2})^{\mathrm{T}},\\
&\mathbf{F}_{s_0 s_1 \ldots s_k 3} = (\mathbf{F}^{(1)}_{s_0 s_1 \ldots s_k 3}, \mathbf{F}^{(2)}_{s_0 s_1 \ldots s_k 3}, \ldots, \mathbf{F}^{(N_{k+1})}_{s_0 s_1 \ldots s_k 3})^{\mathrm{T}}\\
&\text{for } N_{k+1} = 1, 2, \ldots, \infty;
\end{aligned} \tag{4.184}
$$

and

$$
\begin{aligned}
&\mathbf{F}_{s_0 s_1 \ldots s_k 1}(\mathbf{a}_{s_0 s_1 \ldots s_{k+1}}, \mathbf{a}^{\tau}_{s_0 s_1 \ldots s_{k+1}}, \dot{\mathbf{a}}_{s_0 s_1 \ldots s_{k+1}}, \dot{\mathbf{a}}^{\tau}_{s_0 s_1 \ldots s_{k+1}}, \mathbf{p})\\
&\quad = \frac{1}{2\pi}\int_0^{2\pi} \mathbf{g}_{s_0 s_1 \ldots s_k}(\mathbf{a}^{*}_{s_0 s_1 \ldots s_k}, \mathbf{a}^{\tau *}_{s_0 s_1 \ldots s_k}, \dot{\mathbf{a}}^{*}_{s_0 s_1 \ldots s_k}, \dot{\mathbf{a}}^{\tau *}_{s_0 s_1 \ldots s_k}, \mathbf{p}) d\theta_{k+1};\\
&\mathbf{F}^{(l_{k+1})}_{s_0 s_1 \ldots s_k 2}(\mathbf{a}_{s_0 s_1 \ldots s_{k+1}}, \mathbf{a}^{\tau}_{s_0 s_1 \ldots s_{k+1}}, \dot{\mathbf{a}}_{s_0 s_1 \ldots s_{k+1}}, \dot{\mathbf{a}}^{\tau}_{s_0 s_1 \ldots s_{k+1}}, \mathbf{p})\\
&\quad = \frac{1}{\pi}\int_0^{2\pi} \mathbf{g}_{s_0 s_1 \ldots s_k}(\mathbf{a}^{*}_{s_0 s_1 \ldots s_k}, \mathbf{a}^{\tau *}_{s_0 s_1 \ldots s_k}, \dot{\mathbf{a}}^{*}_{s_0 s_1 \ldots s_k}, \dot{\mathbf{a}}^{\tau *}_{s_0 s_1 \ldots s_k}, \mathbf{p}) \cos(l_{k+1}\theta_{k+1}) d\theta_{k+1},\\
&\mathbf{F}^{(l_{k+1})}_{s_0 s_1 \ldots s_k 3}(\mathbf{a}_{s_0 s_1 \ldots s_{k+1}}, \mathbf{a}^{\tau}_{s_0 s_1 \ldots s_{k+1}}, \dot{\mathbf{a}}_{s_0 s_1 \ldots s_{k+1}}, \dot{\mathbf{a}}^{\tau}_{s_0 s_1 \ldots s_{k+1}}, \mathbf{p})\\
&\quad = \frac{1}{\pi}\int_0^{2\pi} \mathbf{g}_{s_0 s_1 \ldots s_k}(\mathbf{a}^{*}_{s_0 s_1 \ldots s_k}, \mathbf{a}^{\tau *}_{s_0 s_1 \ldots s_k}, \dot{\mathbf{a}}^{*}_{s_0 s_1 \ldots s_k}, \dot{\mathbf{a}}^{\tau *}_{s_0 s_1 \ldots s_k}, \mathbf{p}) \sin(l_{k+1}\theta_{k+1}) d\theta_{k+1}
\end{aligned} \tag{4.185}
$$

for $l_{k+1} = 1, 2, \ldots, N_{k+1}$.

3. *Equation (4.182) becomes*

$$
\dot{\mathbf{z}}_{s_0 s_1 \ldots s_{k+1}} = \mathbf{f}_{s_0 s_1 \ldots s_{k+1}}(\mathbf{z}_{s_0 s_1 \ldots s_{k+1}}, \mathbf{z}^{\tau}_{s_0 s_1 \ldots s_{k+1}}) \tag{4.186}
$$

where

$$
\begin{aligned}
\mathbf{z}_{s_0 s_1 \ldots s_{k+1}} &= (\mathbf{a}_{s_0 s_1 \ldots s_{k+1}}, \dot{\mathbf{a}}_{s_0 s_1 \ldots s_{k+1}})^{\mathrm{T}},\\
\mathbf{z}^{\tau}_{s_0 s_1 \ldots s_{k+1}} &= (\mathbf{a}^{\tau}_{s_0 s_1 \ldots s_{k+1}}, \dot{\mathbf{a}}^{\tau}_{s_0 s_1 \ldots s_{k+1}})^{\mathrm{T}},\\
\mathbf{f}_{s_0 s_1 \ldots s_{k+1}} &= (\dot{\mathbf{a}}_{s_0 s_1 \ldots s_{k+1}}, \mathbf{g}_{s_0 s_1 \ldots s_{k+1}})^{\mathrm{T}}
\end{aligned} \tag{4.187}
$$

If equilibrium $\mathbf{z}^{*}_{s_0 s_1 \ldots s_{k+1}} = \mathbf{z}^{\tau *}_{s_0 s_1 \ldots s_{k+1}}$ *of Equation (4.186) (i.e.,* $\mathbf{f}_{s_0 s_1 \ldots s_{k+1}}(\mathbf{z}^{*}_{s_0 s_1 \ldots s_{k+1}}, \mathbf{z}^{\tau *}_{s_0 s_1 \ldots s_{k+1}}) = \mathbf{0}$*) exists, then the approximate solution of the periodic motion of* k*th generalized coordinates for the period-*m *motion exists as in Equation (4.177). In the vicinity of equilibrium* $\mathbf{z}^{*}_{s_0 s_1 \ldots s_{k+1}} = \mathbf{z}^{\tau *}_{s_0 s_1 \ldots s_{k+1}}$*, with*

$$
\begin{aligned}
\mathbf{z}_{s_0 s_1 \ldots s_{k+1}} &= \mathbf{z}^{*}_{s_0 s_1 \ldots s_{k+1}} + \Delta\mathbf{z}_{s_0 s_1 \ldots s_{k+1}},\\
\mathbf{z}^{\tau}_{s_0 s_1 \ldots s_{k+1}} &= \mathbf{z}^{\tau *}_{s_0 s_1 \ldots s_{k+1}} + \Delta\mathbf{z}^{\tau}_{s_0 s_1 \ldots s_{k+1}}
\end{aligned} \tag{4.188}
$$

the linearized equation of Equation (4.186) is

$$
\Delta\dot{\mathbf{z}}_{s_0 s_1 \ldots s_{k+1}} = D_{\mathbf{z}_{s_0 s_1 \ldots s_{k+1}}}\mathbf{f}_{s_0 s_1 \ldots s_{k+1}}\Delta\mathbf{z}_{s_0 s_1 \ldots s_{k+1}} + D_{\mathbf{z}^{\tau}_{s_0 s_1 \ldots s_{k+1}}}\mathbf{f}_{s_0 s_1 \ldots s_{k+1}}\Delta\mathbf{z}^{\tau}_{s_0 s_1 \ldots s_{k+1}} \tag{4.189}
$$

and the eigenvalue analysis of equilibrium $\mathbf{z}^*$ *is given by*

$$|D_{\mathbf{z}_{s_0s_1\ldots s_{k+1}}}\mathbf{f}_{s_0s_1\ldots s_{k+1}} + e^{\lambda\tau}D_{\mathbf{z}^{\tau}_{s_0s_1\ldots s_{k+1}}}\mathbf{f}_{s_0s_1\ldots s_{k+1}} - \lambda\mathbf{I}_{2n_k(2N_{k+1}+1)\times 2n_k(2N_{k+1}+1)}| = 0 \quad (4.190)$$

where

$$D_{\mathbf{z}_{s_0s_1\ldots s_{k+1}}}\mathbf{f}_{s_0s_1\ldots s_{k+1}} = \left.\frac{\partial \mathbf{f}_{s_0s_1\ldots s_{k+1}}\left(\mathbf{z}_{s_0s_1\ldots s_{k+1}}, \mathbf{z}^{\tau}_{s_0s_1\ldots s_{k+1}}\right)}{\partial \mathbf{z}_{s_0s_1\ldots s_{k+1}}}\right|_{(\mathbf{z}^*_{s_0s_1\ldots s_{k+1}}, \mathbf{z}^{\tau*}_{s_0s_1\ldots s_{k+1}})},$$

$$D_{\mathbf{z}^{\tau}_{s_0s_1\ldots s_{k+1}}}\mathbf{f}_{s_0s_1\ldots s_{k+1}} = \left.\frac{\partial \mathbf{f}_{s_0s_1\ldots s_{k+1}}\left(\mathbf{z}_{s_0s_1\ldots s_{k+1}}, \mathbf{z}^{\tau}_{s_0s_1\ldots s_{k+1}}\right)}{\partial \mathbf{z}_{s_0s_1\ldots s_{k+1}}}\right|_{(\mathbf{z}^*_{s_0s_1\ldots s_{k+1}}, \mathbf{z}^{\tau*}_{s_0s_1\ldots s_{k+1}})}. \quad (4.191)$$

The stability and bifurcation of such a periodic motion of the kth *generalized coordinates can be classified by the eigenvalues of Equation (4.189) at equilibrium* $\mathbf{z}^*_{s_0s_1\ldots s_{k+1}} = \mathbf{z}^{\tau*}_{s_0s_1\ldots s_{k+1}}$ *with*

$$(n_1, n_2, n_3 \mid n_4, n_5, n_6). \quad (4.192)$$

a. *If all eigenvalues of the equilibrium possess negative real parts, the approximate quasi-periodic solution is stable.*
b. *If at least one of the eigenvalues of the equilibrium possesses a positive real part, the approximate quasi-periodic solution is unstable.*
c. *The boundaries between stable and unstable equilibriums with higher order singularity give bifurcation and stability conditions with higher order singularity.*

4. *For the* kth *order Hopf bifurcation of period-*m *motion, a relation exists as*

$$p_k\omega_k = \omega_{k-1}. \quad (4.193)$$

a. *If* p_k *is an irrational number, the* kth*-order Hopf bifurcation of the period-*m *motion is called the quasi-period-*p_k *Hopf bifurcation, and the corresponding solution of the* kth *generalized coordinates is* p_k*-quasi-periodic to the system of the* $(k-1)$th *generalized coordinates.*
b. *If* $p_k = 2$, *the* kth*-order Hopf bifurcation of the period-*m *motion is called a period-doubling Hopf bifurcation (or a period-2 Hopf bifurcation), and the corresponding solution of the* kth *generalized coordinates is period-doubling to the system of the* $(k-1)$th *generalized coordinates.*
c. *If* $p_k = q$ *with an integer* q, *the* kth*-order Hopf bifurcation of the period-*m *motion is called a period-*q *Hopf bifurcation, and the corresponding solution of the* kth *generalized coordinates is of* q*-times period to the system of the* $(k-1)$th *generalized coordinates.*
d. *If* $p_k = p/q$ *(*p, q *are irreducible integers), the* kth*-order Hopf bifurcation of the period-*m *motion is called a period-*p/q *Hopf bifurcation, and the corresponding solution of the* kth *generalized coordinates is* p/q*-periodic to the system of the* $(k-1)$th *generalized coordinates.*

Proof. The proof of this theorem is similar to the proof of Theorem 3.14. ■

Consider periodically forced, time-delayed, nonlinear vibration systems, and the analytical solution of quasi-periodic motion relative to period-m flow in such a vibration system is stated from the following theorem.

Theorem 4.8 *Consider a periodically forced,* time-delayed, *nonlinear vibration system as*

$$\ddot{\mathbf{x}} = \mathbf{F}(\mathbf{x}, \dot{\mathbf{x}}; \mathbf{x}^{\tau}, \dot{\mathbf{x}}^{\tau}, t, \mathbf{p}) \in \mathscr{R}^n \tag{4.194}$$

where $\mathbf{F}(\mathbf{x}, \dot{\mathbf{x}}; \mathbf{x}^{\tau}, \dot{\mathbf{x}}^{\tau}, t, \mathbf{p})$ *is a* C^r*-continuous nonlinear function vector* ($r \geq 1$) *with forcing period* $T = 2\pi/\Omega$.

1. *If such a time-delayed, vibration system has a period-*m *motion* $\mathbf{x}^{(m)}(t)$ *with finite norm* $||\mathbf{x}^{(m)}||$, *there is a generalized coordinate transformation with* $\theta = \Omega t$ *for the period-*m *flow of Equation (4.194) in the form of*

$$\begin{aligned}
\mathbf{x}^{(m)}(t) &= \mathbf{a}_0^{(m)}(t) + \sum_{k=1}^{\infty} \mathbf{b}_{k/m}(t) \cos\left(\frac{k}{m}\theta\right) + \mathbf{c}_{k/m}(t) \sin\left(\frac{k}{m}\theta\right);\\
\mathbf{x}^{\tau(m)}(t) &= \mathbf{a}_0^{\tau(m)}(t) + \sum_{k=1}^{\infty} \mathbf{b}_{k/m}^{\tau}(t) \cos\left[\frac{k}{m}(\theta - \theta^{\tau})\right] + \mathbf{c}_{k/m}^{\tau}(t) \sin\left[\frac{k}{m}(\theta - \theta^{\tau})\right]
\end{aligned} \tag{4.195}$$

with $\mathbf{a}_0^{\tau(m)} = \mathbf{a}_0^{(m)}(t-\tau), \mathbf{b}_k^{\tau(m)} = \mathbf{b}_k^{(m)}(t-\tau), \mathbf{c}_k^{\tau(m)} = \mathbf{c}_k^{(m)}(t-\tau), \theta^{\tau} = \Omega\tau$ *and*

$$\begin{aligned}
\mathbf{a}_1^{(0)} &\equiv \mathbf{a}_0^{(m)} = (a_{01}^{(m)}, a_{02}^{(m)}, \ldots, a_{0n}^{(m)})^{\mathrm{T}},\\
\mathbf{a}_2^{(k)} &\equiv \mathbf{b}_{k/m} = (b_{k/m1}, b_{k/m2}, \ldots, b_{k/mn})^{\mathrm{T}},\\
\mathbf{a}_3^{(k)} &\equiv \mathbf{c}_{k/m} = (c_{k/m1}, c_{k/m2}, \ldots, c_{k/mn})^{\mathrm{T}};\\
\mathbf{a}_1^{\tau(0)} &\equiv \mathbf{a}_0^{\tau(m)} = (a_{01}^{\tau(m)}, a_{02}^{\tau(m)}, \ldots, a_{0n}^{\tau(m)})^{\mathrm{T}},\\
\mathbf{a}_2^{\tau(k)} &\equiv \mathbf{b}_{k/m}^{\tau} = (b_{k/m1}^{\tau}, b_{k/m2}^{\tau}, \ldots, b_{k/mn}^{\tau})^{\mathrm{T}},\\
\mathbf{a}_3^{\tau(k)} &\equiv \mathbf{c}_{k/m}^{\tau} = (c_{k/m1}^{\tau}, c_{k/m2}^{\tau}, \ldots, c_{k/mn}^{\tau})^{\mathrm{T}}
\end{aligned} \tag{4.196}$$

which, under $||\mathbf{x}^{(m)}(t) - \mathbf{x}^{(m)*}(t)|| < \varepsilon$ *and* $||\mathbf{x}^{\tau(m)}(t) - \mathbf{x}^{\tau(m)*}(t)|| < \varepsilon^{\tau}$ *with prescribed small* $\varepsilon > 0$ *and* $\varepsilon^{\tau} > 0$, *can be approximated by a finite term transformation* $\mathbf{x}^{(m)*}(t)$ *as*

$$\begin{aligned}
\mathbf{x}^{(m)*}(t) &= \mathbf{a}_0^{(m)}(t) + \sum_{k=1}^{N_0} \mathbf{b}_{k/m}(t) \cos\left(\frac{k}{m}\theta\right) + \mathbf{c}_{k/m}(t) \sin\left(\frac{k}{m}\theta\right);\\
\mathbf{x}^{\tau(m)*}(t) &= \mathbf{a}_0^{\tau(m)}(t) + \sum_{k=1}^{N_0} \mathbf{b}_{k/m}^{\tau}(t) \cos\left[\frac{k}{m}(\theta - \theta^{\tau})\right] + \mathbf{c}_{k/m}^{\tau}(t) \sin\left[\frac{k}{m}(\theta - \theta^{\tau})\right]
\end{aligned} \tag{4.197}$$

and the generalized coordinates are determined by

$$\ddot{\mathbf{a}}_{s_0} = \mathbf{g}_{s_0}(\mathbf{a}_{s_0}, \mathbf{a}_{s_0}^{\tau}, \dot{\mathbf{a}}_{s_0}, \dot{\mathbf{a}}_{s_0}^{\tau}, \mathbf{p}) \tag{4.198}$$

where

$$
\begin{aligned}
&\mathbf{k}_0^{(1)} = diag(\mathbf{I}_{n\times n}, 2\mathbf{I}_{n\times n}, \dots, N_0\mathbf{I}_{n\times n}),\\
&\mathbf{k}_0^{(2)} = diag(\mathbf{I}_{n\times n}, 2^2\mathbf{I}_{n\times n}, \dots, N_0^2\mathbf{I}_{n\times n});\\
&\mathbf{a}_1^{(0)} \equiv \mathbf{a}_0^{(m)}, \mathbf{a}_2^{(k)} \equiv \mathbf{b}_{k/m}, \mathbf{a}_3^{(k)} \equiv \mathbf{c}_{k/m};\\
&\mathbf{a}_1^{\tau(0)} \equiv \mathbf{a}_0^{\tau(m)}, \mathbf{a}_2^{\tau(k)} \equiv \mathbf{b}_{k/m}^{\tau}, \mathbf{a}_3^{(k)} \equiv \mathbf{c}_{k/m}^{\tau};\\
&\mathbf{a}_1 = \mathbf{a}_1^{(0)},\\
&\mathbf{a}_2 = (\mathbf{a}_2^{(1)}, \mathbf{a}_2^{(2)}, \dots, \mathbf{a}_2^{(N_0)})^{\mathrm{T}} \equiv \mathbf{b}^{(m)},\\
&\mathbf{a}_3 = (\mathbf{a}_3^{(1)}, \mathbf{a}_3^{(2)}, \dots, \mathbf{a}_3^{(N_0)})^{\mathrm{T}} \equiv \mathbf{c}^{(m)},\\
&\mathbf{a}_1^{\tau} = \mathbf{a}_1^{\tau(0)},\\
&\mathbf{a}_2^{\tau} = (\mathbf{a}_2^{\tau(1)}, \mathbf{a}_2^{\tau(2)}, \dots, \mathbf{a}_2^{\tau(N_0)})^{\mathrm{T}} \equiv \mathbf{b}^{\tau(m)},\\
&\mathbf{a}_3^{\tau} = (\mathbf{a}_3^{\tau(1)}, \mathbf{a}_3^{\tau(2)}, \dots, \mathbf{a}_3^{\tau(N_0)})^{\mathrm{T}} \equiv \mathbf{c}^{\tau(m)},\\
&\mathbf{F}_1 = \mathbf{F}_0^{(m)},\\
&\mathbf{F}_2 = (\mathbf{F}_{11}^{(m)}, \mathbf{F}_{12}^{(m)}, \dots, \mathbf{F}_{1N_0}^{(m)})^{\mathrm{T}},\\
&\mathbf{F}_3 = (\mathbf{F}_{21}^{(m)}, \mathbf{F}_{22}^{(m)}, \dots, \mathbf{F}_{2N_0}^{(m)})^{\mathrm{T}};\\
&\mathbf{a}_{s_0} = (\mathbf{a}_1, \mathbf{a}_2, \mathbf{a}_3)^{\mathrm{T}},\\
&\mathbf{a}_{s_0}^{\tau} = (\mathbf{a}_1^{\tau}, \mathbf{a}_2^{\tau}, \mathbf{a}_3^{\tau})^{\mathrm{T}},\\
&\mathbf{g}_{s_0} = \left(\mathbf{F}_1, -2\frac{\Omega}{m}\mathbf{k}_0^{(1)}\dot{\mathbf{a}}_3 + \frac{\Omega^2}{m^2}\mathbf{k}_0^{(2)}\mathbf{a}_2 + \mathbf{F}_2, 2\frac{\Omega}{m}\mathbf{k}_0^{(1)}\dot{\mathbf{a}}_2 + \frac{\Omega^2}{m^2}\mathbf{k}_0^{(2)}\mathbf{a}_3 + \mathbf{F}_3\right)^{\mathrm{T}}\\
&\text{for } N_0 = 1, 2, \dots, \infty;
\end{aligned}
\tag{4.199}
$$

and

$$
\begin{aligned}
&\mathbf{F}_0^{(m)}(\mathbf{a}_{s_0}, \mathbf{a}_{s_0}^{\tau}, \dot{\mathbf{a}}_{s_0}, \dot{\mathbf{a}}_{s_0}^{\tau}) = \frac{1}{2m\pi}\int_0^{2m\pi} \mathbf{f}(\mathbf{x}^{(m)*}, \mathbf{x}^{\tau(m)*}, \dot{\mathbf{x}}^{(m)*}, \dot{\mathbf{x}}^{\tau(m)*}, \mathbf{p})d\theta;\\
&\mathbf{F}_{1k}^{(m)}(\mathbf{a}_{s_0}, \mathbf{a}_{s_0}^{\tau}, \dot{\mathbf{a}}_{s_0}, \dot{\mathbf{a}}_{s_0}^{\tau}) = \frac{1}{m\pi}\int_0^{2m\pi} \mathbf{f}(\mathbf{x}^{(m)*}, \mathbf{x}^{\tau(m)*}, \dot{\mathbf{x}}^{(m)*}, \dot{\mathbf{x}}^{\tau(m)*}, \mathbf{p}) \cos\left(\frac{k}{m}\theta\right)d\theta,\\
&\mathbf{F}_{2k}^{(m)}(\mathbf{a}_{s_0}, \mathbf{a}_{s_0}^{\tau}, \dot{\mathbf{a}}_{s_0}, \dot{\mathbf{a}}_{s_0}^{\tau}) = \frac{1}{m\pi}\int_0^{2m\pi} \mathbf{f}(\mathbf{x}^{(m)*}, \mathbf{x}^{\tau(m)*}, \dot{\mathbf{x}}^{(m)*}, \dot{\mathbf{x}}^{\tau(m)*}, \mathbf{p}) \sin\left(\frac{k}{m}\theta\right)d\theta\\
&\text{for } k = 1, 2, \dots, N_0.
\end{aligned}
\tag{4.200}
$$

2. *After the* kth *Hopf bifurcation with* $p_k\omega_k = \omega_{k-1}(k = 1, 2, \dots)$ *and* $\omega_0 = \Omega/m$, *there is a dynamical system of coefficients as*

$$
\ddot{\mathbf{a}}_{s_0s_1\dots s_k} = \mathbf{g}_{s_0s_1\dots s_k}(\mathbf{a}_{s_0s_1\dots s_k}, \mathbf{a}_{s_0s_1\dots s_k}^{\tau}, \dot{\mathbf{a}}_{s_0s_1\dots s_k}, \dot{\mathbf{a}}_{s_0s_1\dots s_k}^{\tau}, \mathbf{p}) \tag{4.201}
$$

where

$$\begin{aligned}
\mathbf{a}_{s_0 s_1 \ldots s_k} &= (\mathbf{a}_{s_0 s_1 \ldots s_{k-1} 1}, \mathbf{a}_{s_0 s_1 \ldots s_{k-1} 2}, \mathbf{a}_{s_0 s_1 \ldots s_{k-1} 3})^{\mathrm{T}}, \\
\mathbf{a}^{\tau}_{s_0 s_1 \ldots s_k} &= (\mathbf{a}^{\tau}_{s_0 s_1 \ldots s_{k-1} 1}, \mathbf{a}^{\tau}_{s_0 s_1 \ldots s_{k-1} 2}, \mathbf{a}^{\tau}_{s_0 s_1 \ldots s_{k-1} 3})^{\mathrm{T}} \\
\mathbf{g}_{s_0 s_1 \ldots s_k} &= (\mathbf{F}_{s_0 s_1 \ldots s_{k-1} 1}, -2\omega_k \mathbf{k}^{(1)}_k \dot{\mathbf{a}}_{s_0 s_1 \ldots s_{k-1} 3} + \omega_k^2 \mathbf{k}^{(2)}_k \mathbf{a}_{s_0 s_1 \ldots s_{k-1} 2} + \mathbf{F}_{s_0 s_1 \ldots s_{k-1} 2}, \\
&\quad - 2\omega_k \mathbf{k}^{(1)}_k \dot{\mathbf{a}}_{s_0 s_1 \ldots s_{k-1} 2} + \omega_k^2 \mathbf{k}^{(2)}_k \mathbf{a}_{s_0 s_1 \ldots s_{k-1} 3} + \mathbf{F}_{s_0 s_1 \ldots s_{k-1} 3})^{\mathrm{T}}, \\
\mathbf{k}^{(1)}_k &= diag(\mathbf{I}_{n_{k-1} \times n_{k-1}}, 2\mathbf{I}_{n_{k-1} \times n_{k-1}}, \ldots, N_k \mathbf{I}_{n_{k-1} \times n_{k-1}}), \\
\mathbf{k}^{(2)}_k &= diag(\mathbf{I}_{n_{k-1} \times n_{k-1}}, 2^2\mathbf{I}_{n_{k-1} \times n_{k-1}}, \ldots, N_k^2 \mathbf{I}_{n_{k-1} \times n_{k-1}}) \\
n_k &= n(2N_0 + 1)(2N_1 + 1) \ldots (2N_{k-1} + 1)
\end{aligned} \tag{4.202}$$

with a periodic solution as

$$\begin{aligned}
\mathbf{a}_{s_0 s_1 \ldots s_k} &= \mathbf{a}^{(0)}_{s_0 s_1 \ldots s_k 1}(t) + \sum_{l_{k+1}=1}^{\infty} \mathbf{a}^{(l_{k+1})}_{s_0 s_1 \ldots s_k 2}(t) \cos(l_{k+1}\theta_{k+1}) \\
&\quad + \mathbf{a}^{(l_{k+1})}_{s_0 s_1 \ldots s_k 3}(t) \sin(l_{k+1}\theta_{k+1}) \\
\mathbf{a}^{\tau}_{s_0 s_1 \ldots s_k} &= \mathbf{a}^{\tau(0)}_{s_0 s_1 \ldots s_k 1}(t) + \sum_{l_{k+1}=1}^{\infty} \mathbf{a}^{\tau(l_{k+1})}_{s_0 s_1 \ldots s_k 2}(t) \cos[l_{k+1}(\theta_{k+1} - \theta^{\tau}_{k+1})] \\
&\quad + \mathbf{a}^{\tau(l_{k+1})}_{s_0 s_1 \ldots s_k 3}(t)[l_{k+1}(\theta_{k+1} - \theta^{\tau}_{k+1})]
\end{aligned} \tag{4.203}$$

with

$$\begin{aligned}
\mathbf{a}^{\tau(0)}_{s_0 s_1 \ldots s_k 1}(t) &= \mathbf{a}^{(0)}_{s_0 s_1 \ldots s_k 1}(t - \tau), \\
\mathbf{a}^{\tau(l_{k+1})}_{s_0 s_1 \ldots s_k 2}(t) &= \mathbf{a}^{\tau(l_{k+1})}_{s_0 s_1 \ldots s_k 2}(t - \tau), \\
\mathbf{a}^{\tau(l_{k+1})}_{s_0 s_1 \ldots s_k 3}(t) &= \mathbf{a}^{\tau(l_{k+1})}_{s_0 s_1 \ldots s_k 3}(t - \tau), \\
\theta^{\tau}_{k+1} &= \omega_{k+1}\tau
\end{aligned} \tag{4.204}$$

and

$$\begin{aligned}
& s_i = 1, 2, 3 \; (i = 0, 1, 2, \ldots, k), \\
& \mathbf{a}_{s_0 s_1 \ldots s_k 1} = \mathbf{a}^{(0)}_{s_0 s_1 \ldots s_k 1} \\
& \mathbf{a}_{s_0 s_1 \ldots s_k 2} = (\mathbf{a}^{(1)}_{s_0 s_1 \ldots s_k 2}, \mathbf{a}^{(2)}_{s_0 s_1 \ldots s_k 2}, \ldots, \mathbf{a}^{(N_{k+1})}_{s_0 s_1 \ldots s_k 2})^{\mathrm{T}}, \\
& \mathbf{a}_{s_0 s_1 \ldots s_k 3} = (\mathbf{a}^{(1)}_{s_0 s_1 \ldots s_k 3}, \mathbf{a}^{(2)}_{s_0 s_1 \ldots s_k 3}, \ldots, \mathbf{a}^{(N_{k+1})}_{s_0 s_1 \ldots s_k 3})^{\mathrm{T}}; \\
& \mathbf{a}_{s_0 s_1 \ldots s_{k-1} 1} = \mathbf{a}^{(0)}_{s_0 s_1 \ldots s_{k-1} 1} \\
& \mathbf{a}_{s_0 s_1 \ldots s_{k-1} 2} = (\mathbf{a}^{(1)}_{s_0 s_1 \ldots s_{k-1} 2}, \mathbf{a}^{(2)}_{s_0 s_1 \ldots s_{k-1} 2}, \ldots, \mathbf{a}^{(N_k)}_{s_0 s_1 \ldots s_{k-1} 2})^{\mathrm{T}},
\end{aligned}$$

$$\mathbf{a}_{s_0 s_1 \ldots s_{k-1} 3} = (\mathbf{a}^{(1)}_{s_0 s_1 \ldots s_{k-1} 3}, \mathbf{a}^{(2)}_{s_0 s_1 \ldots s_{k-1} 3}, \ldots, \mathbf{a}^{(N_k)}_{s_0 s_1 \ldots s_{k-1} 3})^{\mathrm{T}};$$

$$\vdots$$

$$\mathbf{a}_1 = \mathbf{a}^{(0)}_1,$$

$$\mathbf{a}_2 = (\mathbf{a}^{(1)}_2, \mathbf{a}^{(2)}_2, \ldots, \mathbf{a}^{(N_0)}_2)^{\mathrm{T}},$$

$$\mathbf{a}_3 = (\mathbf{a}^{(1)}_3, \mathbf{a}^{(2)}_3, \ldots, \mathbf{a}^{(N_0)}_3)^{\mathrm{T}}; \tag{4.205}$$

and

$$s_i = 1, 2, 3 \; (i = 0, 1, 2, \ldots, k),$$

$$\mathbf{a}^{\tau}_{s_0 s_1 \ldots s_k 1} = \mathbf{a}^{\tau(0)}_{s_0 s_1 \ldots s_k 1},$$

$$\mathbf{a}^{\tau}_{s_0 s_1 \ldots s_k 2} = (\mathbf{a}^{\tau(1)}_{s_0 s_1 \ldots s_k 2}, \mathbf{a}^{\tau(2)}_{s_0 s_1 \ldots s_k 2}, \ldots, \mathbf{a}^{\tau(N_{k+1})}_{s_0 s_1 \ldots s_k 2})^{\mathrm{T}},$$

$$\mathbf{a}^{\tau}_{s_0 s_1 \ldots s_k 2} = (\mathbf{a}^{\tau(1)}_{s_0 s_1 \ldots s_k 3}, \mathbf{a}^{\tau(2)}_{s_0 s_1 \ldots s_k 3}, \ldots, \mathbf{a}^{\tau(N_{k+1})}_{s_0 s_1 \ldots s_k 3})^{\mathrm{T}};$$

$$\mathbf{a}^{\tau}_{s_0 s_1 \ldots s_{k-1} 1} = \mathbf{a}^{\tau(0)}_{s_0 s_1 \ldots s_{k-1} 1},$$

$$\mathbf{a}^{\tau}_{s_0 s_1 \ldots s_{k-1} 2} = (\mathbf{a}^{\tau(1)}_{s_0 s_1 \ldots s_{k-1} 2}, \mathbf{a}^{\tau(2)}_{s_0 s_1 \ldots s_{k-1} 2}, \ldots, \mathbf{a}^{\tau(N_k)}_{s_0 s_1 \ldots s_{k-1} 2})^{\mathrm{T}},$$

$$\mathbf{a}^{\tau}_{s_0 s_1 \ldots s_{k-1} 3} = (\mathbf{a}^{\tau(1)}_{s_0 s_1 \ldots s_{k-1} 3}, \mathbf{a}^{\tau(2)}_{s_0 s_1 \ldots s_{k-1} 3}, \ldots, \mathbf{a}^{\tau(N_k)}_{s_0 s_1 \ldots s_{k-1} 3})^{\mathrm{T}};$$

$$\vdots$$

$$\mathbf{a}^{\tau}_1 = \mathbf{a}^{\tau(0)}_1,$$

$$\mathbf{a}^{\tau}_2 = (\mathbf{a}^{\tau(1)}_2, \mathbf{a}^{\tau(2)}_2, \ldots, \mathbf{a}^{\tau(N_0)}_2)^{\mathrm{T}},$$

$$\mathbf{a}^{\tau}_3 = (\mathbf{a}^{\tau(1)}_3, \mathbf{a}^{\tau(2)}_3, \ldots, \mathbf{a}^{\tau(N_0)}_3)^{\mathrm{T}}; \tag{4.206}$$

which, under $\|\mathbf{a}_{s_0 s_1 \ldots s_k}(t) - \mathbf{a}^{*}_{s_0 s_1 \ldots s_k}(t)\| < \varepsilon$ *and* $\|\mathbf{a}^{\tau}_{s_0 s_1 \ldots s_k}(t) - \mathbf{a}^{\tau *}_{s_0 s_1 \ldots s_k}(t)\| < \varepsilon_{\tau}$ *with a prescribed small* $\varepsilon > 0$ *and* $\varepsilon_{\tau} > 0$*, can be approximated by a finite term transformation, that is,*

$$\begin{aligned}
\mathbf{a}^{*}_{s_0 s_1 \ldots s_k} &= \mathbf{a}^{(0)}_{s_0 s_1 \ldots s_k 1}(t) + \sum_{l_{k+1}=1}^{N_{k+1}} \mathbf{a}^{(l_{k+1})}_{s_0 s_1 \ldots s_k 2}(t) \cos(l_{k+1}\theta_{k+1}) \\
&\quad + \mathbf{a}^{(l_{k+1})}_{s_0 s_1 \ldots s_k 3}(t) \sin(l_{k+1}\theta_{k+1}), \\
\mathbf{a}^{\tau *}_{s_0 s_1 \ldots s_k} &= \mathbf{a}^{\tau(0)}_{s_0 s_1 \ldots s_k 1}(t) + \sum_{l_{k+1}=1}^{N_{k+1}} \mathbf{a}^{\tau(l_{k+1})}_{s_0 s_1 \ldots s_k 2}(t) \cos[l_{k+1}(\theta_{k+1} - \theta^{\tau}_{k+1})] \\
&\quad + \mathbf{a}^{\tau(l_{k+1})}_{s_0 s_1 \ldots s_k 3}(t) \sin[l_{k+1}(\theta_{k+1} - \theta^{\tau}_{k+1})]
\end{aligned} \tag{4.207}$$

and the generalized coordinates are determined by

$$\ddot{\mathbf{a}}_{s_0 s_1 \ldots s_{k+1}} = \mathbf{g}_{s_0 s_1 \ldots s_{k+1}}(\mathbf{a}_{s_0 s_1 \ldots s_{k+1}}, \mathbf{a}^{\tau}_{s_0 s_1 \ldots s_{k+1}}, \dot{\mathbf{a}}_{s_0 s_1 \ldots s_{k+1}}, \dot{\mathbf{a}}^{\tau}_{s_0 s_1 \ldots s_{k+1}}, \mathbf{p}) \tag{4.208}$$

where

$$\begin{aligned}
\mathbf{a}_{s_0s_1\ldots s_{k+1}} &= (\mathbf{a}_{s_0s_1\ldots s_k1}, \mathbf{a}_{s_0s_1\ldots s_k2}, \mathbf{a}_{s_0s_1\ldots s_k3})^{\mathrm{T}},\\
\mathbf{a}^{\tau}_{s_0s_1\ldots s_{k+1}} &= (\mathbf{a}^{\tau}_{s_0s_1\ldots s_k1}, \mathbf{a}^{\tau}_{s_0s_1\ldots s_k2}, \mathbf{a}^{\tau}_{s_0s_1\ldots s_k3})^{\mathrm{T}},\\
\mathbf{g}_{s_0s_1\ldots s_{k+1}} &= (\mathbf{F}_{s_0s_1\ldots s_k1}, -2\omega_{k+1}\mathbf{k}^{(1)}_{k+1}\dot{\mathbf{a}}_{s_0s_1\ldots s_k3} + \omega^2_{k+1}\mathbf{k}^{(2)}_{k+1}\mathbf{a}_{s_0s_1\ldots s_k2} + \mathbf{F}_{s_0s_1\ldots s_k2},\\
&\quad 2\omega_{k+1}\mathbf{k}^{(1)}_{k+1}\dot{\mathbf{a}}_{s_0s_1\ldots s_k2} + \omega^2_{k+1}\mathbf{k}^{(2)}_{k+1}\mathbf{a}_{s_0s_1\ldots s_k3} + \mathbf{F}_{s_0s_1\ldots s_k3})^{\mathrm{T}}
\end{aligned}\tag{4.209}$$

and

$$\begin{aligned}
&\mathbf{k}^{(1)}_{k+1} = diag(\mathbf{I}_{n_k\times n_k}, 2\mathbf{I}_{n_k\times n_k}, \ldots, N_{k+1}\mathbf{I}_{n_k\times n_k}),\\
&\mathbf{k}^{(2)}_{k+1} = diag(\mathbf{I}_{n_k\times n_k}, 2^2\mathbf{I}_{n_k\times n_k}, \ldots, N^2_{k+1}\mathbf{I}_{n_k\times n_k})\\
&n_k = n(2N_0+1)(2N_1+1)\ldots(2N_k+1);\\
&\mathbf{a}_{s_0s_1\ldots s_k1} = \mathbf{a}^{(0)}_{s_0s_1\ldots s_k1},\\
&\mathbf{a}_{s_0s_1\ldots s_k2} = (\mathbf{a}^{(1)}_{s_0s_1\ldots s_k2}, \mathbf{a}^{(2)}_{s_0s_1\ldots s_k2}, \ldots, \mathbf{a}^{(N_{k+1})}_{s_0s_1\ldots s_k2})^{\mathrm{T}},\\
&\mathbf{a}_{s_0s_1\ldots s_k3} = (\mathbf{a}^{(1)}_{s_0s_1\ldots s_k3}, \mathbf{a}^{(2)}_{s_0s_1\ldots s_k3}, \ldots, \mathbf{a}^{(N_{k+1})}_{s_0s_1\ldots s_k3})^{\mathrm{T}};\\
&\mathbf{a}^{\tau}_{s_1s_0\ldots s_k1} = \mathbf{a}^{\tau(0)}_{s_0s_1\ldots s_k1},\\
&\mathbf{a}^{\tau}_{s_0s_1\ldots s_k2} = (\mathbf{a}^{\tau(1)}_{s_0s_1\ldots s_k2}, \mathbf{a}^{\tau(2)}_{s_0s_1\ldots s_k2}, \ldots, \mathbf{a}^{\tau(N_{k+1})}_{s_0s_1\ldots s_k2})^{\mathrm{T}},\\
&\mathbf{a}^{\tau}_{s_0s_1\ldots s_k3} = (\mathbf{a}^{\tau(1)}_{s_0s_1\ldots s_k3}, \mathbf{a}^{\tau(2)}_{s_0s_1\ldots s_k3}, \ldots, \mathbf{a}^{\tau(N_{k+1})}_{s_0s_1\ldots s_k3})^{\mathrm{T}};\\
&\mathbf{F}_{s_0s_1\ldots s_k1} = \mathbf{F}^{(0)}_{s_0s_1\ldots s_k1},\\
&\mathbf{F}_{s_0s_1\ldots s_k2} = (\mathbf{F}^{(1)}_{s_0s_1\ldots s_k2}, \mathbf{F}^{(2)}_{s_0s_1\ldots s_k2}, \ldots, \mathbf{F}^{(N_{k+1})}_{s_0s_1\ldots s_k2})^{\mathrm{T}},\\
&\mathbf{F}_{s_0s_1\ldots s_k3} = (\mathbf{F}^{(1)}_{s_0s_1\ldots s_k3}, \mathbf{F}^{(2)}_{s_0s_1\ldots s_k3}, \ldots, \mathbf{F}^{(N_{k+1})}_{s_0s_1\ldots s_k3})^{\mathrm{T}}\\
&\text{for } N_{k+1} = 1, 2, \ldots, \infty;
\end{aligned}\tag{4.210}$$

and

$$\begin{aligned}
&\mathbf{F}_{s_0s_1\ldots s_k1}(\mathbf{a}_{s_0s_1\ldots s_{k+1}}, \mathbf{a}^{\tau}_{s_0s_1\ldots s_{k+1}}, \dot{\mathbf{a}}_{s_0s_1\ldots s_{k+1}}, \dot{\mathbf{a}}^{\tau}_{s_0s_1\ldots s_{k+1}}, \mathbf{p})\\
&\quad = \frac{1}{2\pi}\int_0^{2\pi} \mathbf{g}_{s_0s_1\ldots s_k}(\mathbf{a}^*_{s_0s_1\ldots s_k}, \mathbf{a}^{\tau*}_{s_0s_1\ldots s_k}, \dot{\mathbf{a}}^*_{s_0s_1\ldots s_k}, \dot{\mathbf{a}}^{\tau*}_{s_0s_1\ldots s_k}, \mathbf{p})d\theta_{k+1};\\
&\mathbf{F}^{(l_{k+1})}_{s_0s_1\ldots s_k2}(\mathbf{a}_{s_0s_1\ldots s_{k+1}}, \mathbf{a}^{\tau}_{s_0s_1\ldots s_{k+1}}, \dot{\mathbf{a}}_{s_0s_1\ldots s_{k+1}}, \dot{\mathbf{a}}^{\tau}_{s_0s_1\ldots s_{k+1}}, \mathbf{p})\\
&\quad = \frac{1}{\pi}\int_0^{2\pi} \mathbf{g}_{s_0s_1\ldots s_k}(\mathbf{a}^*_{s_0s_1\ldots s_k}, \mathbf{a}^{\tau*}_{s_0s_1\ldots s_k}, \dot{\mathbf{a}}^*_{s_0s_1\ldots s_k}, \dot{\mathbf{a}}^{\tau*}_{s_0s_1\ldots s_k}, \mathbf{p})\cos(l_{k+1}\theta_{k+1})d\theta_{k+1},
\end{aligned}$$

$$\begin{aligned}
&\mathbf{F}^{(l_{k+1})}_{s_0 s_1 \ldots s_k 3}(\mathbf{a}_{s_0 s_1 \ldots s_{k+1}}, \mathbf{a}^{\tau}_{s_0 s_1 \ldots s_{k+1}}, \dot{\mathbf{a}}_{s_0 s_1 \ldots s_{k+1}}, \dot{\mathbf{a}}^{\tau}_{s_0 s_1 \ldots s_{k+1}}, \mathbf{p}) \\
&= \frac{1}{\pi}\int_0^{2\pi} \mathbf{g}_{s_0 s_1 \ldots s_k}(\mathbf{a}^{*}_{s_0 s_1 \ldots s_k}, \mathbf{a}^{\tau *}_{s_0 s_1 \ldots s_k}, \dot{\mathbf{a}}^{*}_{s_0 s_1 \ldots s_k}, \dot{\mathbf{a}}^{\tau *}_{s_0 s_1 \ldots s_k}, \mathbf{p}) \sin(l_{k+1}\theta_{k+1}) d\theta_{k+1} \\
&\text{for } l_{k+1} = 1, 2, \ldots, N_{k+1}.
\end{aligned} \tag{4.211}$$

3. *Equation (4.208) becomes*

$$\dot{\mathbf{z}}_{s_0 s_1 \ldots s_{k+1}} = \mathbf{f}_{s_0 s_1 \ldots s_{k+1}}(\mathbf{z}_{s_0 s_1 \ldots s_{k+1}}, \mathbf{z}^{\tau}_{s_0 s_1 \ldots s_{k+1}}) \tag{4.212}$$

where

$$\begin{aligned}
\mathbf{z}_{s_0 s_1 \ldots s_{k+1}} &= (\mathbf{a}_{s_0 s_1 \ldots s_{k+1}}, \dot{\mathbf{a}}_{s_0 s_1 \ldots s_{k+1}})^{\mathrm{T}}, \\
\mathbf{z}^{\tau}_{s_0 s_1 \ldots s_{k+1}} &= (\mathbf{a}^{\tau}_{s_0 s_1 \ldots s_{k+1}}, \dot{\mathbf{a}}^{\tau}_{s_0 s_1 \ldots s_{k+1}})^{\mathrm{T}}, \\
\mathbf{f}_{s_0 s_1 \ldots s_{k+1}} &= (\dot{\mathbf{a}}_{s_0 s_1 \ldots s_{k+1}}, \mathbf{g}_{s_0 s_1 \ldots s_{k+1}})^{\mathrm{T}}.
\end{aligned} \tag{4.213}$$

If equilibrium $\mathbf{z}^{*}_{s_0 s_1 \ldots s_{k+1}} = \mathbf{z}^{\tau *}_{s_0 s_1 \ldots s_{k+1}}$ of Equation (4.212) (i.e., $\mathbf{f}_{s_0 s_1 \ldots s_{k+1}}(\mathbf{z}^{*}_{s_0 s_1 \ldots s_{k+1}}, \mathbf{z}^{\tau *}_{s_0 s_1 \ldots s_{k+1}}) = \mathbf{0}$) exists, then the approximate solution of the periodic motion of kth *generalized coordinates for the period-*m *motion exists as in Equation (4.203). In the vicinity of equilibrium $\mathbf{z}^{*}_{s_0 s_1 \ldots s_{k+1}} = \mathbf{z}^{\tau *}_{s_0 s_1 \ldots s_{k+1}}$, with*

$$\begin{aligned}
\mathbf{z}_{s_0 s_1 \ldots s_{k+1}} &= \mathbf{z}^{*}_{s_0 s_1 \ldots s_{k+1}} + \Delta \mathbf{z}_{s_0 s_1 \ldots s_{k+1}}, \\
\mathbf{z}^{\tau}_{s_0 s_1 \ldots s_{k+1}} &= \mathbf{z}^{\tau *}_{s_0 s_1 \ldots s_{k+1}} + \Delta \mathbf{z}^{\tau}_{s_0 s_1 \ldots s_{k+1}}
\end{aligned} \tag{4.214}$$

the linearized equation of Equation (4.212) is

$$\Delta \dot{\mathbf{z}}_{s_0 s_1 \ldots s_{k+1}} = D_{\mathbf{z}_{s_0 s_1 \ldots s_{k+1}}} \mathbf{f}_{s_0 s_1 \ldots s_{k+1}} \Delta \mathbf{z}_{s_0 s_1 \ldots s_{k+1}} + D_{\mathbf{z}^{\tau}_{s_0 s_1 \ldots s_{k+1}}} \mathbf{f}_{s_0 s_1 \ldots s_{k+1}} \Delta \mathbf{z}^{\tau}_{s_0 s_1 \ldots s_{k+1}} \tag{4.215}$$

and the eigenvalue analysis of equilibrium $\mathbf{z}^{*}$ is given by

$$|D_{\mathbf{z}_{s_0 s_1 \ldots s_{k+1}}} \mathbf{f}_{s_0 s_1 \ldots s_{k+1}} + e^{\lambda \tau} D_{\mathbf{z}^{\tau}_{s_0 s_1 \ldots s_{k+1}}} \mathbf{f}_{s_0 s_1 \ldots s_{k+1}} - \lambda \mathbf{I}_{2n_k(2N_{k+1}+1) \times 2n_k(2N_{k+1}+1)}| = 0 \tag{4.216}$$

where

$$\begin{aligned}
D_{\mathbf{z}_{s_0 s_1 \ldots s_{k+1}}} \mathbf{f}_{s_0 s_1 \ldots s_{k+1}} &= \left. \frac{\partial \mathbf{f}_{s_0 s_1 \ldots s_{k+1}}\left(\mathbf{z}_{s_0 s_1 \ldots s_{k+1}}, \mathbf{z}^{\tau}_{s_0 s_1 \ldots s_{k+1}}\right)}{\partial \mathbf{z}_{s_0 s_1 \ldots s_{k+1}}} \right|_{(\mathbf{z}^{*}_{s_0 s_1 \ldots s_{k+1}}, \mathbf{z}^{\tau *}_{s_0 s_1 \ldots s_{k+1}})}, \\
D_{\mathbf{z}^{\tau}_{s_0 s_1 \ldots s_{k+1}}} \mathbf{f}_{s_0 s_1 \ldots s_{k+1}} &= \left. \frac{\partial \mathbf{f}_{s_0 s_1 \ldots s_{k+1}}\left(\mathbf{z}_{s_0 s_1 \ldots s_{k+1}}, \mathbf{z}^{\tau}_{s_0 s_1 \ldots s_{k+1}}\right)}{\partial \mathbf{z}_{s_0 s_1 \ldots s_{k+1}}} \right|_{(\mathbf{z}^{*}_{s_0 s_1 \ldots s_{k+1}}, \mathbf{z}^{\tau *}_{s_0 s_1 \ldots s_{k+1}})}.
\end{aligned} \tag{4.217}$$

The stability and bifurcation of such a periodic flow of the kth *generalized coordinates can be classified by the eigenvalues of Equation (4.215) at equilibrium $\mathbf{z}^{*}_{s_0 s_1 \ldots s_{k+1}} = \mathbf{z}^{\tau *}_{s_0 s_1 \ldots s_{k+1}}$ with*

$$(n_1, n_2, n_3 \,|\, n_4, n_5, n_6). \tag{4.218}$$

a. *If all eigenvalues of the equilibrium possess negative real parts, the approximate quasi-periodic solution is stable.*
b. *If at least one of the eigenvalues of the equilibrium possesses a positive real part, the approximate quasi-periodic solution is unstable.*
c. *The boundaries between stable and unstable equilibriums with higher order singularity give bifurcation and stability conditions with higher order singularity.*

4. *For the* kth *order Hopf bifurcation of period-*m *motion, a relation exists as*

$$p_k \omega_k = \omega_{k-1}. \tag{4.219}$$

a. *If p_k is an irrational number, the* kth*-order Hopf bifurcation of the period-*m *motion is called the quasi-period-p_k Hopf bifurcation, and the corresponding solution of the* kth *generalized coordinates is p_k-quasi-periodic to the system of the* $(k-1)$th *generalized coordinates.*
b. *If $p_k = 2$, the* kth*-order Hopf bifurcation of the period-*m *motion is called a period-doubling Hopf bifurcation (or a period-2 Hopf bifurcation), and the corresponding solution of the* kth *generalized coordinates is period-doubling to the system of the* $(k-1)$th *generalized coordinates.*
c. *If $p_k = q$ with an integer q, the* kth*-order Hopf bifurcation of the period-*m *motion is called a period-q Hopf bifurcation, and the corresponding solution of the* kth *generalized coordinates is of q-times period to the system of the* $(k-1)$th *generalized coordinates.*
d. *If $p_k = p/q$ (p, q are irreducible integers), the* kth*-order Hopf bifurcation of the period-*m *motion is called a period-p/q Hopf bifurcation, and the corresponding solution of the* kth *generalized coordinates is p/q-periodic to the system of the* $(k-1)$th *generalized coordinates.*

Proof. The proof of this theorem is similar to the proof of Theorem 3.16. ■

5

Quadratic Nonlinear Oscillators

In this chapter, analytical solutions for period-m motions in a periodically forced, quadratic nonlinear oscillator will be presented through the Fourier series solutions with finite harmonic terms, and the stability and bifurcation analyses of the corresponding period-1 motions will be carried out. There are many period-1 motions in such a nonlinear oscillator, and the parameter map for excitation amplitude and frequency will be developed for different period-1 motions. For each period-1 motion branch, analytical bifurcation trees of period-1 motions to chaos will be presented. For a better understanding of complex period-m motions in such a quadratic nonlinear oscillator, trajectories, and amplitude spectrums will be illustrated numerically.

5.1 Period-1 Motions

In this section, period-1 motions in a periodically forced, quadratic nonlinear oscillator will be discussed. The analytical solutions with only two harmonic terms in the Fourier series expressions will be discussed first as an introduction. The appropriate analytical solutions will be presented with finite harmonic terms based on the prescribed accuracy of harmonic amplitudes. From appropriate solutions, the analytical bifurcation trees for period-1 motions to chaos can be found. Infinite, countable period-1 motions that exist in such an oscillator will be presented, and the corresponding parameter maps will be presented, and complex period-1 motions will be illustrated.

5.1.1 Analytical Solutions

Consider a periodically forced, nonlinear oscillator

$$\ddot{x} + \delta\dot{x} + \alpha x + \beta x^2 = Q_0 \cos \Omega t \tag{5.1}$$

where δ is the linear damping coefficient. α and β are linear and quadratic spring coefficients, respectively. Q_0 and Ω are excitation amplitude and frequency, respectively. In Luo (2012a), the standard form of Equation (5.1) can be written as

$$\ddot{x} = F(x, \dot{x}, t) \tag{5.2}$$

where

$$F(\dot{x}, x, t) = -\delta\dot{x} - \alpha x - \beta x^2 + Q_0 \cos \Omega t. \tag{5.3}$$

Toward Analytical Chaos in Nonlinear Systems, First Edition. Albert C. J. Luo.
© 2014 John Wiley & Sons, Ltd. Published 2014 by John Wiley & Sons, Ltd.

The analytical solution of period-1 motion for the above equation is

$$x^*(t) = a_0(t) + \sum_{k=1}^{N} b_k(t)\cos(k\theta) + c_k(t)\sin(k\theta) \tag{5.4}$$

where $a_0(t)$, $b_k(t)$ and $c_k(t)$ vary with time and $\theta = \Omega t$. The first and second order of derivatives of $x^*(t)$ are

$$\dot{x}^*(t) = \dot{a}_0(t) + \sum_{k=1}^{N}(\dot{b}_k + k\Omega c_k)\cos(k\theta) + (\dot{c}_k - k\Omega b_k)\sin(k\theta), \tag{5.5}$$

$$\begin{aligned}\ddot{x}^*(t) = \ddot{a}_0(t) + \sum_{k=1}^{N}[\ddot{b}_k + 2(k\Omega)\dot{c}_k - (k\Omega)^2 b_k]\cos(k\theta)\\ + [\ddot{c}_k - 2(k\Omega)\dot{b}_k - (k\Omega)^2 c_k]\sin(k\theta).\end{aligned} \tag{5.6}$$

Substitution of Equations (5.4)–(5.6) into Equation (5.1) and application of the virtual work principle for a basis of constant, $\cos(k\theta)$ and $\sin(k\theta)$ ($k = 1, 2, \ldots$) as a set of virtual displacements gives

$$\begin{aligned}\ddot{a}_0 &= F_0(a_0, \mathbf{b}, \mathbf{c}, \dot{a}_0, \dot{\mathbf{b}}, \dot{\mathbf{c}}),\\ \ddot{b}_k + 2(k\Omega)\dot{c}_k - (k\Omega)^2 b_k &= F_{1k}(a_0, \mathbf{b}, \mathbf{c}, \dot{a}_0, \dot{\mathbf{b}}, \dot{\mathbf{c}}),\\ \ddot{c}_k - 2(k\Omega)\dot{b}_k - (k\Omega)^2 c_k &= F_{2k}(a_0, \mathbf{b}, \mathbf{c}, \dot{a}_0, \dot{\mathbf{b}}, \dot{\mathbf{c}})\\ k &= 1, 2, \ldots, N\end{aligned} \tag{5.7}$$

where

$$\begin{aligned}F_0(a_0, \mathbf{b}, \mathbf{c}, \dot{a}_0, \dot{\mathbf{b}}, \dot{\mathbf{c}}) &= \frac{1}{T}\int_0^T F(x^*, \dot{x}^*, t)dt\\ &= -\delta\dot{a}_0 - \alpha a_0 - \beta a_0{}^2 - \frac{\beta}{2}\sum_{l=1}^{N}(b_l{}^2 + c_l{}^2),\\ F_{1k}(a_0, \mathbf{b}, \mathbf{c}, \dot{a}_0, \dot{\mathbf{b}}, \dot{\mathbf{c}}) &= \frac{2}{T}\int_0^T F(x^*, \dot{x}^*, t)\cos(k\Omega t)dt\\ &= -\delta(\dot{b}_k + c_k k\Omega) - \alpha b_k - 2\beta a_0 b_k + f_{1k} + Q_0\delta_k^1,\\ F_{2k}(a_0, \mathbf{b}, \mathbf{c}, \dot{a}_0, \dot{\mathbf{b}}, \dot{\mathbf{c}}) &= \frac{2}{T}\int_0^T F(x^*, \dot{x}^*, t)\sin(k\Omega t)dt\\ &= -\delta(\dot{c}_k - b_k k\Omega) - \alpha c_k - 2\beta a_0 c_k + f_{2k}\end{aligned} \tag{5.8}$$

and

$$\begin{aligned}f_{1k} &= -\beta\sum_{l=1}^{N}\sum_{j=1}^{N}[(b_l b_j + c_l c_j)\delta_{j-l}^k + \frac{1}{2}(b_l b_j - c_l c_j)\delta_{j+l}^k],\\ f_{2k} &= -\beta\sum_{l=1}^{N}\sum_{j=1}^{N} b_l c_j(\delta_{j+l}^k + \delta_{j-l}^k - \delta_{l-j}^k).\end{aligned} \tag{5.9}$$

Define

$$\begin{aligned}\mathbf{z} &\triangleq (a_0, \mathbf{b}^{\mathrm{T}}, \mathbf{c}^{\mathrm{T}})^{\mathrm{T}} \\ &= (a_0, b_1, \ldots, b_N, c_1, \ldots, c_N)^{\mathrm{T}} \equiv (z_0, z_1, \ldots, z_{2N})^{\mathrm{T}}, \\ \mathbf{z}_1 &= \dot{\mathbf{z}} = (\dot{a}_0, \dot{\mathbf{b}}^{\mathrm{T}}, \dot{\mathbf{c}}^{\mathrm{T}})^{\mathrm{T}} \\ &= (\dot{a}_0, \dot{b}_1, \ldots, \dot{b}_N, \dot{c}_1, \ldots, \dot{c}_N)^{\mathrm{T}} \equiv (\dot{z}_0, \dot{z}_1, \ldots, \dot{z}_{2N})^{\mathrm{T}}\end{aligned} \tag{5.10}$$

where

$$\mathbf{b} = (b_1, b_2, \ldots, b_N)^{\mathrm{T}} \text{ and } \mathbf{c} = (c_1, c_2, \ldots, c_N)^{\mathrm{T}}. \tag{5.11}$$

Equation (5.7) can be expressed in the form of vector field as

$$\dot{\mathbf{z}} = \mathbf{z}_1 \text{ and } \dot{\mathbf{z}}_1 = \mathbf{g}(\mathbf{z}, \mathbf{z}_1) \tag{5.12}$$

where

$$\mathbf{g}(\mathbf{z}, \mathbf{z}_1) = \begin{pmatrix} F_0(\mathbf{z}, \mathbf{z}_1) \\ \mathbf{F}_1(\mathbf{z}, \mathbf{z}_1) - 2\mathbf{k}_1 \Omega \dot{\mathbf{c}} + \mathbf{k}_2 \Omega^2 \mathbf{b} \\ \mathbf{F}_2(\mathbf{z}, \mathbf{z}_1) + 2\mathbf{k}_1 \Omega \dot{\mathbf{b}} + \mathbf{k}_2 \Omega^2 \mathbf{c} \end{pmatrix} \tag{5.13}$$

and

$$\begin{aligned}\mathbf{k}_1 &= diag(1, 2, \ldots, N), \\ \mathbf{k}_2 &= diag(1, 2^2, \ldots, N^2), \\ \mathbf{F}_1 &= (F_{11}, F_{12}, \ldots, F_{1N})^{\mathrm{T}}, \\ \mathbf{F}_2 &= (F_{21}, F_{22}, \ldots, F_{2N})^{\mathrm{T}} \\ &\text{for } N = 1, 2, \ldots, \infty.\end{aligned} \tag{5.14}$$

Introducing

$$\mathbf{y} \equiv (\mathbf{z}, \mathbf{z}_1) \text{ and } \mathbf{f} = (\mathbf{z}_1, \mathbf{g})^{\mathrm{T}}, \tag{5.15}$$

Equation (5.12) becomes

$$\dot{\mathbf{y}} = \mathbf{f}(\mathbf{y}). \tag{5.16}$$

The steady-state solutions for periodic motion in Equation (5.1) can be obtained by setting $\dot{\mathbf{y}} = \mathbf{0}$, that is,

$$\begin{aligned}F_0(a_0^*, \mathbf{b}^*, \mathbf{c}^*, 0, \mathbf{0}, \mathbf{0}) &= 0, \\ \mathbf{F}_1(a_0^*, \mathbf{b}^*, \mathbf{c}^*, 0, \mathbf{0}, \mathbf{0}) + \Omega^2 \mathbf{k}_2 \mathbf{b}^* &= \mathbf{0}, \\ \mathbf{F}_2(a_0^*, \mathbf{b}^*, \mathbf{c}^*, 0, \mathbf{0}, \mathbf{0}) + \Omega^2 \mathbf{k}_2 \mathbf{c}^* &= \mathbf{0}.\end{aligned} \tag{5.17}$$

The $(2N + 1)$ nonlinear equations in Equation (5.17) are solved by the Newton-Raphson method. In Luo (2012a), the linearized equation at the equilibrium point $\mathbf{y}^* = (\mathbf{z}^*, \mathbf{0})^{\mathrm{T}}$ is given by

$$\Delta \dot{\mathbf{y}} = D\mathbf{f}(\mathbf{y}^*) \Delta \mathbf{y} \tag{5.18}$$

where

$$D\mathbf{f}(\mathbf{y}^*) = \partial \mathbf{f}(\mathbf{y}) / \partial \mathbf{y}|_{\mathbf{y}^*} = \begin{bmatrix} \mathbf{0}_{(2N+1)\times(2N+1)} & \mathbf{I}_{(2N+1)\times(2N+1)} \\ \mathbf{G}_{(2N+1)\times(2N+1)} & \mathbf{H}_{(2N+1)\times(2N+1)} \end{bmatrix} \tag{5.19}$$

and

$$\mathbf{G} = \frac{\partial \mathbf{g}}{\partial \mathbf{z}} = (\mathbf{G}^{(0)}, \mathbf{G}^{(c)}, \mathbf{G}^{(s)})^{\mathrm{T}}; \tag{5.20}$$

$$\begin{aligned}
\mathbf{G}^{(0)} &= (G_0^{(0)}, G_1^{(0)}, \ldots, G_{2N}^{(0)}),\\
\mathbf{G}^{(c)} &= (\mathbf{G}_1^{(c)}, \mathbf{G}_2^{(c)}, \ldots, \mathbf{G}_N^{(c)})^{\mathrm{T}},\\
\mathbf{G}^{(s)} &= (\mathbf{G}_1^{(s)}, \mathbf{G}_2^{(s)}, \ldots, \mathbf{G}_N^{(s)})^{\mathrm{T}}
\end{aligned} \tag{5.21}$$

for $N = 1, 2, \ldots \infty$ with

$$\begin{aligned}
\mathbf{G}_k^{(c)} &= (G_{k0}^{(c)}, G_{k1}^{(c)}, \ldots, G_{k(2N)}^{(c)}),\\
\mathbf{G}_k^{(s)} &= (G_{k0}^{(s)}, G_{k1}^{(s)}, \ldots, G_{k(2N)}^{(s)})
\end{aligned} \tag{5.22}$$

for $k = 1, 2, \ldots N$. The corresponding components are

$$\begin{aligned}
G_r^{(0)} &= -\alpha\delta_0^r - \beta g_{2r}^{(0)},\\
G_{kr}^{(c)} &= (k\Omega)^2\delta_k^r - \alpha\delta_k^r - \delta k\Omega\delta_{k+N}^r - \beta g_{2kr}^{(c)},\\
G_{kr}^{(s)} &= (k\Omega)^2\delta_{k+N}^r + \delta k\Omega\delta_k^r - \alpha\delta_{k+N}^r - \beta g_{2kr}^{(s)}
\end{aligned} \tag{5.23}$$

where

$$g_{2r}^{(0)} = 2a_0\delta_0^r + b_k\delta_k^r + c_k\delta_{k+N}^r, \tag{5.24}$$

$$\begin{aligned}
g_{2kr}^{(c)} = {} & 2b_k\delta_r^0 + 2a_0\delta_k^r + \sum_{i=1}^{N}\sum_{j=1}^{N}[b_j(\delta_{j-i}^k + \delta_{i-j}^k + \delta_{i+j}^k)\delta_i^r\\
& + c_j(\delta_{j-i}^k + \delta_{i-j}^k - \delta_{i+j}^k)\delta_{i+N}^r],
\end{aligned} \tag{5.25}$$

$$\begin{aligned}
g_{2kr}^{(s)} = {} & 2c_k\delta_0^r + 2a_0\delta_{k+N}^r + \sum_{i=1}^{N}\sum_{j=1}^{N}[c_j(\delta_{i+j}^k + \delta_{j-i}^k - \delta_{i-j}^k)\delta_i^r\\
& + b_i(\delta_{i+j}^k + \delta_{j-i}^k - \delta_{i-j}^k)\delta_{j+N}^r]
\end{aligned} \tag{5.26}$$

for $r = 0, 1, \ldots 2N$.

$$\mathbf{H} = \frac{\partial \mathbf{g}}{\partial \mathbf{z}_1} = (\mathbf{H}^{(0)}, \mathbf{H}^{(c)}, \mathbf{H}^{(s)})^{\mathrm{T}} \tag{5.27}$$

where

$$\begin{aligned}
\mathbf{H}^{(0)} &= (H_0^{(0)}, H_1^{(0)}, \ldots, H_{2N}^{(0)}),\\
\mathbf{H}^{(c)} &= (\mathbf{H}_1^{(c)}, \mathbf{H}_2^{(c)}, \ldots, \mathbf{H}_N^{(c)})^{\mathrm{T}},\\
\mathbf{H}^{(s)} &= (\mathbf{H}_1^{(s)}, \mathbf{H}_2^{(s)}, \ldots, \mathbf{H}_N^{(s)})^{\mathrm{T}}
\end{aligned} \tag{5.28}$$

for $N = 1, 2, \ldots \infty$, with

$$\begin{aligned}
\mathbf{H}_k^{(c)} &= (H_{k0}^{(c)}, H_{k1}^{(c)}, \ldots, H_{k(2N)}^{(c)}),\\
\mathbf{H}_k^{(s)} &= (H_{k0}^{(s)}, H_{k1}^{(s)}, \ldots, H_{k(2N)}^{(s)})
\end{aligned} \tag{5.29}$$

for $k = 1, 2, \ldots N$. The corresponding components are

$$
\begin{aligned}
H_r^{(0)} &= -\delta\delta_0^r, \\
H_{kr}^{(c)} &= -2k\Omega\delta_{k+N}^r - \delta\delta_k^r, \\
H_{kr}^{(s)} &= 2k\Omega\delta_k^r - \delta\delta_{k+N}^r
\end{aligned} \tag{5.30}
$$

for $r = 0, 1, \ldots, 2N$.

The corresponding eigenvalues are determined by

$$
|D\mathbf{f}(\mathbf{y}^*) - \lambda\mathbf{I}_{2(2N+1)\times 2(2N+1)}| = 0. \tag{5.31}
$$

From Luo (2012a), the eigenvalues of $D\mathbf{f}(\mathbf{y}^*)$ are classified as

$$
(n_1, n_2, n_3 | n_4, n_5, n_6) \tag{5.32}
$$

where n_1 is the total number of negative real eigenvalues, n_2 is the total number of positive real eigenvalues, n_3 is the total number of negative zero eigenvalues; n_4 is the total pair number of complex eigenvalues with negative real parts, n_5 is the total pair number of complex eigenvalues with positive real parts, n_6 is the total pair number of complex eigenvalues with zero real parts. If $\mathrm{Re}(\lambda_k) < 0$ $(k = 1, 2, \ldots, 2(2N + 1))$, the approximate steady-state solution $\mathbf{y}^*$ with truncation of $\cos(N\Omega t)$ and $\sin(N\Omega t)$ is stable. If $\mathrm{Re}(\lambda_k) > 0$ $(k \in \{1, 2, \ldots, 2(2N + 1)\})$, the truncated approximate steady-state solution is unstable. The corresponding boundary between the stable and unstable solution is given by the saddle-node bifurcation (SN) and Hopf bifurcation (HB).

5.1.2 Frequency-Amplitude Characteristics

The exact steady-state solutions of periodic motions in the nonlinear oscillator should be obtained through the infinite harmonic terms. Unfortunately, it is impossible to compute the exact solution of periodic motions in such an oscillator. Thus, one uses the truncated solutions to obtain the approximate solutions of the nonlinear oscillator with enough precision $(A_N \le \varepsilon)$ where the number N is the total number of harmonic terms used in the approximate solution and ε is a prescribed precision (i.e., $\varepsilon = 10^{-8}$). If more terms are used in the Fourier series solution of periodic motions, the better prediction of the periodic motions can be obtained. However, the computational workload will dramatically increase. It is very important that the suitable precision ε is selected. The eigenvalue analysis of such approximate, analytical solutions can be done through dynamics of time-varying coefficients in the Fourier series expression of periodic motion, and the stability and bifurcation analysis can be completed. The equilibrium solution of Equation (5.12) can be obtained from Equation (5.17) by using the Newton-Raphson method, and the stability analysis will be discussed.

The harmonic amplitudes varying with excitation frequency Ω is illustrated. The harmonic amplitude and phase are defined by

$$
A_k \equiv \sqrt{b_k^2 + c_k^2} \text{ and } \varphi_k = \arctan\frac{c_k}{b_k}. \tag{5.33}
$$

The corresponding solution in Equation (5.4) becomes

$$
x^*(t) = a_0 + \sum_{k=1}^{N} A_k \cos(k\Omega t - \varphi_k). \tag{5.34}
$$

Consider system parameters as

$$\delta = 0.05,\ \alpha = 10.0,\ \beta = 5.0, Q_0 = 4.5 \tag{5.35}$$

Without losing generality, as in Luo and Yu (2013a), the analytical approximate solutions for periodic motion based on two harmonic terms (HB2) are presented first. The constant term a_0, the first and second harmonic amplitudes A_1 and A_2 are presented in Figure 5.1(a)–(c), respectively. In Figure 5.1, the stable and unstable solutions of period-1 motion for the quadratic nonlinear oscillator with periodic excitation are predicted analytically. The period-1 motion possesses four branches of solutions for different frequency ranges. The stability and bifurcation analysis are completed. For the HB2 analytical solutions, eigenvalue analysis provides the possible conditions of stability and bifurcation. The acronyms "HB", "SN," and "UHB" are used to represent the Hopf bifurcation, saddle-node bifurcation, and unstable Hopf bifurcation, respectively. Solid and dashed curves represent the stable and unstable period-1 motions, respectively. The corresponding phase angles versus excitation frequency are presented in Figure 5.2(a)–(c). The corresponding stability and bifurcation points are labeled. To consider effects of excitation amplitude, $Q_0 = 1.5, 2.5, 3.5, 4.0, 5.0$ are employed to show the behavior of harmonic frequency-amplitude curves. The stability ranges for period-1 motions can be observed clearly in Figure 5.3(a)–(c). The red dashed curves and the black solid curves are unstable and stable periodic solutions. The arrow direction represents how the curves change with excitation frequency. In Figure 5.3(b), for the second harmonic term, $A_2 < 10^{-3}$ for $\Omega > 6$. For $\Omega < 6$, more harmonic terms should be considered to get a good prediction of period-1 motion. Thus, the 30 harmonic terms (HB30) for period-1 motion will be considered.

As in Luo and Yu (2013b), the analytical prediction of period-1 motions based on 30 harmonic terms (HB30) are presented in Figure 5.4(i)–(vi). In Figure 5.4(i), the constant term of period-1 motion is presented. For $\Omega < 2.0$, there are many branches of period-1 motions and bifurcations, and they have similar structures. The unstable Hopf bifurcation is observed, which is also called the subcritical Hopf bifurcation. The stable Hopf bifurcation is also called the supercritical Hopf bifurcation. For $\Omega > 1.0$, the curves of constant terms varying excitation frequency is very simple. In Figure 5.4(ii), the harmonic amplitude A_1 versus excitation frequency Ω is presented, and the frequency-amplitude curves for $\Omega > 1.0$ are very clearly presented. However, for $\Omega < 2.0$, many branches of period-1 solutions are crowded, but the similar structures for each branch of period-1 motion are observed. In Figure 5.4(iii), the harmonic amplitude A_2 is presented. For $\Omega > 5.0$, $A_2 < 3 \times 10^{-3}$ from the zoomed window and for $\Omega \in (2.0, 5.0)$, $A_2 \in (10^{-4}, 0.3)$ is observed. For $\Omega \in (0, 2.0)$, $A_2 \sim 10^0$ with many branches of period-1 motion and they become more crowded. The aforementioned three plots are based on a linear scale system. Once the amplitude quantity level changes with the power laws, it is very difficult to present the changes of amplitude with excitation frequency. For $\Omega \in (0, 2.0)$, the harmonic amplitudes in the backbone curves of period-1 motions are more crowded. Thus, the common logarithm scale is used to plot the harmonic amplitude. The overview of amplitude quantity levels can clearly be observed. In Figure 5.4(iv), the harmonic amplitude A_3 is plotted through the common logarithm because quantity level changes are too big for $\Omega \in (0.0, 7.0)$. The branches of period-1 motions in $\Omega \in (0, 2.0)$ are obviously presented and the zoomed window for $\Omega \in (2.0, 7.0)$ shows that the harmonic amplitude quantity level change is very clear with $A_3 \in (10^{-7}, 10^{-2})$. Similarly, the harmonic amplitude A_4 varying with excitation frequency is presented in Figure 5.4(v), and for $\Omega \in (2.0, 7.0)$ the harmonic amplitude A_4 lies in the range of $A_4 \in (10^{-10}, 10^{-3})$. For $\Omega < 2$, the harmonic

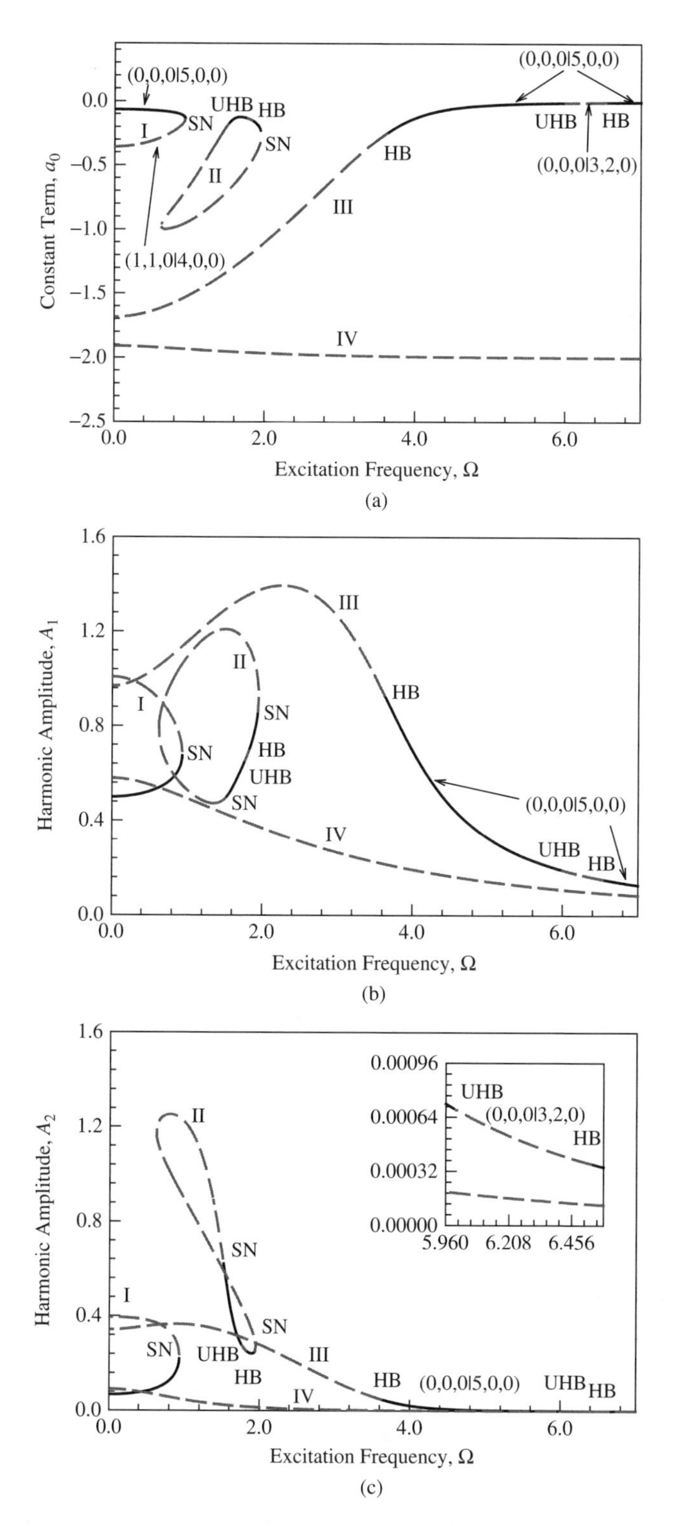

Figure 5.1 Analytical prediction of period-1 motions from two harmonic terms (HB2): (a) a_0, (b) A_1, and (c) A_2. Parameters: ($\delta = 0.05, \alpha = 10.0, \beta = 5.0, Q_0 = 4.5$)

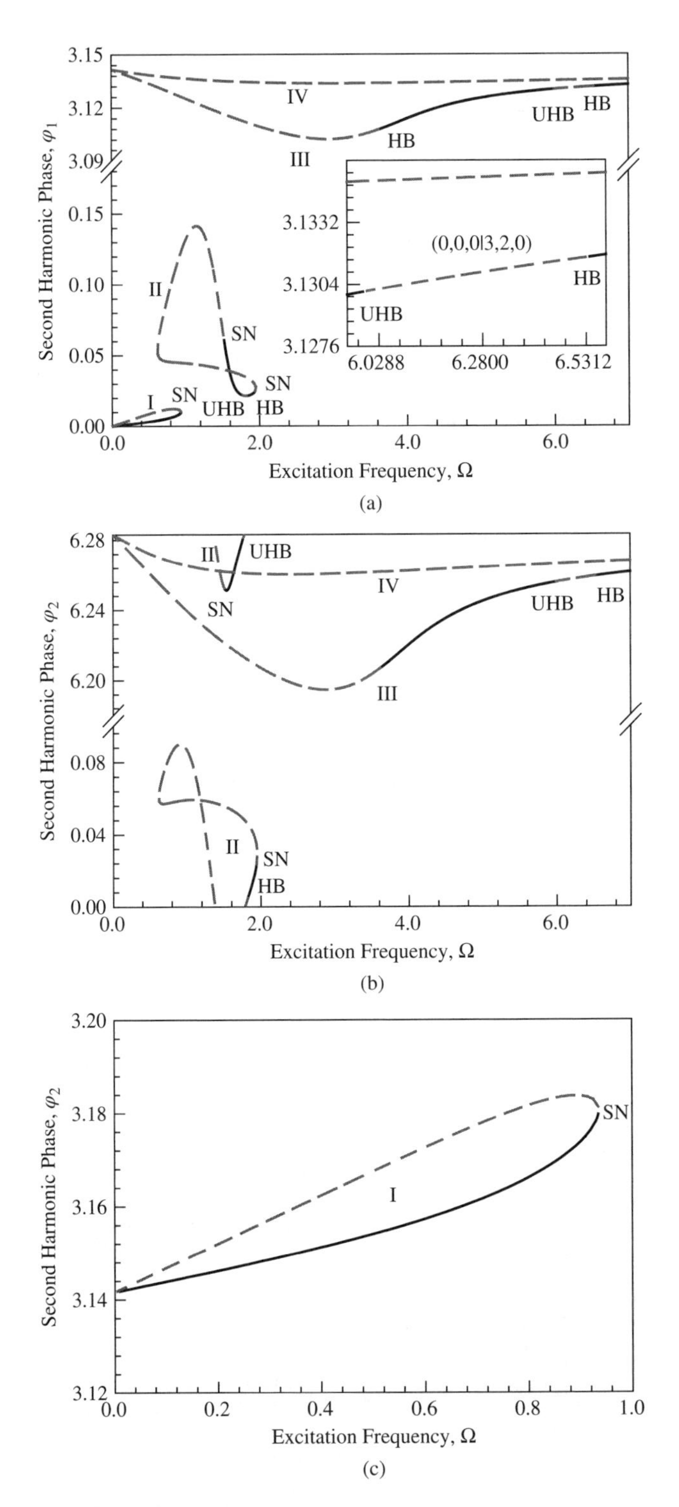

Figure 5.2 Analytical prediction of period-1 motions from two harmonic terms (HB2): (a) φ_1, (b) φ_2, and (c) a zoomed view of φ_2. Parameters: ($\delta = 0.05, \alpha = 10.0, \beta = 5.0, Q_0 = 4.5$)

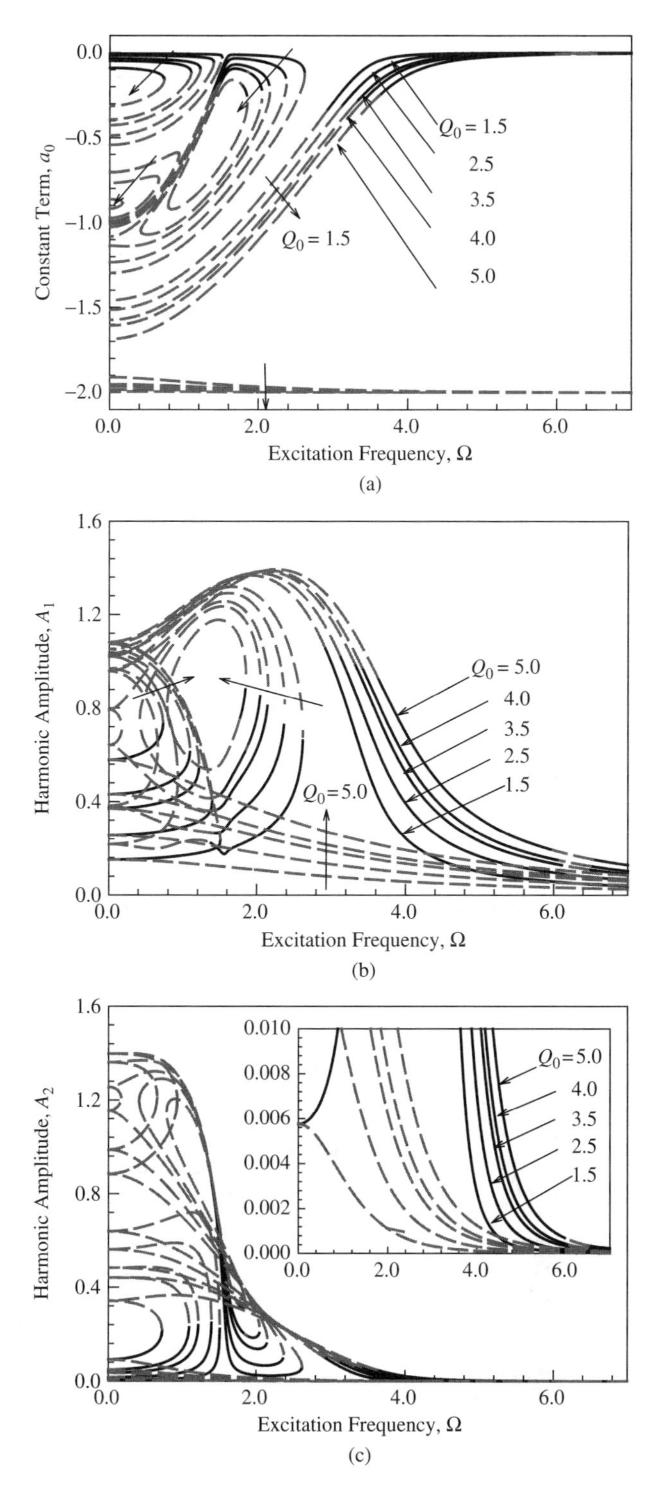

Figure 5.3 Excitation amplitudes effects on periodic motions from two harmonic terms (HB2) for $Q_0 = 1.5, 2.5, 3.5, 4.0, 5.0$: (a) a_0, (b) A_1, and (c) A_2. Parameters: ($\delta = 0.05, \alpha = 10.0, \beta = 5.0, Q_0 = 4.5$)

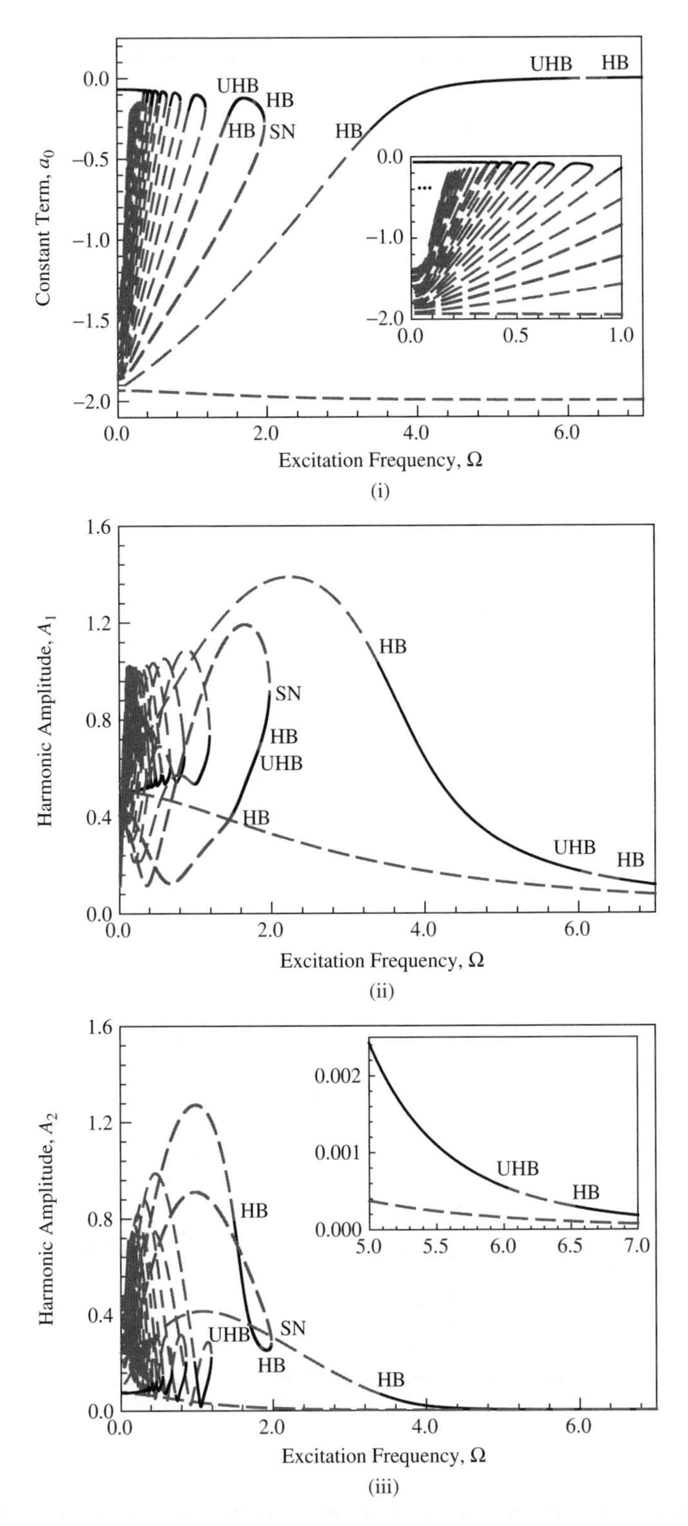

Figure 5.4 Analytical prediction of period-1 motions based on 30 harmonic terms (HB30): (i) a_0 and (ii)–(vi) A_k ($k = 1, 2, \ldots, 4, 30$). Parameters: ($\delta = 0.05, \alpha = 10.0, \beta = 5.0, Q_0 = 4.5$)

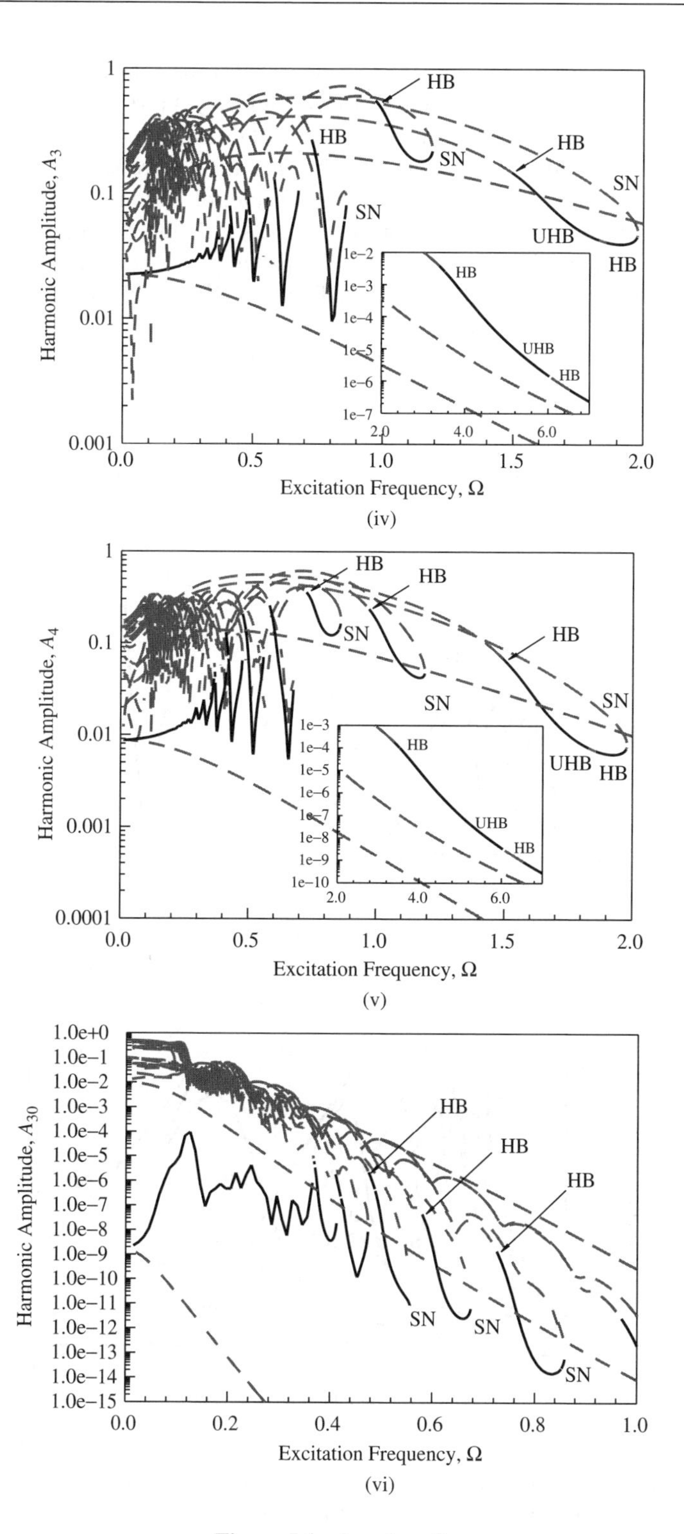

Figure 5.4 *(continued)*

amplitude A_4 is $A_4 \in (10^{-4}, 10^0)$. To avoid abundent illustrions, the harmonic amplitude A_{30} for $\Omega \in (0.0, 1.0)$ are presented in Figure 5.4(vi) with $A_{20} \in (10^{-15}, 10^0)$ and for $\Omega > 1.0$, the harmonic amplitude $A_{30} < 10^{-15}$ can be ignored. For excitation frequency close to zero, more harmonic terms should be included for the analytical expression in the Fourier series solution of period-1 motion. To obtain the entire picture of period-1 motion, the parameter map (Ω, Q_0) is presented in Figure 5.5(a),(b). For $\Omega < 2.5$, the parameter map is zoomed, which shows the similar bifurcation and stability patterns. The solid and dashed curves are for stable and unstable Hopf bifurcations (HB and UHB), respectively. The dash-dotted curve is for the saddle-node bifurcation (SN). The notation $\mathrm{S}^m\mathrm{U}^n$ $(m, n = 0, 1, 2, \ldots)$ represents m stable solutions and n unstable solutions for period-1 motions in the corresponding region.

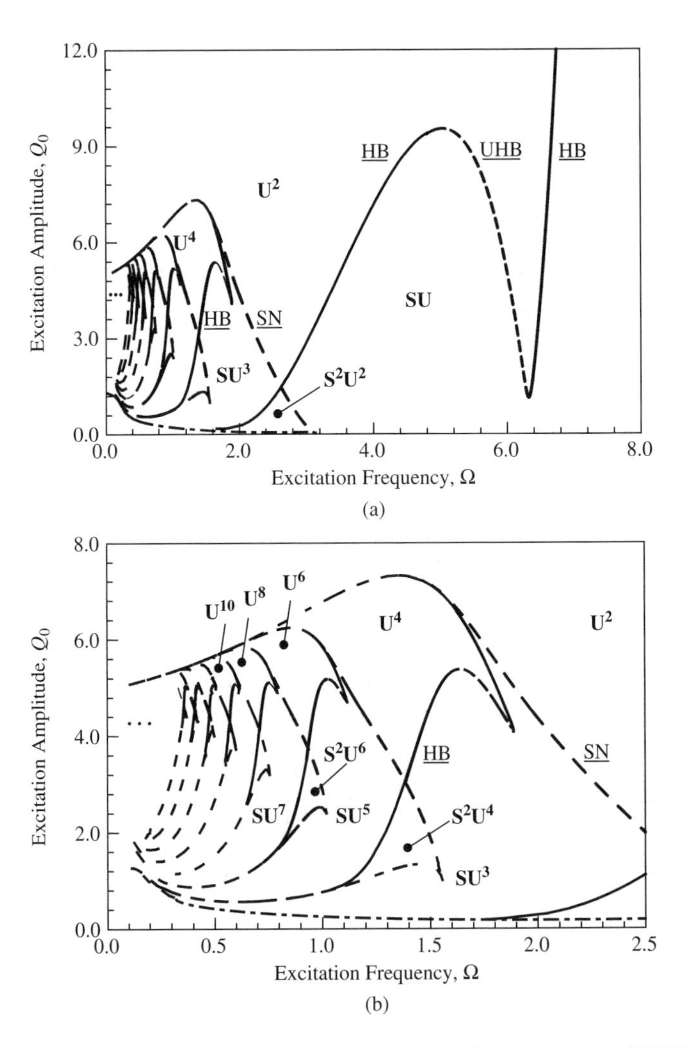

Figure 5.5 A parameter map for period-1 motions based on 30 harmonic terms (HB30): (a) overview and (b) zoomed view. Parameters: $(\delta = 0.05, \alpha = 10.0, \beta = 5.0)$

5.1.3 Numerical Illustrations

To verify the approximate analytical solutions of periodic motion in such a quadratic nonlinear oscillator, numerical simulations are carried out by the midpoint scheme. The initial conditions for numerical simulation are computed from the approximate analytical solutions. The numerical results are depicted by solid curves, but the analytical solutions are given by red circular symbols. The big filled circular symbol is the initial condition.

The displacement, velocity, trajectory, and amplitude spectrum of stable period-1 motion are presented in Figure 5.6 for $\Omega = 5.8$ with initial conditions ($x_0 \approx -0.197945, \dot{x}_0 \approx 0.013202$). This analytical solution is based on four harmonic terms (HB4) in the Fourier series solution of period-1 motion. In Figure 5.6(a),(b), for over 40 periods, the analytical and numerical solutions match very well. In Figure 5.6(c), analytical and numerical trajectories match very well. In Figure 5.6(d), the amplitude spectrum versus the harmonic order is presented. $a_0 \approx -9.032271\text{e-}3$, $A_1 \approx 0.189645$, $A_2 \approx 7.213358\text{e-}4$, $A_3 \approx 2.335642\text{e-}6$, $A_4 \approx 6.654057\text{e-}9$. The second harmonic term ($A_2 \sim 10^{-3}$) and higher order harmonic terms are very small and ignorable. Such an approximate solution with one harmonic term can be easily obtained even if the perturbation method or traditional harmonic balance is used.

Since there are many branches of period-1 motions, only the trajectory and amplitude spectrums are plotted to look into the regularity and complexity of period-1 motions. The input data for numerical simulations are presented in Table 5.1. In Figure 5.7(i)–(vi), the trajectories and harmonic amplitude spectrums of period-1 motions are presented for $\Omega = 3.6, 1.101, 0.98$. In Figure 5.7(i) and (ii), the seven harmonic terms (HB7) are used in the analytical solution of period-1 motion for $\Omega = 3.6$. The analytical and numerical simulations match very well with a cycle. For this periodic motion, the second order harmonic term becomes more important except for the first harmonic term. The harmonic amplitude distributions are $a_0 = -0.2280$, $A_1 \sim 10^0$, $A_2 \sim 5 \times 10^{-2}$, $A_3 \sim 2 \times 10^{-3}$, $A_4 \sim 7 \times 10^{-5}$, $A_5 \sim 2 \times 10^{-6}$, $A_6 \sim 8 \times 10^{-8}$, and $A_7 \sim 2.5 \times 10^{-9}$. The second and third harmonic terms will have relatively important contributions on the analytical solution of such a period-1 motion. In Figure 5.7(iii) and (iv), the 14 harmonic terms (HB14) are employed in the analytical solution of period-1 motion for $\Omega = 1.101$. The analytical and numerical simulations match very well with two cycles, which is on the second solution branch of period-1 motion. The harmonic amplitude distributions are $a_0 \approx -0.106346, A_1 \sim 6 \times 10^{-1}, A_2 \sim 7.6 \times 10^{-2}, A_3 \sim 2 \times 10^{-1}$, $A_4 \sim 6 \times 10^{-2}$, $A_5 \sim 4 \times 10^{-3}$, $A_6 \sim 3 \times 10^{-3}$, $A_7 \sim 1.4 \times 10^{-3}$, $A_8 \sim 2.2 \times 10^{-4}$, $A_9 \sim 2.4 \times 10^{-5}$, $A_{10} \sim 2.2 \times 10^{-5}$, $A_{11} \sim 5.3 \times 10^{-6}$, $A_{12} \sim 2.9 \times 10^{-7}$, $A_{13} \sim 2.5 \times 10^{-7}$,

Table 5.1 Input data for numerical illustrations ($\delta = 0.05, \alpha = 10.0, \beta = 5.0, Q_0 = 4.5$)

Figure no.	Ω	Initial condition ($x_0, \dot{x}_0$)	Type	Harmonics terms
Figure 5.7(i),(ii)	3.6	$(-1.801123, 0.092939)$	P-1	HB7 (stable)
Figure 5.7(iii),(iv)	1.101	$(0.155338, -2.175708\text{e} - 3)$	P-1	HB14 (stable)
Figure 5.7(v),(vi)	0.98	$(-0.245784, 0.098412)$	P-1	HB21 (stable)
Figure 5.8(i),(ii)	0.735	$(0.815054, -0.196194)$	P-1	HB21 (stable)
Figure 5.8(iii),(iv)	0.6	$(0.137234, 0.181059)$	P-1	HB21 (stable)
Figure 5.8(v),(vi)	0.477	$(0.781392, -0.229502)$	P-1	HB32 (stable)

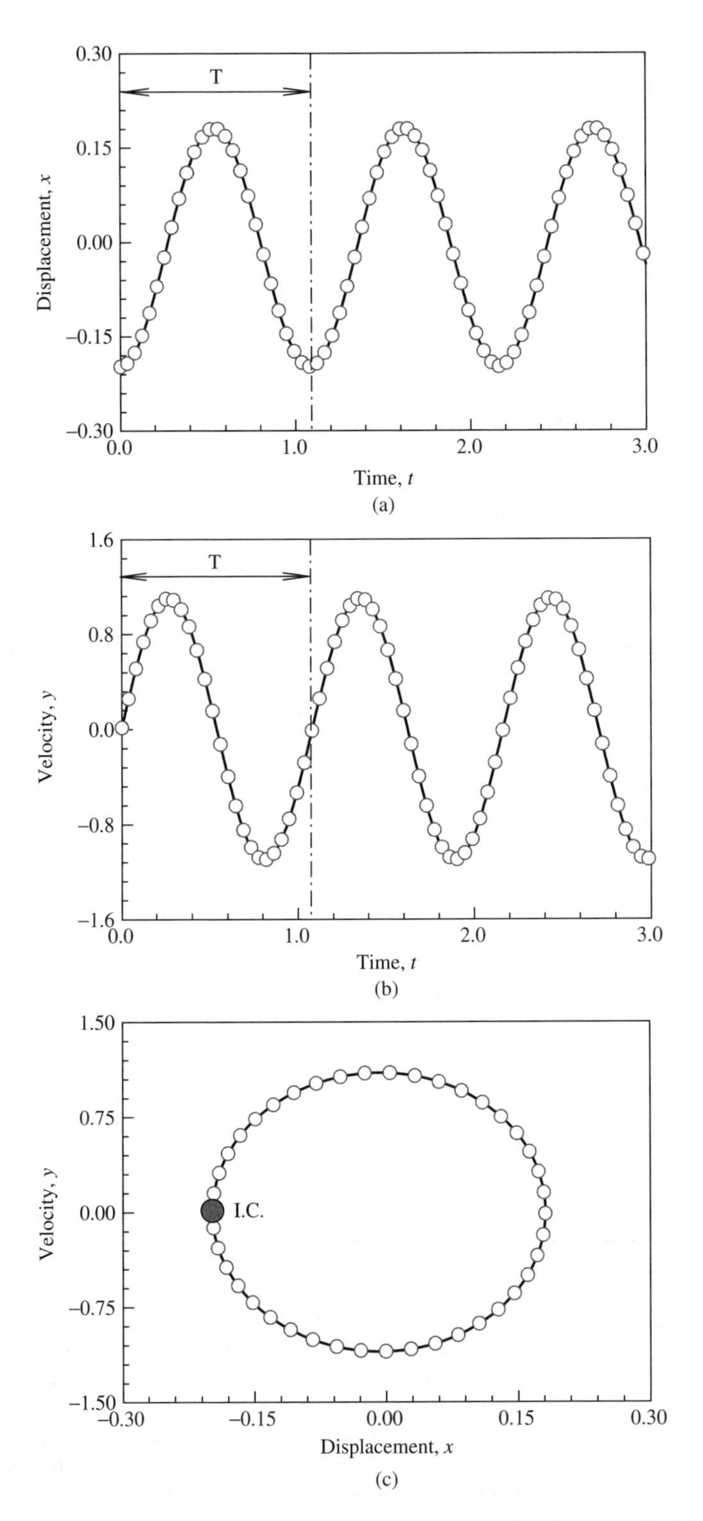

Figure 5.6 Analytical and numerical solutions of stable period-1 motion based on the 4 harmonic terms (HB4): (a) displacement, (b) velocity, (c) phase plane, and (d) amplitude spectrum. Initial condition ($x_0 \approx -0.197945, \dot{x}_0 \approx 0.013202$). Parameters: ($\delta = 0.05, \alpha = 10.0, \beta = 5.0, Q_0 = 4.5, \Omega = 6.5$)

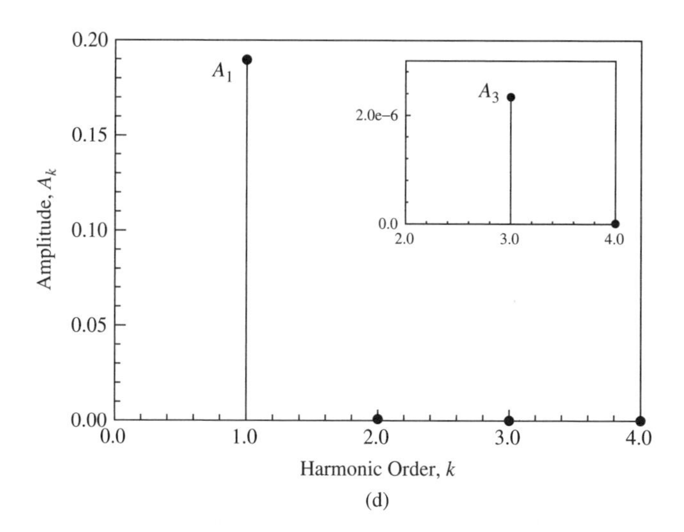

(d)

Figure 5.6 (*continued*)

$A_{14} \sim 9.3 \times 10^{-8}$. In Figure 5.7(v) and (vi), the 21 harmonic terms (HB21) are used in the analytical solution of period-1 motion for $\Omega = 0.98$. The analytical and numerical simulations match very well with three cycles, which is still on the second solution branch of period-1 motions. The harmonic amplitude distributions are $a_0 \approx -0.178142$, $A_{1,3} \sim 5 \times 10^{-1}$, $A_2 \sim 1.8 \times 10^{-1}$, $A_4 \sim 2.2 \times 10^{-1}$, $A_5 \sim 6 \times 10^{-2}$, $A_{6,7} \sim 1.5 \times 10^{-2}$, $A_8 \sim 6.3 \times 10^{-3}$, $A_9 \sim 8.3 \times 10^{-4}$, $A_{10} \sim 4.5 \times 10^{-4}$, $A_{11} \sim 3.5 \times 10^{-4}$, $A_{12} \sim 1.1 \times 10^{-4}$, $A_{13} \sim 8 \times 10^{-6}$, $A_{14} \sim 1.2 \times 10^{-5}$, $A_{15} \sim 6.6 \times 10^{-6}$, $A_{17} \sim 1.7 \times 10^{-8}$, $A_{18} \sim 2.5 \times 10^{-7}$, $A_{19} \sim 1.1 \times 10^{-7}$, $A_{20} \sim 2.3 \times 10^{-8}$, $A_{21} \sim 3.8 \times 10^{-9}$. From the amplitude distribution, it is observed that many higher order harmonic terms contribute significantly on the period-1 motion.

To further look into the complexity of period-1 motion, in Figure 5.8(i)–(vi), the trajectories and harmonic amplitude spectrums of period-1 motions are presented for $\Omega = 0.735, 0.6, 0.477$. For the three period-1 motions are on three different branches of period-1 motions. The analytical and numerical solutions match very well for the three period-1 motions. The trajectory of period-1 motion with 21 harmonic terms (HB21) in the Fourier solution for $\Omega = 0.735$ has four cycles in Figure 5.8(i). The main amplitude distributions for $\Omega = 0.735$ in Figure 5.8(ii) are $a_0 \approx -0.141771$, $A_1 \sim 5.5 \times 10^{-1}$, $A_2 \sim 4.9 \times 10^{-2}$, $A_3 \sim 2.3 \times 10^{-1}$, $A_4 \sim 3.3 \times 10^{-1}$, $A_5 \sim 2.3 \times 10^{-1}$, $A_6 \sim 6.7 \times 10^{-2}$, $A_7 \sim 1.0 \times 10^{-2}$. For higher order harmonic terms, $A_{17} \sim 5.1 \times 10^{-7}$, $A_{18} \sim 2.6 \times 10^{-6}$, $A_{19} \sim 3.9 \times 10^{-6}$, $A_{20} \sim 1.9 \times 10^{-6}$, $A_{21} \sim 4.1 \times 10^{-7}$. In Figure 5.8(iii), the trajectory of period-1 motion with 21 harmonic terms (HB21) in the Fourier solution for $\Omega = 0.6$ experiences five cycles. In Figure 5.8(iv), the main amplitude distributions for $\Omega = 0.6$ are $a_0 \approx -0.095182$, $A_1 \sim 5.4 \times 10^{-1}$, $A_2 \sim 7.5 \times 10^{-2}$, $A_3 \sim 4.9 \times 10^{-2}$, $A_4 \sim 1.4 \times 10^{-1}$, $A_{5,6} \sim 1.5 \times 10^{-1}$, $A_7 \sim 4.2 \times 10^{-2}$. For higher order harmonic terms, $A_{16} \sim 1.6 \times 10^{-5}$, $A_{17} \sim 2.9 \times 10^{-5}$, $A_{18} \sim 2.4 \times 10^{-5}$, $A_{19} \sim 8.5 \times 10^{-6}$, $A_{20} \sim 2.1 \times 10^{-6}$, $A_{21} \sim 9.6 \times 10^{-7}$. In Figure 5.8(v), the trajectory of period-1 motion with 32 harmonic terms (HB32) in the Fourier solution for $\Omega = 0.477$ possesses six cycles. In Figure 5.8(vi), the main amplitude distributions

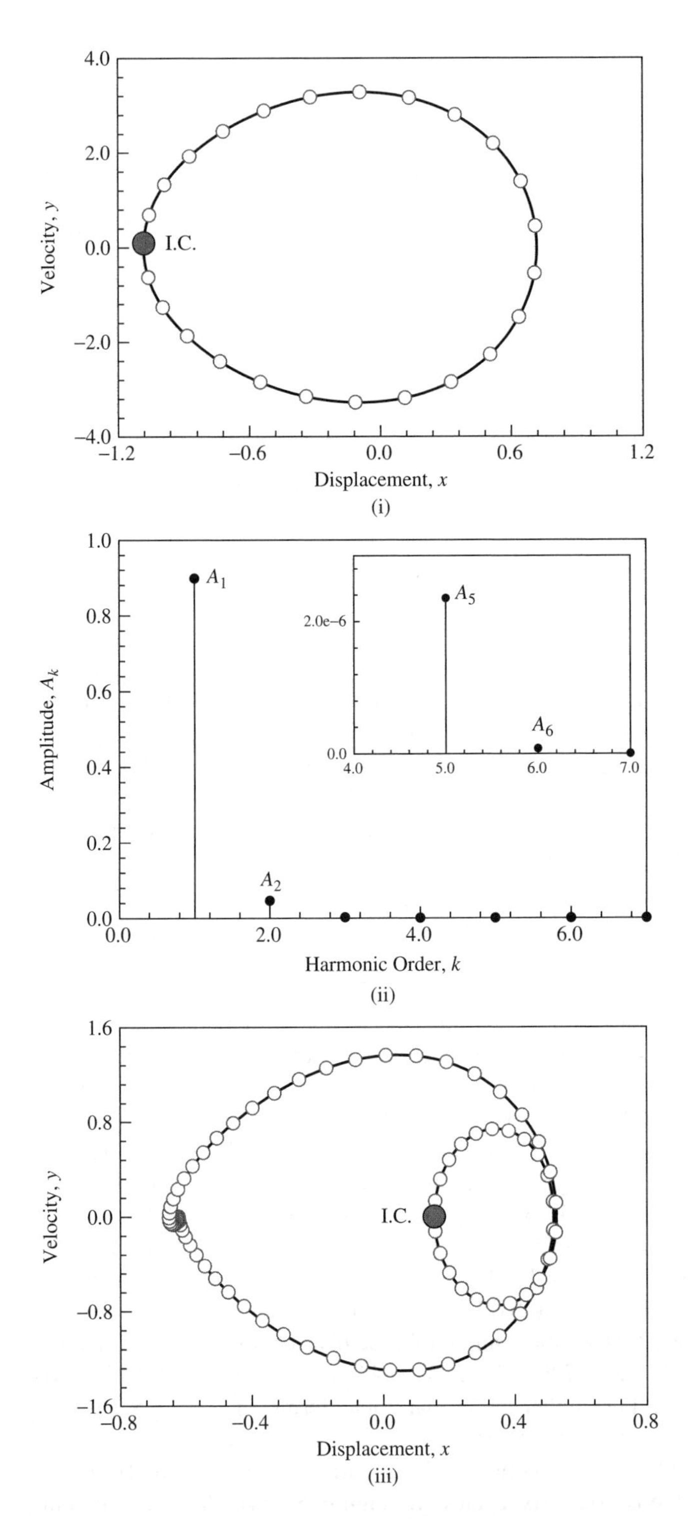

Figure 5.7 Phase plane and amplitude spectrums of stable period-1 motions, (i,ii): $\Omega = 3.6$ with ($x_0 \approx -1.081123$, $\dot{x}_0 \approx 0.092939$, HB7). (iii,iv) $\Omega = 1.101$ with ($x_0 \approx 0.155338$, $\dot{x}_0 \approx -2.175708\text{e-}3$, HB14) and (v,vi): $\Omega = 0.98$ with ($x_0 \approx -0.245784$, $\dot{x}_0 \approx 0.098412$, HB21). Parameters: ($\delta = 0.05$, $\alpha = 10.0$, $\beta = 5.0$, $Q_0 = 4.5$)

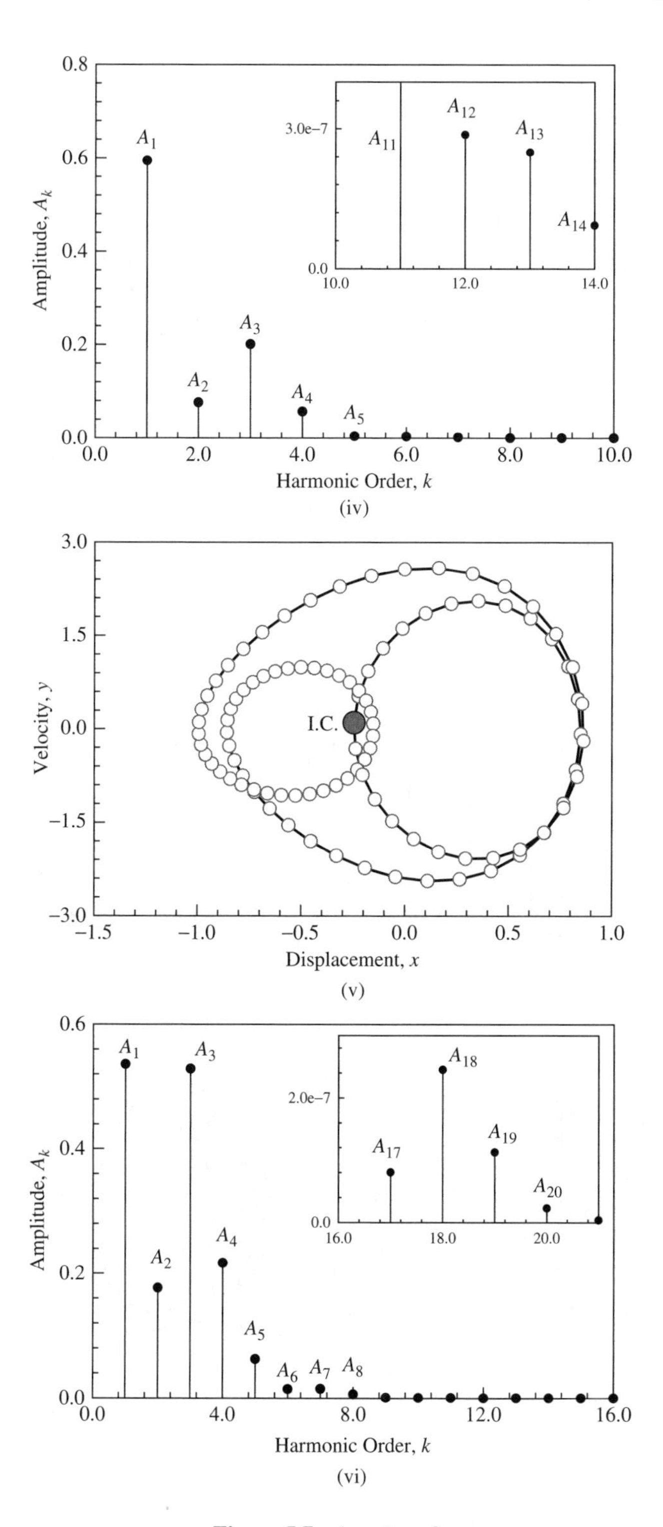

Figure 5.7 (*continued*)

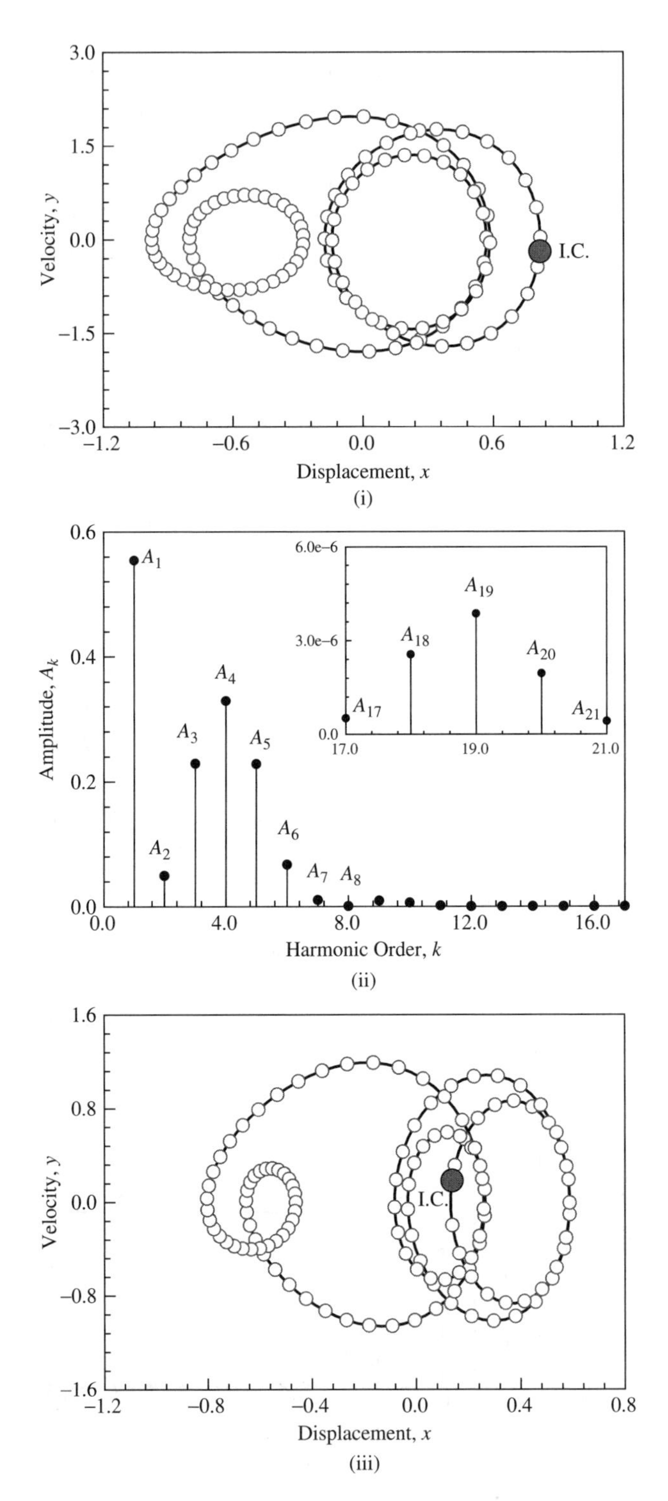

Figure 5.8 Phase plane and amplitude spectrums of stable period-1 motions for lower frequency: (i,ii): $\Omega = 0.735$ with ($x_0 \approx 0.815054$, $\dot{x}_0 \approx -0.196194$, HB21). (iii,iv) $\Omega = 0.6$ with ($x_0 \approx 0.137234$, $\dot{x}_0 \approx 0.181059$, HB21) and (v,vi): $\Omega = 0.477$ with ($x_0 \approx 0.781392$, $\dot{x}_0 \approx -0.229502$, HB32). Parameters: ($\delta = 0.05, \alpha = 10.0, \beta = 5.0, Q_0 = 4.5$)

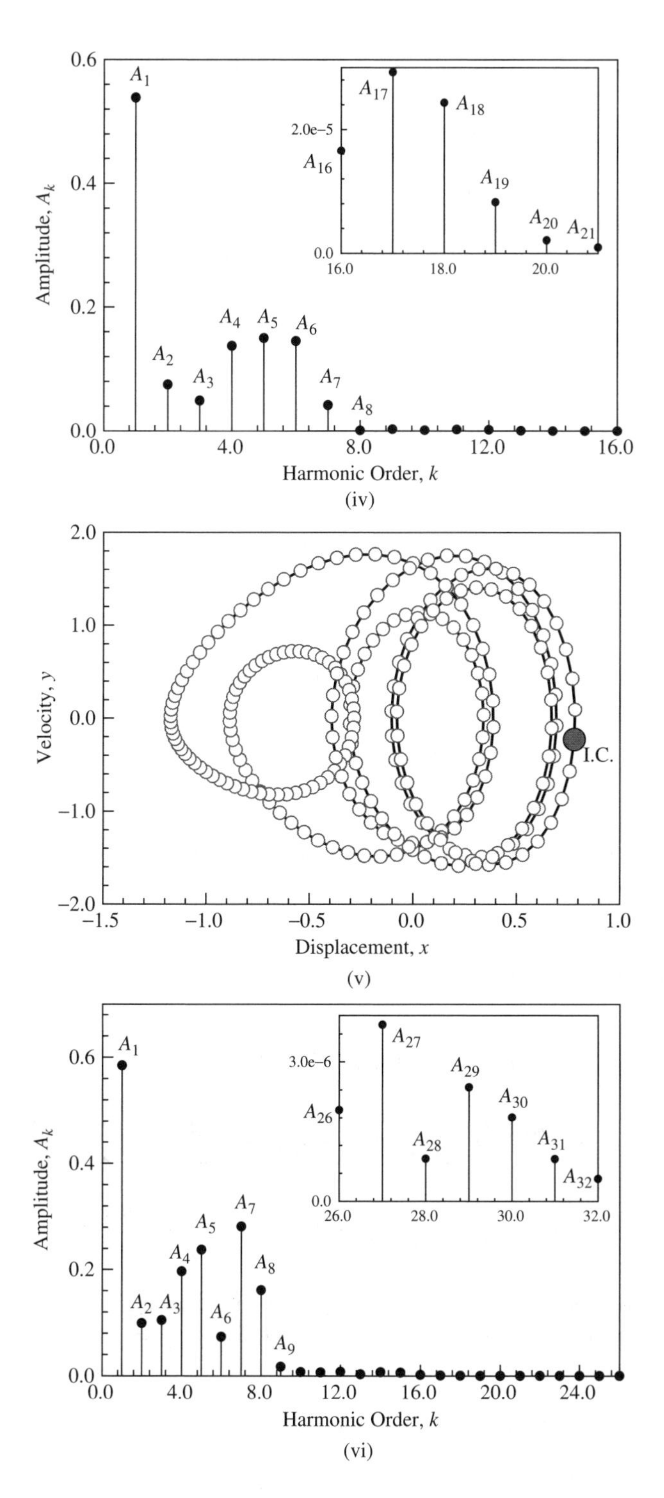

Figure 5.8 (*continued*)

for $\Omega = 0.477$ are $a_0 \approx -0.075150$, $A_1 \sim 5.8 \times 10^{-1}$, $A_2 \sim 9.8 \times 10^{-2}$, $A_3 \sim 1.0 \times 10^{-1}$, $A_4 \sim 1.9 \times 10^{-1}$, $A_5 \sim 2.4 \times 10^{-1}$, $A_6 \sim 7.3 \times 10^{-2}$, $A_7 \sim 2.8 \times 10^{-1}$, $A_8 \sim 1.6 \times 10^{-1}$, $A_9 \sim 1.7 \times 10^{-2}$. For the zoomed window with higher order harmonic amplitudes, $A_{26} \sim 2 \times 10^{-6}$, $A_{27} \sim 3.8 \times 10^{-6}$, $A_{28} \sim 9.1 \times 10^{-5}$, $A_{29} \sim 2.5 \times 10^{-6}$, $A_{30} \sim 1.8 \times 10^{-6}$, $A_{31} \sim 9.1 \times 10^{-7}$, $A_{32} \sim 4.8 \times 10^{-7}$. The complexity of period-1 motions is strongly dependent on the harmonic amplitude contributions. With reduction of excitation frequency, more harmonic terms should be included in the Fourier series solution of period-1 motion.

5.2 Period-*m* Motions

In this section, period-*m* motions in a periodically forced, quadratic nonlinear oscillator will be discussed. The appropriate analytical solutions will also be presented with finite harmonic terms based on the prescribed accuracy of harmonic amplitudes. The analytical bifurcation tree for period-1 motion to chaos will be determined. Period-2 and period-4 motions will be illustrated.

5.2.1 Analytical Solutions

In Luo (2012a), the analytical solution of period-*m* motion in Equation (5.1) is

$$x^{(m)*}(t) = a_0^{(m)}(t) + \sum_{k=1}^{N} b_{k/m}(t)\cos\left(\frac{k}{m}\theta\right) + c_{k/m}(t)\sin\left(\frac{k}{m}\theta\right). \tag{5.36}$$

where $a_0^{(m)}(t)$, $b_{k/m}(t)$ and $c_{k/m}(t)$ vary with time and $\theta = \Omega t$. The first and second order of derivatives of $x^*(t)$ are

$$\begin{aligned} \dot{x}^{(m)*}(t) = \dot{a}_0^{(m)} + \sum_{k=1}^{N} &\left(\dot{b}_{k/m} + \frac{k\Omega}{m}c_{k/m}\right)\cos\left(\frac{k\theta}{m}\right) \\ &+ \left(\dot{c}_{k/m} - \frac{k\Omega}{m}b_{k/m}\right)\sin\left(\frac{k\theta}{m}\right), \end{aligned} \tag{5.37}$$

$$\begin{aligned} \ddot{x}^{(m)*}(t) = \ddot{a}_0^{(m)} + \sum_{k=1}^{N} &\left(\ddot{b}_{k/m} + 2\frac{k\Omega}{m}\dot{c}_{k/m} - \left(\frac{k\Omega}{m}\right)^2 b_{k/m}\right)\cos\left(\frac{k\theta}{m}\right) \\ &+ \left(\ddot{c}_{k/m} - 2\frac{k\Omega}{m}\dot{b}_{k/m} - \left(\frac{k\Omega}{m}\right)^2 c_{k/m}\right)\sin\left(\frac{k\theta}{m}\right). \end{aligned} \tag{5.38}$$

Substitution of Equations (5.36)–(5.38) into Equation (5.1) and application of the virtual work principle for a basis of constant, $\cos(k\theta/m)$ and $\sin(k\theta/m)$ $(k = 1, 2, \ldots)$ as a set of virtual displacements gives

$$\begin{aligned} &\ddot{a}_0^{(m)} = F_0^{(m)}(a_0^{(m)}, \mathbf{b}^{(m)}, \mathbf{c}^{(m)}, \dot{a}_0^{(m)}, \dot{\mathbf{b}}^{(m)}, \dot{\mathbf{c}}^{(m)}), \\ &\ddot{b}_{k/m} + 2\frac{k\Omega}{m}\dot{c}_{k/m} - \left(\frac{k\Omega}{m}\right)^2 b_{k/m} = F_{1k}^{(m)}(a_0^{(m)}, \mathbf{b}^{(m)}, \mathbf{c}^{(m)}, \dot{a}_0^{(m)}, \dot{\mathbf{b}}^{(m)}, \dot{\mathbf{c}}^{(m)}), \end{aligned}$$

$$\ddot{c}_{k/m} - 2\frac{k\Omega}{m}\dot{b}_{k/m} - \left(\frac{k\Omega}{m}\right)^2 c_{k/m} = F_{2k}^{(m)}(a_0^{(m)}, \mathbf{b}^{(m)}, \mathbf{c}^{(m)}, \dot{a}_0^{(m)}, \dot{\mathbf{b}}^{(m)}, \dot{\mathbf{c}}^{(m)})$$
$$\text{for } k = 1, 2, \ldots, N \tag{5.39}$$

where

$$\begin{aligned}
&F_0^{(m)}(a_0^{(m)}, \mathbf{b}^{(m)}, \mathbf{c}^{(m)}, \dot{a}_0^{(m)}, \dot{\mathbf{b}}^{(m)}, \dot{\mathbf{c}}^{(m)}) \\
&\quad = \frac{1}{mT}\int_0^{mT} F(x^{(m)*}, \dot{x}^{(m)*}, t)dt \\
&\quad = -\delta\dot{a}_0^{(m)} - \alpha a_0^{(m)} - \beta(a_0^{(m)})^2 - \frac{\beta}{2}\sum_{i=1}^{N}(b_{i/m}^2 + c_{i/m}^2), \\
&F_{1k}^{(m)}(a_0^{(m)}, \mathbf{b}^{(m)}, \mathbf{c}^{(m)}, \dot{a}_0^{(m)}, \dot{\mathbf{b}}^{(m)}, \dot{\mathbf{c}}^{(m)}) \\
&\quad = \frac{2}{mT}\int_0^{mT} F(x^{(m)*}, \dot{x}^{(m)*}, t)\cos\left(\frac{k}{m}\Omega t\right) dt \\
&\quad = -\delta\left(\dot{b}_{k/m} + c_{k/m}\frac{k\Omega}{m}\right) - \alpha b_{k/m} - 2\beta a_0^{(m)} b_{k/m} - f_{1k/m} + Q_0\delta_k^m, \\
&F_{2k}^{(m)}(a_0^{(m)}, \mathbf{b}^{(m)}, \mathbf{c}^{(m)}, \dot{a}_0^{(m)}, \dot{\mathbf{b}}^{(m)}, \dot{\mathbf{c}}^{(m)}) \\
&\quad = \frac{2}{mT}\int_0^{mT} F(x^{(m)*}, \dot{x}^{(m)*}, t)\sin\left(\frac{k}{m}\Omega t\right) dt \\
&\quad = -\delta\left(\dot{c}_{k/m} - b_{k/m}\frac{k\Omega}{m}\right) - \alpha c_{k/m} - 2\beta a_0^{(m)} c_{k/m} - f_{2k/m};
\end{aligned} \tag{5.40}$$

and

$$\begin{aligned}
f_{1k/m} &= \beta\sum_{i=1}^{N}\sum_{j=1}^{N}\Big[\left(b_{i/m}b_{j/m} + c_{i/m}c_{j/m}\right)\delta_{j-i}^k \\
&\quad + \frac{1}{2}\left(b_{i/m}b_{j/m} - c_{i/m}c_{j/m}\right)\delta_{i+j}^k\Big], \\
f_{2k/m} &= \beta\sum_{i=1}^{N}\sum_{j=1}^{N} b_{i/m}c_{j/m}(\delta_{i+j}^k + \delta_{j-i}^k - \delta_{i-j}^k).
\end{aligned} \tag{5.41}$$

Define

$$\begin{aligned}
\mathbf{z}^{(m)} &\triangleq (a_0^{(m)}, \mathbf{b}^{(m)}, \mathbf{c}^{(m)})^{\mathrm{T}} \\
&= (a_0^{(m)}, b_{1/m}, \ldots, b_{N/m}, c_{1/m}, \ldots, c_{N/m})^{\mathrm{T}} \\
&\equiv (z_0^{(m)}, z_1^{(m)}, \ldots, z_{2N}^{(m)})^{\mathrm{T}},
\end{aligned}$$

$$\begin{aligned}\mathbf{z}_1 &\triangleq \dot{\mathbf{z}} = (\dot{a}_0^{(m)}, \dot{\mathbf{b}}^{(m)}, \dot{\mathbf{c}}^{(m)})^{\mathrm{T}} \\ &= (\dot{a}_0^{(m)}, \dot{b}_{1/m}, \ldots, \dot{b}_{N/m}, \dot{c}_{1/m}, \ldots, \dot{c}_{N/m})^{\mathrm{T}} \\ &\equiv (\dot{z}_0^{(m)}, \dot{z}_1^{(m)}, \ldots, \dot{z}_{2N}^{(m)})^{\mathrm{T}}\end{aligned} \tag{5.42}$$

where

$$\begin{aligned}\mathbf{b}^{(m)} &= (b_{1/m}, \ldots, b_{N/m})^{\mathrm{T}}, \\ \mathbf{c}^{(m)} &= (c_{1/m}, \ldots, c_{N/m})^{\mathrm{T}}.\end{aligned} \tag{5.43}$$

Equation (5.39) can be expressed in the form of vector field as

$$\dot{\mathbf{z}}^{(m)} = \mathbf{z}_1^{(m)} \text{ and } \dot{\mathbf{z}}_1^{(m)} = \mathbf{g}^{(m)}(\mathbf{z}^{(m)}, \mathbf{z}_1^{(m)}) \tag{5.44}$$

where

$$\mathbf{g}^{(m)}(\mathbf{z}^{(m)}, \mathbf{z}_1^{(m)}) = \begin{pmatrix} F_0^{(m)}(\mathbf{z}^{(m)}, \mathbf{z}_1^{(m)}) \\ \mathbf{F}_1^{(m)}(\mathbf{z}^{(m)}, \mathbf{z}_1^{(m)}) - 2\mathbf{k}_1\left(\dfrac{\Omega}{m}\right)\dot{\mathbf{c}}^{(m)} + \mathbf{k}_2\left(\dfrac{\Omega}{m}\right)^2\mathbf{b}^{(m)} \\ \mathbf{F}_2^{(m)}(\mathbf{z}^{(m)}, \mathbf{z}_1^{(m)}) + 2\mathbf{k}_1\left(\dfrac{\Omega}{m}\right)\dot{\mathbf{b}}^{(m)} + \mathbf{k}_2\left(\dfrac{\Omega}{m}\right)^2\mathbf{c}^{(m)} \end{pmatrix} \tag{5.45}$$

and

$$\begin{aligned}\mathbf{k}_1 &= diag(1, 2, \ldots, N), \\ \mathbf{k}_2 &= diag(1, 2^2, \ldots, N^2), \\ \mathbf{F}_1^{(m)} &= (F_{11}^{(m)}, F_{12}^{(m)}, \ldots, F_{1N}^{(m)})^{\mathrm{T}}, \\ \mathbf{F}_2^{(m)} &= (F_{21}^{(m)}, F_{22}^{(m)}, \ldots, F_{2N}^{(m)})^{\mathrm{T}} \\ &\text{for } N = 1, 2, \ldots, \infty.\end{aligned} \tag{5.46}$$

Introducing

$$\mathbf{y}^{(m)} \equiv (\mathbf{z}^{(m)}, \mathbf{z}_1^{(m)}) \text{ and } \mathbf{f}^{(m)} = (\mathbf{z}_1^{(m)}, \mathbf{g}^{(m)})^{\mathrm{T}}, \tag{5.47}$$

Equation (5.44) becomes

$$\dot{\mathbf{y}}^{(m)} = \mathbf{f}^{(m)}(\mathbf{y}^{(m)}). \tag{5.48}$$

The steady-state solutions for periodic motion in Equation (5.1) can be obtained by setting $\dot{\mathbf{y}}^{(m)} = \mathbf{0}$, that is,

$$\begin{aligned} F_0^{(m)}(\mathbf{z}^{(m)}, \mathbf{0}) &= 0, \\ \mathbf{F}_1^{(m)}(\mathbf{z}^{(m)}, \mathbf{0}) - \mathbf{k}_2\left(\frac{\Omega}{m}\right)^2\mathbf{b}^{(m)} &= \mathbf{0}, \\ \mathbf{F}_2^{(m)}(\mathbf{z}^{(m)}, \mathbf{0}) - \mathbf{k}_2\left(\frac{\Omega}{m}\right)^2\mathbf{c}^{(m)} &= \mathbf{0}. \end{aligned} \tag{5.49}$$

The $(2N + 1)$ nonlinear equations in Equation (5.49) are solved by the Newton-Raphson method. As in Luo (2012a) (or Chapter 3), the linearized equation at equilibrium point $\mathbf{y}^* = (\mathbf{z}^*, \mathbf{0})^{\mathrm{T}}$ is given by

$$\Delta\dot{\mathbf{y}}^{(m)} = D\mathbf{f}(\mathbf{y}^{(m)*})\Delta\mathbf{y}^{(m)} \tag{5.50}$$

where

$$D\mathbf{f}(\mathbf{y}^{(m)*}) = \partial\mathbf{f}(\mathbf{y}^{(m)})/\partial\mathbf{y}^{(m)}|_{\mathbf{y}^{(m)*}}. \tag{5.51}$$

The corresponding eigenvalues are determined by

$$|D\mathbf{f}(\mathbf{y}^{(m)*}) - \lambda\mathbf{I}_{2(2N+1)\times 2(2N+1)}| = 0, \tag{5.52}$$

where

$$D\mathbf{f}(\mathbf{y}^{(m)*}) = \begin{bmatrix} \mathbf{0}_{(2N+1)\times(2N+1)} & \mathbf{I}_{(2N+1)\times(2N+1)} \\ \mathbf{G}_{(2N+1)\times(2N+1)} & \mathbf{H}_{(2N+1)\times(2N+1)} \end{bmatrix} \tag{5.53}$$

and

$$\mathbf{G} = \frac{\partial\mathbf{g}^{(m)}}{\partial\mathbf{z}^{(m)}} = (\mathbf{G}^{(0)}, \mathbf{G}^{(c)}, \mathbf{G}^{(s)})^{\mathrm{T}}; \tag{5.54}$$

$$\begin{aligned} \mathbf{G}^{(0)} &= (G_0^{(0)}, G_1^{(0)}, \ldots, G_{2N}^{(0)}), \\ \mathbf{G}^{(c)} &= (\mathbf{G}_1^{(c)}, \mathbf{G}_2^{(c)}, \ldots, \mathbf{G}_N^{(c)})^{\mathrm{T}}, \\ \mathbf{G}^{(s)} &= (\mathbf{G}_1^{(s)}, \mathbf{G}_2^{(s)}, \ldots, \mathbf{G}_N^{(s)})^{\mathrm{T}} \end{aligned} \tag{5.55}$$

for $N = 1, 2, \ldots \infty$ with

$$\begin{aligned} \mathbf{G}_k^{(c)} &= (G_{k0}^{(c)}, G_{k1}^{(c)}, \ldots, G_{k(2N)}^{(c)}), \\ \mathbf{G}_k^{(s)} &= (G_{k0}^{(s)}, G_{k1}^{(s)}, \ldots, G_{k(2N)}^{(s)}) \end{aligned} \tag{5.56}$$

for $k = 1, 2, \ldots N$. The corresponding components are

$$\begin{aligned} G_r^{(0)} &= -\alpha\delta_0^r - \beta g_{2r}^{(0)}, \\ G_{kr}^{(c)} &= \left(\frac{k\Omega}{m}\right)^2 \delta_k^r - \alpha\delta_k^r - \delta\left(\frac{k\Omega}{m}\right)\delta_{k+N}^r - \beta g_{2kr}^{(c)}, \\ G_{kr}^{(s)} &= \left(\frac{k\Omega}{m}\right)^2 \delta_{k+N}^r + \delta\left(\frac{k\Omega}{m}\right)\delta_k^r - \alpha\delta_{k+N}^r - \beta g_{2kr}^{(s)} \end{aligned} \tag{5.57}$$

where

$$g_{2r}^{(0)} = 2a_0^{(m)}\delta_0^r + b_{k/m}\delta_k^r + c_{k/m}\delta_{k+N}^r, \tag{5.58}$$

$$\begin{aligned} g_{2kr}^{(c)} &= 2b_{k/m}\delta_r^0 + 2a_0^{(m)}\delta_k^r + \sum_{i=1}^{N}\sum_{j=1}^{N}[b_{j/m}(\delta_{j-i}^k + \delta_{i-j}^k + \delta_{i+j}^k)\delta_i^r \\ &\quad + c_{j/m}(\delta_{j-i}^k + \delta_{i-j}^k - \delta_{i+j}^k)\delta_{i+N}^r], \end{aligned} \tag{5.59}$$

$$\begin{aligned} g_{2kr}^{(s)} &= 2c_{k/m}\delta_0^r + 2a_0^{(m)}\delta_{k+N}^r + \sum_{i=1}^{N}\sum_{j=1}^{N} c_{j/m}(\delta_{i+j}^k + \delta_{j-i}^k - \delta_{i-j}^k)\delta_i^r \\ &\quad + b_{i/m}(\delta_{i+j}^k + \delta_{j-i}^k - \delta_{i-j}^k)\delta_{j+N}^r \end{aligned} \tag{5.60}$$

for $r = 0, 1, \ldots, 2N$.

$$\mathbf{H} = \frac{\partial \mathbf{g}^{(m)}}{\partial \mathbf{z}_1^{(m)}} = (\mathbf{H}^{(0)}, \mathbf{H}^{(c)}, \mathbf{H}^{(s)})^{\mathrm{T}} \tag{5.61}$$

where

$$\begin{aligned}
\mathbf{H}^{(0)} &= (H_0^{(0)}, H_1^{(0)}, \ldots, H_{2N}^{(0)}),\\
\mathbf{H}^{(c)} &= (\mathbf{H}_1^{(c)}, \mathbf{H}_2^{(c)}, \ldots, \mathbf{H}_N^{(c)})^{\mathrm{T}},\\
\mathbf{H}^{(s)} &= (\mathbf{H}_1^{(s)}, \mathbf{H}_2^{(s)}, \ldots, \mathbf{H}_N^{(s)})^{\mathrm{T}}
\end{aligned} \tag{5.62}$$

for $N = 1, 2, \ldots \infty$, with

$$\begin{aligned}
\mathbf{H}_k^{(c)} &= (H_{k0}^{(c)}, H_{k1}^{(c)}, \ldots, H_{k(2N)}^{(c)}),\\
\mathbf{H}_k^{(s)} &= (H_{k0}^{(s)}, H_{k1}^{(s)}, \ldots, H_{k(2N)}^{(s)})
\end{aligned} \tag{5.63}$$

for $k = 1, 2, \ldots N$. The corresponding components are

$$\begin{aligned}
H_r^{(0)} &= -\delta\delta_0^r,\\
H_{kr}^{(c)} &= -2\frac{k\Omega}{m}\delta_{k+N}^r - \delta\delta_k^r,\\
H_{kr}^{(s)} &= 2\frac{k\Omega}{m}\delta_k^r - \delta\delta_{k+N}^r
\end{aligned} \tag{5.64}$$

for $r = 0, 1, \ldots, 2N$.

From Luo (2012a), the eigenvalues of $D\mathbf{f}(\mathbf{y}^{(m)*})$ are classified as

$$(n_1, n_2, n_3 | n_4, n_5, n_6) \tag{5.65}$$

where n_1 is the total number of negative real eigenvalues, n_2 is the total number of positive real eigenvalues, n_3 is the total number of negative zero eigenvalues; n_4 is the total pair number of complex eigenvalues with negative real parts, n_5 is the total pair number of complex eigenvalues with positive real parts, n_6 is the total pair number of complex eigenvalues with zero real parts. If $\mathrm{Re}(\lambda_k) < 0$ $(k = 1, 2, \ldots, 2(2N + 1))$, the approximate steady-state solution $\mathbf{y}^{(m)*}$ with truncation of $\cos(N\Omega t/m)$ and $\sin(N\Omega t/m)$ is stable. If $\mathrm{Re}(\lambda_k) > 0$ $(k \in \{1, 2, \ldots, 2(2N + 1)\})$, the truncated approximate steady-state solution is unstable. The corresponding boundary between the stable and unstable solution is given by the saddle-node bifurcation and Hopf bifurcation.

5.2.2 *Analytical Bifurcation Trees*

The exact steady-state solutions of periodic motions in the nonlinear oscillator should be obtained through the infinite harmonic terms. Unfortunately, it is impossible to compute the exact solution of periodic motions in such an oscillator. Thus, as in Luo and Yu (2013c), one uses the truncated solutions to obtain the approximate solutions of the nonlinear oscillator with enough precision ($A_{N/m} \leq \varepsilon$) where the number N is the total number of harmonic terms used in the approximate solution and ε is a prescribed precision (i.e., $\varepsilon = 10^{-8}$). The more terms are used in the Fourier series solution of periodic motions, the better prediction of the periodic motions can be obtained. However, the computational workload will dramatically increase. It is very important that a suitable precision ε is selected. Similarly, the eigenvalue analysis of such approximate, analytical solutions can be done through dynamics of time-varying coefficients in the Fourier series expression of periodic motion, and the stability and bifurcation analysis can be completed as well. The equilibrium solution of Equation (5.44) can be obtained from Equation (5.49) by using the Newton-Raphson method, and the stability analysis will be discussed.

The harmonic amplitude varying with excitation frequency Ω are illustrated. The harmonic amplitude and phase are defined by

$$A_{k/m} \equiv \sqrt{b_{k/m}^2 + c_{k/m}^2} \text{ and } \varphi_{k/m} = \arctan \frac{c_{k/m}}{b_{k/m}}. \tag{5.66}$$

The corresponding solution in Equation (5.36) becomes

$$x^*(t) = a_0^{(m)} + \sum_{k=1}^{N} A_{k/m} \cos\left(\frac{k}{m}\Omega t - \varphi_{k/m}\right). \tag{5.67}$$

Consider system parameters as

$$\delta = 0.05,\ \alpha = 10.0,\ \beta = 5.0, Q_0 = 4.5. \tag{5.68}$$

In Section 5.1, multiple period-1 motions were discovered. Since the Hopf bifurcation of the period-1 motion exists, the period-doubling of the period-1 motion for such an oscillator occurs and the corresponding period-2 motions will be obtained. If such a period-2 motion possesses the Hopf bifurcation, then the period-doubling bifurcation of period-2 motion occurs and the corresponding period-4 motion will be obtained. Continuously, the bifurcation tree of period-1 motion to chaos for such an oscillator will be obtained. To understand the analytical bifurcation tree of period-1 motion to chaos, the overview of the analytical prediction of period-1 motions to period-4 motions are presented in Figure 5.9. The detailed, zoomed views of analytical prediction of period-1 motion to period-4 motions are arranged in Figures 5.10–5.12. For low frequency $\Omega < 1.2$, because thc ranges of stable period-4 motions are very short, only period-1 motion to period-2 motions in the fourth zoomed view for the bifurcation routes are presented in Figure 5.13.

In Figure 5.9(i), the constant term $a_0^{(m)}$ for period-1 motion to period-4 motion is presented with 120 harmonic terms (HB120). For $\Omega > 2$, such a bifurcation tree is clearly illustrated. The Hopf bifurcations and unstable Hopf bifurcation are observed. With excitation frequency increase, no saddle-node bifurcation is observed. However, for the next near bifurcation tree, the Hopf bifurcations, unstable Hopf bifurcation, and saddle-node bifurcation are observed. For $\Omega < 1$, the bifurcation tree is crowded, which will be presented later with the zoomed view in detail. In Figure 5.9(ii), the harmonic amplitude $A_{1/4}$ of period-4 motions is presented for $\Omega > 1$ only. The saddle-node bifurcations are observed, which is for the period-4 appearance, which is the same as the Hopf bifurcation of period-2 motion. The Hopf bifurcations for the appearance of period-8 motion. In Figure 5.9(iii), the harmonic amplitude $A_{1/2}$ of period-2 motions are presented. The saddle-node bifurcations of period-2 motions for the appearance of period-2 motion, which is the same as the Hopf bifurcations of the period-1 motions. For $\Omega < 1$, the range for stable period-4 motions is very small and the analytical prediction for the solutions is very crowded. The unstable saddle-node bifurcation (USN) for period-2 motion is observed, which is the same as the unstable Hopf bifurcation (or subcritical Hopf bifurcation). It is obvious that the Hope bifurcation of period-2 motion, the period-4 motion appears. In Figure 5.9(iv), the harmonic amplitude $A_{1/4}$ of period-4 motions is presented for $\Omega > 1$ only,

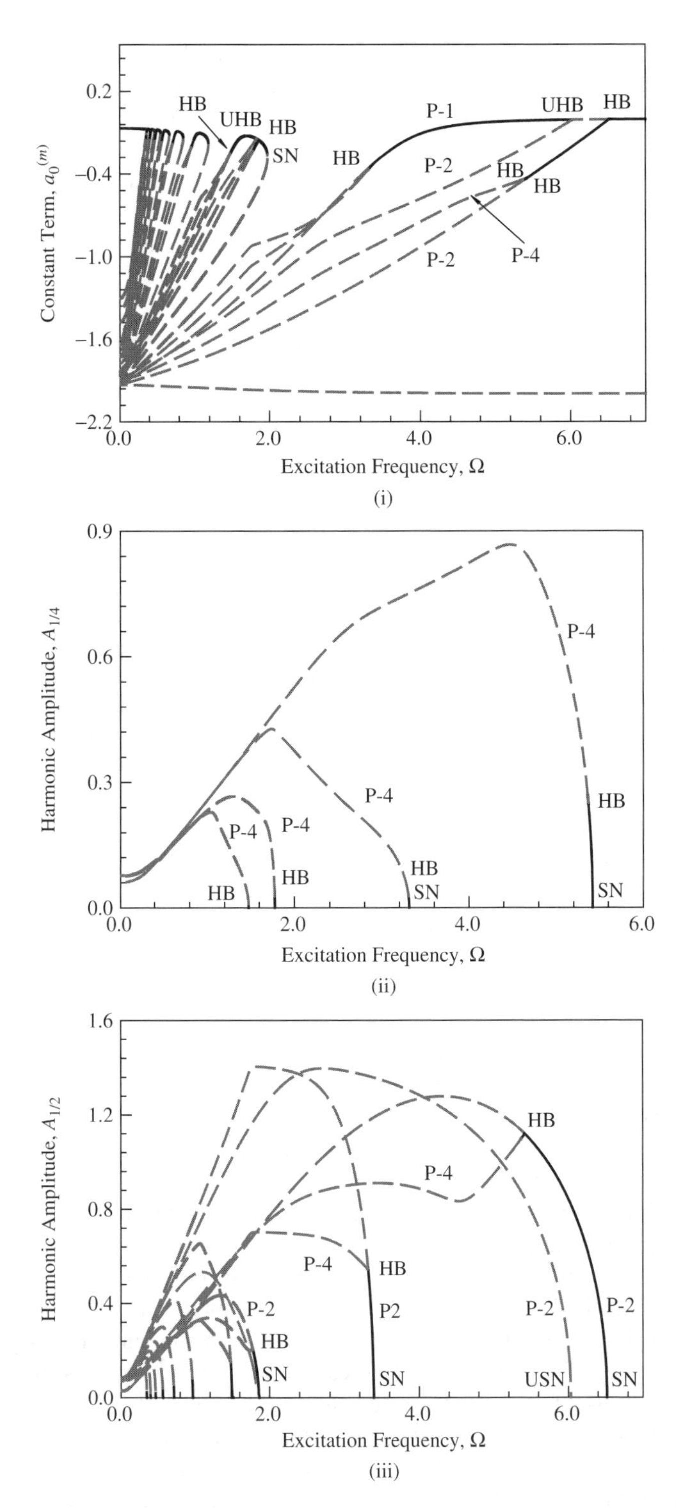

Figure 5.9 An overview for the analytical prediction of period-1 to period-4 motions based on the 30 harmonic terms (HB120): (i) $a_0^{(m)}$ and (ii)–(vi) $A_{k/m}$ $(k = 1, 2, \ldots, 4, 8, m = 4)$. Parameters: $(\delta = 0.05, \alpha = 10.0, \beta = 5.0, Q_0 = 4.5)$

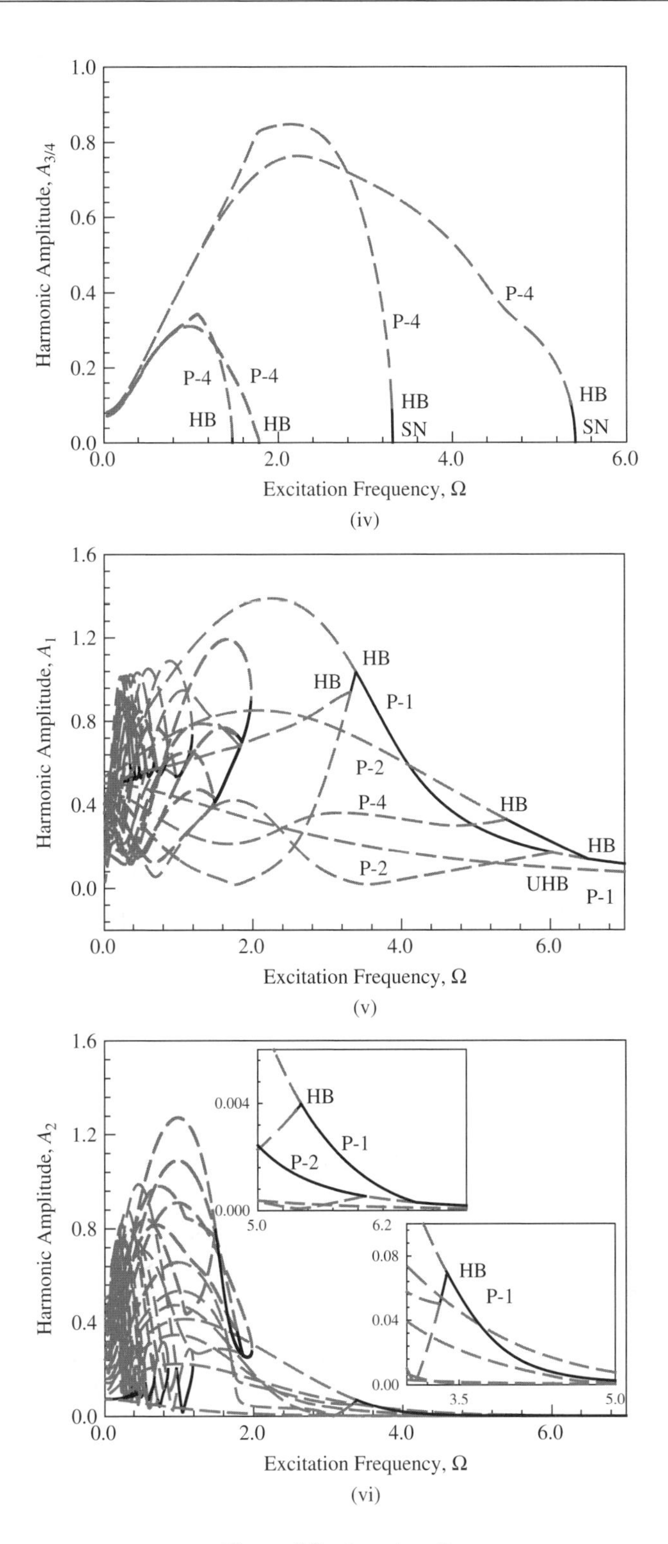

Figure 5.9 *(continued)*

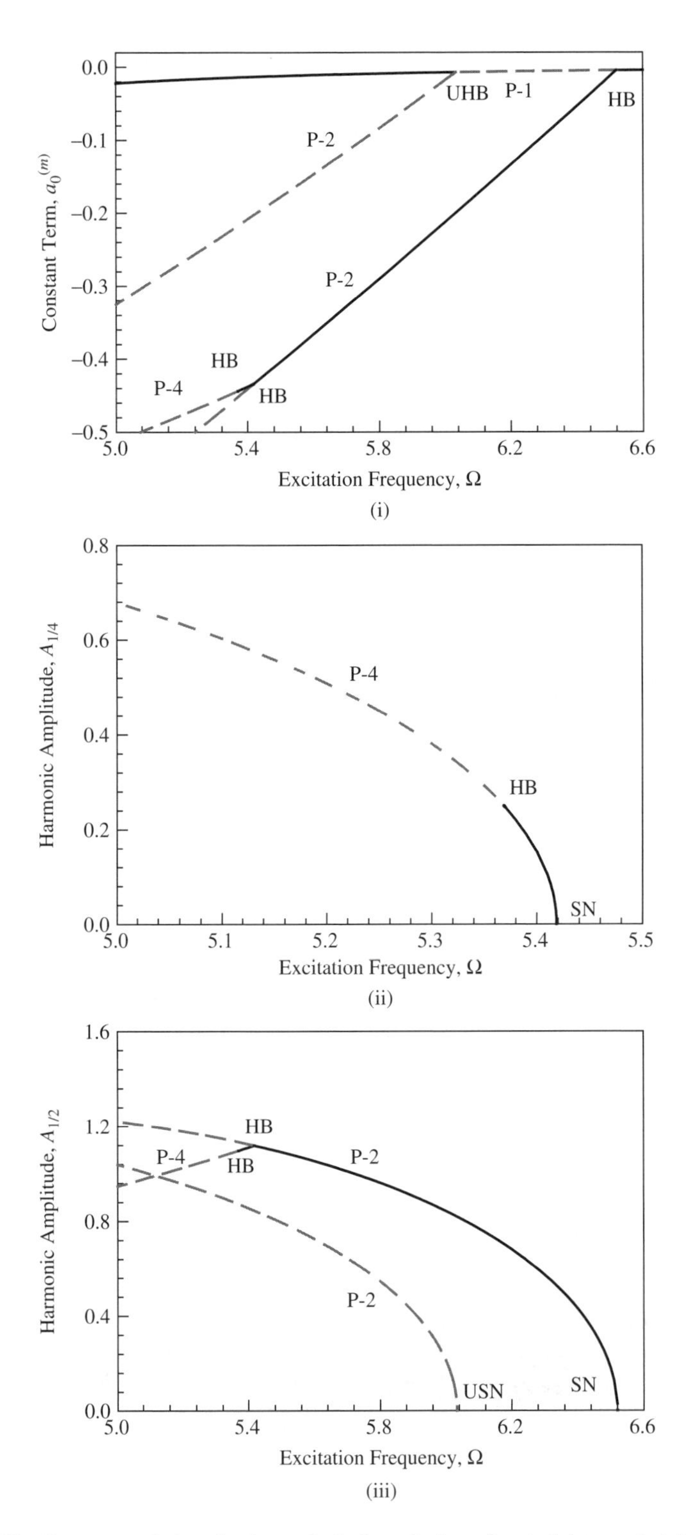

Figure 5.10 The first zoomed view for the analytical prediction of period-1 to period-4 motions based on 120 harmonic terms (HB120): (i) $a_0^{(m)}$ and (ii)–(xii) $A_{k/m}$ ($k = 1, 2, \ldots, 4, 6, \ldots, 14, 15, 16, m = 4$). Parameters: ($\delta = 0.05, \alpha = 10.0, \beta = 5.0, Q_0 = 4.5$)

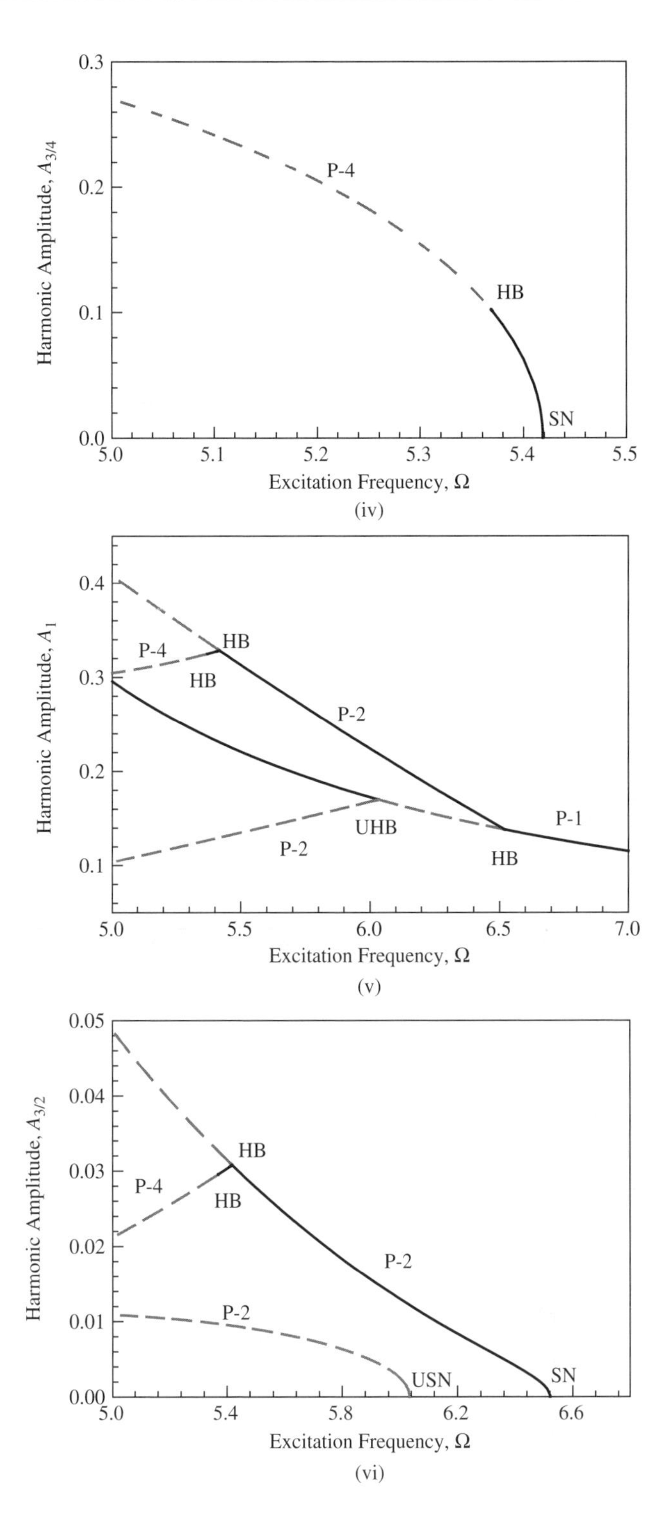

Figure 5.10 (*continued*)

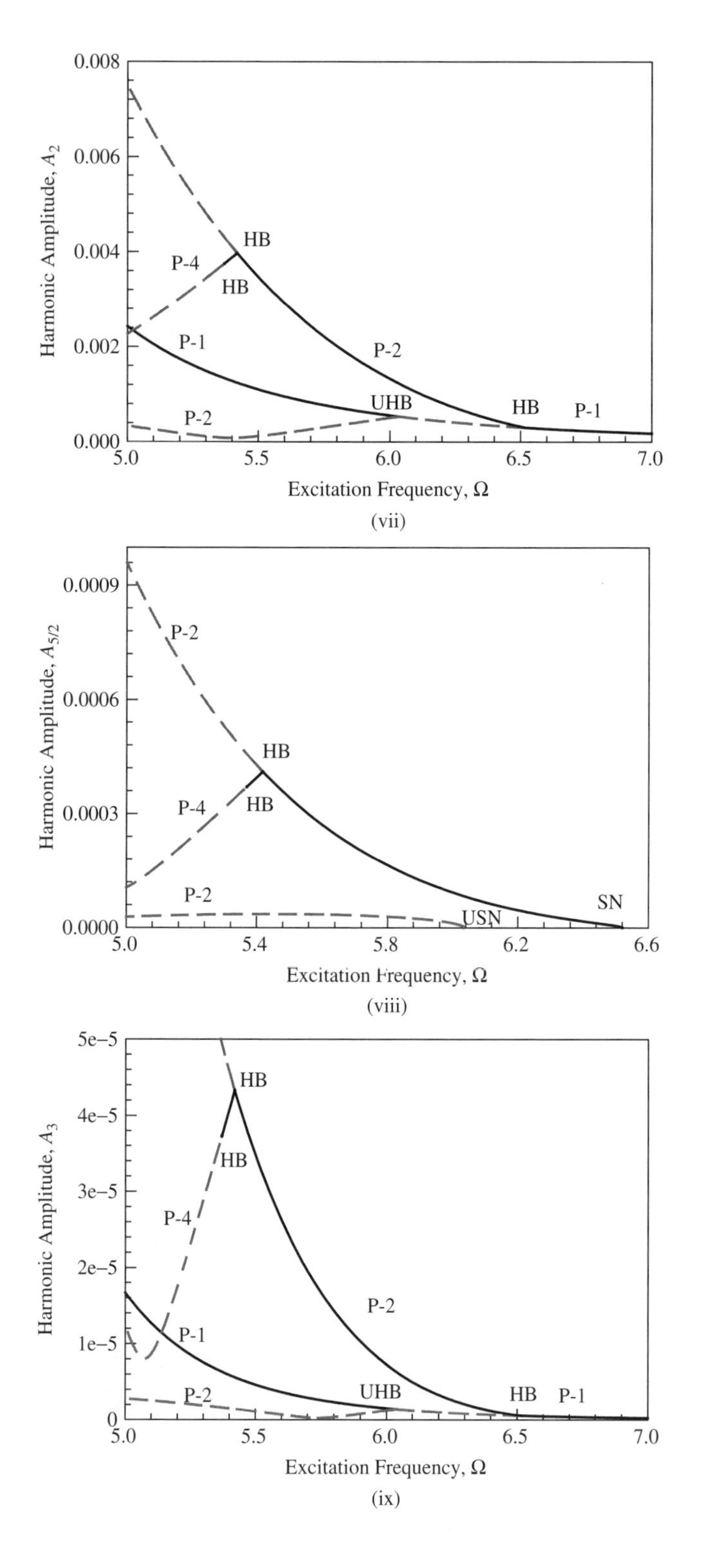

Figure 5.10 *(continued)*

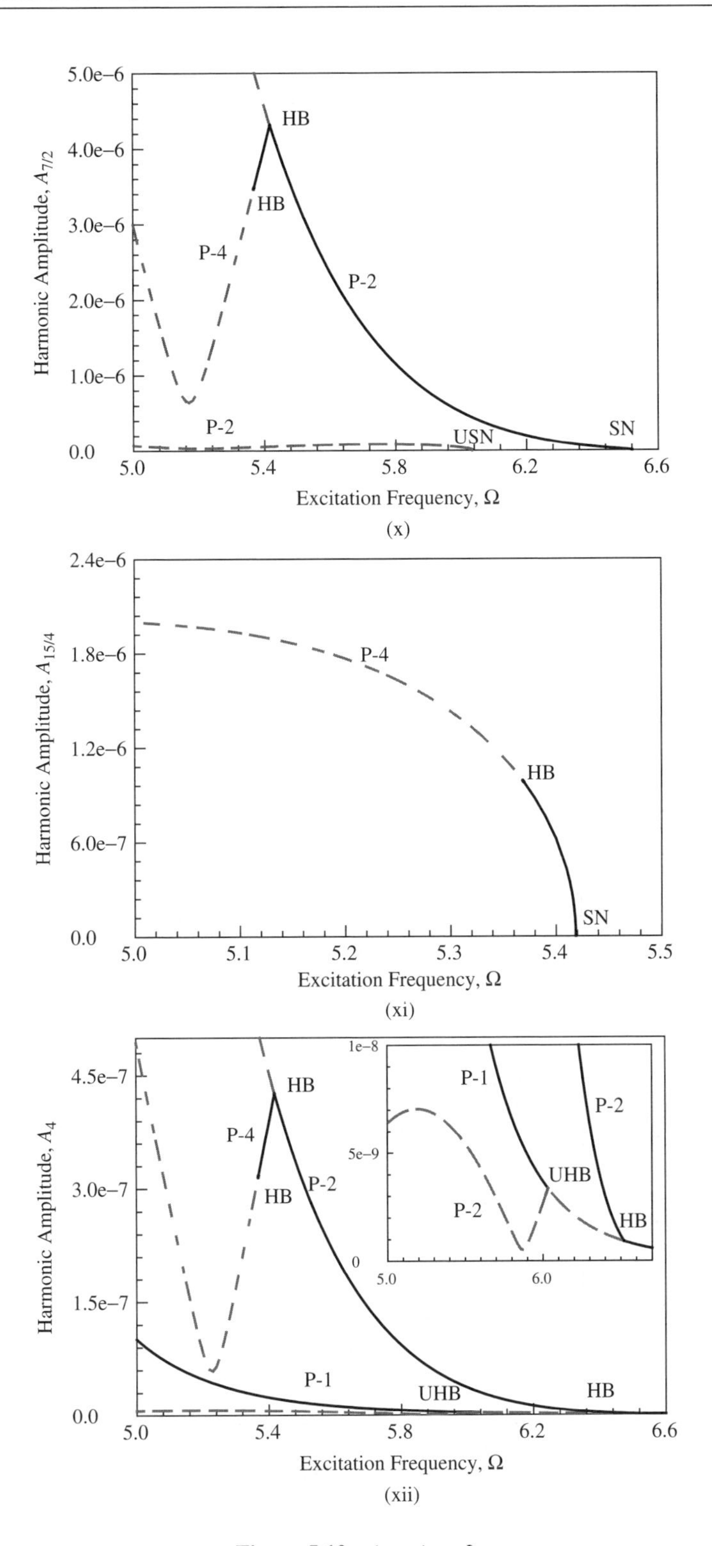

Figure 5.10 *(continued)*

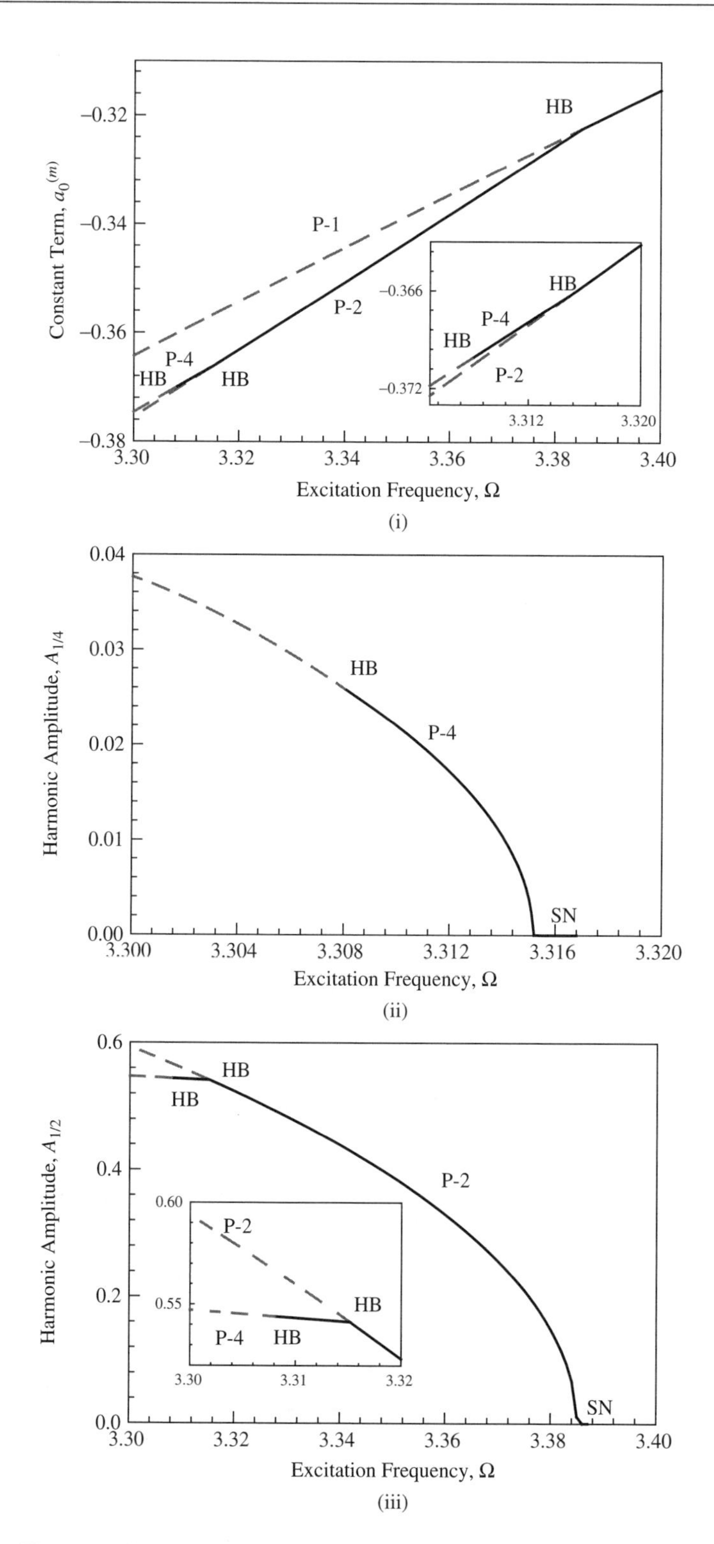

Figure 5.11 The second zoomed view for the analytical prediction of period-1 to period-4 motions based on the 120 harmonic terms (HB120): (i) $a_0^{(m)}$ and (ii)–(xii) $A_{k/m}$ $(k = 1, 2, \ldots, 4, 6, 8, 12, \ldots, 20, 22, 24, m = 4)$. Parameters: $(\delta = 0.05, \alpha = 10.0, \beta = 5.0, Q_0 = 4.5)$

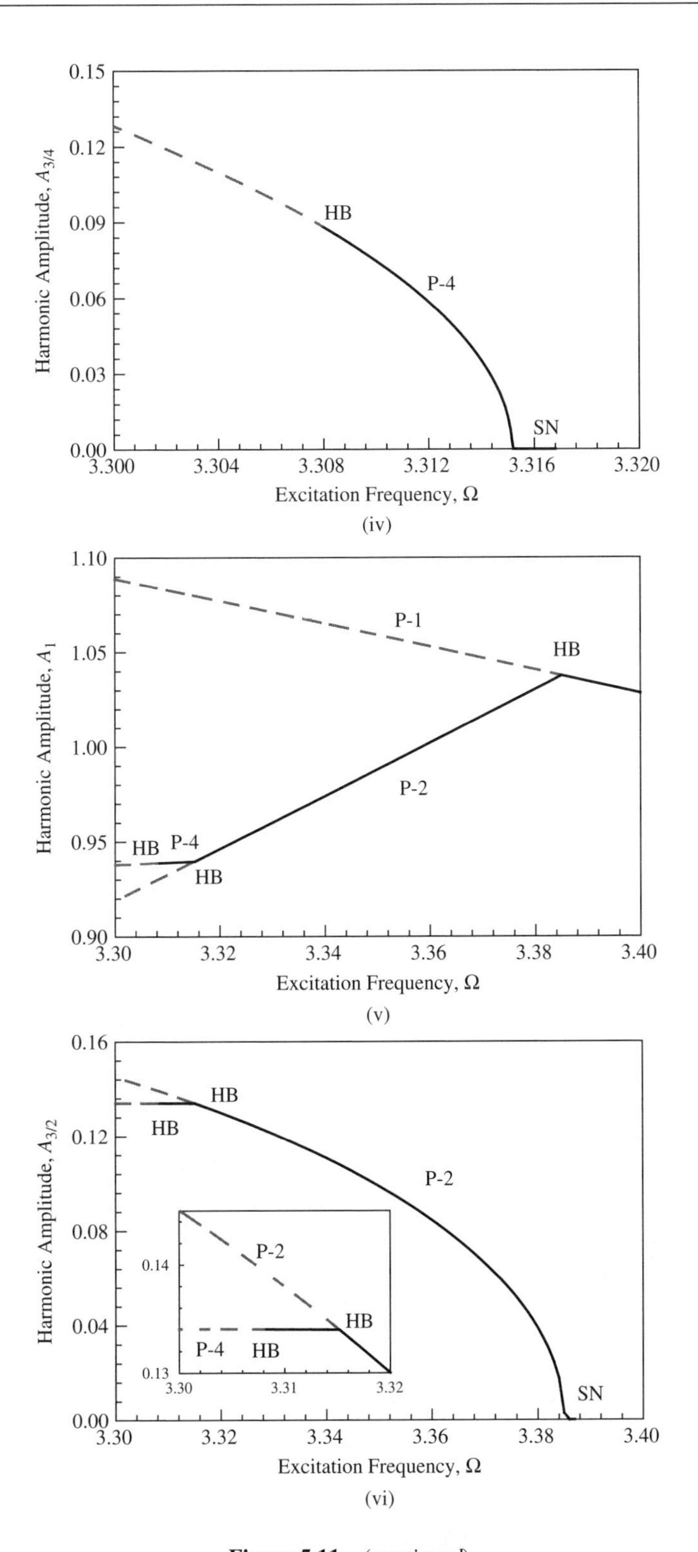

Figure 5.11 *(continued)*

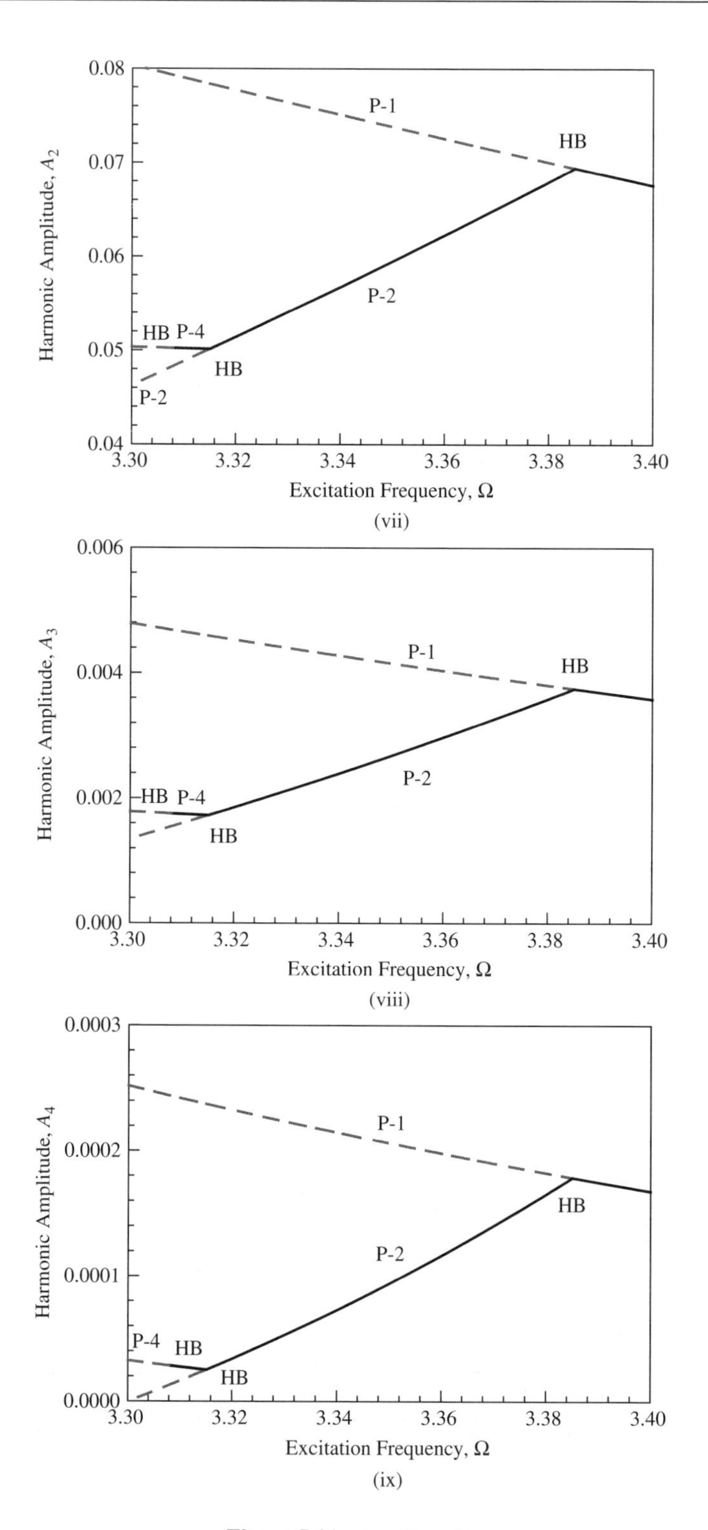

Figure 5.11 *(continued)*

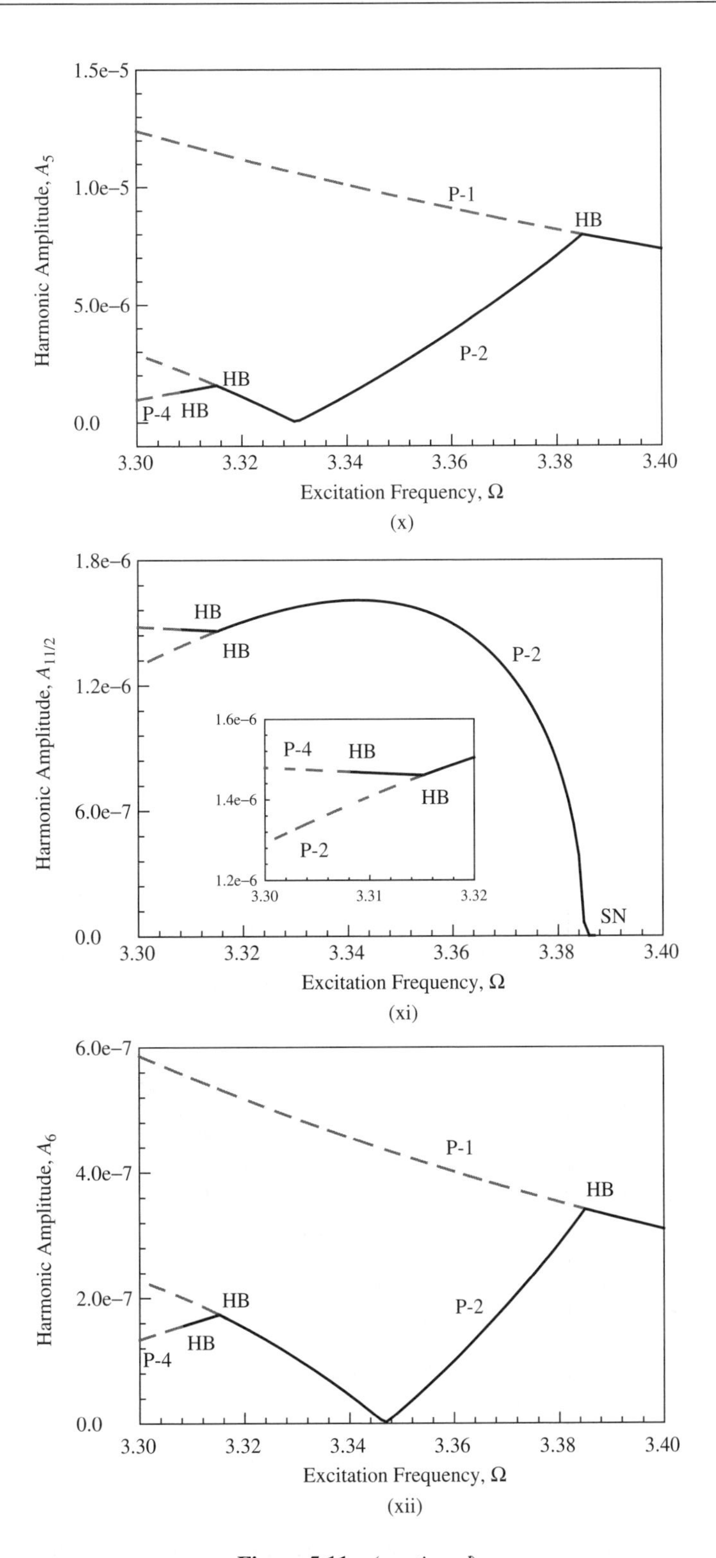

Figure 5.11 *(continued)*

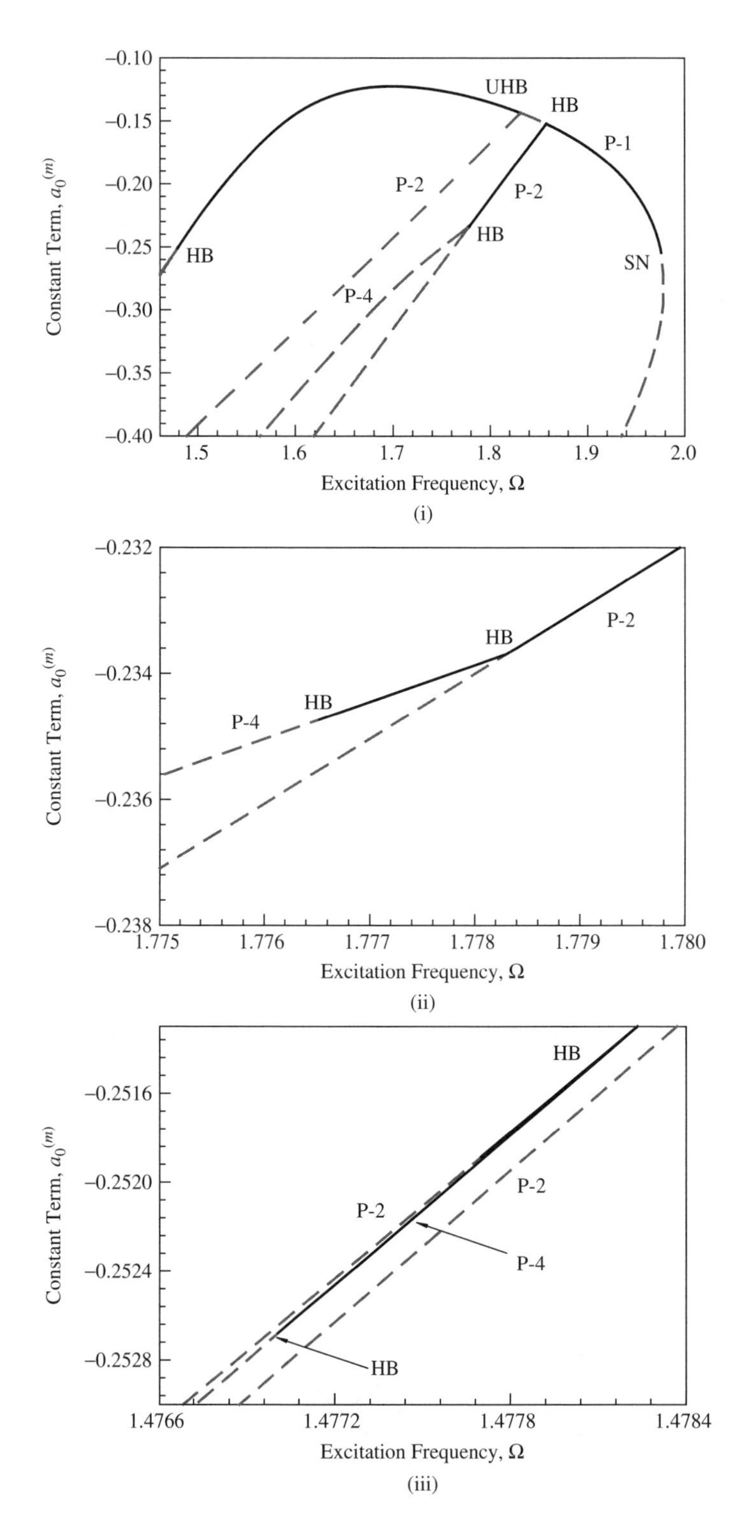

Figure 5.12 The third zoomed view for the analytical prediction of period-1 to period-4 motions based on the 120 harmonic terms (HB120): (i)–(iii) $a_0^{(m)}$ and (iv)–(xii) $A_{k/m}$ ($k = 1, 2, \ldots, 4, 6, \ldots, 12, 28, 28$, $m = 4$). Parameters: ($\delta = 0.05, \alpha = 10.0, \beta = 5.0, Q_0 = 4.5$)

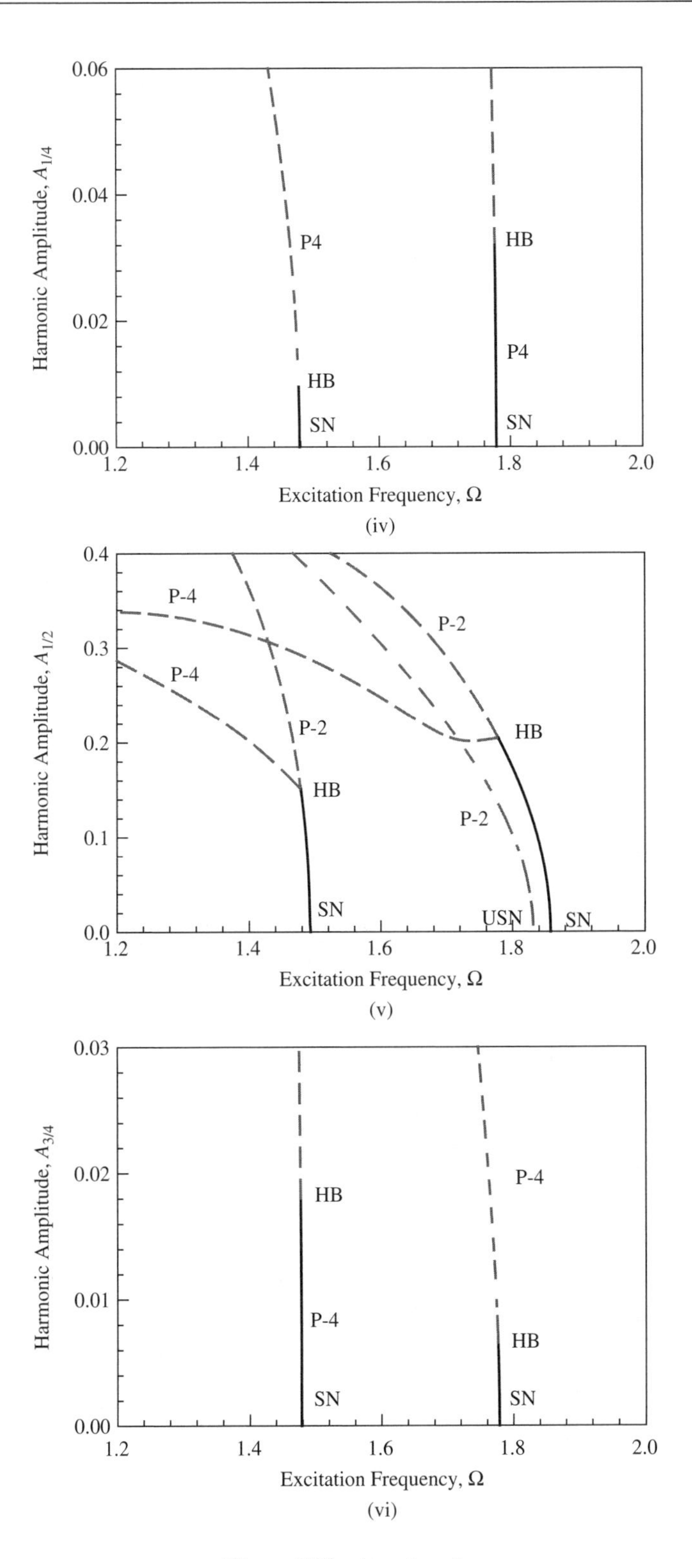

Figure 5.12 *(continued)*

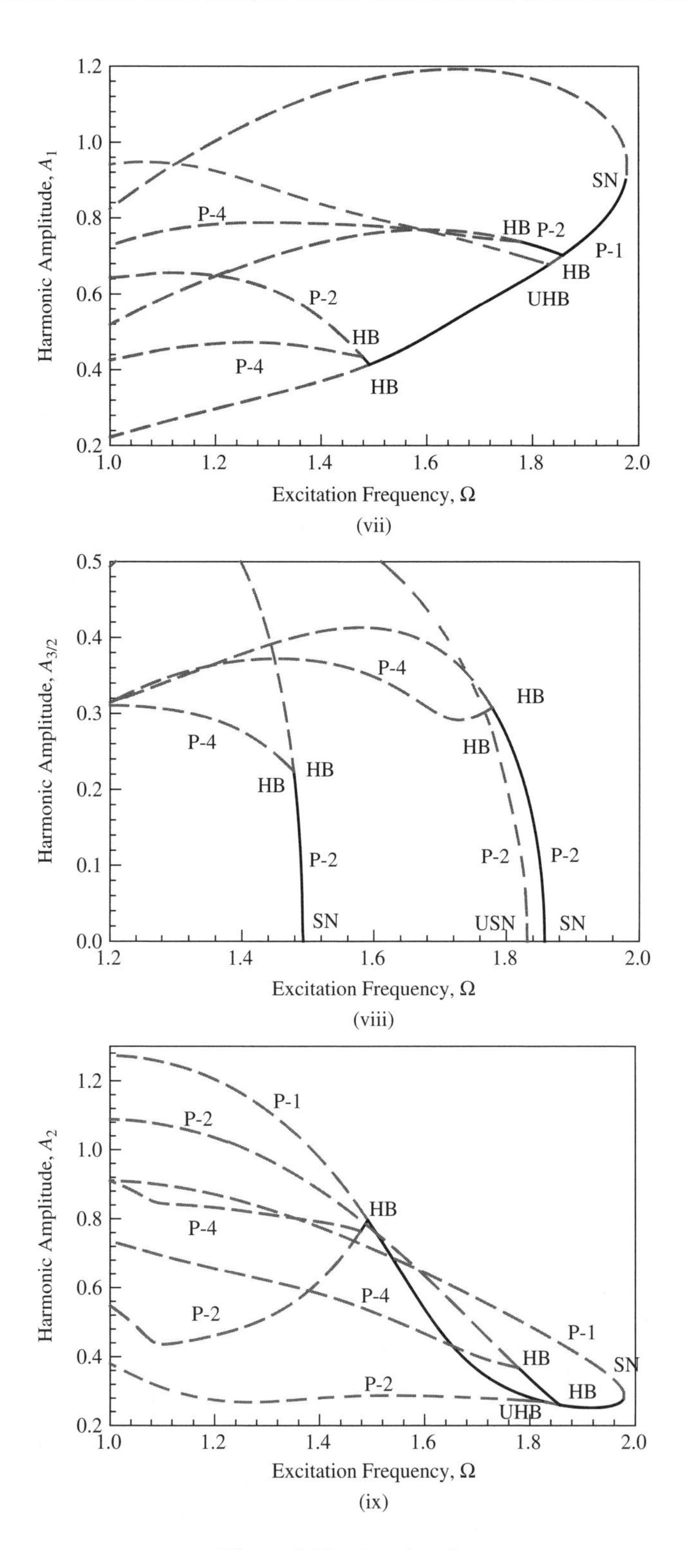

Figure 5.12 *(continued)*

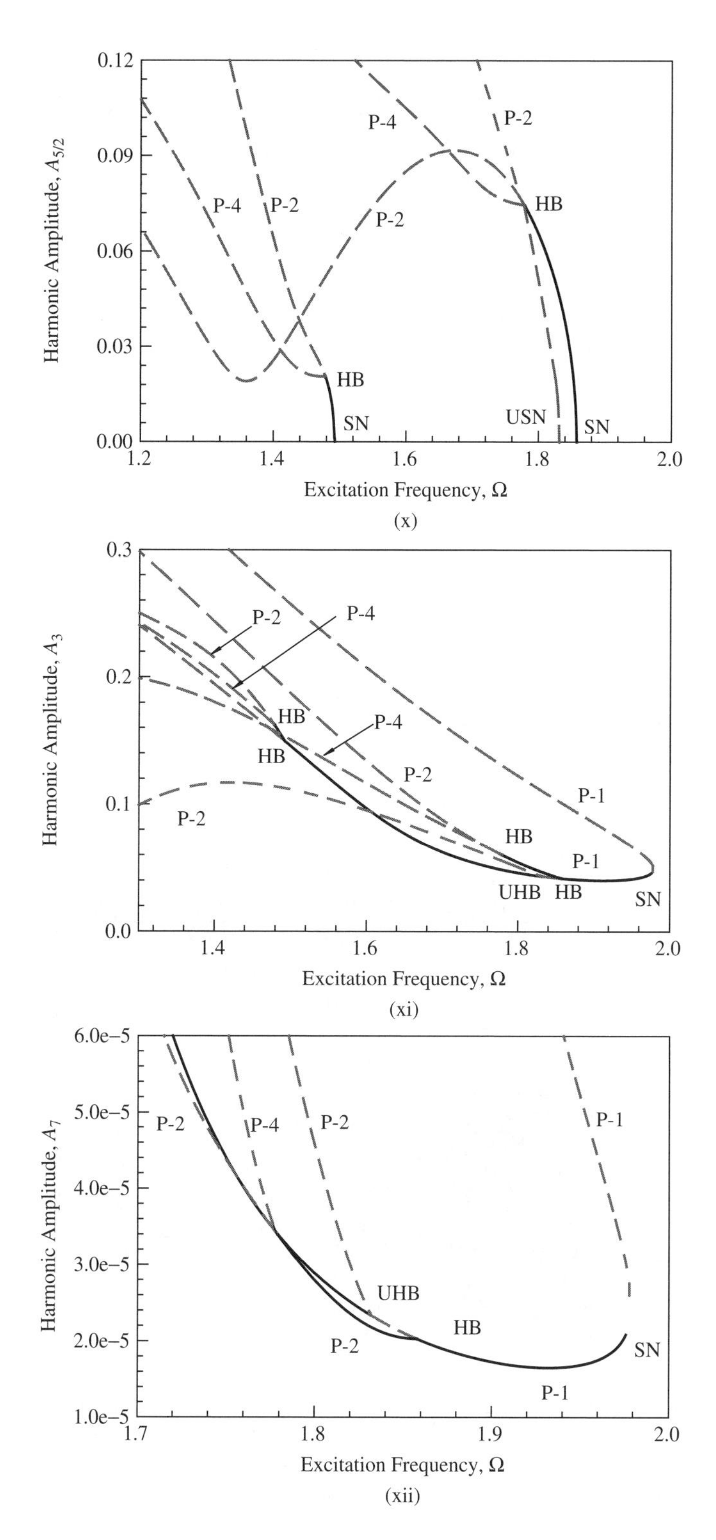

Figure 5.12 (*continued*)

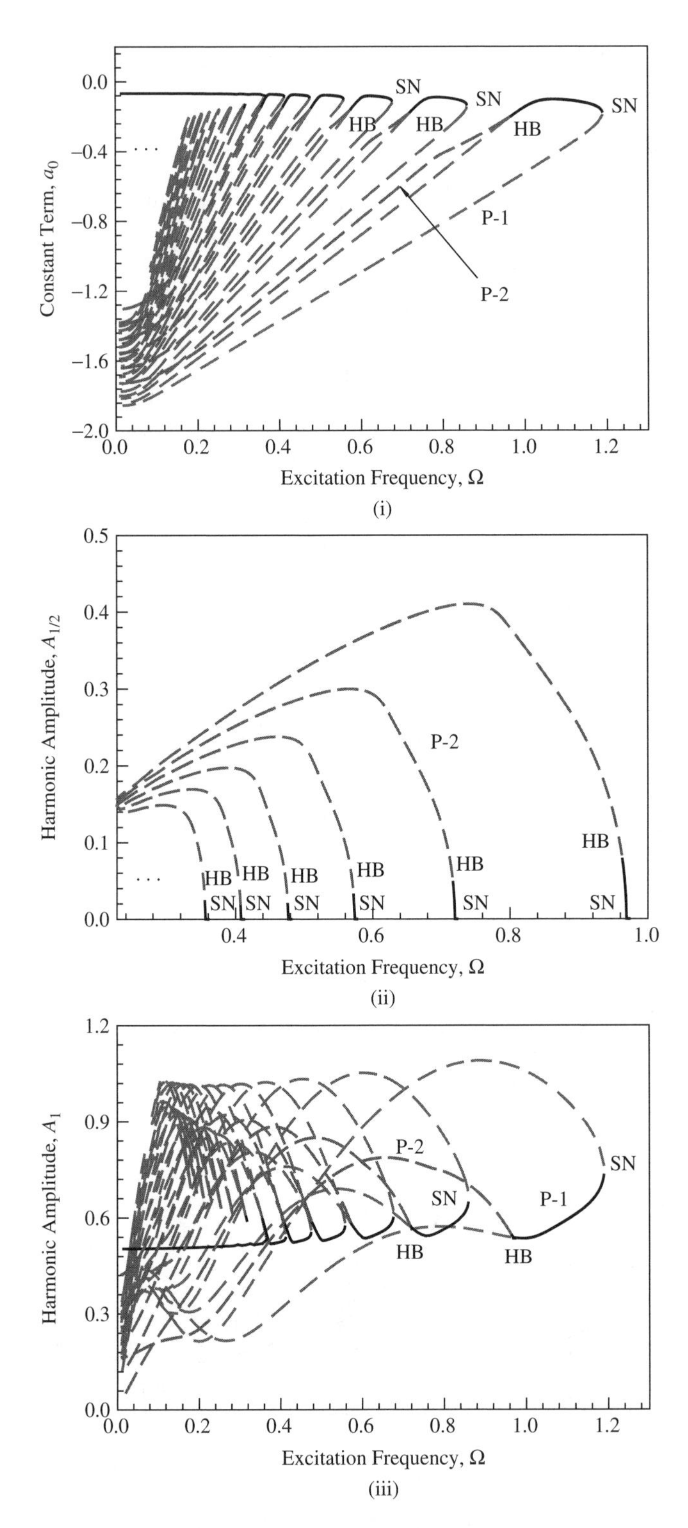

Figure 5.13 The fourth zoomed view for the analytical prediction of period-1 to period-2 motions based on the 60 harmonic terms (HB60): (i) $a_0^{(m)}$ and (ii)–(xii) $A_{k/m}$ ($k = 1, 2, \ldots, 4, 6, \ldots, 14, 15, 16, m = 2$). Parameters: ($\delta = 0.05, \alpha = 10.0, \beta = 5.0, Q_0 = 4.5$)

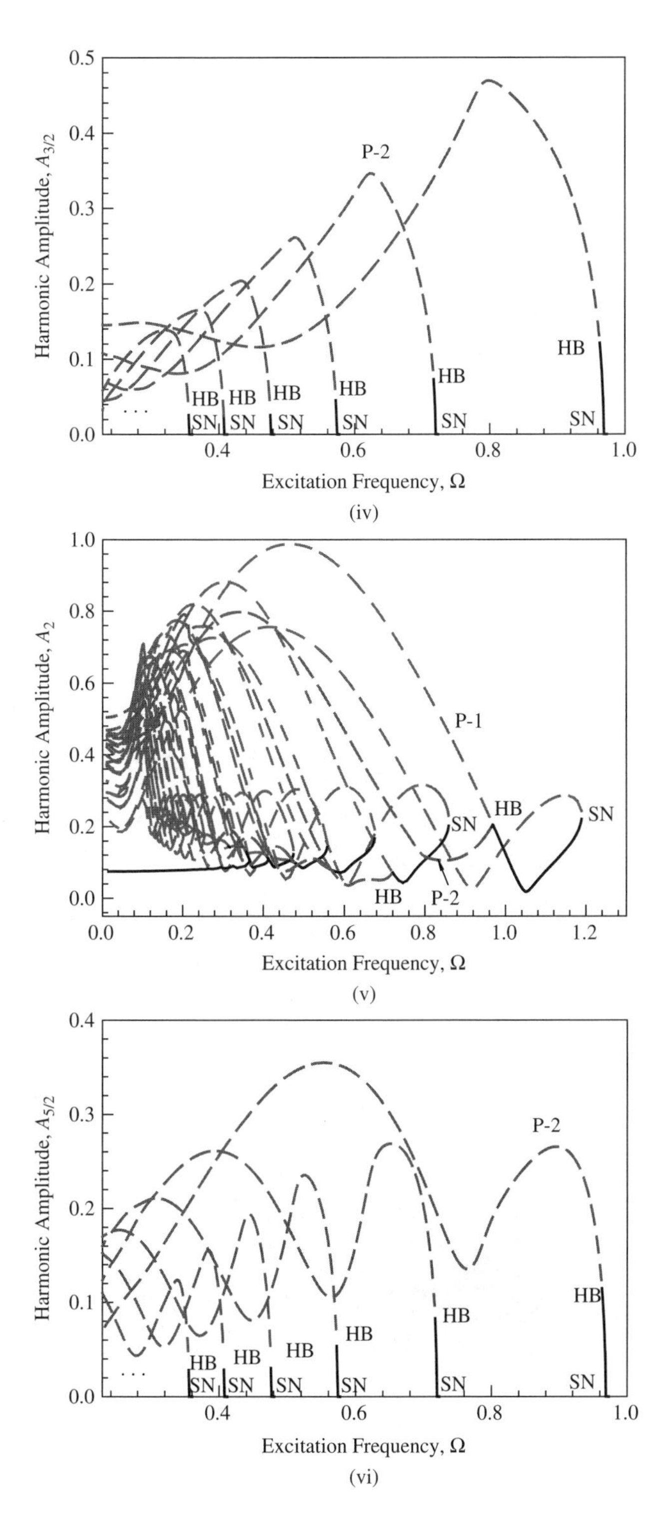

Figure 5.13 *(continued)*

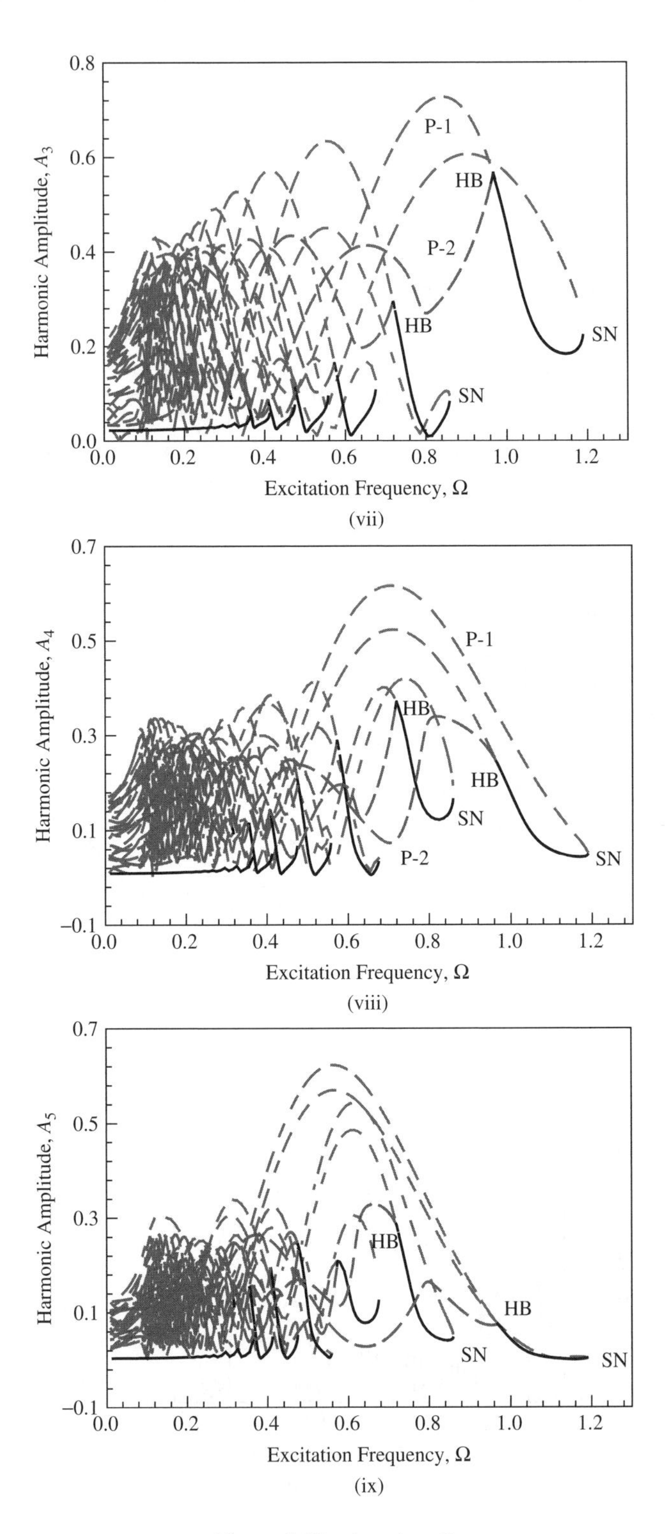

Figure 5.13 (*continued*)

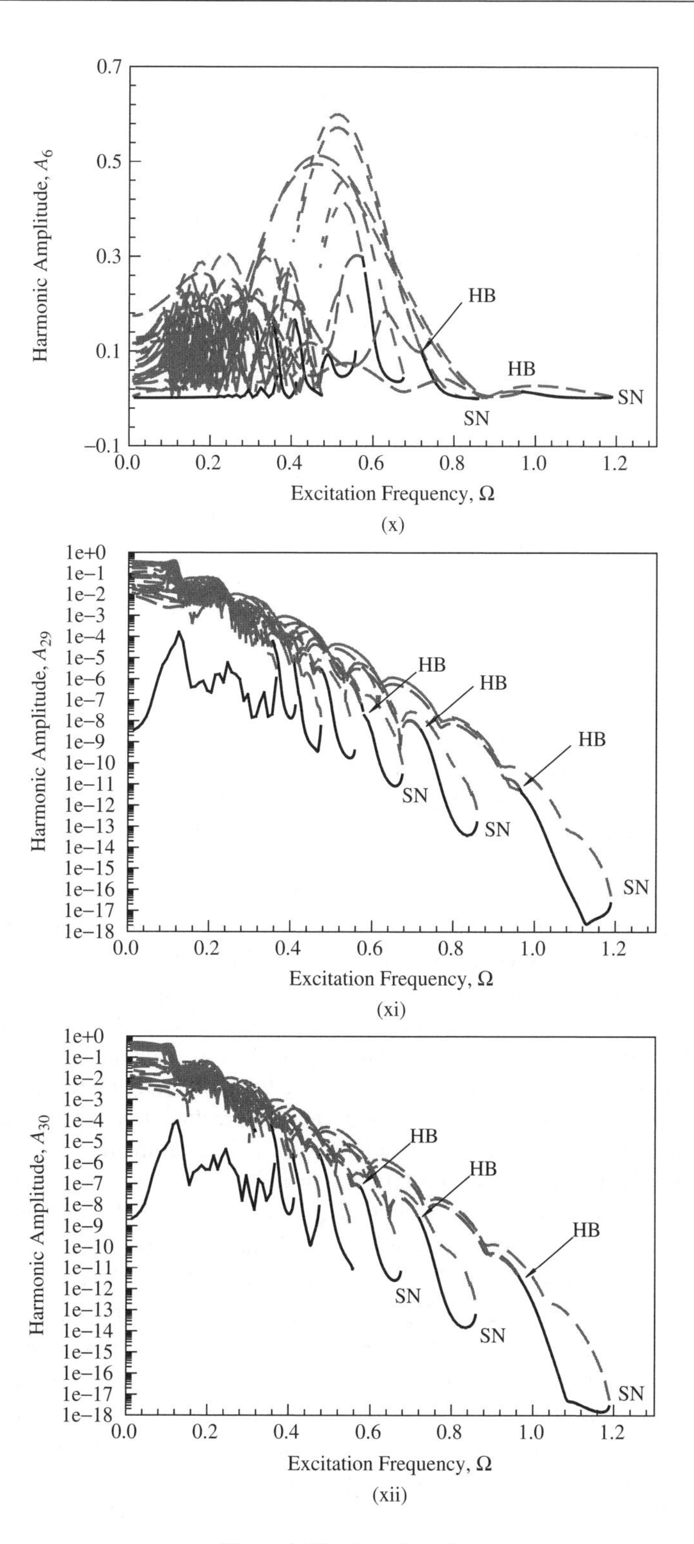

Figure 5.13 *(continued)*

similar to the results in Figure 5.9(ii). In Figure 5.9(v), the harmonic amplitude A_1 varying with excitation frequency is presented. For this amplitude, all period-1 motion through period-4 motion is included. For $\Omega > 1$, the bifurcation tree is clear, but for $\Omega < 1$, the bifurcation tree is very crowded, which means many bifurcation trees exist in such a frequency range. Similarly, the harmonic amplitude A_2 varying with excitation frequency is presented, and all period-1 motion through period-4 motion is included in such a plot. Since the zoomed view will be presented for each bifurcation tree. It is not necessary to give an overview.

From the overview of bifurcation trees for period-1 motion to period-4 motion in $\Omega \in (0, 7.0)$, it is very difficult to know the detail of each bifurcation tree. The detailed view and discussions should be presented for the analytical bifurcation tree. In Figure 5.10(i)–(xii), the analytical prediction of the bifurcation tree of period-1 motion to period-4 motion based on 120 harmonic terms is carried out for the first zoomed view for $\Omega \in (5.0, 6.0)$. In Figure 5.10(i), the constant $a_0^{(m)}$ ($m = 4$) versus excitation frequency is presented. The stable Hopf bifurcation of the period-1 motion is at $\Omega \approx 6.53$, and the period-2 motion appears, and the stable period-2 motion lies in $\Omega \in (5.418, 6.53)$. The stable Hopf bifurcation of period-2 motion occurs at $\Omega \approx 5.418$, and the period-4 motion appears. The stable period-4 motion is in $\Omega \in (5.3685, 5.418)$. The stable Hopf bifurcation of period-4 motion occurs at $\Omega \approx 5.3685$, and the period-8 motion will appear. Because the stable period-8 motion possesses a very small range of excitation frequency, it should not be computed herein. Other unstable periodic motions go to $\Omega \approx 0$. The unstable Hopf bifurcation (or subcritical Hopf bifurcation) of the period-1 motion is at $\Omega \approx 5.4180$, and the entire unstable period-2 motion from the unstable Hopf bifurcation is predicted. The analytical prediction of period-1 and period-2 motions is very clearly illustrated. The stable period-4 motion is not clear because of the small range of excitation frequency. In Figure 5.10(ii), the harmonic amplitude $A_{1/4}$ is presented. The saddle-node bifurcation of period-4 motion occurs at $\Omega \approx 5.418$ for the period-4 motion appearance, and the Hopf bifurcation of the period-4 motion at $\Omega \approx 5.3685$ is clearly illustrated. $A_{1/4} = 0$ for all stable and unstable period-1 and period-2 motions. In Figure 5.10(iii), the harmonic amplitude $A_{1/2}$ is presented for period-2 and period-4 motions, and $A_{1/2} = 0$ for the period-1 motion.

The stable and unstable saddle-node bifurcations of period-2 motions are observed, and the bifurcation of period-2 to period-4 motion is shown. In addition, $A_{1/2}$ for the unstable period-2 motion only is depicted via dashed curves, started from the unstable saddle-node bifurcation. In Figure 5.10(iv), the harmonic amplitude $A_{3/4}$ is depicted, which is similar to the harmonic amplitude $A_{1/4}$. In Figure 5.10(v), the harmonic amplitude A_1 for all period-1, period-2, and period-4 motions is presented. The bifurcation tree relative to A_1 is more complicated than the bifurcation tree relative to $A_{1/2}$. The bifurcation tree for period-1 to period-4 motion is developed. So the bifurcation trees relative to $A_{k/4}$ ($\mathrm{mod}(k, 4) \neq 0$ and $\mathrm{mod}(k, 2) \neq 0$) are very simple, as similar to Figure 5.10(ii) and (iv), and they will not be presented further herein. The harmonic amplitudes $A_{k/4}$ ($k = 6, 8, \ldots, 14, 15, 16$) are presented in Figure 5.10(vi)–(xii). $A_{3/2} \sim 5 \times 10^{-2}$, $A_2 \sim 8 \times 10^{-3}$, $A_{5/2} \sim 9 \times 10^{-4}$, $A_3 \sim 5 \times 10^{-5}$, $A_{7/2} \sim 5 \times 10^{-6}$, $A_{15/4} \sim 2.5 \times 10^{-6}$, $A_4 \sim 5 \times 10^{-7}$. The harmonic amplitudes $A_{k/4} \leq 10^{-7}$ ($k = 17, 18, \ldots, 120$) will not be presented. For such a range of excitation frequency, $A_{30} \sim 10^{-80}$. For this branch of period-1 motion, there is another bifurcation tree and the second zoomed view for $\Omega \in (3.30, 3.32)$ will be presented to show the bifurcation detail.

In Figure 5.11(i), the constant $a_0^{(m)}$ ($m = 4$) versus excitation frequency is presented but the period-4 motion and period-2 motion are too close. The further zoomed window is presented. In Figure 5.11(ii), the harmonic amplitude $A_{1/4}$ is presented. In Figure 5.11(iii), the harmonic amplitude $A_{1/2}$ is presented, and the further zoomed window is to make the bifurcation tree for period-2 and period-4 motion clear. In Figure 5.11(iv), the harmonic amplitude $A_{3/4}$ is presented, as similar to $A_{1/4}$. In Figure 5.11(v), the bifurcation tree of the harmonic amplitude A_1 is presented for period-1 motion to period-4 motion. To save space, only the harmonic amplitudes $A_{k/4}$ $(k = 6, 8, 12 \ldots, 20, 22, 24)$ are presented in Figure 5.11(vi)–(xii). $A_{3/2} \sim 1.6 \times 10^{-1}$, $A_2 \sim 8 \times 10^{-2}$, $A_3 \sim 6 \times 10^{-3}$, $A_4 \sim 3 \times 10^{-4}$, $A_5 \sim 1.5 \times 10^{-5}, A_{11/2} \sim 1.8 \times 10^{-6}, A_6 \sim 6.0 \times 10^{-7}$. The harmonic amplitudes $A_{k/4} \leq 10^{-7}$ $(k = 25, 26, \ldots, 120)$ will not be presented.

The bifurcation tree for the second branch of period-1 motion is presented for $\Omega \in (1.46, 2.0)$ in Figure 5.12(i)–(xii) through the third zoomed view. In Figure 5.12(i), the constant $a_0^{(m)}$ $(m = 4)$ versus excitation frequency is presented. This branch is different from the first branch of the bifurcation tree for period-1 motion to period-4 motion. The saddle-node bifurcation of period-1 motion is at $\Omega \approx 1.9760$, the unstable Hopf bifurcation, and stable Hopf bifurcations of period-1 motions are at $\Omega \approx 1.8580, 1.8310, 1.4930$, respectively. The Hopf bifurcations for two stable period-2 motions are at $\Omega \approx 1.7785, 1.4790$. Since the stable period-4 is very short, two further zoomed views for constants are presented in Figure 5.12(ii),(iii). The Hopf bifurcations for two period-4 motions are at $\Omega \approx 1.7765, 1.4770$. In Figure 5.12(iv), the harmonic amplitudes $A_{1/4}$ for period-4 motions are presented. The saddle-node bifurcations are for period-4 motions. In Figure 5.12(v), two bifurcation trees of harmonic amplitude $A_{1/2}$ for stable period-2 and period-4 motions are presented, and the unstable period-2 motion only is shown with unstable saddle-node bifurcation. In Figure 5.12(vi), the harmonic amplitudes $A_{3/4}$ for period-4 motions are presented, which are different from the harmonic amplitude $A_{1/4}$. The bifurcation trees of the harmonic amplitude A_1 for period-1, period-2, and period-4 motions are presented in Figure 5.12(vii) and the bifurcation trees become more complicated. As in $A_{1/2}$, the two bifurcation trees of harmonic amplitude $A_{3/2}$ are presented in Figure 5.12(viii), and the unstable period-2 motion only is given as well. In Figure 5.12(ix), the bifurcation trees of the harmonic amplitude A_2 for period-1, period-2, and period-4 motions are presented, which are quite different from the harmonic amplitude A_1. In Figure 5.12(x), the two bifurcation trees of harmonic amplitude $A_{5/2}$ are presented, which are similar to the harmonic amplitudes $A_{1/2}$ and $A_{3/2}$. To know change of harmonic amplitudes, the bifurcation trees of the harmonic amplitude A_3 for period-1, period-2, and period-4 motions are presented in Figure 5.12(xi), and with increasing excitation frequency, the harmonic amplitude decreases. However, the quantity level from $a_0^{(m)}$ to A_3 is still between $10^0 \sim 10^{-1}$. To avoid abundent illustrations, the bifurcation trees of harmonic amplitude $A_7 \in (2 \times 10^{-5}, 2 \times 10^{-3})$ are presented for such a range of excitation frequency, and the bifurcation trees only for $\Omega \in (1.7, 2.0)$ are presented in Figure 5.12(xii).

With decreasing excitation frequency, there are many branches of period-1 motions and the corresponding bifurcation trees are induced. Thus, the fourth zoomed view of the bifurcation tree for $\Omega \in (0, 1.2)$ is presented in Figure 5.13. Since the frequency ranges of stable period-4 motions are very small, only period-1 and period-2 motions in the bifurcation tree are presented. In Figure 5.13(i), the constant $a_0^{(m)} \in (-2.0, 0.0)$ versus excitation frequency are presented and six branches of bifurcation trees are observed. The corresponding saddle-node (SN) and Hopf (HB) bifurcations are observed with similar patterns. In Figure 5.13(ii), the harmonic amplitude $A_{1/2}$ is presented. The saddle-node and Hopf bifurcations are observed and the curves of $A_{1/2}$ for different branches of bifurcation trees are illustrated clearly. In Figure 5.13(iii), the harmonic amplitudes A_1 for different bifurcation trees are presented, but the bifurcation tree is very crowded. In Figure 5.13(iv), the harmonic amplitudes $A_{3/2}$ for different bifurcation trees are presented, which are different from the amplitude $A_{1/2}$. The harmonic amplitudes A_2 for different branches of the bifurcation trees are shown in

Figure 5.13(v), which are distinguished from the corresponding amplitude A_1. To look into the higher order contributions on period-2 motion, the harmonic amplitudes $A_{5/2}$ for different bifurcation trees are presented in Figure 5.13(vi). $A_{k/2} \sim 5 \times 10^{-1}$ $(k = 1, 3, 5)$ still holds and they have similar patterns. To avoid abundent illustrations, the harmonic amplitudes $A_{k/2}$ $(k = 4, 8, 10, 12)$ for period-1 and period-2 motions are presented in Figure 5.13(vii)–(x). $A_{k/2} \sim 10^0$ $(k = 2, 4, \ldots, 12)$. For higher harmonics orders, the amplitudes with the larger excitation frequency become smaller. To further confirm the reduction of the harmonic amplitudes A_{29} and A_{30} are presented in Figure 5.13(xi),(xii). The range of the two harmonic amplitudes for period-1 and period-2 motions is from 10^0 to 10^{-18}. With increasing excitation frequency, the harmonic amplitudes experience exponential reduction.

5.2.3 *Numerical Illustrations*

To verify the approximate analytical solutions of periodic motion in such a quadratic nonlinear oscillator, numerical simulations are carried out by the symplectic scheme. The initial conditions for numerical simulation are computed from the approximate analytical solutions. The numerical results are depicted by solid curves, but the analytical solutions are given by red circular symbols. The big filled circular symbols are initial conditions.

The displacement, velocity, trajectory, and amplitude spectrum of stable period-4 motion are presented in Figure 5.14 for $\Omega = 1.7778$ with initial conditions $(x_0 \approx 1.001540, \dot{x}_0 \approx 1.120421)$. This analytical solution is based on 48 harmonic terms (HB48) in the Fourier series solution of period-4 motion. In Figure 5.14(a),(b), for over 40 periods, the analytical and numerical solutions match very well. In Figure 5.14(c), analytical and numerical trajectories match very well. It is obviously observed that the period-4 motion is generated just off the Hopf bifurcation of period-2 motion. This special Hopf bifurcation of the dynamical systems coefficients of period-2 motion gives the so called period-doubling bifurcation of the period-2 motion. The doubled trajectories of period-2 motion are very close to the trajectory of period-4 motion. In Figure 5.14(d), the amplitude spectrums are $a_0^{(4)} \approx -0.233986$, $A_{1/4} \approx 0.017290$, $A_{1/2} \approx 0.204460$, $A_{3/4} \approx 3.381562\text{e-}3$, $A_1 \approx 0.737368$, $A_{5/4} \approx 0.017580$, $A_{3/2} \approx 0.307271$, $A_{7/4} \approx 0.014590$, $A_2 \approx 0.368200$, $A_{9/4} \approx 4.273747\text{e-}3$, $A_{5/2} \approx 0.074591$, $A_{11/4} \approx 2.251820\text{e-}3$, $A_3 \approx 0.060604$, $A_{13/4} \approx 2.340337\text{e-}3$, $A_{7/2} \approx 0.026013$, $A_{15/4} \approx 7.143238\text{e-}4$, $A_4 \approx 0.011351$. The harmonic amplitudes for the zoomed window are $A_{10} \sim 3 \times 10^{-8}$, $A_{41/4} \sim 1.9 \times 10^{-7}$, $A_{21/2} \sim 1.7 \times 10^{-8}$, $A_{43/4} \sim 6.8 \times 10^{-8}$, $A_{11} \sim 4.3 \times 10^{-9}$, $A_{45/4} \sim 2.8 \times 10^{-8}$, $A_{23/2} \sim 3.4 \times 10^{-9}$, $A_{47/4} \sim 1.3 \times 10^{-8}$, and $A_{12} \sim 6.0 \times 10^{-10}$.

From the bifurcation trees, period-2 and period-4 motions will be presented through trajectories and amplitude spectrums. The input data for numerical simulations are presented in Table 5.2. Since the Hopf bifurcation of period-1 motion occurs, the period-2 motion will appear. The numerical illustrations of period-2 motions are presented in Figure 5.15 for $\Omega = 6.0, 3.36, 1.8$. In Figure 5.15(i), the trajectory of period-2 motion for $\Omega = 6.0$ is presented with 10 harmonic terms (HB10) in the Fourier series solution, which looks like period-1 motion. In the vicinity of period-2 motion, the period-1 motion can be approximated by five harmonic terms in the Fourier series solution and such an approximate solution gives a good approximation of period-1 motion. The numerical and analytical solutions are in good agreement. The harmonic amplitudes in Figure 5.15(ii) are $a_0^{(2)} \approx -0.212815$, $A_{1/2} \approx 0.842766$, $A_1 \approx 0.224171$, $A_{3/2} \approx 0.012995$, $A_2 \sim 1.3 \times 10^{-3}$, $A_{5/2} \sim 9.3 \times 10^{-5}$, $A_3 \sim 7.3 \times 10^{-7}$, $A_{7/2} \sim 5.1 \times 10^{-7}$, $A_4 \sim 3.6 \times 10^{-8}$, $A_{9/2} \sim 2.5 \times 10^{-9}$, and $A_5 \sim 1.7\times\ 10^{-10}$.

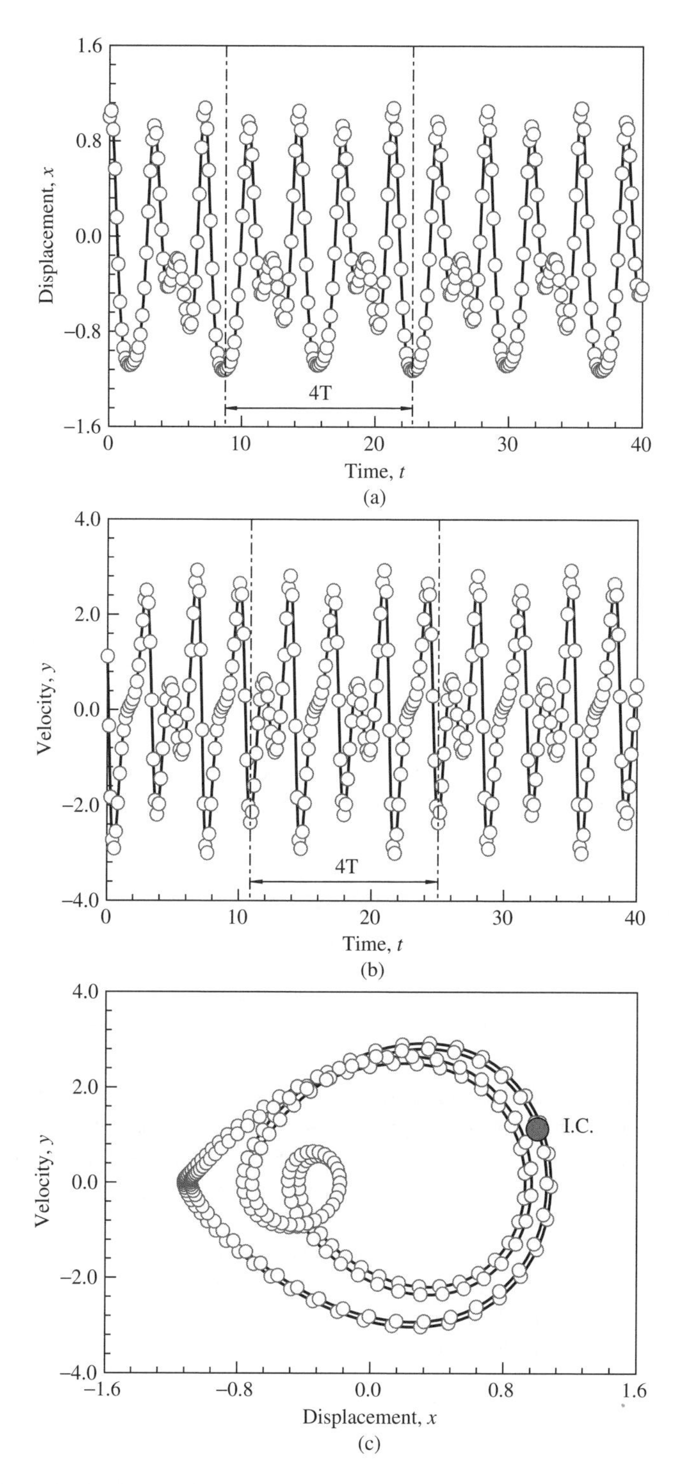

Figure 5.14 Analytical and numerical solutions of stable period-4 motion based on 48 harmonic terms (HB48): (a) displacement, (b) velocity, (c) phase plane, and (d) amplitude spectrum. Initial condition ($x_0 \approx 1.001540, \dot{x}_0 \approx 1.120421$). Parameters: ($\delta = 0.05, \alpha = 10.0, \beta = 5.0, Q_0 = 4.5, \Omega = 6.5$)

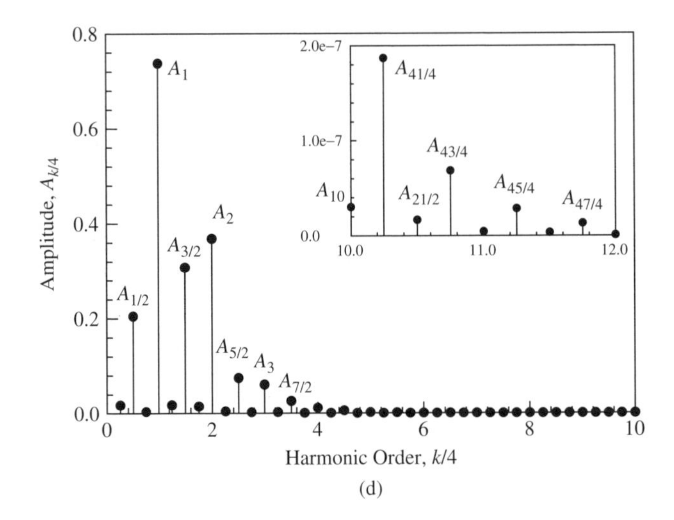

(d)

Figure 5.14 (*continued*)

Table 5.2 Input data for numerical illustrations ($\delta = 0.05, \alpha = 10.0, \beta = 5.0, Q_0 = 4.5$)

Figure no.	Ω	Initial condition $(x_0, \dot{x}_0)$	Types	Harmonics terms
Figure 5.15(i),(ii)	6.0	(−0.352338, 2.489982)	P-2	HB10 (stable)
Figure 5.15(iii),(iv)	3.36	(−1.533046, 0.085869)	P-2	HB16 (stable)
Figure 5.15(v),(vi)	1.8	(0.990675, 0.981511)	P-2	HB24 (stable)
Figure 5.16(i),(ii)	5.4	(−0.820573, 2.809580)	P-4	HB20 (stable)
Figure 5.16(iii),(iv)	3.3136	(−0.788258, 0.116180)	P-4	HB32 (stable)
Figure 5.16(v),(vi)	1.775	(0.980357, 1.122485)	P-4	HB52 (stable)
Figure 5.17(i),(ii)	1.779	(1.092702, 0.220018)	P-2	HB26 (unstable)
Figure 5.17(iii),(iv)	1.772	(0.968350, 1.122528)	P-4	HB48 (unstable)
Figure 5.17(v),(vi)	3.305	(−0.711542, 0.113098)	P-4	HB32 (unstable)

For this period-2 motion, the harmonic terms of $A_{1/2}$, A_1, and $A_{3/2}$ play an important role in the period-2 motion. For a rough approximation, one can consider three harmonic terms to determine such a period-2 motion. In Figure 5.15(iii), the trajectory of a period-2 motion for $\Omega = 3.36$ is illustrated with 16 harmonic terms (HB16) in the Fourier series solution. In the vicinity of a period-2 motion, the period-1 motion is very well approximated by eight harmonic terms in the Fourier series solution. The period-1 motion near such a period-2 motion has one cycle. Thus, the period-2 motion becomes two cycles and the two cycles are too similar, which is different from the cycles in period-1 motions with low frequency. In Figure 5.15(iv), the main harmonic amplitudes are $a_0^{(2)} \approx -0.338140$, $A_{1/2} \approx 0.329618$, $A_1 \approx 1.002062$, $A_{3/2} \approx 0.084585$, $A_2 \approx 0.062260$, $A_{5/2} \sim 8.2 \times 10^{-3}$, $A_3 \sim 3.9 \times 10^{-3}$, $A_{7/2} \sim 5.5 \times 10^{-4}$, $A_4 \sim 1.2 \times 10^{-4}$, $A_{9/2} \sim 3.0 \times 10^{-5}$, and $A_5 \sim 3.8 \times 10^{-6}$. The harmonic amplitudes in the zoomed window are $A_6 \sim 1.0 \times 10^{-8}$, $A_{13/2} \sim 6.7 \times 10^{-8}$, $A_7 \sim 1.3 \times 10^{-9}$, $A_{15/2} \sim 2.8 \times 10^{-9}$, and $A_8 \sim 6.9 \times 10^{-11}$. For the afore-discussed period-2 motions, their

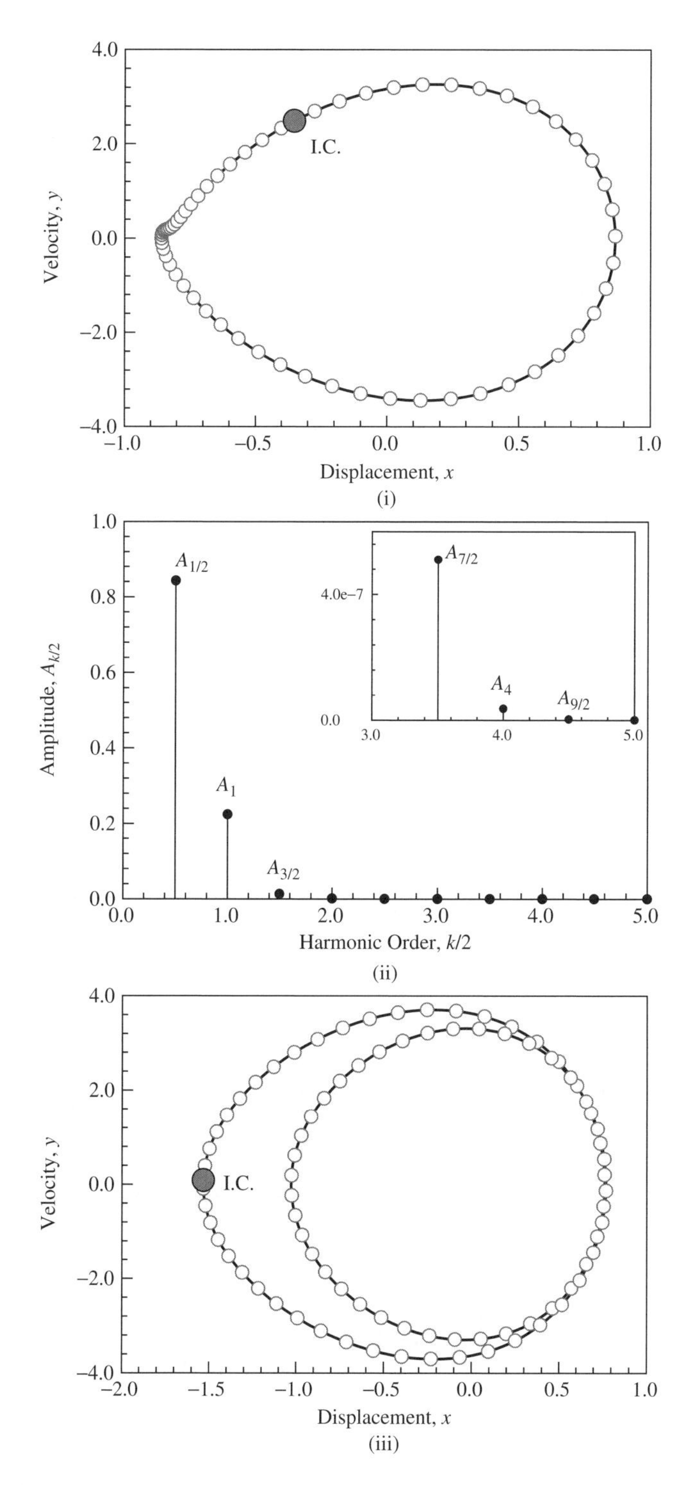

Figure 5.15 Phase plane and amplitude spectrums of period-2 motions: (i,ii): $\Omega = 6.0$ with ($x_0 \approx -0.352338, \dot{x}_0 \approx 2.489982$, HB10). (iii,iv): $\Omega = 3.36$ with ($x_0 \approx -1.533046$, $\dot{x}_0 \approx 0.085869$, HB16). (v,vi): $\Omega = 1.8$ with ($x_0 \approx 0.990675$, $\dot{x}_0 \approx 0.981511$, HB24). Parameters: ($\delta = 0.05, \alpha = 10.0, \beta = 5.0$, $Q_0 = 4.5$)

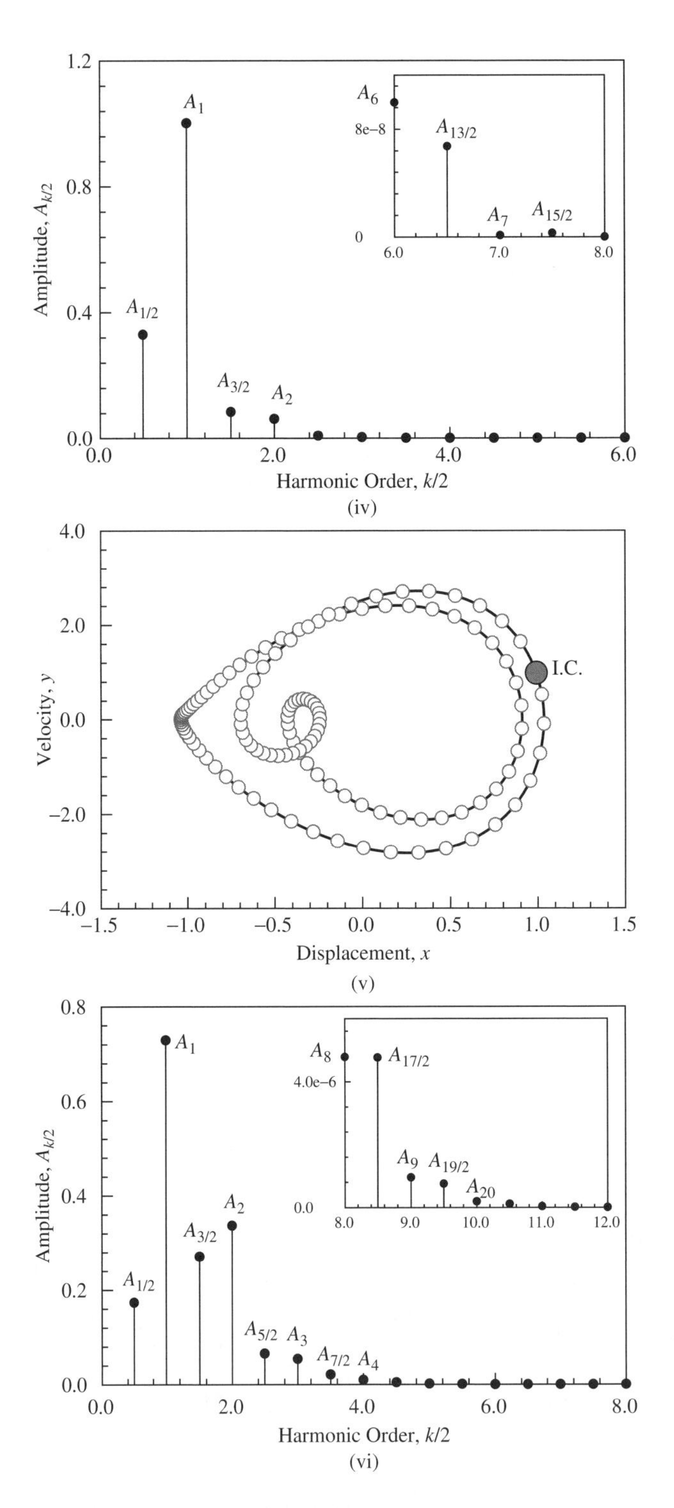

Figure 5.15 (*continued*)

harmonic amplitudes are completely different. In Figure 5.15(v), the trajectory of period-2 motion for $\Omega = 1.8$ is illustrated with 24 harmonic terms (HB24). In the vicinity of such a period-2 motion, the period-1 motion can be approximated by 12 harmonic terms in the Fourier series solution. The trajectory is different from the period-2 motion trajectory of $\Omega = 3.36$. In Figure 5.15(vi), the harmonic amplitude distribution is completely different from the afore-discussed two period-2 motions. The main harmonic amplitudes are $a_0^{(2)} \approx -0.211375$, $A_{1/2} \approx 0.173351$, $A_1 \approx 0.729050$, $A_{3/2} \approx 0.271045$, $A_2 \approx 0.336619$, $A_{5/2} \approx 0.065433$, $A_3 \approx 0.054397$, $A_{7/2} \approx 0.021060$, $A_4 \sim 9.7 \times 10^{-3}$, $A_{9/2} \sim 4.4 \times 10^{-3}$, and $A_5 \sim 1.4 \times 10^{-3}$. The harmonic amplitudes in the zoomed window are $A_8 \sim 4.8 \times 10^{-6}$, $A_{17/2} \sim 4.8 \times 10^{-6}$, $A_9 \sim 9.6 \times 10^{-7}$, $A_{19/2} \sim 7.5 \times 10^{-7}$, $A_{10} \sim 1.9 \times 10^{-7}$, $A_{21/2} \sim 1.4 \times 10^{-7}$, $A_{11} \sim 3.7 \times 10^{-8}$, $A_{23/2} \sim 1.7 \times 10^{-8}$, and $A_{12} \sim 6.8 \times 10^{-9}$.

For the Hopf bifurcation of period-2 motions, the corresponding period-4 motion will appear. In the vicinity of excitation frequencies of the above-discussed period-2 motions, the analytical and numerical solutions of period-4 motions are presented in Figure 5.16 for $\Omega = 5.4, 3.3136, 1.775$. In Figure 5.16(i), the period-4 motion for $\Omega = 5.4$ is based on the 20 harmonic terms (HB20), and such a period-4 motion possesses two cycles because the relative period-2 motion has one cycle. In the neighborhood of such a period-4 motion, the period-2 motion is very well approximated by 10 harmonic terms in the Fourier series solution. Thus, the period-4 motion can be approximately determined by 20 harmonic terms in the corresponding Fourier series solution. In Figure 5.16(ii), harmonic amplitude distribution gives $a_0^{(4)} \approx -0.437875$, $A_{1/4} \approx 0.153046$, $A_{1/2} \approx 1.110241$, $A_{3/4} \approx 0.062656$, $A_1 \approx 0.326945$, $A_{5/4} \approx 0.014690$, $A_{3/2} \approx 0.030307$, $A_{7/4} \sim 2.5 \times 10^{-3}$, $A_2 \sim 3.9 \times 10^{-3}$, $A_{9/4} \sim 3.5 \times 10^{-4}$, $A_{5/2} \sim 4.0 \times 10^{-4}$, $A_{11/4} \sim 4.5 \times 10^{-5}$, $A_3 \sim 4.1 \times 10^{-5}$, $A_{13/4} \sim 5.4 \times 10^{-6}$, $A_{7/2} \sim 4.0 \times 10^{-6}$, $A_{15/4} \sim 6.2 \times 10^{-7}$, $A_4 \sim 3.8 \times 10^{-7}$, $A_{17/4} \sim 6.8 \times 10^{-8}$, $A_{9/2} \sim 3.6 \times 10^{-8}$, $A_{9/4} \sim 7.3 \times 10^{-9}$, and $A_5 \sim 3.3 \times 10^{-9}$. From such results, the harmonic amplitudes $A_{k/4}$ ($\text{mod}(k, 4) \neq 0$ and $\text{mod}(k, 2) \neq 0$) play an important role in the period-4 motion. In Figure 5.16(iii), the trajectory of a period-4 motion in phase plane is presented for $\Omega = 3.3136$ and the analytical solution is given by the 32 harmonic terms (HB32) because the period-2 motion is well approximated by 16 harmonic terms (HB16) in the vicinity of such a period-4 motion. The period-4 motion has four cycles, and two cycles are very close because of the period-doubling. In Figure 5.16(iv), the main harmonic amplitude are $a_0^{(4)} \approx -0.367029$, $A_{1/4} \approx 0.012090$, $A_{1/2} \approx 0.541932$, $A_{3/4} \approx 0.041072$, $A_1 \approx 0.939406$, $A_{5/4} \approx$ 4.060521e-3, $A_{3/2} \approx 0.133993$, $A_{7/4} \approx$ 6.970779e-3, $A_2 \approx 0.050155$, $A_{9/4} \approx$ 1.245390e-3, $A_{5/2} \approx 0.012243$, $A_{11/4} \sim 4.9 \times 10^{-4}$, $A_3 \sim 1.7 \times 10^{-3}$, $A_{13/4} \sim 1.4 \times 10^{-4}$, $A_{7/2} \sim 7.5 \times 10^{-4}$, $A_{15/4} \sim 2.0 \times 10^{-5}$, $A_4 \sim 2.6 \times 10^{-5}$, $A_{17/4} \sim 1.0 \times 10^{-5}$, $A_{9/2} \sim 3.6 \times 10^{-5}$, and $A_5 \sim 1.5 \times 10^{-6}$. The harmonic amplitudes in the zoomed window are $A_6 \sim 1.7 \times 10^{-7}$, $A_{25/4} \sim 2.7 \times 10^{-8}$, $A_{13/2} \sim 4.9 \times 10^{-8}$, $A_{27/4} \sim 3.6 \times 10^{-9}$, $A_7 \sim 1.0 \times 10^{-8}$, $A_{29/4} \sim 1.0 \times 10^{-9}$, $A_{15/2} \sim 1.3 \times 10^{-9}$, $A_{29/4} \sim 2.5 \times 10^{-10}$, and $A_8 \sim 5.2 \times 10^{-10}$. For $\Omega = 1.775$, the trajectory of period-4 motion is presented in Figure 5.16(v) with the analytical solution possessing 52 harmonic terms (HB52) because the period-2 motion is well approximated by the 26 harmonic terms (HB26). The harmonic amplitude for analytical solution of such a period-4 motion is plotted in Figure 5.16(vi). The main harmonic amplitudes are $a_0^{(4)} \approx -0.235623$, $A_{1/4} \approx 0.043745$, $A_{1/2} \approx 0.204120$, $A_{3/4} \approx$ 8.638851e-3, $A_1 \approx 0.737768$, $A_{5/4} \approx 0.044368$, $A_{3/2} \approx 0.305761$, $A_{7/4} \approx 0.036892$, $A_2 \approx 0.369171$, $A_{9/4} \approx 0.010847$, $A_{5/2} \approx 0.074685$, $A_{11/4} \approx$ 5.739027e-3, $A_3 \approx 0.061447$, $A_{13/4} \approx$ 5.916787e-3, $A_{7/2} \approx$

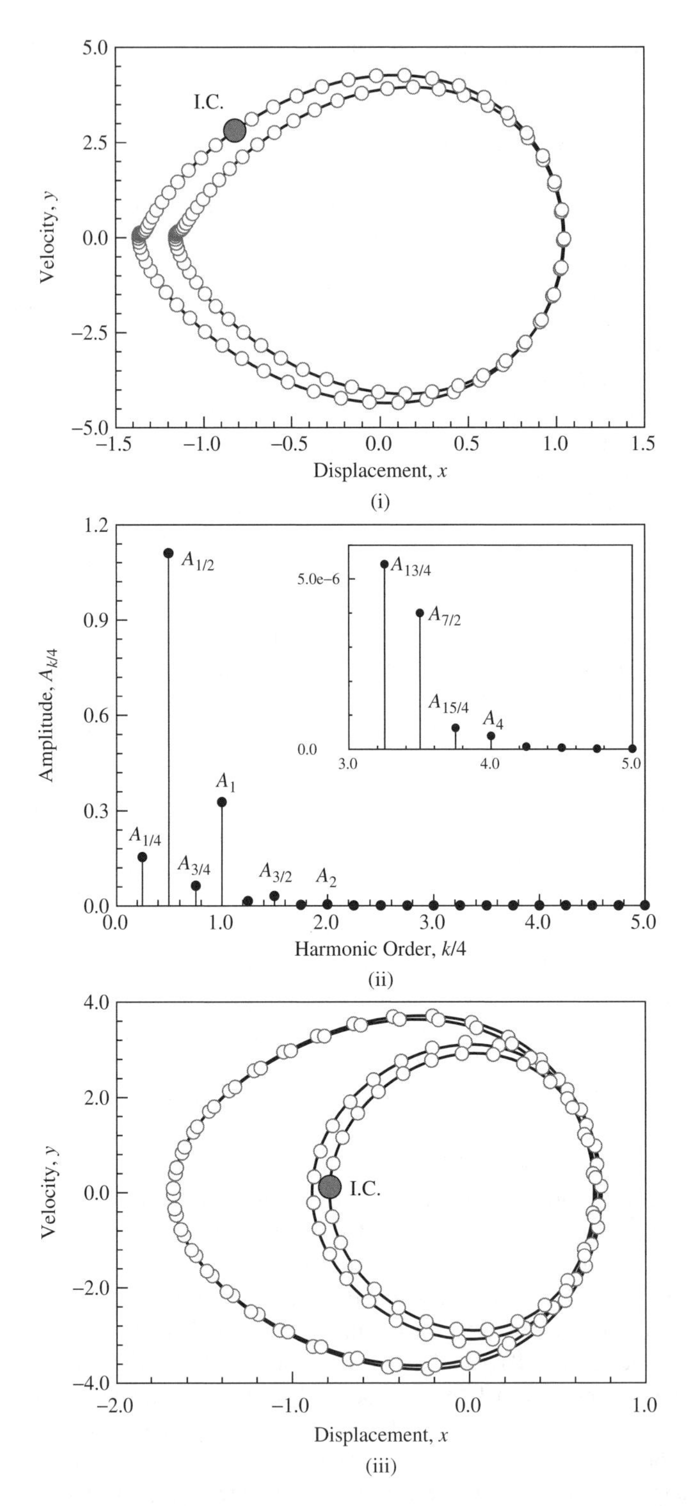

Figure 5.16 Phase plane and amplitude spectrums of stable period-4 motions: (i,ii) $\Omega = 5.4$ with ($x_0 \approx -0.820573$, $\dot{x}_0 \approx 2.809580$, HB20), (iii,iv): $\Omega = 3.3136$ with ($x_0 \approx -0.788258$, $\dot{x}_0 \approx 0.116180$, HB32), and (v,iv): $\Omega = 1.775$ with ($x_0 \approx 0.980357$, $\dot{x}_0 \approx 1.122485$, HB52). Parameters: ($\delta = 0.05, \alpha = 10.0, \beta = 5.0, Q_0 = 4.5$)

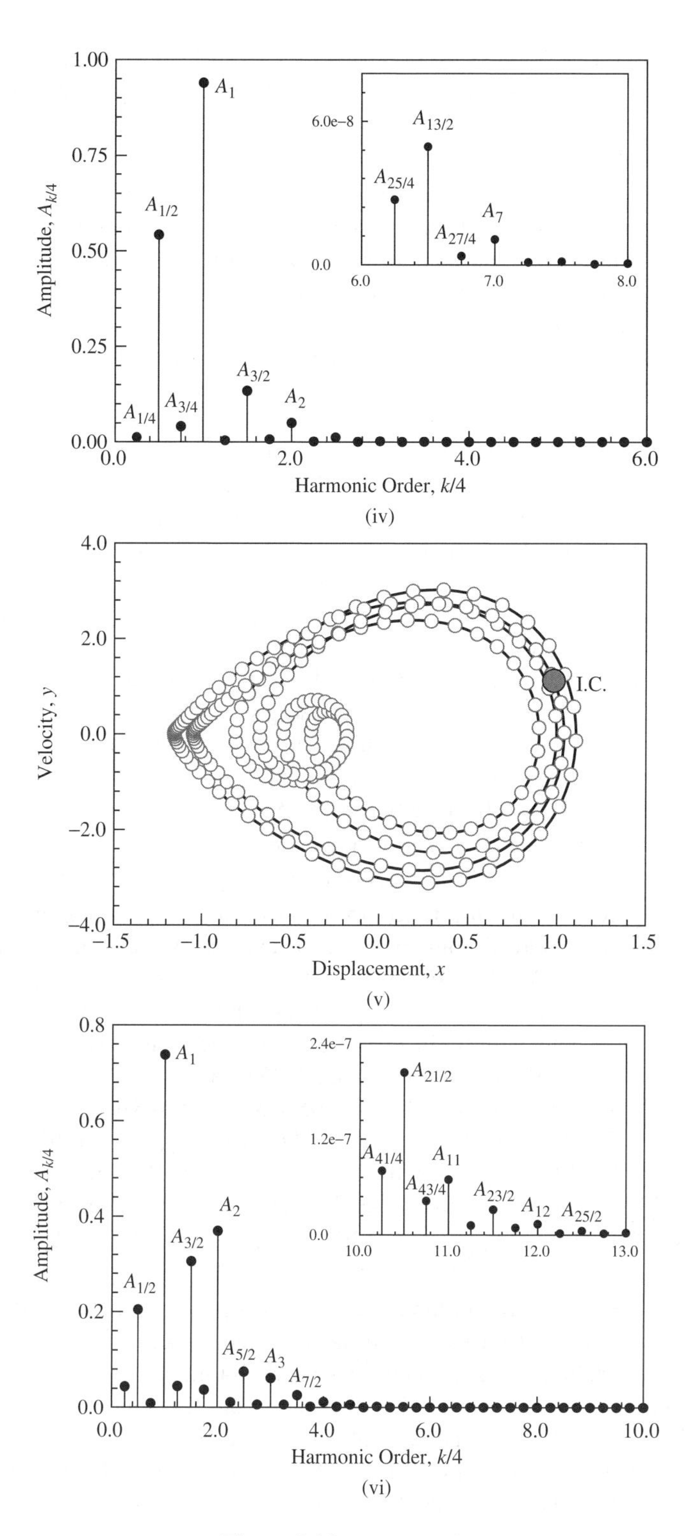

Figure 5.16 (*continued*)

0.026001, $A_{15/4} \approx 1.821890\text{e-}3$, $A_4 \approx 0.011594$, $A_{17/4} \approx 1.452601\text{e-}3$, $A_{9/2} \approx 5.734844\text{e-}3$, $A_{15/4} \approx 1.973087\text{e-}4$, and $A_5 \approx 1.710901\text{e-}3$. The harmonic amplitudes in the zoomed window are $A_{10} \sim 4.8 \times 10^{-6}$, $A_{41/4} \sim 8.0 \times 10^{-8}$, $A_{21/2} \sim 2.0 \times 10^{-7}$, $A_{43/4} \sim 4.3 \times 10^{-8}$, $A_{11} \sim 6.9 \times 10^{-8}$, $A_{45/4} \sim 1.2 \times 10^{-8}$, $A_{23/2} \sim 3.2 \times 10^{-8}$, $A_{47/4} \sim 8.7 \times 10^{-9}$, $A_{12} \sim 1.4 \times 10^{-8}$, $A_{49/4} \sim 1.7 \times 10^{-9}$, $A_{25/2} \sim 4.8 \times 10^{-9}$, $A_{51/4} \sim 1.7 \times 10^{-9}$, and $A_{13} \sim 2.5 \times 10^{-9}$.

To verify the accuracy of the analytical solutions of periodic motions, the unstable periodic motion should be illustrated. If the analytical solution is exact and the numerical simulation is without computational error, the numerical and analytical solution should match very well and the numerical solution of the unstable motion will not move away to a stable periodic motion or chaotic state. If the analytical solution of the unstable motion is not accurate, the numerical solution of unstable periodic motion with the initial condition from the analytical solution will move away very soon. In other words, if the analytical solution of the unstable motion is very accurate and the computational error of numerical simulation is very small, then, the numerical simulation of the unstable motion with the initial condition from the analytical solution will match with the analytical solution for quite a long time. The analytical and numerical solutions of unstable period-2 and period-4 motions are presented in Figure 5.17 for $\Omega = 1.779, 1.772, 3.305$. In Figure 5.17(i), the trajectory of an unstable period-2 motion is plotted for $\Omega = 1.779$, and the analytical solution of such an unstable period-2 motion is based on 26 harmonic terms (HB26). For the first 30 periods, the analytical and numerical solutions of period-2 motion have a good agreement. After the 30 periods, the numerical solutions moves away and goes to infinity outside of the corresponding separatrix. In Figure 5.17(ii), the main harmonic amplitudes for the analytical solution of period-2 motion are $a_0^{(2)} \approx -0.232980$, $A_{1/2} \approx 0.133982$, $A_1 \approx 0.699304$, $A_{3/2} \approx 0.274961$, $A_2 \approx 0.274593$, $A_{5/2} \approx 0.071686$, $A_3 \approx 0.055582$, $A_{7/2} \approx 0.019680$, $A_4 \approx 0.011054$, $A_{9/2} \sim 4.3 \times 10^{-3}$, and $A_5 \sim 2.1 \times 10^{-3}$. The harmonic amplitudes in the zoomed window are $A_{10} \sim 4.4 \times 10^{-7}$, $A_{21/2} \sim 1.8 \times 10^{-7}$, $A_{11} \sim 7.7 \times 10^{-8}$, $A_{23/2} \sim 3.2 \times 10^{-8}$, $A_{12} \sim 1.4 \times 10^{-8}$, $A_{25/2} \sim 5.7 \times 10^{-9}$, and $A_{13} \sim 2.4 \times 10^{-9}$. In Figure 5.17(iii), the trajectory of unstable period-4 motion for $\Omega = 1.772$ is presented, and the analytical solution of such unstable period-4 motion possesses 48 harmonic terms (HB48). For the first 50 periods, the analytical and numerical solutions have a good match. After 50 periods, the numerical solution moves away from the analytical solution and goes into infinity. In Figure 5.17(iv), the main harmonic amplitudes are $a_0^{(4)} \approx -0.237380$, $A_{1/4} \approx 0.060278$, $A_{1/2} \approx 0.203765$, $A_{3/4} \approx 0.012037$, $A_1 \approx 0.738190$, $A_{5/4} \approx 0.060972$, $A_{3/2} \approx 0.304194$, $A_{7/4} \approx 0.050804$, $A_2 \approx 0.370219$, $A_{9/4} \approx 0.015001$, $A_{5/2} \approx 0.074795$, $A_{11/4} \approx 7.969294\text{e-}3$, $A_3 \approx 0.062351$, $A_{13/4} \approx 8.146255\text{e-}3$, $A_{7/2} \approx 0.025992$, $A_{15/4} \approx 2.531950\text{e-}3$, $A_4 \approx 0.011857$, $A_{17/4} \approx 2.011493\text{e-}3$, $A_{9/2} \approx 5.780753\text{e-}3$, $A_{15/4} \approx 2.813292\text{e-}4$, and $A_5 \approx 1.784252\text{e-}3$. The harmonic amplitudes in the zoomed window are $A_{10} \sim 3.3 \times 10^{-7}$, $A_{41/4} \sim 1.6 \times 10^{-7}$, $A_{21/2} \sim 2.2 \times 10^{-7}$, $A_{43/4} \sim 6.0 \times 10^{-8}$, $A_{11} \sim 7.0 \times 10^{-8}$, $A_{45/4} \sim 1.8 \times 10^{-8}$, $A_{23/2} \sim 3.5 \times 10^{-8}$, $A_{47/4} \sim 1.2 \times 10^{-9}$, and $A_{12} \sim 1.4 \times 10^{-9}$. For the two unstable period-2 and period-4 motions, their excitation frequencies are very close. Consider an unstable period-4 motion with a large excitation frequency (i.e., $\Omega = 3.305$). For such an unstable period-4 motion, its trajectory is presented in Figure 5.17(v) and the analytical solution is based on 32 harmonic terms. For the first 30 periods, the analytical and numerical solutions match very well. After 30 periods, the numerical solution moves away from the analytical solutions and goes to infinity. The main harmonic amplitudes of the analytical solutions for the unstable period-4 motions in

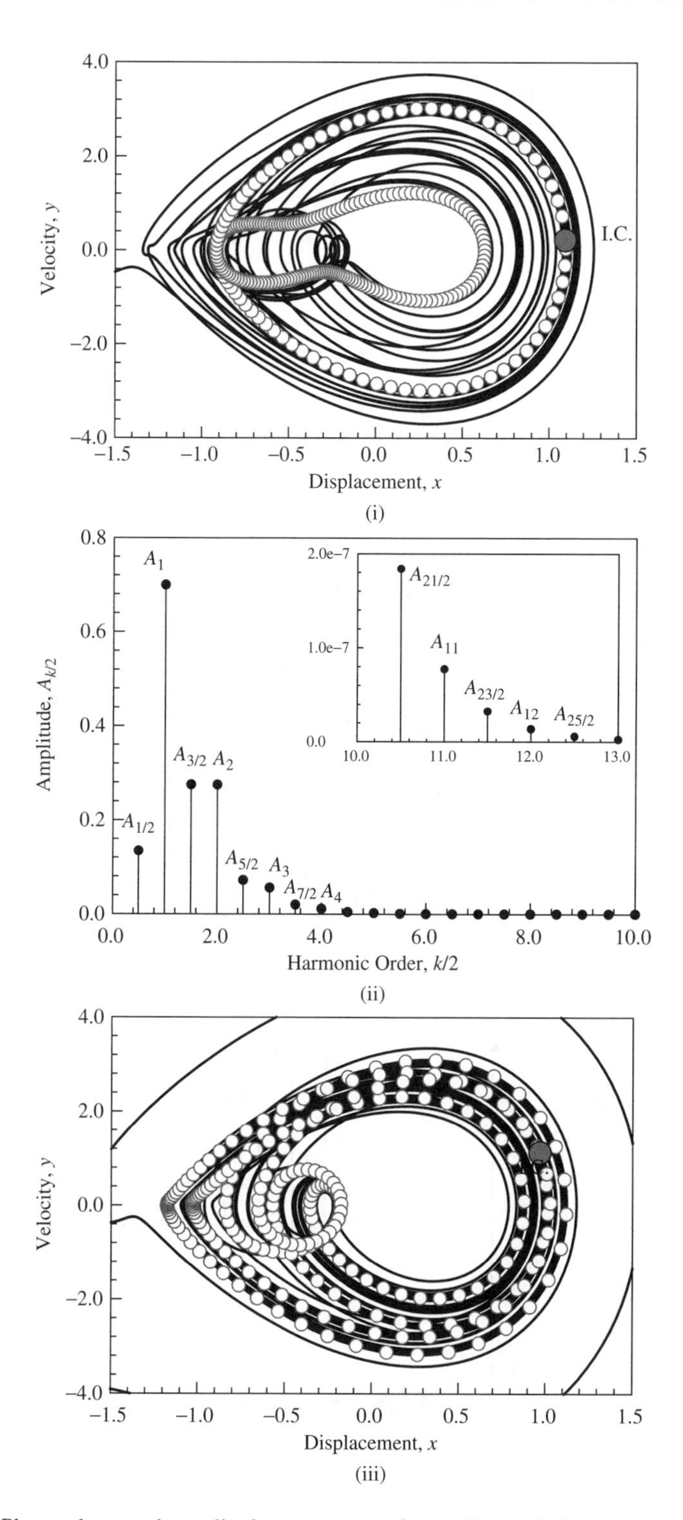

Figure 5.17 Phase plane and amplitude spectrums of unstable periodic motions: (i,ii) period-2 motion: $\Omega = 1.779$ with ($x_0 \approx 1.092702$, $\dot{x}_0 \approx 0.220018$, HB26). (iii,iv) Period-4 motion: $\Omega = 1.772$ with ($x_0 \approx 0.968350$, $\dot{x}_0 \approx 1.122528$, HB48). (v,vi) Period-4 motion: $\Omega = 3.305$ with ($x_0 \approx -0.711542$, $\dot{x}_0 \approx 0.113098$, HB32). Parameters: ($\delta = 0.05, \alpha = 10.0, \beta = 5.0, Q_0 = 4.5$)

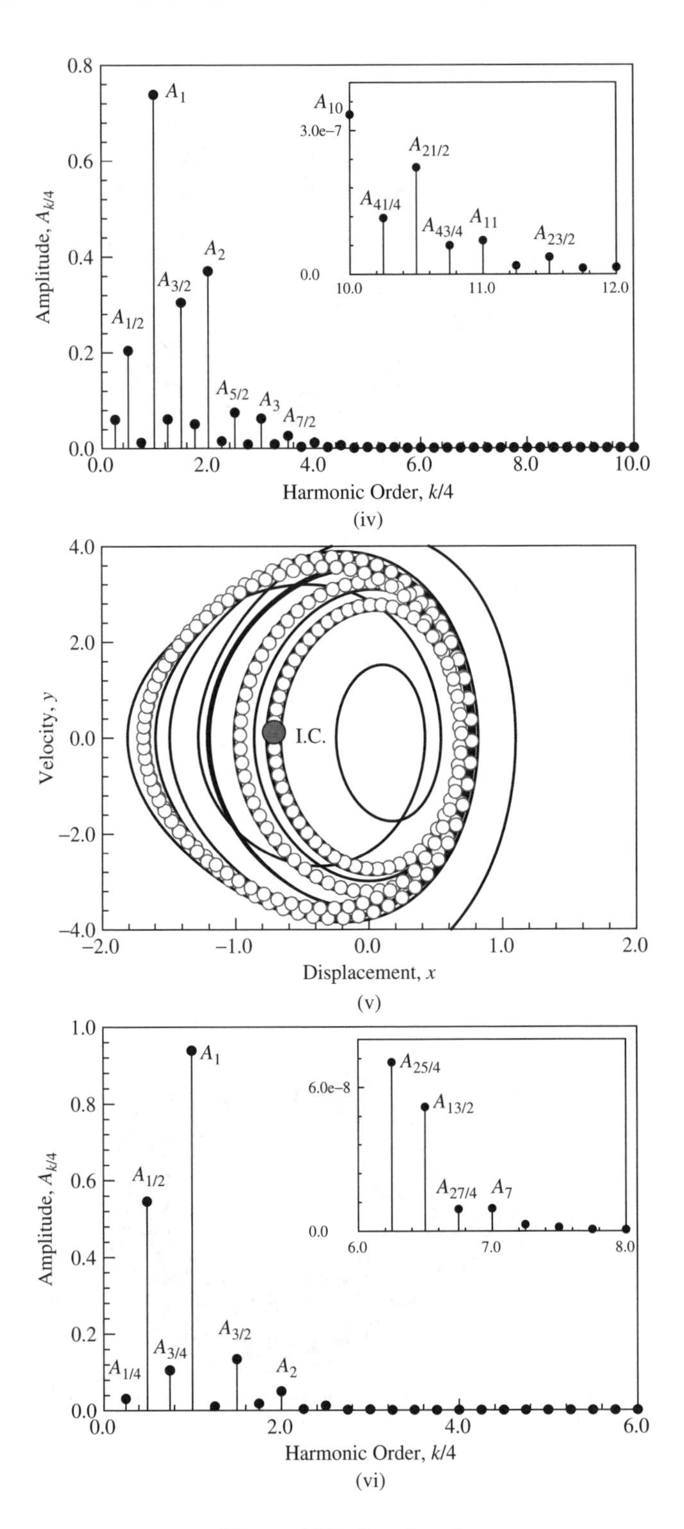

Figure 5.17 (*continued*)

Figure 5.17(vi) are $a_0^{(4)} \approx -0.371836$, $A_{1/4} \approx 0.030860$, $A_{1/2} \approx 0.545223$, $A_{3/4} \approx 0.104964$, $A_1 \approx 0.938345$, $A_{5/4} \approx 0.010593$, $A_{3/2} \approx 0.134037$, $A_{7/4} \approx 0.017856$, $A_2 \approx 0.050273$, $A_{9/4} \approx 3.226846\text{e-}3$, $A_{5/2} \approx 0.012186$, $A_{11/4} \sim 1.3\times10^{-3}$, $A_3 \approx 1.8\times10^{-3}$, $A_{13/4} \sim 3.6\times10^{-4}$, $A_{7/2} \sim 7.4\times10^{-4}$, $A_{15/4} \sim 5.3\times10^{-5}$, $A_4 \sim 3.0\times10^{-5}$, $A_{17/4} \sim 2.6\times10^{-5}$, $A_{9/2} \sim 3.6\times10^{-5}$, $A_{19/4} \sim 8.3\times10^{-7}$, and $A_5 \sim 1.2\times10^{-6}$. The harmonic amplitudes in the zoomed window are $A_6 \sim 1.5\times10^{-7}$, $A_{25/4} \sim 7.0\times10^{-8}$, $A_{13/2} \sim 5.2\times10^{-8}$, $A_{27/4} \sim 9.1\times10^{-9}$, $A_7 \sim 9.4\times10^{-9}$, $A_{29/4} \sim 2.7\times10^{-9}$, $A_{15/2} \sim 1.5\times10^{-9}$, $A_{29/4} \sim 6.3\times10^{-10}$, and $A_8 \sim 4.7\times10^{-10}$.

5.3 Arbitrary Periodical Forcing

Consider a quadratic nonlinear oscillator with arbitrary periodical forcing as

$$\ddot{x} + \delta\dot{x} + \alpha x + \beta x^2 = f(t) \tag{5.69}$$

where the arbitrary periodical forcing is given by

$$f(t) = Q_0^{(0)} + \sum_{n=1}^{N_1} Q_{0n}^{(c)} \cos(n\Omega t) + Q_{0n}^{(s)} \sin(n\Omega t) \tag{5.70}$$

An analytical solution of period-m motion in Equation (5.69) is assumed as

$$x^{(m)*}(t) = a_0^{(m)}(t) + \sum_{k=1}^{N} b_{k/m}(t)\cos\left(\frac{k}{m}\theta\right) + c_{k/m}(t)\sin\left(\frac{k}{m}\theta\right). \tag{5.71}$$

where $a_0^{(m)}(t)$, $b_{k/m}(t)$ and $c_{k/m}(t)$ vary with time and $\theta = \Omega t$. The first and second order of derivatives of $x^*(t)$ are

$$\begin{aligned}\dot{x}^{(m)*}(t) &= \dot{a}_0^{(m)} + \sum_{k=1}^{N}\left(\dot{b}_{k/m} + \frac{k\Omega}{m}c_{k/m}\right)\cos\left(\frac{k\theta}{m}\right)\\ &+ \left(\dot{c}_{k/m} - \frac{k\Omega}{m}b_{k/m}\right)\sin\left(\frac{k\theta}{m}\right)\end{aligned} \tag{5.72}$$

$$\begin{aligned}\ddot{x}^{(m)*}(t) &= \ddot{a}_0^{(m)} + \sum_{k=1}^{N}\left(\ddot{b}_{k/m} + 2\frac{k\Omega}{m}\dot{c}_{k/m} - \left(\frac{k\Omega}{m}\right)^2 b_{k/m}\right)\cos\left(\frac{k\theta}{m}\right)\\ &+ \left(\ddot{c}_{k/m} - 2\frac{k\Omega}{m}\dot{b}_{k/m} - \left(\frac{k\Omega}{m}\right)^2 c_{k/m}\right)\sin\left(\frac{k\theta}{m}\right)\end{aligned} \tag{5.73}$$

Substitution of Equations (5.71)–(5.73) into Equation (5.69) and application of the virtual work principle for a basis of constant, $\cos(k\theta/m)$ and $\sin(k\theta/m)$ $(k = 1, 2, \ldots)$ as a set of virtual displacements gives

$$\begin{aligned}\ddot{a}_0^{(m)} &= F_0^{(m)}(a_0^{(m)}, \mathbf{b}^{(m)}, \mathbf{c}^{(m)}, \dot{a}_0^{(m)}, \dot{\mathbf{b}}^{(m)}, \dot{\mathbf{c}}^{(m)})\\ \ddot{b}_{k/m} &+ 2\frac{k\Omega}{m}\dot{c}_{k/m} - \left(\frac{k\Omega}{m}\right)^2 b_{k/m}\\ &= F_{1k}^{(m)}(a_0^{(m)}, \mathbf{b}^{(m)}, \mathbf{c}^{(m)}, \dot{a}_0^{(m)}, \dot{\mathbf{b}}^{(m)}, \dot{\mathbf{c}}^{(m)})\end{aligned}$$

$$\begin{aligned}
&\ddot{c}_{k/m} - 2\frac{k\Omega}{m}\dot{b}_{k/m} - \left(\frac{k\Omega}{m}\right)^2 c_{k/m} \\
&\quad = F_{2k}^{(m)}(a_0^{(m)}, \mathbf{b}^{(m)}, \mathbf{c}^{(m)}, \dot{a}_0^{(m)}, \dot{\mathbf{b}}^{(m)}, \dot{\mathbf{c}}^{(m)}) \\
&\text{for } k = 1, 2, \ldots, N
\end{aligned} \tag{5.74}$$

where

$$\begin{aligned}
&F_0^{(m)}(a_0^{(m)}, \mathbf{b}^{(m)}, \mathbf{c}^{(m)}, \dot{a}_0^{(m)}, \dot{\mathbf{b}}^{(m)}, \dot{\mathbf{c}}^{(m)}) \\
&\quad = \frac{1}{mT}\int_0^{mT} F(x^{(m)*}, \dot{x}^{(m)}, t)dt \\
&\quad = -\delta\dot{a}_0^{(m)} - \alpha a_0^{(m)} - \beta(a_0^{(m)})^2 - \frac{\beta}{2}\sum_{i=1}^{N}(b_{i/m}^2 + c_{i/m}^2) + Q_0^{(0)} \\
&F_{1k}^{(m)}(a_0^{(m)}, \mathbf{b}^{(m)}, \mathbf{c}^{(m)}, \dot{a}_0^{(m)}, \dot{\mathbf{b}}^{(m)}, \dot{\mathbf{c}}^{(m)}) \\
&\quad = \frac{2}{mT}\int_0^{mT} F(x^{(m)*}, \dot{x}^{(m)}, t)\cos\left(\frac{k}{m}\Omega t\right)dt \\
&\quad = -\delta\left(\dot{b}_{k/m} + c_{k/m}\frac{k\Omega}{m}\right) - \alpha b_{k/m} - 2\beta a_0^{(m)} b_{k/m} - f_{1k/m} + \sum_{n=1}^{N_1} Q_{0n}^{(c)}\delta_k^{nm} \\
&F_{2k}^{(m)}(a_0^{(m)}, \mathbf{b}^{(m)}, \mathbf{c}^{(m)}, \dot{a}_0^{(m)}, \dot{\mathbf{b}}^{(m)}, \dot{\mathbf{c}}^{(m)}) \\
&\quad = \frac{2}{mT}\int_0^{mT} F(x^{(m)*}, \dot{x}^{(m)}, t)\sin\left(\frac{k}{m}\Omega t\right)dt \\
&\quad = -\delta\left(\dot{c}_{k/m} - b_{k/m}\frac{k\Omega}{m}\right) - \alpha c_{k/m} - 2\beta a_0^{(m)} c_{k/m} - f_{2k/m} + \sum_{n=1}^{N_1} Q_{0n}^{(s)}\delta_k^{nm}
\end{aligned} \tag{5.75}$$

and

$$\begin{aligned}
f_{1k/m} &= \beta\sum_{i=1}^{N}\sum_{j=1}^{N}[(b_{i/m}b_{j/m} + c_{i/m}c_{j/m})\delta_{j-i}^k \\
&\quad + \frac{1}{2}\left(b_{i/m}b_{j/m} - c_{i/m}c_{j/m}\right)\delta_{i+j}^k\Big] \\
f_{2k/m} &= \beta\sum_{i=1}^{N}\sum_{j=1}^{N} b_{i/m}c_{j/m}(\delta_{i+j}^k + \delta_{j-i}^k - \delta_{i-j}^k).
\end{aligned} \tag{5.76}$$

As in Section 5.2, using Equations (5.42)–(5.65), the period-m solutions for a quadratic nonlinear oscillator with an arbitrary periodic forcing can be determined and the corresponding stability and bifurcation can be determined.

6

Time-Delayed Nonlinear Oscillators

In this chapter, analytical solutions for period-m motions in a time-delayed, nonlinear oscillator will be presented through the Fourier series, and the stability and bifurcation analyses of the corresponding periodic motions are presented through the eigenvalue analysis. Analytical bifurcation trees of periodic motions to chaos will be presented through the frequency-amplitude curves. Trajectories and amplitude spectrums of periodic motions in such a time-delayed nonlinear system are illustrated numerically for a better understanding of time-delayed nonlinear dynamical systems.

6.1 Analytical Solutions

In this section, the analytical solutions of periodic motions in time-delayed, nonlinear systems will be developed through finite Fourier series. Consider a periodically forced, time-delayed, nonlinear oscillator as

$$\begin{aligned}\ddot{x} + \delta_1\dot{x} - \delta_2\dot{x}^{\tau} + \alpha_1 x - \alpha_2 x^{\tau} + \beta_1 x^2 - \beta_2(x^{\tau})^2 + \gamma_1 x^3 - \gamma_2(x^{\tau})^3 \\ = Q_0 \cos\Omega t \end{aligned} \tag{6.1}$$

where $x^{\tau} = x(t-\tau)$ and $\dot{x}^{\tau} = \dot{x}(t-\tau)$. Coefficients in Equation (6.1) are δ_1 and δ_2 for linear damping, α_1 and α_2 for linear springs and delay, β_1 and β_2 for quadratic nonlinearity and delay, γ_1 and γ_2 for the cubic nonlinearity and delay, Q_0 and Ω for excitation amplitude and frequency, respectively. In Luo (2012a), the standard form of Equation (6.1) can be written as

$$\ddot{x} = F(x, \dot{x}, x^{\tau}, \dot{x}^{\tau}, t) \tag{6.2}$$

where

$$\begin{aligned}F(x, \dot{x}, x^{\tau}, \dot{x}^{\tau}, t) = -\delta_1\dot{x} + \delta_2\dot{x}^{\tau} - \alpha_1 x + \alpha_2 x^{\tau} - \beta_1 x^2 + \beta_2(x^{\tau})^2 \\ - \gamma_1 x^3 + \gamma_2(x^{\tau})^3 + Q_0\cos(\Omega t). \end{aligned} \tag{6.3}$$

The analytical solution of period-m motion for the above equation is

$$x^{(m)*} = a_0^{(m)}(t) + \sum_{k=1}^{N} b_{k/m}(t)\cos\left(\frac{k}{m}\theta\right) + c_{k/m}(t)\sin\left(\frac{k}{m}\theta\right),$$

Toward Analytical Chaos in Nonlinear Systems, First Edition. Albert C. J. Luo.
© 2014 John Wiley & Sons, Ltd. Published 2014 by John Wiley & Sons, Ltd.

$$x^{\tau(m)*} = a_0^{\tau(m)}(t) + \sum_{k=1}^{N} \left[b_{k/m}^{\tau}(t) \cos\left(\frac{k}{m}\theta^{\tau}\right) - c_{k/m}^{\tau}(t) \sin\left(\frac{k}{m}\theta^{\tau}\right) \right] \cos\left(\frac{k}{m}\theta\right)$$
$$+ \left[b_{k/m}^{\tau}(t) \sin\left(\frac{k}{m}\theta^{\tau}\right) + c_{k/m}^{\tau}(t) \cos\left(\frac{k}{m}\theta^{\tau}\right) \right] \sin\left(\frac{k}{m}\theta\right) \tag{6.4}$$

where $a_0^{\tau(m)}(t) = a_0^{(m)}(t-\tau), b_{k/m}^{\tau}(t) = b_{k/m}(t-\tau), c_{k/m}^{\tau}(t) = c_{k/m}(t-\tau)$ $\theta = \Omega t$ and $\theta^{\tau} = \Omega \tau$. The coefficients $a_0^{(m)}(t)$, $b_{k/m}(t)$, $c_{k/m}(t)$ vary with time, and the derivatives of the foregoing equations are

$$\dot{x}^{(m)*} = \dot{a}_0^{(m)}(t) + \sum_{k=1}^{N} \left[\dot{b}_{k/m}(t) + \frac{k\Omega}{m} c_{k/m}(t) \right] \cos\left(\frac{k}{m}\theta\right)$$
$$+ \left[\dot{c}_{k/m}(t) - \frac{k\Omega}{m} b_{k/m}(t) \right] \sin\left(\frac{k}{m}\theta\right),$$
$$\dot{x}^{\tau(m)*} = \dot{a}_0^{\tau(m)}(t) + \sum_{k=1}^{N} \left\{ \left[\dot{b}_{k/m}^{\tau}(t) + \frac{k\Omega}{m} c_{k/m}^{\tau}(t) \right] \cos\left(\frac{k}{m}\theta^{\tau}\right) \right.$$
$$\left. - \left[\dot{c}_{k/m}^{\tau}(t) - \frac{k\Omega}{m} b_{k/m}^{\tau}(t) \right] \sin\left(\frac{k}{m}\theta^{\tau}\right) \right\} \cos\left(\frac{k}{m}\theta\right)$$
$$+ \left\{ \left[\dot{b}_{k/m}^{\tau}(t) + \frac{k\Omega}{m} c_{k/m}^{\tau}(t) \right] \sin\left(\frac{k}{m}\theta^{\tau}\right) \right.$$
$$\left. + \left[\dot{c}_{k/m}^{\tau}(t) - \frac{k\Omega}{m} b_{k/m}^{\tau}(t) \right] \cos\left(\frac{k}{m}\theta^{\tau}\right) \right\} \sin\left(\frac{k}{m}\theta\right). \tag{6.5}$$

$$\ddot{x}^{(m)*} = \ddot{a}_0^{(m)}(t) + \sum_{k=1}^{N} \left[\ddot{b}_{k/m}(t) + 2\frac{k\Omega}{m} \dot{c}_{k/m}(t) - \left(\frac{k\Omega}{m}\right)^2 b_{k/m}(t) \right] \cos\left(\frac{k}{m}\theta\right)$$
$$+ \left[\ddot{c}_{k/m}(t) - 2\frac{k\Omega}{m} \dot{b}_{k/m}(t) - \left(\frac{k\Omega}{m}\right)^2 c_{k/m}(t) \right] \sin\left(\frac{k}{m}\theta\right). \tag{6.6}$$

Substitution of Equations (6.4)–(6.6) into Equation (6.1) and application of the virtual work principle for a basis of constant, $\cos(k\theta/m)$ and $\sin(k\theta/m)$ $(k = 1, 2, \ldots)$ as a set of virtual displacements gives

$$\ddot{a}_0^{(m)} = F_0^{(m)}(\mathbf{z}^{(m)}, \dot{\mathbf{z}}^{(m)}; \mathbf{z}^{\tau(m)}, \dot{\mathbf{z}}^{\tau(m)}),$$
$$\ddot{b}_{k/m} + 2\frac{k\Omega}{m}\dot{c}_{k/m} - \left(\frac{k\Omega}{m}\right)^2 b_{k/m} = F_{1k}^{(m)}(\mathbf{z}^{(m)}, \dot{\mathbf{z}}^{(m)}; \mathbf{z}^{\tau(m)}, \dot{\mathbf{z}}^{\tau(m)}),$$
$$\ddot{c}_{k/m} - 2\frac{k\Omega}{m}\dot{b}_{k/m} - \left(\frac{k\Omega}{m}\right)^2 c_{k/m} = F_{2k}^{(m)}(\mathbf{z}^{(m)}, \dot{\mathbf{z}}^{(m)}; \mathbf{z}^{\tau(m)}, \dot{\mathbf{z}}^{\tau(m)})$$
$$k = 1, 2, \ldots, N \tag{6.7}$$

where

$$\mathbf{z}^{(m)} = (a_0^{(m)}, \mathbf{b}^{(m)}, \mathbf{c}^{(m)})^{\mathrm{T}} \text{ and } \dot{\mathbf{z}}^{(m)} = (\dot{a}_0^{(m)}, \dot{\mathbf{b}}^{(m)}, \dot{\mathbf{c}}^{(m)})^{\mathrm{T}},$$
$$\mathbf{z}^{\tau(m)} = (a_0^{\tau(m)}, \mathbf{b}^{\tau(m)}, \mathbf{c}^{\tau(m)})^{\mathrm{T}} \text{ and } \dot{\mathbf{z}}^{\tau(m)} = (\dot{a}_0^{\tau(m)}, \dot{\mathbf{b}}^{\tau(m)}, \dot{\mathbf{c}}^{\tau(m)})^{\mathrm{T}};$$

$$\mathbf{b}^{(m)} = (b_1^{(m)}, b_2^{(m)}, \ldots, b_N^{(m)})^{\mathrm{T}} \text{ and } \mathbf{b}^{\tau(m)} = (b_1^{\tau(m)}, b_2^{\tau(m)}, \ldots, b_N^{\tau(m)})^{\mathrm{T}},$$
$$\mathbf{c}^{(m)} = (c_1^{(m)}, c_2^{(m)}, \ldots, c_N^{(m)})^{\mathrm{T}} \text{ and } \mathbf{c}^{\tau(m)} = (c_1^{\tau(m)}, c_2^{\tau(m)}, \ldots, c_N^{\tau(m)})^{\mathrm{T}};$$
$$\dot{\mathbf{b}}^{(m)} = (\dot{b}_1^{(m)}, \dot{b}_2^{(m)}, \ldots, \dot{b}_N^{(m)})^{\mathrm{T}} \text{ and } \dot{\mathbf{b}}^{\tau(m)} = (\dot{b}_1^{\tau(m)}, \dot{b}_2^{\tau(m)}, \ldots, \dot{b}_N^{\tau(m)})^{\mathrm{T}},$$
$$\dot{\mathbf{c}}^{(m)} = (\dot{c}_1^{(m)}, \dot{c}_2^{(m)}, \ldots, \dot{c}_N^{(m)})^{\mathrm{T}} \text{ and } \dot{\mathbf{c}}^{\tau(m)} = (\dot{c}_1^{\tau(m)}, \dot{c}_2^{\tau(m)}, \ldots, \dot{c}_N^{\tau(m)})^{\mathrm{T}}; \tag{6.8}$$

$$\begin{aligned}
&F_0^{(m)}(\mathbf{z}^{(m)*}, \dot{\mathbf{z}}^{(m)}; \mathbf{z}^{\tau(m)}, \dot{\mathbf{z}}^{\tau(m)}) \\
&\quad = \frac{1}{mT}\int_0^{mT} F(x^{(m)*}, \dot{x}^{(m)*}, x^{(m)\tau*}, \dot{x}^{(m)\tau*}, t)dt, \\
&F_{1k}^{(m)}(\mathbf{z}^{(m)}, \dot{\mathbf{z}}^{(m)}; \mathbf{z}^{\tau(m)}, \dot{\mathbf{z}}^{\tau(m)}) \\
&\quad = \frac{2}{mT}\int_0^{mT} F(x^{(m)*}, \dot{x}^{(m)*}, x^{(m)\tau*}, \dot{x}^{(m)\tau*}, t)\cos\left(\frac{k}{m}\Omega t\right)dt, \\
&F_{2k}^{(m)}(\mathbf{z}^{(m)}, \dot{\mathbf{z}}^{(m)}; \mathbf{z}^{\tau(m)}, \dot{\mathbf{z}}^{\tau(m)}) \\
&\quad = \frac{2}{mT}\int_0^{mT} F(x^{(m)*}, \dot{x}^{(m)*}, x^{\tau(m)*}, \dot{x}^{\tau(m)*}, t)\sin\left(\frac{k}{m}\Omega t\right)dt \\
&\text{for } k = 1, 2, \ldots N.
\end{aligned} \tag{6.9}$$

Therefore, the coefficients of constant, $\cos(k\theta/m)$ and $\sin(k\theta/m)$ for the function of $f(x, \dot{x}, t)$ can be obtained. The constant term is given by

$$\begin{aligned}
F_0^{(m)}(\mathbf{z}^{(m)}, \dot{\mathbf{z}}^{(m)}; \mathbf{z}^{\tau(m)}, \dot{\mathbf{z}}^{\tau(m)}) = &-\delta_1\dot{a}_0^{(m)} + \delta_2\dot{a}_0^{\tau(m)} - \alpha_1 a_0^{(m)} + \alpha_2 a_0^{\tau(m)} \\
&- \beta_1 f_1^{(0)} + \beta_2 f_1^{\tau(0)} - \gamma_1 f_2^{(0)} + \gamma_2 f_2^{\tau(0)}
\end{aligned} \tag{6.10}$$

The constants caused by quadratic nonlinearity are

$$\begin{aligned}
f_1^{(0)} &= (a_0^{(m)})^2 + \sum_{i=1}^{N}\frac{1}{2}(b_{i/m}^2 + c_{i/m}^2), \\
f_1^{\tau(0)} &= (a_0^{\tau(m)})^2 + \sum_{i=1}^{N}\frac{1}{2}[(b_{i/m}^{\tau})^2 + (c_{i/m}^{\tau})^2]
\end{aligned} \tag{6.11}$$

The constants caused by cubic nonlinearity are

$$f_2^{(0)} = (a_0^{(m)})^3 + \sum_{q=1}^{3}\sum_{i=1}^{N}\sum_{j=1}^{N}\sum_{l=1}^{N} f_2^{(0)}(i, j, l, q) \tag{6.12}$$

with

$$f_2^{(0)}(i, j, l, 1) = \frac{1}{2N^2}(3a_0^{(m)}b_{i/m}^2 + 3a_0^{(m)}c_{i/m}^2),$$

$$
\begin{aligned}
f_2^{(0)}(i,j,l,2) &= \frac{1}{4} b_{i/m} b_{j/m} b_{l/m} (\delta_{i+j}^l + \delta_{i+l}^j + \delta_{j+l}^i), \\
f_2^{(0)}(i,j,l,3) &= \frac{3}{4} b_{i/m} c_{j/m} c_{l/m} (\delta_{i+j}^l + \delta_{i+l}^j - \delta_{j+l}^i).
\end{aligned} \tag{6.13}
$$

The time-delay related constants, caused by cubic nonlinearity, are

$$
f_2^{\tau(0)} = (a_0^{\tau(m)})^3 + \sum_{q=1}^{11} \sum_{i=1}^{N} \sum_{j=1}^{N} \sum_{l=1}^{N} f_2^{\tau(0)}(i,j,l,q) \tag{6.14}
$$

with

$$
\begin{aligned}
f_2^{\tau(0)}(i,j,l,1) &= \frac{3}{2N^2} a_0^{\tau(m)} [(b_i^\tau)^2 + (c_i^\tau)^2], \\
f_2^{\tau(0)}(i,j,l,2) &= \frac{1}{4} b_{i/m}^\tau b_{j/m}^\tau b_{l/m}^\tau \cos\left(\frac{i}{m}\Omega\tau\right) \cos\left(\frac{j}{m}\Omega\tau\right) \cos\left(\frac{l}{m}\Omega\tau\right) \Delta_1^{(0)}, \\
f_2^{\tau(0)}(i,j,l,3) &= -\frac{1}{4} c_{i/m}^\tau c_{j/m}^\tau c_{l/m}^\tau \sin\left(\frac{i}{m}\Omega\tau\right) \sin\left(\frac{j}{m}\Omega\tau\right) \sin\left(\frac{l}{m}\Omega\tau\right) \Delta_1^{(0)}, \\
f_2^{\tau(0)}(i,j,l,4) &= -\frac{3}{4} b_{i/m}^\tau b_{j/m}^\tau c_{l/m}^\tau \cos\left(\frac{i}{m}\Omega\tau\right) \cos\left(\frac{j}{m}\Omega\tau\right) \sin\left(\frac{l}{m}\Omega\tau\right) \Delta_1^{(0)}, \\
f_2^{\tau(0)}(i,j,l,5) &= \frac{3}{4} b_{i/m}^\tau c_{j/m}^\tau c_{l/m}^\tau \cos\left(\frac{i}{m}\Omega\tau\right) \sin\left(\frac{j}{m}\Omega\tau\right) \sin\left(\frac{l}{m}\Omega\tau\right) \Delta_1^{(0)}, \\
f_2^{\tau(0)}(i,j,l,6) &= \frac{3}{4} b_{i/m}^\tau c_{j/m}^\tau c_{l/m}^\tau \cos\left(\frac{i}{m}\Omega\tau\right) \cos\left(\frac{j}{m}\Omega\tau\right) \cos\left(\frac{l}{m}\Omega\tau\right) \Delta_2^{(0)}, \\
f_2^{\tau(0)}(i,j,l,7) &= \frac{3}{4} b_{i/m}^\tau b_{j/m}^\tau b_{l/m}^\tau \cos\left(\frac{i}{m}\Omega\tau\right) \sin\left(\frac{j}{m}\Omega\tau\right) \sin\left(\frac{l}{m}\Omega\tau\right) \Delta_2^{(0)}, \\
f_2^{\tau(0)}(i,j,l,8) &= -\frac{3}{4} c_{i/m}^\tau c_{j/m}^\tau c_{l/m}^\tau \sin\left(\frac{i}{m}\Omega\tau\right) \cos\left(\frac{j}{m}\Omega\tau\right) \cos\left(\frac{l}{m}\Omega\tau\right) \Delta_2^{(0)}, \\
f_2^{\tau(0)}(i,j,l,9) &= -\frac{3}{4} c_{i/m}^\tau b_{j/m}^\tau b_{l/m}^\tau \sin\left(\frac{i}{m}\Omega\tau\right) \sin\left(\frac{j}{m}\Omega\tau\right) \sin\left(\frac{l}{m}\Omega\tau\right) \Delta_2^{(0)}, \\
f_2^{\tau(0)}(i,j,l,10) &= \frac{3}{2} b_{i/m}^\tau c_{j/m}^\tau b_{l/m}^\tau \cos(\frac{i}{m}\Omega\tau) \cos(\frac{j}{m}\Omega\tau) \sin(\frac{l}{m}\Omega\tau) \Delta_2^{(0)}, \\
f_2^{\tau(0)}(i,j,l,11) &= -\frac{3}{2} c_{i/m}^\tau b_{j/m}^\tau c_{l/m}^\tau \sin\left(\frac{i}{m}\Omega\tau\right) \sin\left(\frac{j}{m}\Omega\tau\right) \cos\left(\frac{l}{m}\Omega\tau\right) \Delta_2^{(0)},
\end{aligned} \tag{6.15}
$$

where

$$
\begin{aligned}
\Delta_1^{(0)} &= \delta_{i+j}^l + \delta_{i+l}^j + \delta_{j+l}^i, \\
\Delta_2^{(0)} &= \delta_{i+j}^l + \delta_{i+l}^j - \delta_{j+l}^i.
\end{aligned} \tag{6.16}
$$

The cosine term is given by

$$
\begin{aligned}
& F_{1k}^{(m)}(\mathbf{z}^{(m)}, \dot{\mathbf{z}}^{(m)}; \mathbf{z}^{\tau(m)} \dot{\mathbf{z}}^{\tau(m)}) \\
&= -\delta_1 \left(\dot{b}_{k/m} + \frac{k\Omega}{m} c_{k/m} \right) + \delta_2 \left[\left(\dot{b}_{k/m}^\tau + \frac{k\Omega}{m} c_{k/m}^\tau \right) \cos\left(\frac{k}{m}\Omega\tau\right) \right.
\end{aligned}
$$

$$+\left(-\dot{c}^{\tau}_{k/m}+\frac{k\Omega}{m}b^{\tau}_{k/m}\right)\sin\left(\frac{k}{m}\Omega\tau\right)\Big]-\alpha_1 b_{k/m}$$
$$+\alpha_2\left[b^{\tau}_{k/m}\cos\left(\frac{k}{m}\Omega\tau\right)-c^{\tau}_{k/m}\sin\left(\frac{k}{m}\Omega\tau\right)\right]$$
$$-\beta_1 f^{(c)}_{1k}+\beta_2 f^{\tau(c)}_{1k}-\gamma_1 f^{(c)}_{2k}+\gamma_2 f^{\tau(c)}_{2k}+Q_0\delta^m_k. \quad (6.17)$$

The cosine terms, caused by the quadratic nonlinear terms, are

$$f^{(c)}_{1k}=2a_0^{(m)}b_{k/m}+\sum_{i=1}^{N}\sum_{j=1}^{N}\frac{1}{2}b_{i/m}b_{j/m}\Delta^{1(c)}_{1k}+\frac{1}{2}c_{i/m}c_{j/m}\Delta^{2(c)}_{1k}, \quad (6.18)$$

$$f^{\tau(c)}_{1k}=2a_0^{\tau(m)}\left[b^{\tau}_{k/m}\cos\left(\frac{k}{m}\Omega\tau\right)-c^{\tau}_{k/m}\sin\left(\frac{k}{m}\Omega\tau\right)\right]+\sum_{q=1}^{6}\sum_{i=1}^{N}\sum_{j=1}^{N}f^{\tau(c)}_{1k}(i,j,q), \quad (6.19)$$

with

$$f^{\tau(c)}_{1k}(i,j,1)=\frac{1}{2}b^{\tau}_{i/m}b^{\tau}_{j/m}\cos\left(\frac{i}{m}\Omega\tau\right)\cos\left(\frac{j}{m}\Omega\tau\right)\Delta^{1(c)}_{1k},$$
$$f^{\tau(c)}_{1k}(i,j,2)=\frac{1}{2}c^{\tau}_{i/m}c^{\tau}_{j/m}\sin\left(\frac{i}{m}\Omega\tau\right)\sin\left(\frac{j}{m}\Omega\tau\right)\Delta^{1(c)}_{1k},$$
$$f^{\tau(c)}_{1k}(i,j,3)=-b^{\tau}_{i/m}c^{\tau}_{j/m}\cos\left(\frac{i}{m}\Omega\tau\right)\sin\left(\frac{j}{m}\Omega\tau\right)\Delta^{1(c)}_{1k};$$
$$f^{\tau(c)}_{1k}(i,j,4)=\frac{1}{2}c^{\tau}_{i/m}c^{\tau}_{j/m}\cos\left(\frac{i}{m}\Omega\tau\right)\cos\left(\frac{j}{m}\Omega\tau\right)\Delta^{2(c)}_{1k},$$
$$f^{\tau(c)}_{1k}(i,j,5)=\frac{1}{2}b^{\tau}_{i/m}b^{\tau}_{j/m}\sin\left(\frac{i}{m}\Omega\tau\right)\sin\left(\frac{j}{m}\Omega\tau\right)\Delta^{2(c)}_{1k},$$
$$f^{\tau(c)}_{1k}(i,j,6)=c^{\tau}_{i/m}b^{\tau}_{j/m}\cos\left(\frac{i}{m}\Omega\tau\right)\sin\left(\frac{j}{m}\Omega\tau\right)\Delta^{2(c)}_{1k}; \quad (6.20)$$

where

$$\Delta^{1(c)}_{1k}=\delta^k_{i+j}+\delta^k_{j-i}+\delta^k_{i-j},$$
$$\Delta^{2(c)}_{1k}=\delta^k_{j-i}-\delta^k_{i+j}+\delta^k_{i-j}. \quad (6.21)$$

The cosine terms, caused by the cubic nonlinearity, are given by

$$f^{(c)}_{2k}=3(a_0^{(m)})^2 b_{k/m}+\sum_{q=1}^{5}\sum_{i=1}^{N}\sum_{j=1}^{N}\sum_{l=1}^{N}f^{(c)}_{2k}(i,j,l,q) \quad (6.22)$$

with

$$f^{(c)}_{2k}(i,j,l,1)=\frac{3}{2N}a_0^{(m)}b_{i/m}b_{j/m}\Delta^{1(c)}_{2k},$$

$$
\begin{aligned}
f_{2k}^{(c)}(i,j,l,2) &= \frac{3}{2N}a_0^{(m)}b_{i/m}b_{j/m}\Delta_{2k}^{2(c)},\\
f_{2k}^{(c)}(i,j,l,3) &= \frac{1}{4}b_{i/m}b_{j/m}b_{l/m}\Delta_{2k}^{3(c)},\\
f_{2k}^{(c)}(i,j,l,4) &= \frac{3}{4}b_{i/m}c_{j/m}c_{l/m}\Delta_{2k}^{3(c)};
\end{aligned}
\tag{6.23}
$$

where

$$
\begin{aligned}
\Delta_{2k}^{1(c)} &= \delta_{i+j}^{k} + \delta_{j-i}^{k} + \delta_{i-j}^{k},\\
\Delta_{2k}^{2(c)} &= \delta_{j-i}^{k} - \delta_{i+j}^{k} + \delta_{i-j}^{k};\\
\Delta_{2k}^{3(c)} &= \delta_{i+j-l}^{k} + \delta_{i+l-j}^{k} + \delta_{i-j-l}^{k} + \delta_{j+l-i}^{k} + \delta_{i+j+l}^{k} + \delta_{l-i-j}^{k} + \delta_{j-i-l}^{k},\\
\Delta_{2k}^{4(c)} &= \delta_{i+j-l}^{k} + \delta_{i+l-j}^{k} - \delta_{j+l-i}^{k} - \delta_{i-j-l}^{k} - \delta_{i+j+l}^{k} + \delta_{l-i-j}^{k} + \delta_{j-i-l}^{k}.
\end{aligned}
\tag{6.24}
$$

The time-delayed cosine terms, caused by the cubic nonlinearity, are given by

$$
\begin{aligned}
f_{2k}^{\tau(c)} &= 3(a_0^{\tau(m)})^2\left[b_{k/m}^{\tau}\cos\left(\frac{k}{m}\Omega\tau\right) - c_{k/m}^{\tau}\sin\left(\frac{k}{m}\Omega\tau\right)\right]\\
&\quad + \sum_{q=1}^{16}\sum_{i=1}^{N}\sum_{j=1}^{N}\sum_{l=1}^{N} f_{2k}^{\tau(c)}(i,j,l,q)
\end{aligned}
\tag{6.25}
$$

with

$$
\begin{aligned}
f_{2k}^{\tau(c)}(i,j,l,1) &= \frac{3}{2N}a_0^{\tau(m)}b_{i/m}^{\tau}b_{j/m}^{\tau}\cos\left(\frac{i}{m}\Omega\tau\right)\cos\left(\frac{j}{m}\Omega\tau\right)\Delta_{2k}^{1(c)},\\
f_{2k}^{\tau(c)}(i,j,l,2) &= \frac{3}{2N}a_0^{\tau(m)}c_{i/m}^{\tau}c_{j/m}^{\tau}\sin\left(\frac{i}{m}\Omega\tau\right)\sin\left(\frac{j}{m}\Omega\tau\right)\Delta_{2k}^{1(c)},\\
f_{2k}^{\tau(c)}(i,j,l,3) &= -\frac{3}{N}a_0^{\tau(m)}b_{i/m}^{\tau}c_{j/m}^{\tau}\cos\left(\frac{i}{m}\Omega\tau\right)\sin\left(\frac{j}{m}\Omega\tau\right)\Delta_{2k}^{1(c)};\\
f_{2k}^{\tau(c)}(i,j,l,4) &= \frac{3}{2N}a_0^{\tau(m)}c_{i/m}^{\tau}c_{j/m}^{\tau}\cos\left(\frac{i}{m}\Omega\tau\right)\cos\left(\frac{j}{m}\Omega\tau\right)\Delta_{2k}^{2(c)},\\
f_{2k}^{\tau(c)}(i,j,l,5) &= \frac{3}{2N}a_0^{\tau(m)}b_{i/m}^{\tau}b_{j/m}^{\tau}\sin\left(\frac{i}{m}\Omega\tau\right)\sin\left(\frac{j}{m}\Omega\tau\right)\Delta_{2k}^{2(c)},\\
f_{2k}^{\tau(c)}(i,j,l,6) &= \frac{3}{N}a_0^{\tau(m)}b_{i/m}^{\tau}c_{j/m}^{\tau}\sin\left(\frac{i}{m}\Omega\tau\right)\cos\left(\frac{j}{m}\Omega\tau\right)\Delta_{2k}^{2(c)};\\
f_{2k}^{\tau(c)}(i,j,l,7) &= \frac{1}{4}b_{i/m}^{\tau}b_{j/m}^{\tau}b_{l/m}^{\tau}\cos\left(\frac{i}{m}\Omega\tau\right)\cos\left(\frac{j}{m}\Omega\tau\right)\cos\left(\frac{l}{m}\Omega\tau\right)\Delta_{2k}^{3(c)},\\
f_{2k}^{\tau(c)}(i,j,l,8) &= -\frac{1}{4}c_{i/m}^{\tau}c_{j/m}^{\tau}c_{l/m}^{\tau}\sin\left(\frac{i}{m}\Omega\tau\right)\sin\left(\frac{j}{m}\Omega\tau\right)\sin\left(\frac{l}{m}\Omega\tau\right)\Delta_{2k}^{3(c)},\\
f_{2k}^{\tau(c)}(i,j,l,9) &= -\frac{3}{4}b_{i/m}^{\tau}b_{j/m}^{\tau}c_{l/m}^{\tau}\cos\left(\frac{i}{m}\Omega\tau\right)\cos\left(\frac{j}{m}\Omega\tau\right)\sin\left(\frac{l}{m}\Omega\tau\right)\Delta_{2k}^{3(c)},
\end{aligned}
$$

$$f_{2k}^{\tau(c)}(i,j,l,10) = \frac{3}{4} b_{i/m}^{\tau} c_{j/m}^{\tau} c_{l/m}^{\tau} \cos\left(\frac{i}{m}\Omega\tau\right) \sin\left(\frac{j}{m}\Omega\tau\right) \sin\left(\frac{l}{m}\Omega\tau\right) \Delta_{2k}^{3(c)},$$

$$f_{2k}^{\tau(c)}(i,j,l,11) = \frac{3}{4} b_{i/m}^{\tau} c_{j/m}^{\tau} c_{l/m}^{\tau} \cos\left(\frac{i}{m}\Omega\tau\right) \cos\left(\frac{j}{m}\Omega\tau\right) \cos\left(\frac{l}{m}\Omega\tau\right) \Delta_{2k}^{2(c)},$$

$$f_{2k}^{\tau(c)}(i,j,l,12) = \frac{3}{4} b_{i/m}^{\tau} b_{j/m}^{\tau} b_{l/m}^{\tau} \cos\left(\frac{i}{m}\Omega\tau\right) \sin\left(\frac{j}{m}\Omega\tau\right) \sin\left(\frac{l}{m}\Omega\tau\right) \Delta_{2k}^{4(c)},$$

$$f_{2k}^{\tau(c)}(i,j,l,13) = -\frac{3}{4} c_{i/m}^{\tau} c_{j/m}^{\tau} c_{l/m}^{\tau} \sin\left(\frac{i}{m}\Omega\tau\right) \cos\left(\frac{j}{m}\Omega\tau\right) \cos\left(\frac{l}{m}\Omega\tau\right) \Delta_{2k}^{4(c)},$$

$$f_{2k}^{\tau(c)}(i,j,l,14) = -\frac{3}{4} c_{i/m}^{\tau} b_{j/m}^{\tau} b_{l/m}^{\tau} \sin\left(\frac{i}{m}\Omega\tau\right) \sin\left(\frac{j}{m}\Omega\tau\right) \sin\left(\frac{l}{m}\Omega\tau\right) \Delta_{2k}^{4(c)},$$

$$f_{2k}^{\tau(c)}(i,j,l,15) = -\frac{3}{2} b_{i/m}^{\tau} c_{j/m}^{\tau} b_{l/m}^{\tau} \cos\left(\frac{i}{m}\Omega\tau\right) \cos\left(\frac{j}{m}\Omega\tau\right) \sin\left(\frac{l}{m}\Omega\tau\right) \Delta_{2k}^{4(c)},$$

$$f_{2k}^{\tau(c)}(i,j,l,16) = -\frac{3}{2} c_{i/m}^{\tau} b_{j/m}^{\tau} c_{l/m}^{\tau} \sin\left(\frac{i}{m}\Omega\tau\right) \sin\left(\frac{j}{m}\Omega\tau\right) \cos\left(\frac{l}{m}\Omega\tau\right) \Delta_{2k}^{4(c)}. \tag{6.26}$$

The sine term is given by

$$\begin{aligned}
&F_{2k}^{(m)}(\mathbf{z}^{(m)}, \dot{\mathbf{z}}^{(m)}; \mathbf{z}^{\tau(m)} \dot{\mathbf{z}}^{\tau(m)}) \\
&= -\delta_1 \left(\dot{c}_{k/m} - \frac{k\Omega}{m} b_{k/m}\right) + \delta_2 \left[\left(\dot{b}_{k/m}^{\tau} + \frac{k\Omega}{m} c_{k/m}^{\tau}\right) \sin\left(\frac{k}{m}\Omega\tau\right)\right. \\
&\quad \left. + \left(\dot{c}_{k/m}^{\tau} - \frac{k\Omega}{m} b_{k/m}^{\tau}\right) \cos\left(\frac{k}{m}\Omega\tau\right)\right] - \alpha_1 c_{k/m} \\
&\quad + \alpha_2 \left[c_{k/m} \cos\left(\frac{k}{m}\Omega\tau\right) + b_{k/m} \sin\left(\frac{k}{m}\Omega\tau\right)\right] \\
&\quad - \beta_1 f_{1k}^{(s)} + \beta_2 f_{1k}^{\tau(s)} - \gamma_1 f_{2k}^{(s)} + \gamma_2 f_{2k}^{\tau(s)}.
\end{aligned} \tag{6.27}$$

The sine term, caused by the quadratic nonlinearity, is given by

$$f_{1k}^{(s)} = 2a_0^{(m)} c_{k/m} + \sum_{i=1}^{N} \sum_{j=1}^{N} b_{i/m} c_{j/m} \Delta_{1k}^{2(s)}, \tag{6.28}$$

$$f_{1k}^{\tau(s)} = 2a_0^{\tau(m)} \left[c_{k/m}^{\tau} \cos\left(\frac{k}{m}\Omega\tau\right) + b_{k/m}^{\tau} \sin\left(\frac{k}{m}\Omega\tau\right)\right] + \sum_{q=1}^{4} \sum_{i=1}^{N} \sum_{j=1}^{N} f_{1k}^{\tau(s)}(i,j,q) \tag{6.29}$$

with

$$f_{1k}^{\tau(s)}(i,j,1) = b_{i/m}^{\tau} c_{j/m}^{\tau} \cos\left(\frac{i}{m}\Omega\tau\right) \cos\left(\frac{j}{m}\Omega\tau\right) \Delta_{1k}^{2(s)},$$

$$f_{1k}^{\tau(s)}(i,j,2) = b_{i/m}^{\tau} b_{j/m}^{\tau} \cos\left(\frac{i}{m}\Omega\tau\right) \sin\left(\frac{j}{m}\Omega\tau\right) \Delta_{1k}^{2(s)},$$

$$f_{1k}^{\tau(s)}(i,j,3) = -c_{i/m}^{\tau} c_{j/m}^{\tau} \sin\left(\frac{i}{m}\Omega\tau\right) \cos\left(\frac{j}{m}\Omega\tau\right) \Delta_{1k}^{2(s)},$$

$$f_{1k}^{\tau(s)}(i,j,4) = -c_{i/m}^{\tau} c_{j/m}^{\tau} \sin\left(\frac{i}{m}\Omega\tau\right) \cos\left(\frac{j}{m}\Omega\tau\right) \Delta_{1k}^{2(s)} \tag{6.30}$$

where

$$\Delta_{1k}^{2(s)} = \delta_{i+j}^{k} + \delta_{j-i}^{k} - \delta_{i-j}^{k}. \tag{6.31}$$

The sine term, caused by the cubic nonlinearity, is given by

$$f_{2k}^{(s)} = 3(a_0^{(m)})^2 c_{k/m} + \sum_{q=1}^{3}\sum_{i=1}^{N}\sum_{j=1}^{N}\sum_{l=1}^{N} f_{2k}^{(s)}(i,j,l,q) \tag{6.32}$$

with

$$f_{2k}^{(s)}(i,j,l,1) = \frac{3}{N} a_0^{(m)} b_{i/m} c_{j/m} \Delta_{2k}^{2(s)},$$

$$f_{2k}^{(s)}(i,j,l,2) = \frac{1}{4} c_{i/m} c_{j/m} c_{l/m} \Delta_{2k}^{3(s)},$$

$$f_{2k}^{(s)}(i,j,l,3) = \frac{3}{4} b_{i/m} b_{j/m} c_{l/m} \Delta_{2k}^{4(s)}; \tag{6.33}$$

where

$$\Delta_{2k}^{3(s)} = (\delta_{i+j-l}^{k} + \delta_{i+l-j}^{k} + \delta_{j+l-i}^{k} - \delta_{i-j-l}^{k} - \delta_{i+j+l}^{k} - \delta_{l-i-j}^{k} - \delta_{j-i-l}^{k}),$$

$$\Delta_{2k}^{4(s)} = (\delta_{i+l-j}^{k} - \delta_{i+j-l}^{k} + \delta_{j+l-i}^{k} - \delta_{i-j-l}^{k} + \delta_{i+j+l}^{k} + \delta_{l-i-j}^{k} - \delta_{j-i-l}^{k}). \tag{6.34}$$

The time-delayed sine term, caused by the cubic nonlinearity, is given by

$$f_{2k}^{\tau(s)} = 3(a_0^{\tau(m)})^2 \left[c_k^{\tau} \cos\left(\frac{k}{m}\Omega\tau\right) + b_k^{\tau} \sin\left(\frac{k}{m}\Omega\tau\right)\right] + \sum_{q=1}^{14}\sum_{i=1}^{N}\sum_{j=1}^{N}\sum_{l=1}^{N} f_{2k}^{\tau(s)}(i,j,l,q), \tag{6.35}$$

with

$$f_{2k}^{\tau(s)}(i,j,l,1) = \frac{3}{N} a_0^{\tau(m)} b_{i/m}^{\tau} c_{j/m}^{\tau} \cos\left(\frac{i}{m}\Omega\tau\right) \cos\left(\frac{j}{m}\Omega\tau\right) \Delta_{2k}^{2(s)},$$

$$f_{2k}^{\tau(s)}(i,j,l,2) = \frac{3}{N} a_0^{\tau(m)} b_{i/m}^{\tau} b_{j/m}^{\tau} \cos\left(\frac{i}{m}\Omega\tau\right) \sin\left(\frac{j}{m}\Omega\tau\right) \Delta_{2k}^{2(s)},$$

$$f_{2k}^{\tau(s)}(i,j,l,3) = -\frac{3}{N} a_0^{\tau(m)} c_{i/m}^{\tau} c_{j/m}^{\tau} \sin\left(\frac{i}{m}\Omega\tau\right) \cos\left(\frac{j}{m}\Omega\tau\right) \Delta_{2k}^{2(s)},$$

$$f_{2k}^{\tau(s)}(i,j,l,4) = -\frac{3}{N} a_0^{\tau(m)} c_{i/m}^{\tau} b_{j/m}^{\tau} \sin\left(\frac{i}{m}\Omega\tau\right) \sin\left(\frac{j}{m}\Omega\tau\right) \Delta_{2k}^{2(s)};$$

$$f_{2k}^{\tau(s)}(i,j,l,5) = \frac{1}{4} c_{i/m}^{\tau} c_{j/m}^{\tau} c_{l/m}^{\tau} \cos\left(\frac{i}{m}\Omega\tau\right) \cos\left(\frac{j}{m}\Omega\tau\right) \cos\left(\frac{l}{m}\Omega\tau\right) \Delta_{2k}^{3(s)},$$

$$f_{2k}^{\tau(s)}(i,j,l,6) = \frac{1}{4} b_{i/m}^{\tau} b_{j/m}^{\tau} b_{l/m}^{\tau} \sin\left(\frac{i}{m}\Omega\tau\right) \sin\left(\frac{j}{m}\Omega\tau\right) \sin\left(\frac{l}{m}\Omega\tau\right) \Delta_{2k}^{3(s)},$$

$$f_{2k}^{\tau(s)}(i,j,l,7) = \frac{3}{4}c_{i/m}^{\tau}c_{j/m}^{\tau}b_{l/m}^{\tau}\cos\left(\frac{i}{m}\Omega\tau\right)\cos\left(\frac{j}{m}\Omega\tau\right)\sin\left(\frac{l}{m}\Omega\tau\right)\Delta_{2k}^{3(s)},$$

$$f_{2k}^{\tau(s)}(i,j,l,8) = \frac{3}{4}c_{i/m}^{\tau}b_{j/m}^{\tau}b_{l/m}^{\tau}\cos\left(\frac{i}{m}\Omega\tau\right)\sin\left(\frac{j}{m}\Omega\tau\right)\sin\left(\frac{l}{m}\Omega\tau\right)\Delta_{2k}^{3(s)};$$

$$f_{2k}^{\tau(s)}(i,j,l,9) = \frac{3}{4}b_{i/m}^{\tau}b_{j/m}^{\tau}c_{l/m}^{\tau}\cos\left(\frac{i}{m}\Omega\tau\right)\cos\left(\frac{j}{m}\Omega\tau\right)\cos\left(\frac{l}{m}\Omega\tau\right)\Delta_{2k}^{4(s)},$$

$$f_{2k}^{\tau(s)}(i,j,l,10) = \frac{3}{4}b_{i/m}^{\tau}b_{j/m}^{\tau}b_{l/m}^{\tau}\cos\left(\frac{i}{m}\Omega\tau\right)\cos\left(\frac{j}{m}\Omega\tau\right)\sin\left(\frac{l}{m}\Omega\tau\right)\Delta_{2k}^{4(s)},$$

$$f_{2k}^{\tau(s)}(i,j,l,11) = \frac{3}{4}c_{i/m}^{\tau}c_{j/m}^{\tau}c_{l/m}^{\tau}\sin\left(\frac{i}{m}\Omega\tau\right)\sin\left(\frac{j}{m}\Omega\tau\right)\cos\left(\frac{l}{m}\Omega\tau\right)\Delta_{2k}^{4(s)},$$

$$f_{2k}^{\tau(s)}(i,j,l,12) = \frac{3}{4}c_{i/m}^{\tau}c_{j/m}^{\tau}b_{l/m}^{\tau}\sin\left(\frac{i}{m}\Omega\tau\right)\sin\left(\frac{j}{m}\Omega\tau\right)\sin\left(\frac{l}{m}\Omega\tau\right)\Delta_{2k}^{4(s)},$$

$$f_{2k}^{\tau(s)}(i,j,l,13) = -\frac{3}{2}b_{i/m}^{\tau}c_{j/m}^{\tau}c_{l/m}^{\tau}\cos\left(\frac{i}{m}\Omega\tau\right)\sin\left(\frac{j}{m}\Omega\tau\right)\cos\left(\frac{l}{m}\Omega\tau\right)\Delta_{2k}^{4(s)},$$

$$f_{2k}^{\tau(s)}(i,j,l,14) = -\frac{3}{2}b_{i/m}^{\tau}c_{j/m}^{\tau}b_{l/m}^{\tau}\cos\left(\frac{i}{m}\Omega\tau\right)\sin\left(\frac{j}{m}\Omega\tau\right)\sin\left(\frac{l}{m}\Omega\tau\right)\Delta_{2k}^{4(s)}. \tag{6.36}$$

Define

$$\begin{aligned}
\mathbf{z}^{(m)} &\triangleq (a_0^{(m)}, \mathbf{b}^{(m)}, \mathbf{c}^{(m)})^{\mathrm{T}} \\
&= (a_0^{(m)}, b_{1/m}, \ldots, b_{N/m}, c_{1/m}, \ldots, c_{N/m})^{\mathrm{T}} \equiv (z_0^{(m)}, z_1^{(m)}, \ldots, z_{2N}^{(m)})^{\mathrm{T}}, \\
\mathbf{z}_1^{(m)} &\triangleq \dot{\mathbf{z}}^{(m)} = (\dot{a}_0^{(m)}, \dot{\mathbf{b}}^{(m)}, \dot{\mathbf{c}}^{(m)})^{\mathrm{T}} \\
&= (\dot{a}_0^{(m)}, \dot{b}_{1/m}, \ldots, \dot{b}_{N/m}, \dot{c}_{1/m}, \ldots, \dot{c}_{N/m})^{\mathrm{T}} \equiv (\dot{z}_0^{(m)}, \dot{z}_1^{(m)}, \ldots, \dot{z}_{2N}^{(m)})^{\mathrm{T}}, \\
\mathbf{z}^{\tau(m)} &\triangleq (a_0^{\tau(m)}, \mathbf{b}^{\tau(m)}, \mathbf{c}^{\tau(m)})^{\mathrm{T}} \\
&= (a_0^{\tau(m)}, b_{1/m}^{\tau}, \ldots, b_{N/m}^{\tau}, c_{1/m}^{\tau}, \ldots, c_{2N/m}^{\tau})^{\mathrm{T}} \equiv (z_0^{\tau(m)}, z_1^{\tau(m)}, \ldots, z_{2N}^{\tau(m)})^{\mathrm{T}}, \\
\mathbf{z}_1^{\tau(m)} &\triangleq \dot{\mathbf{z}}^{\tau(m)} = (\dot{a}_0^{\tau(m)}, \dot{\mathbf{b}}^{\tau(m)}, \dot{\mathbf{c}}^{\tau(m)})^{\mathrm{T}} \\
&= (\dot{a}_0^{\tau(m)}, \dot{b}_{1/m}^{\tau}, \ldots, \dot{b}_{N/m}^{\tau}, \dot{c}_{1/m}^{\tau}, \ldots, \dot{c}_{2N/m}^{\tau})^{\mathrm{T}} \equiv (\dot{z}_0^{\tau(m)}, \dot{z}_1^{\tau(m)}, \ldots, \dot{z}_{2N}^{\tau(m)})^{\mathrm{T}}.
\end{aligned} \tag{6.37}$$

Equation (6.7) can be expressed in the form of vector field as

$$\dot{\mathbf{z}}^{(m)} = \mathbf{z}_1^{(m)} \text{ and } \dot{\mathbf{z}}_1^{(m)} = \mathbf{g}^{(m)}(\mathbf{z}^{(m)}, \mathbf{z}_1^{(m)}, \mathbf{z}^{\tau(m)}, \mathbf{z}_1^{\tau(m)}) \tag{6.38}$$

where

$$\begin{aligned}
&\mathbf{g}^{(m)}(\mathbf{z}^{(m)}, \mathbf{z}_1^{(m)}, \mathbf{z}^{\tau(m)}, \mathbf{z}_1^{\tau(m)}) \\
&= \begin{pmatrix} F_0^{(m)}(\mathbf{z}^{(m)}, \mathbf{z}_1^{(m)}, \mathbf{z}^{\tau(m)}, \mathbf{z}_1^{\tau(m)}) \\ \mathbf{F}_1^{(m)}(\mathbf{z}^{(m)}, \mathbf{z}_1^{(m)}, \mathbf{z}^{\tau(m)}, \mathbf{z}_1^{\tau(m)}) - 2\mathbf{k}_1 \frac{\Omega}{m}\dot{\mathbf{c}}^{(m)} + \mathbf{k}_2\left(\frac{\Omega}{m}\right)^2 \mathbf{b}^{(m)} \\ \mathbf{F}_2^{(m)}(\mathbf{z}^{(m)}, \mathbf{z}_1^{(m)}, \mathbf{z}^{\tau(m)}, \mathbf{z}_1^{\tau(m)}) + 2\mathbf{k}_1 \frac{\Omega}{m}\dot{\mathbf{b}}^{(m)} + \mathbf{k}_2\left(\frac{\Omega}{m}\right)^2 \mathbf{c}^{(m)} \end{pmatrix}
\end{aligned} \tag{6.39}$$

and

$$\begin{aligned}
\mathbf{k}_1 &= diag(1, 2, \ldots, N),\\
\mathbf{k}_2 &= diag(1, 2^2, \ldots, N^2),\\
\mathbf{F}_1^{(m)} &= (F_{11}^{(m)}, F_{12}^{(m)}, \ldots, F_{1N}^{(m)})^{\mathrm{T}},\\
\mathbf{F}_2^{(m)} &= (F_{21}^{(m)}, F_{22}^{(m)}, \ldots, F_{2N}^{(m)})^{\mathrm{T}}\\
&\text{for } N = 1, 2, \ldots, \infty.
\end{aligned} \tag{6.40}$$

Introducing

$$\mathbf{y}^{(m)} \equiv (\mathbf{z}^{(m)}, \mathbf{z}_1^{(m)}), \quad \mathbf{y}^{\tau(m)} \equiv (\mathbf{z}^{\tau(m)}, \mathbf{z}_1^{\tau(m)}) \text{ and } \mathbf{f}^{(m)} = (\mathbf{z}_1^{(m)}, \mathbf{g}^{(m)})^{\mathrm{T}} \tag{6.41}$$

Equation (6.38) becomes

$$\dot{\mathbf{y}}^{(m)} = \mathbf{f}^{(m)}(\mathbf{y}^{(m)}, \mathbf{y}^{\tau(m)}). \tag{6.42}$$

The steady-state solutions for periodic motion in Equation (6.1) can be obtained by setting $\dot{\mathbf{y}}^{(m)} = \mathbf{0}$ and $\dot{\mathbf{y}}^{\tau(m)} = \mathbf{0}$, that is,

$$\begin{aligned}
F_0^{(m)}(a_0^{(m)*}, \mathbf{b}^{(m)*}, \mathbf{c}^{(m)*}, 0, \mathbf{0}, \mathbf{0}, a_0^{(m)*}, \mathbf{b}^{(m)*}, \mathbf{c}^{(m)*}, 0, \mathbf{0}, \mathbf{0}) = 0,\\
\mathbf{F}_1^{(m)}(a_0^{(m)*}, \mathbf{b}^{(m)*}, \mathbf{c}^{(m)*}, 0, \mathbf{0}, \mathbf{0}, a_0^{(m)*}, \mathbf{b}^{(m)*}, \mathbf{c}^{(m)*}, 0, \mathbf{0}, \mathbf{0}) - \left(\frac{\Omega}{m}\right)^2 \mathbf{k}_2 \mathbf{b}^{(m)*} = \mathbf{0},\\
\mathbf{F}_2^{(m)}(a_0^{(m)*}, \mathbf{b}^{(m)*}, \mathbf{c}^{(m)*}, 0, \mathbf{0}, \mathbf{0}, a_0^{(m)*}, \mathbf{b}^{(m)*}, \mathbf{c}^{(m)*}, 0, \mathbf{0}, \mathbf{0}) - \left(\frac{\Omega}{m}\right)^2 \mathbf{k}_2 \mathbf{c}^{(m)*} = \mathbf{0}.
\end{aligned} \tag{6.43}$$

The $(2N + 1)$ nonlinear equations in Equation (6.43) are solved by the Newton-Raphson method. In Luo (2012a), the linearized equation at equilibrium $\mathbf{y}^{(m)*} = (\mathbf{z}^{(m)*}, \mathbf{0})^{\mathrm{T}}$ and $\mathbf{y}^{\tau(m)*} = (\mathbf{z}^{(m)*}, \mathbf{0})^{\mathrm{T}}$ is given by

$$\Delta\dot{\mathbf{y}}^{(m)} = \mathbf{A}\Delta\mathbf{y}^{(m)} + \mathbf{B}\Delta\mathbf{y}^{(m)\tau} \tag{6.44}$$

where

$$\begin{aligned}
\mathbf{A} &= \partial\mathbf{f}^{(m)}(\mathbf{y}^{(m)}, \mathbf{y}^{\tau(m)})/\partial\mathbf{y}^{(m)}|_{(\mathbf{y}^{(m)*}, \mathbf{y}^{\tau(m)*})},\\
\mathbf{B} &= \partial\mathbf{f}^{(m)}(\mathbf{y}^{(m)}, \mathbf{y}^{\tau(m)})/\partial\mathbf{y}^{\tau(m)}|_{(\mathbf{y}^{(m)*}, \mathbf{y}^{\tau(m)*})}.
\end{aligned} \tag{6.45}$$

The Jacobian matrices are

$$\begin{aligned}
\mathbf{A} &= \begin{bmatrix} \mathbf{0}_{(2N+1)\times(2N+1)} & \mathbf{I}_{(2N+1)\times(2N+1)} \\ \mathbf{G} & \mathbf{H} \end{bmatrix},\\
\mathbf{B} &= \begin{bmatrix} \mathbf{0}_{(2N+1)\times(2N+1)} & \mathbf{I}_{(2N+1)\times(2N+1)} \\ \mathbf{G}^{\tau} & \mathbf{H}^{\tau} \end{bmatrix};
\end{aligned} \tag{6.46}$$

and

$$\begin{aligned}
\mathbf{G} &= \frac{\partial\mathbf{g}^{(m)}}{\partial\mathbf{z}^{(m)}} = (\mathbf{G}^{(0)}, \mathbf{G}^{(c)}, \mathbf{G}^{(s)})^{\mathrm{T}},\\
\mathbf{G}^{\tau} &= \frac{\partial\mathbf{g}^{(m)}}{\partial\mathbf{z}^{\tau(m)}} = (\mathbf{G}^{\tau(0)}, \mathbf{G}^{\tau(c)}, \mathbf{G}^{\tau(s)})^{\mathrm{T}};
\end{aligned} \tag{6.47}$$

$$\begin{aligned}\mathbf{G}^{(0)} &= (G_0^{(0)}, G_1^{(0)}, \ldots, G_{2N}^{(0)}),\\ \mathbf{G}^{(c)} &= (\mathbf{G}_1^{(c)}, \mathbf{G}_2^{(c)}, \ldots, \mathbf{G}_N^{(c)})^{\mathrm{T}},\\ \mathbf{G}^{(s)} &= (\mathbf{G}_1^{(s)}, \mathbf{G}_2^{(s)}, \ldots, \mathbf{G}_N^{(s)})^{\mathrm{T}};\\ \mathbf{G}^{\tau(0)} &= (G_0^{\tau(0)}, G_1^{\tau(0)}, \ldots, G_{2N}^{\tau(0)}),\\ \mathbf{G}^{\tau(c)} &= (\mathbf{G}_1^{\tau(c)}, \mathbf{G}_2^{\tau(c)}, \ldots, \mathbf{G}_N^{\tau(c)})^{\mathrm{T}},\\ \mathbf{G}^{\tau(s)} &= (\mathbf{G}_1^{\tau(s)}, \mathbf{G}_2^{\tau(s)}, \ldots, \mathbf{G}_N^{\tau(s)})^{\mathrm{T}}\end{aligned} \tag{6.48}$$

for $N = 1, 2, \ldots, \infty$ with

$$\begin{aligned}\mathbf{G}_k^{(c)} &= (G_{k0}^{(c)}, G_{k1}^{(c)}, \ldots, G_{k(2N)}^{(c)}),\\ \mathbf{G}_k^{(s)} &= (G_{k0}^{(s)}, G_{k1}^{(s)}, \ldots, G_{k(2N)}^{(s)});\\ \mathbf{G}_k^{\tau(c)} &= (G_{k0}^{\tau(c)}, G_{k1}^{\tau(c)}, \ldots, G_{k(2N)}^{\tau(c)}),\\ \mathbf{G}_k^{\tau(s)} &= (G_{k0}^{\tau(s)}, G_{k1}^{\tau(s)}, \ldots, G_{k(2N)}^{\tau(s)})\end{aligned} \tag{6.49}$$

for $k = 1, 2, \ldots, N$. The corresponding components for constants are

$$G_r^{(0)} = -\alpha_1 \delta_r^0 - \beta_1 g_{1r}^{(0)} - \gamma_1 g_{2r}^{(0)} \tag{6.50}$$

where for $r = 0, 1, \ldots, 2N$

$$g_{1r}^{(0)} = 2a_0^{(m)} \delta_r^0 + \sum_{i=1}^{N} (b_{i/m} \delta_i^r + c_{i/m} \delta_{i+N}^r), \tag{6.51}$$

$$g_{2r}^{(0)} = 3(a_0^{(m)})^2 \delta_r^0 + \sum_{q=1}^{4} \sum_{i=1}^{N} \sum_{j=1}^{N} \sum_{l=1}^{N} g_{2r}^{(0)}(i,j,l,q) \tag{6.52}$$

with

$$\begin{aligned}g_{2r}^{(0)}(i,j,l,1) &= \frac{3}{2N^2}(b_{i/m}^2 + c_{i/m}^2)\delta_r^0,\\ g_{2r}^{(0)}(i,j,l,2) &= \frac{3}{N^2} a_0^{(m)} (b_{i/m} \delta_i^r + c_{i/m} \delta_{i+N}^r),\\ g_{2r}^{(0)}(i,j,l,3) &= \frac{3}{4} b_{j/m} b_{l/m} \delta_i^r \Delta_1^{(0)},\\ g_{2r}^{(0)}(i,j,l,4) &= \frac{3}{4}(c_{j/m} c_{l/m} \delta_i^r + b_{i/m} c_{l/m} \delta_{j+N}^r)\Delta_2^{(0)}.\end{aligned} \tag{6.53}$$

The corresponding components for cosine terms are

$$G_{kr}^{(c)} = \left(\frac{k\Omega}{m}\right)^2 \delta_k^r - \delta_1 \frac{k\Omega}{m} \delta_{k+N}^r - \alpha_1 \delta_k^r - \beta_1 g_{2r}^{(c)} - \gamma_1 g_{2r}^{(c)} \tag{6.54}$$

where

$$g_{1r}^{(c)} = 2(b_{k/m}\delta_r^0 + a_0^{(m)}\delta_k^r) + \sum_{q=1}^{2}\sum_{i=1}^{N}\sum_{j=1}^{N} g_{1r}^{(c)}(i,j,q) \tag{6.55}$$

with

$$\begin{aligned} g_{1r}^{(c)}(i,j,1) &= b_{j/m}\delta_i^r \Delta_{1k}^{1(c)}, \\ g_{1r}^{(c)}(i,j,2) &= c_{i/m}\delta_{i+N}^r \Delta_{1k}^{2(c)}; \end{aligned} \tag{6.56}$$

and

$$g_{1kr}^{(c)} = 3a_0^{(m)}(2b_{k/m}\delta_r^0 + a_0^{(m)}\delta_k^r) + \sum_{q=1}^{6}\sum_{i=1}^{N}\sum_{j=1}^{N}\sum_{l=1}^{N} g_{1kr}^{(c)}(i,j,l,q) \tag{6.57}$$

with

$$\begin{aligned} g_{1kr}^{(c)}(i,j,l,1) &= \frac{3}{2N} b_{i/m} b_{j/m}\delta_r^0 \Delta_{1k}^{1(c)}, \\ g_{1kr}^{(c)}(i,j,l,2) &= \frac{3}{N} a_0^{(m)} b_{j/m}\delta_i^r \Delta_{1k}^{1(c)}, \\ g_{1kr}^{(c)}(i,j,l,3) &= \frac{3}{2N} a_0^{(m)} b_{j/m}\delta_i^r \Delta_{1k}^{2(c)}, \\ g_{1kr}^{(c)}(i,j,l,4) &= \frac{3}{4} b_{j/m} b_{l/m}\delta_i^r \Delta_{2k}^{3(c)}, \\ g_{1kr}^{(c)}(i,j,l,5) &= \frac{3}{4} c_{j/m} c_{l/m}\delta_i^r \Delta_{2k}^{4(c)}, \\ g_{1kr}^{(c)}(i,j,l,6) &= \frac{3}{2} b_{i/m} c_{l/m}\delta_{j+N}^r \Delta_{2k}^{4(c)}. \end{aligned} \tag{6.58}$$

The corresponding components for sine terms are

$$G_{kr}^{(s)} = \left(\frac{k\Omega}{m}\right)^2 \delta_{k+N}^r + \delta_1 \frac{k\Omega}{m}\delta_k^r - \alpha_1\delta_{k+N}^r - \beta_1 g_{2kr}^{(s)} - \gamma_1 g_{2kr}^{(s)} \tag{6.59}$$

where

$$g_{1kr}^{(s)} = 2(c_{k/m}\delta_r^0 + a_0^{(m)}\delta_{k+N}^r) + \sum_{q=1}^{2}\sum_{i=1}^{N}\sum_{j=1}^{N} g_{1kr}^{(s)}(i,j,q) \tag{6.60}$$

with

$$\begin{aligned} g_{1kr}^{(s)}(i,j,1) &= c_{j/m}\delta_i^r \Delta_{1k}^{2(s)}, \\ g_{1kr}^{(s)}(i,j,2) &= b_{i/m}\delta_{j+N}^r \Delta_{1k}^{2(s)} \end{aligned} \tag{6.61}$$

and

$$g_{2kr}^{(s)} = 6a_0^{(m)} c_{k/m}\delta_r^0 + \sum_{q=1}^{6}\sum_{i=1}^{N}\sum_{j=1}^{N}\sum_{l=1}^{N} g_{2kr}^{(s)}(i,j,l,q) \tag{6.62}$$

with

$$g_{2kr}^{(s)}(i,j,l,1)=\frac{3}{N}b_{i/m}c_{j/m}\delta_r^0\Delta_{1k}^{2(s)},$$
$$g_{2kr}^{(s)}(i,j,l,2)=\frac{3}{N}a_0^{(m)}c_{j/m}\delta_i^r\Delta_{1k}^{2(s)},$$
$$g_{2kr}^{(s)}(i,j,l,3)=\frac{3}{N}a_0^{(m)}b_{i/m}\delta_{j+N}^r\Delta_{1k}^{2(s)},$$
$$g_{2kr}^{(s)}(i,j,l,5)=\frac{3}{4}c_{j/m}c_{l/m}\delta_i^r\Delta_{2k}^{3(s)},$$
$$g_{2kr}^{(s)}(i,j,l,6)=\frac{3}{2}b_{j/m}c_{l/m}\delta_i^r\Delta_{2k}^{4(s)},$$
$$g_{2kr}^{(s)}(i,j,l,7)=\frac{3}{4}b_{i/m}b_{j/m}\delta_{l+N}^r\Delta_{2k}^{4(s)}. \tag{6.63}$$

The components relative to time-delay for constants are for $r=0,1,\ldots,2N$

$$G_r^{\tau(0)}=\alpha_2\delta_r^0+\beta_2 g_{1r}^{\tau(0)}+\gamma_2 g_{2r}^{\tau(0)} \tag{6.64}$$

where

$$g_{1r}^{\tau(0)}=2a_0^{\tau(m)}\delta_r^0+\sum_{i=1}^{N}(b_{i/m}^{\tau}\delta_i^{\tau}+c_{i/m}^{\tau}\delta_{i+N}^{\tau}) \tag{6.65}$$

and

$$g_{2r}^{\tau(0)}=3(a_0^{\tau(m)})^2\delta_0^r+\sum_{q=1}^{12}\sum_{i=1}^{N}\sum_{j=1}^{N}\sum_{l=1}^{N}g_{2r}^{\tau(0)}(i,j,l,q) \tag{6.66}$$

with

$$g_{2r}^{\tau(0)}(i,j,l,1)=\frac{3}{2N^2}[(b_{i/m}^{\tau})^2+(c_{i/m}^{\tau})^2]\delta_0^r,$$
$$g_{2r}^{\tau(0)}(i,j,l,2)=\frac{3}{2N^2}a_0^{\tau(m)}(b_{i/m}^{\tau}\delta_i^r+c_{i/m}^{\tau}\delta_{i+N}^r),$$
$$g_{2r}^{\tau(0)}(i,j,l,3)=\frac{3}{4}b_{j/m}^{\tau}b_{l/m}^{\tau}\delta_i^r\cos\left(\frac{i}{m}\Omega\tau\right)\cos\left(\frac{j}{m}\Omega\tau\right)\cos\left(\frac{l}{m}\Omega\tau\right)\Delta_1^{(0)},$$
$$g_{2r}^{\tau(0)}(i,j,l,4)=-\frac{3}{4}c_{j/m}^{\tau}c_{l/m}^{\tau}\delta_{i+N}^r\sin\left(\frac{i}{m}\Omega\tau\right)\sin\left(\frac{j}{m}\Omega\tau\right)\sin\left(\frac{l}{m}\Omega\tau\right)\Delta_1^{(0)},$$
$$g_{2r}^{\tau(0)}(i,j,l,5)=-\frac{3}{4}(2b_{j/m}^{\tau}c_{l/m}^{\tau}\delta_i^r+b_{i/m}^{\tau}b_{j/m}^{\tau}\delta_{l+N}^r)$$
$$\times\cos\left(\frac{i}{m}\Omega\tau\right)\cos\left(\frac{j}{m}\Omega\tau\right)\sin\left(\frac{l}{m}\Omega\tau\right)\Delta_1^{(0)},$$
$$g_{2r}^{\tau(0)}(i,j,l,6)=\frac{3}{4}(c_{j/m}^{\tau}c_{l/m}^{\tau}\delta_i^r+b_{i/m}^{\tau}c_{l/m}^{\tau}\delta_{j+N}^r)$$
$$\times\cos\left(\frac{i}{m}\Omega\tau\right)\sin\left(\frac{j}{m}\Omega\tau\right)\sin\left(\frac{l}{m}\Omega\tau\right)\Delta_1^{(0)},$$

$$g_{2r}^{\tau(0)}(i,j,l,7) = \frac{3}{4}(c_{j/m}^{\tau}c_{l/m}^{\tau}\delta_{i}^{r} + 2b_{i/m}^{\tau}c_{l/m}^{\tau}\delta_{j+N}^{r})$$
$$\times \cos\left(\frac{i}{m}\Omega\tau\right)\cos\left(\frac{j}{m}\Omega\tau\right)\cos\left(\frac{l}{m}\Omega\tau\right)\Delta_{2}^{(0)},$$

$$g_{2r}^{\tau(0)}(i,j,l,8) = \frac{3}{4}(b_{j/m}^{\tau}b_{l/m}^{\tau}\delta_{i}^{r} + 2b_{i/m}^{\tau}b_{l/m}^{\tau}\delta_{j+N}^{r})$$
$$\times \cos\left(\frac{i}{m}\Omega\tau\right)\sin\left(\frac{j}{m}\Omega\tau\right)\sin\left(\frac{l}{m}\Omega\tau\right)\Delta_{2}^{(0)},$$

$$g_{2r}^{\tau(0)}(i,j,l,9) = -\frac{3}{4}(c_{j/m}^{\tau}c_{l/m}^{\tau}\delta_{i+N}^{r} + 2c_{i/m}^{\tau}c_{l/m}^{\tau}\delta_{j+N}^{r})$$
$$\times \sin\left(\frac{i}{m}\Omega\tau\right)\cos\left(\frac{j}{m}\Omega\tau\right)\cos\left(\frac{l}{m}\Omega\tau\right)\Delta_{2}^{(0)},$$

$$g_{2r}^{\tau(0)}(i,j,l,10) = -\frac{3}{4}(b_{j/m}^{\tau}b_{l/m}^{\tau}\delta_{i+N}^{r} + c_{i/m}^{\tau}b_{l/m}^{\tau}\delta_{j}^{r})$$
$$\times \sin\left(\frac{i}{m}\Omega\tau\right)\sin\left(\frac{j}{m}\Omega\tau\right)\sin\left(\frac{l}{m}\Omega\tau\right)\Delta_{2}^{(0)},$$

$$g_{2r}^{\tau(0)}(i,j,l,11) = \frac{3}{2}(c_{j/m}^{\tau}b_{l/m}^{\tau}\delta_{i}^{r} + b_{i/m}^{\tau}b_{l/m}^{\tau}\delta_{j+N}^{r} + b_{i/m}^{\tau}c_{j/m}^{\tau}\delta_{l}^{r})$$
$$\times \cos\left(\frac{i}{m}\Omega\tau\right)\cos\left(\frac{j}{m}\Omega\tau\right)\sin\left(\frac{l}{m}\Omega\tau\right)\Delta_{2}^{(0)},$$

$$g_{2r}^{\tau(0)}(i,j,l,12) = -\frac{3}{2}(b_{j/m}^{\tau}c_{l/m}^{\tau}\delta_{i+N}^{r} + c_{i/m}^{\tau}c_{l/m}^{\tau}\delta_{j}^{r} + c_{i/m}^{\tau}b_{j/m}^{\tau}\delta_{l+N}^{r})$$
$$\times \sin\left(\frac{i}{m}\Omega\tau\right)\sin\left(\frac{j}{m}\Omega\tau\right)\cos\left(\frac{l}{m}\Omega\tau\right)\Delta_{2}^{(0)}. \tag{6.67}$$

The components relative to time-delay for cosine terms are

$$G_{kr}^{\tau(c)} = \delta_2\left[\frac{k\Omega}{m}\delta_{k+N}^{r}\cos\left(\frac{k}{m}\Omega\tau\right) + \frac{k\Omega}{m}\delta_{k}^{r}\sin\left(\frac{k}{m}\Omega\tau\right)\right]$$
$$+ \alpha_2\left[\delta_{k}^{r}\cos\left(\frac{k}{m}\Omega\tau\right) - \delta_{k+N}^{r}\sin\left(\frac{k}{m}\Omega\tau\right)\right] + \beta_2 g_{1rk}^{\tau(c)} + \gamma_2 g_{2rk}^{\tau(c)} \tag{6.68}$$

where

$$g_{1rk}^{\tau(c)} = 2\left[\left(b_{k/m}^{\tau}\delta_{r}^{0} + a_{0}^{\tau(m)}\delta_{k}^{r}\right)\cos\left(\frac{k}{m}\Omega\tau\right) - (c_{k/m}^{\tau}\delta_{r}^{0} + a_{0}^{\tau(m)}\delta_{k+N}^{r})\sin\left(\frac{k}{m}\Omega\tau\right)\right]$$
$$+ \sum_{q=1}^{6}\sum_{i=1}^{N}\sum_{j=1}^{N} g_{1rk}^{\tau(c)}(i,j,q) \tag{6.69}$$

with

$$g_{1rk}^{\tau(c)}(i,j,1) = b_{j/m}^{\tau}\delta_{i}^{r}\cos\left(\frac{i}{m}\Omega\tau\right)\cos\left(\frac{j}{m}\Omega\tau\right)\Delta_{1k}^{1(c)},$$

$$g_{1rk}^{\tau(c)}(i,j,2) = c_{j/m}^{\tau}\delta_{i+N}^{r}\sin\left(\frac{i}{m}\Omega\tau\right)\sin\left(\frac{j}{m}\Omega\tau\right)\Delta_{1k}^{1(c)},$$

$$g_{1rk}^{\tau(c)}(i,j,3) = (c_{j/m}^{\tau}\delta_i^r + b_{i/m}^{\tau}\delta_{j+N}^r)\cos\left(\frac{i}{m}\Omega\tau\right)\sin\left(\frac{j}{m}\Omega\tau\right)\Delta_{1k}^{1(c)},$$

$$g_{1rk}^{\tau(c)}(i,j,4) = c_{j/m}^{\tau}\delta_{i+N}^r\cos\left(\frac{i}{m}\Omega\tau\right)\cos\left(\frac{j}{m}\Omega\tau\right)\Delta_{1k}^{2(c)},$$

$$g_{1rk}^{\tau(c)}(i,j,5) = b_{j/m}^{\tau}\delta_i^r\sin\left(\frac{i}{m}\Omega\tau\right)\sin\left(\frac{j}{m}\Omega\tau\right)\Delta_{1k}^{2(c)},$$

$$g_{1rk}^{\tau(c)}(i,j,6) = (b_{j/m}^{\tau}\delta_{i+N}^r + c_{i/m}^{\tau}\delta_j^r)\cos\left(\frac{i}{m}\Omega\tau\right)\sin\left(\frac{j}{m}\Omega\tau\right)\Delta_{1k}^{2(c)}; \tag{6.70}$$

and

$$\begin{aligned} g_{2kr}^{\tau(c)} = {} & 3\left\{\left[2a_0^{\tau(m)}b_{k/m}^{\tau}\delta_r^0 + (a_0^{\tau(m)})^2\delta_k^r\right]\cos\left(\frac{k}{m}\Omega\tau\right)\right. \\ & \left. - \left[2a_0^{\tau(m)}c_{k/m}^{\tau}\delta_r^0 + (a_0^{\tau(m)})^2\delta_{k+N}^r\right]\sin\left(\frac{k}{m}\Omega\tau\right)\right\} \\ & + \sum_{q=1}^{16}\sum_{i=1}^{N}\sum_{j=1}^{N}\sum_{l=1}^{N} g_{2kr}^{\tau(c)}(i,j,l,q) \end{aligned} \tag{6.71}$$

with

$$g_{2kr}^{\tau(c)}(i,j,l,1) = \frac{3}{2N}\left[\left(b_{i/m}^{\tau}b_{j/m}^{\tau}\delta_r^0 + 2a_0^{\tau(m)}b_{j/m}^{\tau}\delta_i^r\right)\cos\left(\frac{i}{m}\Omega\tau\right)\cos\left(\frac{j}{m}\Omega\tau\right)\Delta_{2k}^{1(c)},\right.$$

$$g_{2kr}^{\tau(c)}(i,j,l,2) = \frac{3}{2N}(c_{i/m}^{\tau}c_{j/m}^{\tau}\delta_r^0 + 2a_0^{\tau(m)}c_{j/m}^{\tau}\delta_{i+N}^r)\sin\left(\frac{i}{m}\Omega\tau\right)\sin\left(\frac{j}{m}\Omega\tau\right)\Delta_{2k}^{1(c)},$$

$$\begin{aligned} g_{2kr}^{\tau(c)}(i,j,l,3) = {} & -\frac{3}{N}(b_{i/m}^{\tau}c_{j/m}^{\tau}\delta_r^0 + a_0^{\tau(m)}c_{j/m}^{\tau}\delta_i^r + a_0^{\tau(m)}b_{i/m}^{\tau}\delta_{j+N}^r) \\ & \times\cos\left(\frac{i}{m}\Omega\tau\right)\sin\left(\frac{j}{m}\Omega\tau\right)\Delta_{2k}^{1(c)}, \end{aligned}$$

$$g_{2kr}^{\tau(c)}(i,j,l,4) = \frac{3}{2N}(c_{i/m}^{\tau}c_{j/m}^{\tau}\delta_r^0 + 2a_0^{\tau(m)}c_{j/m}^{\tau}\delta_{i+N}^r)\cos\left(\frac{i}{m}\Omega\tau\right)\cos\left(\frac{j}{m}\Omega\tau\right)\Delta_{2k}^{2(c)},$$

$$g_{2kr}^{\tau(c)}(i,j,l,5) = \frac{3}{2N}(b_{i/m}^{\tau}b_{j/m}^{\tau}\delta_r^0 + 2a_0^{\tau(m)}b_{j/m}^{\tau}\delta_i^r)\sin\left(\frac{i}{m}\Omega\tau\right)\sin\left(\frac{j}{m}\Omega\tau\right)\Delta_{2k}^{2(c)},$$

$$\begin{aligned} g_{2kr}^{\tau(c)}(i,j,l,6) = {} & \frac{3}{N}(b_{i/m}^{\tau}c_{j/m}^{\tau}\delta_r^0 + a_0^{\tau(m)}c_{j/m}^{\tau}\delta_i^r + a_0^{\tau(m)}b_{i/m}^{\tau}\delta_{j+N}^r) \\ & \times\sin\left(\frac{i}{m}\Omega\tau\right)\cos\left(\frac{j}{m}\Omega\tau\right)\Delta_{2k}^{2(c)}, \end{aligned}$$

$$g_{2kr}^{\tau(c)}(i,j,l,7) = \frac{3}{4}b_{j/m}^{\tau}b_{l/m}^{\tau}\delta_i^r\cos\left(\frac{i}{m}\Omega\tau\right)\cos\left(\frac{j}{m}\Omega\tau\right)\cos\left(\frac{l}{m}\Omega\tau\right)\Delta_{2k}^{3(c)},$$

$$g_{2kr}^{\tau(c)}(i,j,l,8) = -\frac{1}{4}c_{j/m}^{\tau}c_{l/m}^{\tau}\delta_{i+N}^r\sin\left(\frac{i}{m}\Omega\tau\right)\sin\left(\frac{j}{m}\Omega\tau\right)\sin\left(\frac{l}{m}\Omega\tau\right)\Delta_{2k}^{3(c)},$$

$$
\begin{aligned}
g_{2kr}^{\tau(c)}(i,j,l,9) &= -\frac{3}{4}(2b_{j/m}^{\tau}c_{l/m}^{\tau}\delta_i^r + b_{i/m}^{\tau}b_{j/m}^{\tau}\delta_{l+N}^r) \\
&\quad \times \cos\left(\frac{i}{m}\Omega\tau\right)\cos\left(\frac{j}{m}\Omega\tau\right)\sin\left(\frac{l}{m}\Omega\tau\right)\Delta_{2k}^{3(c)}, \\
g_{2kr}^{\tau(c)}(i,j,l,10) &= \frac{3}{4}(c_{j/m}^{\tau}c_{l/m}^{\tau}\delta_i^r + b_{i/m}^{\tau}c_{l/m}^{\tau}\delta_{j+N}^r) \\
&\quad \times \cos\left(\frac{i}{m}\Omega\tau\right)\sin\left(\frac{j}{m}\Omega\tau\right)\sin\left(\frac{l}{m}\Omega\tau\right)\Delta_{2k}^{3(c)}, \\
g_{2kr}^{\tau(c)}(i,j,l,11) &= \frac{3}{4}(c_{j/m}^{\tau}c_{l/m}^{\tau}\delta_i^r + 2b_{i/m}^{\tau}c_{l/m}^{\tau}\delta_{j+N}^r) \\
&\quad \times \cos\left(\frac{i}{m}\Omega\tau\right)\cos\left(\frac{j}{m}\Omega\tau\right)\cos\left(\frac{l}{m}\Omega\tau\right)\Delta_{2k}^{4(c)}, \\
g_{2kr}^{\tau(c)}(i,j,l,12) &= \frac{3}{4}(b_{j/m}^{\tau}b_{l/m}^{\tau}\delta_i^r + 2b_{i/m}^{\tau}b_{l/m}^{\tau}\delta_j^r) \\
&\quad \times \cos\left(\frac{i}{m}\Omega\tau\right)\sin\left(\frac{j}{m}\Omega\tau\right)\sin\left(\frac{l}{m}\Omega\tau\right)\Delta_{2k}^{4(c)}, \\
g_{2kr}^{\tau(c)}(i,j,l,13) &= -\frac{3}{4}(c_{j/m}^{\tau}c_{l/m}^{\tau}\delta_i^r + 2c_{i/m}^{\tau}c_{l/m}^{\tau}\delta_j^r) \\
&\quad \times \sin\left(\frac{i}{m}\Omega\tau\right)\cos\left(\frac{j}{m}\Omega\tau\right)\cos\left(\frac{l}{m}\Omega\tau\right)\Delta_{2k}^{4(c)}, \\
g_{2kr}^{\tau(c)}(i,j,l,14) &= -\frac{3}{4}(b_{j/m}^{\tau}b_{l/m}^{\tau}\delta_{i+N}^r + 2c_{i/m}^{\tau}b_{l/m}^{\tau}\delta_j^r) \\
&\quad \times \sin\left(\frac{i}{m}\Omega\tau\right)\sin\left(\frac{j}{m}\Omega\tau\right)\sin\left(\frac{l}{m}\Omega\tau\right)\Delta_{2k}^{4(c)}, \\
g_{2kr}^{\tau(c)}(i,j,l,15) &= \frac{3}{2}(c_{j/m}^{\tau}b_{l/m}^{\tau}\delta_i^r + b_{i/m}^{\tau}b_{l/m}^{\tau}\delta_{j+N}^r + b_{i/m}^{\tau}c_{j/m}^{\tau}\delta_l^r) \\
&\quad \times \cos\left(\frac{i}{m}\Omega\tau\right)\cos\left(\frac{j}{m}\Omega\tau\right)\sin\left(\frac{l}{m}\Omega\tau\right)\Delta_{2k}^{4(c)}, \\
g_{2kr}^{\tau(c)}(i,j,l,16) &= -\frac{3}{2}(b_{j/m}^{\tau}c_{l/m}^{\tau}\delta_{i+N}^r + c_{i/m}^{\tau}c_{l/m}^{\tau}\delta_j^r + c_{i/m}^{\tau}b_{j/m}^{\tau}\delta_{l+N}^r) \\
&\quad \times \sin\left(\frac{i}{m}\Omega\tau\right)\sin\left(\frac{j}{m}\Omega\tau\right)\cos\left(\frac{l}{m}\Omega\tau\right)\Delta_{2k}^{4(c)}.
\end{aligned} \tag{6.72}
$$

The components relative to time-delay for sine terms are

$$
\begin{aligned}
G_{kr}^{\tau(s)} &= \delta_2\left[\frac{k\Omega}{m}\delta_{k+N}^r\sin\left(\frac{k}{m}\Omega\tau\right) - \frac{k\Omega}{m}\delta_k^r\cos\left(\frac{k}{m}\Omega\tau\right)\right] \\
&\quad + \alpha_2\left[\delta_{k+N}^r\cos\left(\frac{k}{m}\Omega\tau\right) + \delta_k^r\sin\left(\frac{k}{m}\Omega\tau\right)\right] + \beta_2 g_{1kr}^{\tau(s)} + \gamma_2 g_{2kr}^{\tau(s)}
\end{aligned} \tag{6.73}
$$

where

$$g_{1kr}^{\tau(s)} = 2\left[\left(c_{k/m}^{\tau}\delta_r^0 + a_0^{\tau(m)}\delta_{k+N}^r\right)\cos\left(\frac{k}{m}\Omega\tau\right) + \left(b_{k/m}^{\tau}\delta_r^0 + a_0^{\tau(m)}\delta_k^r\right)\sin\left(\frac{k}{m}\Omega\tau\right)\right] + \sum_{q=1}^{4}\sum_{i=1}^{N}\sum_{j=1}^{N} g_{1kr}^{\tau(s)}(i,j,q) \tag{6.74}$$

with

$$\begin{aligned}
g_{1kr}^{\tau(s)}(i,j,1) &= (c_{j/m}^{\tau}\delta_i^r + b_{i/m}^{\tau}\delta_{j+N}^r)\cos\left(\frac{i}{m}\Omega\tau\right)\cos\left(\frac{j}{m}\Omega\tau\right)\Delta_{1k}^{2(s)},\\
g_{1kr}^{\tau(s)}(i,j,2) &= (b_{j/m}^{\tau}\delta_i^r + b_{i/m}^{\tau}\delta_j^r)\cos\left(\frac{i}{m}\Omega\tau\right)\sin\left(\frac{j}{m}\Omega\tau\right)\Delta_{1k}^{2(s)},\\
g_{1kr}^{\tau(s)}(i,j,3) &= -(c_{j/m}^{\tau}\delta_{i+N}^r + c_{i/m}^{\tau}\delta_{j+N}^r)\sin\left(\frac{i}{m}\Omega\tau\right)\cos\left(\frac{j}{m}\Omega\tau\right)\Delta_{1k}^{2(s)},\\
g_{1kr}^{\tau(s)}(i,j,4) &= (b_{j/m}^{\tau}\delta_{i+N}^r + c_{i/m}^{\tau}\delta_j^r)\sin\left(\frac{i}{m}\Omega\tau\right)\sin\left(\frac{j}{m}\Omega\tau\right)\Delta_{1k}^{2(s)};
\end{aligned} \tag{6.75}$$

and

$$g_{2kr}^{\tau(s)} = 3\left\{\left[2a_0^{\tau(m)}c_k^{\tau}\delta_r^0 + (a_0^{\tau(m)})^2\delta_{k+N}^r\right]\cos\left(\frac{k}{m}\Omega\tau\right) + \left[2a_0^{\tau(m)}b_k^{\tau}\delta_r^0 + (a_0^{\tau(m)})^2\delta_k^r\right]\sin\left(\frac{k}{m}\Omega\tau\right)\right\} + \sum_{q=1\}}^{14}\sum_{i=1}^{N}\sum_{j=1}^{N}\sum_{l=1}^{N} g_{2kr}^{\tau(s)}(i,j,l,q) \tag{6.76}$$

with

$$\begin{aligned}
g_{2kr}^{\tau(s)}(i,j,l,1) &= \frac{3}{N}(b_{i/m}^{\tau}c_{j/m}^{\tau}\delta_r^0 + a_0^{\tau(m)}c_{j/m}^{\tau}\delta_i^r + a_0^{\tau(m)}b_{i/m}^{\tau}\delta_{j+N}^r)\\
&\quad\times\cos\left(\frac{i}{m}\Omega\tau\right)\cos\left(\frac{j}{m}\Omega\tau\right)\Delta_{2k}^{2(s)},\\
g_{2kr}^{\tau(s)}(i,j,l,2) &= \frac{3}{N}(b_{i/m}^{\tau}b_{l/m}^{\tau}\delta_r^0 + a_0^{\tau(m)}b_{j/m}^{\tau}\delta_i^r + a_0^{\tau(m)}b_{i/m}^{\tau}\delta_j^r)\\
&\quad\times\cos\left(\frac{i}{m}\Omega\tau\right)\sin\left(\frac{j}{m}\Omega\tau\right)\Delta_{2k}^{2(s)},\\
g_{2kr}^{\tau(s)}(i,j,l,3) &= -\frac{3}{N}(c_{i/m}^{\tau}c_{j/m}^{\tau}\delta_r^0 + a_0^{\tau(m)}c_{j/m}^{\tau}\delta_{i+N}^r + a_0^{\tau(m)}c_{i/m}^{\tau}\delta_{j+N}^r)\\
&\quad\times\sin\left(\frac{i}{m}\Omega\tau\right)\cos\left(\frac{j}{m}\Omega\tau\right)\Delta_{2k}^{2(s)},
\end{aligned}$$

$$g_{2kr}^{\tau(s)}(i,j,l,4) = -\frac{3}{N}(c_{i/m}^{\tau}b_{j/m}^{\tau}\delta_r^0 + a_0^{\tau(m)}c_{i/m}^{\tau}b_{j/m}^{\tau}\delta_{i+N}^r + a_0^{\tau(m)}c_{i/m}^{\tau}b_{j/m}^{\tau}\delta_j^r)$$
$$\times \sin\left(\frac{i}{m}\Omega\tau\right)\sin\left(\frac{j}{m}\Omega\tau\right)\Delta_{2k}^{2(s)},$$

$$g_{2kr}^{\tau(s)}(i,j,l,5) = \frac{3}{4}c_{j/m}^{\tau}c_{l/m}^{\tau}\delta_{i+N}^r\cos\left(\frac{i}{m}\Omega\tau\right)\cos\left(\frac{j}{m}\Omega\tau\right)\cos\left(\frac{l}{m}\Omega\tau\right)\Delta_{2k}^{3(s)},$$

$$g_{2kr}^{\tau(s)}(i,j,l,6) = \frac{3}{4}b_{j/m}^{\tau}b_{l/m}^{\tau}\delta_i^r\sin\left(\frac{i}{m}\Omega\tau\right)\sin\left(\frac{j}{m}\Omega\tau\right)\sin\left(\frac{l}{m}\Omega\tau\right)\Delta_{2k}^{3(s)},$$

$$g_{2kr}^{\tau(s)}(i,j,l,7) = \frac{3}{4}(2c_{j/m}^{\tau}b_{l/m}^{\tau}\delta_{i+N}^r + c_{i/m}^{\tau}c_{j/m}^{\tau}\delta_l^r)$$
$$\times \cos\left(\frac{i}{m}\Omega\tau\right)\cos\left(\frac{j}{m}\Omega\tau\right)\sin\left(\frac{l}{m}\Omega\tau\right)\Delta_{2k}^{3(s)},$$

$$g_{2kr}^{\tau(s)}(i,j,l,8) = \frac{3}{4}(b_{j/m}^{\tau}b_{l/m}^{\tau}\delta_{i+N}^r + 2c_{i/m}^{\tau}b_{l/m}^{\tau}\delta_j^r)$$
$$\times \cos\left(\frac{i}{m}\Omega\tau\right)\sin\left(\frac{j}{m}\Omega\tau\right)\sin\left(\frac{l}{m}\Omega\tau\right)\Delta_{2k}^{3(s)},$$

$$g_{2kr}^{\tau(s)}(i,j,l,9) = \frac{3}{4}(b_{j/m}^{\tau}c_{l/m}^{\tau}\delta_i^r + b_{i/m}^{\tau}c_{l/m}^{\tau}\delta_j^r + b_{i/m}^{\tau}b_{j/m}^{\tau}\delta_{l+N}^r)$$
$$\times \cos\left(\frac{i}{m}\Omega\tau\right)\cos\left(\frac{j}{m}\Omega\tau\right)\cos\left(\frac{l}{m}\Omega\tau\right)\Delta_{2k}^{4(s)},$$

$$g_{2kr}^{\tau(s)}(i,j,l,10) = \frac{3}{4}(b_{j/m}^{\tau}b_{l/m}^{\tau}\delta_i^r + b_{i/m}^{\tau}b_{l/m}^{\tau}\delta_j^r + b_{i/m}^{\tau}b_{j/m}^{\tau}\delta_l^r)$$
$$\times \cos\left(\frac{i}{m}\Omega\tau\right)\cos\left(\frac{j}{m}\Omega\tau\right)\sin\left(\frac{l}{m}\Omega\tau\right)\Delta_{2k}^{4(s)},$$

$$g_{2kr}^{\tau(s)}(i,j,l,11) = \frac{3}{4}(c_{j/m}^{\tau}c_{l/m}^{\tau}\delta_{i+N}^r + c_{i/m}^{\tau}c_{l/m}^{\tau}\delta_{j+N}^r + c_{i/m}^{\tau}c_{j/m}^{\tau}\delta_{l+N}^r)$$
$$\times \sin\left(\frac{i}{m}\Omega\tau\right)\sin\left(\frac{j}{m}\Omega\tau\right)\cos\left(\frac{l}{m}\Omega\tau\right)\Delta_{2k}^{4(s)},$$

$$g_{2kr}^{\tau(s)}(i,j,l,11) = \frac{3}{4}(c_{j/m}^{\tau}c_{l/m}^{\tau}\delta_{i+N}^r + c_{i/m}^{\tau}c_{l/m}^{\tau}\delta_{j+N}^r + c_{i/m}^{\tau}c_{j/m}^{\tau}\delta_{l+N}^r)$$
$$\times \sin\left(\frac{i}{m}\Omega\tau\right)\sin\left(\frac{j}{m}\Omega\tau\right)\cos\left(\frac{l}{m}\Omega\tau\right)\Delta_{2k}^{4(s)},$$

$$g_{2kr}^{\tau(s)}(i,j,l,12) = \frac{3}{4}(c_{j/m}^{\tau}b_{l/m}^{\tau}\delta_{i+N}^r + c_{i/m}^{\tau}b_{l/m}^{\tau}\delta_{j+N}^r + 3c_{i/m}^{\tau}c_{j/m}^{\tau}\delta_l^r)$$
$$\times \sin\left(\frac{i}{m}\Omega\tau\right)\sin\left(\frac{j}{m}\Omega\tau\right)\sin\left(\frac{l}{m}\Omega\tau\right)\Delta_{2k}^{4(s)},$$

$$g_{2kr}^{\tau(s)}(i,j,l,13) = -\frac{3}{2}(c_{j/m}^{\tau}c_{l/m}^{\tau}\delta_i^r + b_{i/m}^{\tau}c_{l/m}^{\tau}\delta_{j+N}^r + b_{i/m}^{\tau}c_{j/m}^{\tau}\delta_{l+N}^r)$$
$$\times \cos\left(\frac{i}{m}\Omega\tau\right)\sin\left(\frac{j}{m}\Omega\tau\right)\cos\left(\frac{l}{m}\Omega\tau\right)\Delta_{2k}^{4(s)},$$

$$g_{2kr}^{\tau(s)}(i,j,l,14) = -\frac{3}{2}(c_{j/m}^{\tau} b_{l/m}^{\tau} \delta_i^r + b_{i/m}^{\tau} b_{l/m}^{\tau} \delta_{j+N}^r + b_{i/m}^{\tau} c_{j/m}^{\tau} \delta_l^r)$$

$$\times \cos\left(\frac{i}{m}\Omega\tau\right) \sin\left(\frac{j}{m}\Omega\tau\right) \sin\left(\frac{l}{m}\Omega\tau\right) \Delta_{2k}^{4(s)}. \tag{6.77}$$

The matrices relative to the velocity are

$$\mathbf{H} = \frac{\partial \mathbf{g}^{(m)}}{\partial \mathbf{z}_1{}^{(m)}} = (\mathbf{H}^{(0)}, \mathbf{H}^{(c)}, \mathbf{H}^{(s)})^{\mathrm{T}},$$

$$\mathbf{H}^{\tau} = \frac{\partial \mathbf{g}^{(m)}}{\partial \mathbf{z}_1^{\tau(m)}} = (\mathbf{H}^{\tau(0)}, \mathbf{H}^{\tau(c)}, \mathbf{H}^{\tau(s)})^{\mathrm{T}} \tag{6.78}$$

where

$$\mathbf{H}^{(0)} = (H_0^{(0)}, H_1^{(0)}, \ldots, H_{2N}^{(0)}),$$

$$\mathbf{H}^{(c)} = (\mathbf{H}_1^{(c)}, \mathbf{H}_2^{(c)}, \ldots, \mathbf{H}_N^{(c)})^{\mathrm{T}},$$

$$\mathbf{H}^{(s)} = (\mathbf{H}_1^{(s)}, \mathbf{H}_2^{(s)}, \ldots, \mathbf{H}_N^{(s)})^{\mathrm{T}};$$

$$\mathbf{H}^{\tau(0)} = (H_0^{\tau(0)}, H_1^{\tau(0)}, \ldots, H_{2N}^{\tau(0)}),$$

$$\mathbf{H}^{\tau(c)} = (\mathbf{H}_1^{\tau(c)}, \mathbf{H}_2^{\tau(c)}, \ldots, \mathbf{H}_N^{\tau(c)})^{\mathrm{T}},$$

$$\mathbf{H}^{\tau(s)} = (\mathbf{H}_1^{\tau(s)}, \mathbf{H}_2^{\tau(s)}, \ldots, \mathbf{H}_N^{\tau(s)})^{\mathrm{T}} \tag{6.79}$$

for $N = 1, 2, \ldots \infty$, with

$$\mathbf{H}_k^{(c)} = (H_{k0}^{(c)}, H_{k1}^{(c)}, \ldots, H_{k(2N)}^{(c)}),$$

$$\mathbf{H}_k^{(s)} = (H_{k0}^{(s)}, H_{k1}^{(s)}, \ldots, H_{k(2N)}^{(s)});$$

$$\mathbf{H}_k^{\tau(c)} = (H_{k0}^{\tau(c)}, H_{k1}^{\tau(c)}, \ldots, H_{k(2N)}^{\tau(c)}),$$

$$\mathbf{H}_k^{\tau(s)} = (H_{k0}^{\tau(s)}, H_{k1}^{\tau(s)}, \ldots, H_{k(2N)}^{\tau(s)}). \tag{6.80}$$

for $k = 1, 2, \ldots N$. The corresponding components are

$$H_r^{(0)} = -\delta_1 \delta_0^r,$$

$$H_{kr}^{(c)} = -2\frac{k}{m}\Omega\delta_{k+N}^r - \delta_1 \delta_k^r,$$

$$H_{kr}^{(s)} = 2\frac{k}{m}\Omega\delta_k^r - \delta_1 \delta_{k+N}^r;$$

$$H_r^{\tau(0)} = \delta_2 \delta_0^r,$$

$$H_{kr}^{\tau(c)} = \delta_2 \left[-\sin\left(\frac{k}{m}\Omega\tau\right) \delta_{k+N}^r + \cos\left(\frac{k}{m}\Omega\tau\right) \delta_k^r\right],$$

$$H_{kr}^{\tau(s)} = \delta_2 \left[\sin\left(\frac{k}{m}\Omega\tau\right) \delta_k^r + \cos\left(\frac{k}{m}\Omega\tau\right) \delta_{k+N}^r\right] \tag{6.81}$$

for $r = 0, 1, \ldots, 2N$.

The corresponding eigenvalues of equilibrium are determined by

$$|\mathbf{A} + \mathbf{B}e^{-\lambda\tau} - \lambda\mathbf{I}_{2(2N+1)\times 2(2N+1)}| = 0. \tag{6.82}$$

From Luo (2012a), the eigenvalues of $D\mathbf{f}(\mathbf{y}^*)$ are classified as

$$(n_1, n_2, n_3 | n_4, n_5, n_6). \tag{6.83}$$

If $\mathrm{Re}(\lambda_k) < 0$ $(k = 1, 2, \ldots, 2(2N + 1))$, the approximate solution of periodic motion relative to $\mathbf{y}^{(m)*} = \mathbf{y}^{\tau(m)*}$ with truncation of $\cos(N\Omega t/m)$ and $\sin(N\Omega t/m)$ is stable. If $\mathrm{Re}(\lambda_k) > 0$ $(k \in \{1, 2, \ldots, 2(2N + 1)\})$, the truncated approximate solution relative to $\mathbf{y}^{(m)*} = \mathbf{y}^{\tau(m)*}$ is unstable. The corresponding boundary between the stable and unstable solutions is given by the saddle-node bifurcation and Hopf bifurcation.

6.2 Analytical Bifurcation Trees

The harmonic amplitude varying with excitation frequency Ω is presented to illustrate the bifurcation tree of period-1 motion to chaos. The harmonic amplitude and phase are defined by

$$A_{k/m} \equiv \sqrt{b_{k/m}^2 + c_{k/m}^2} \text{ and } \varphi_{k/m} = \arctan \frac{c_{k/m}}{b_{k/m}}. \tag{6.84}$$

The corresponding solution in Equation (6.2) becomes

$$x^*(t) = a_0^{(m)} + \sum_{k=1}^{N} A_{k/m} \cos\left(\frac{k}{m}\Omega t - \varphi_{k/m}\right). \tag{6.85}$$

As in Luo and Jin (2013), consider a time-delayed, quadratic nonlinear oscillator under a periodic excitation with system parameters as

$$\begin{aligned} &\delta_1 = 0.05, \alpha_1 = 15.0, \alpha_2 = 5.0, \beta_1 = 5.0, Q_0 = 4.5, \tau = T/4, \\ &\delta_2 = \beta_2 = \gamma_1 = \gamma_2 = 0. \end{aligned} \tag{6.86}$$

From the prescribed system parameters, an overview of the bifurcation tree of period-1 motion to chaos for the time-delayed, quadratic nonlinear oscillator will be presented in Figure 6.1 through the 160 harmonic terms (HB160). In Figure 6.1(i), the constant varying with excitation frequency is presented. For $\Omega > 2$, the period-1 motion exists. In Figure 6.1(ii), the harmonic amplitude $A_{1/8}$ versus excitation frequency is presented for period-8 motion only. Such amplitudes for period-1, period-2, and period-4 motions are zero. The saddle-node and Hopf bifurcations occur at $\Omega \approx 1.8896$ and $\Omega \approx 1.8881$, respectively. Once the Hopf bifurcation occurs, the period-16 motions can be similarly determined by the 320 harmonic terms. Many coexisting unstable period-8 motions are observed. In Figure 6.1(iii), harmonic amplitude $A_{1/4}$ varying with excitation frequency is presented for period-4 and period-8 motions. The saddle-node and Hopf bifurcations of period-4 motion occur at $\Omega \approx 1.8896$ and $\Omega \approx 1.8957$. Such harmonic amplitudes for period-1 and period-2 motions are zero. In Figure 6.1(iv), the harmonic amplitude $A_{3/8}$ versus excitation frequency is presented that is

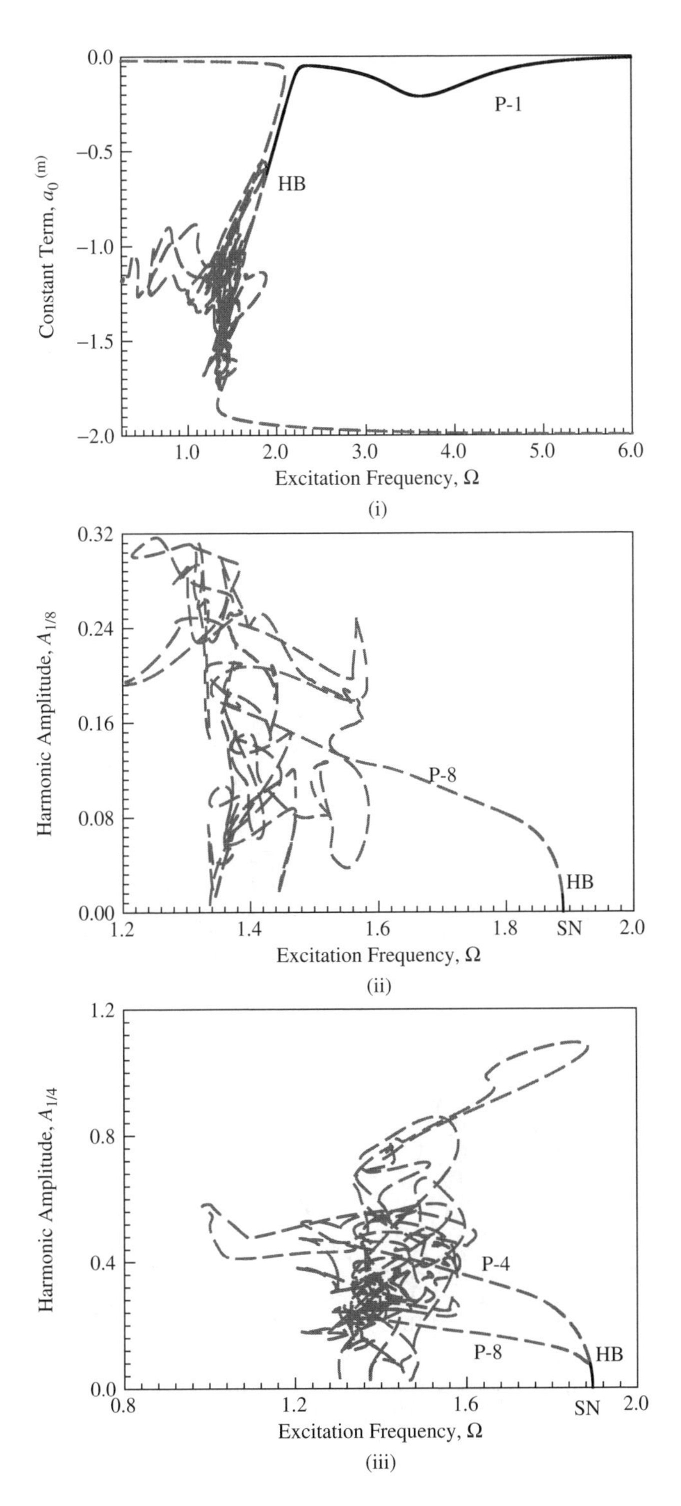

Figure 6.1 An overview for the analytical prediction of period-1 to period-8 motions of the time-delayed, quadratic nonlinear oscillator based on the 160 harmonic terms (HB160): (i) $a_0^{(m)}$ and (ii)–(ix) $A_{k/m}$ ($k = 1, 2, 3, 4; 8, 16, 24; 160, m = 8$). Parameters: ($\delta_1 = 0.05, \alpha_1 = 15.0, \alpha_2 = 5.0, \beta_1 = 5.0, Q_0 = 4.5, \tau = T/4$)

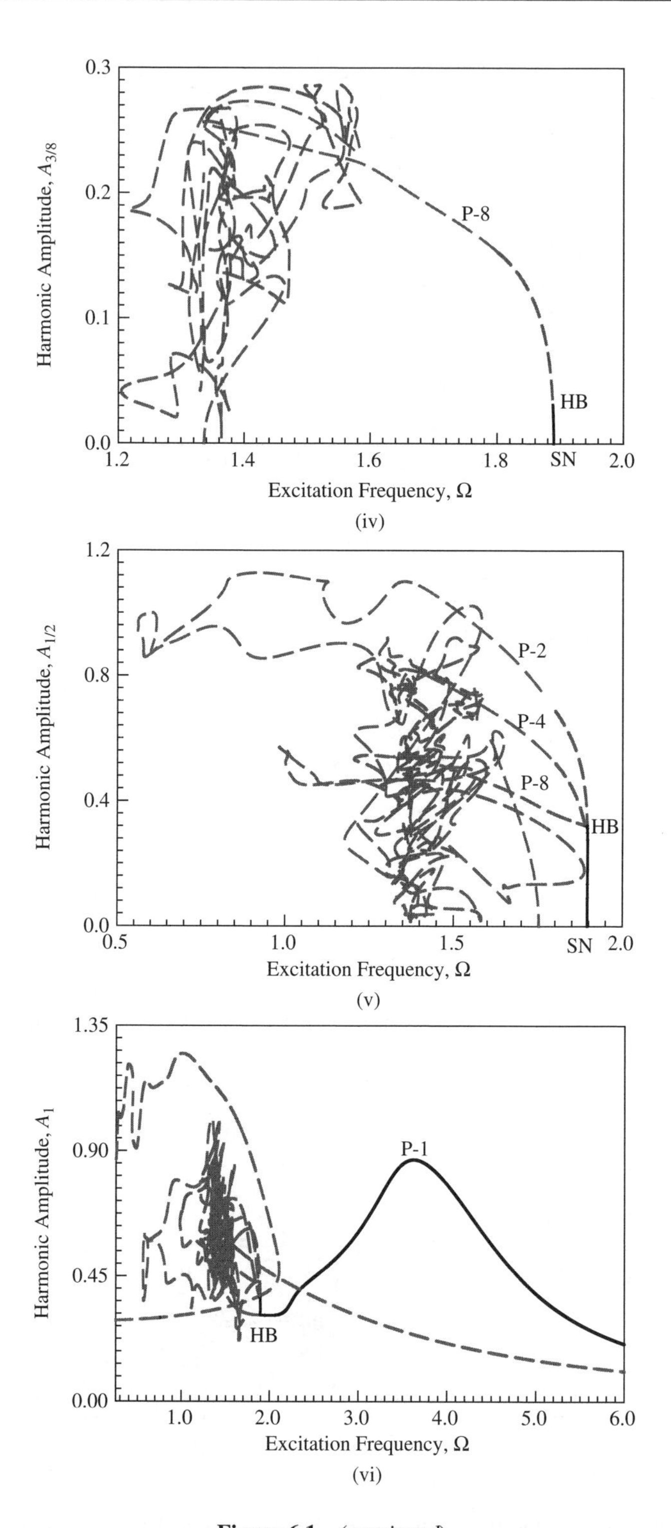

Figure 6.1 *(continued)*

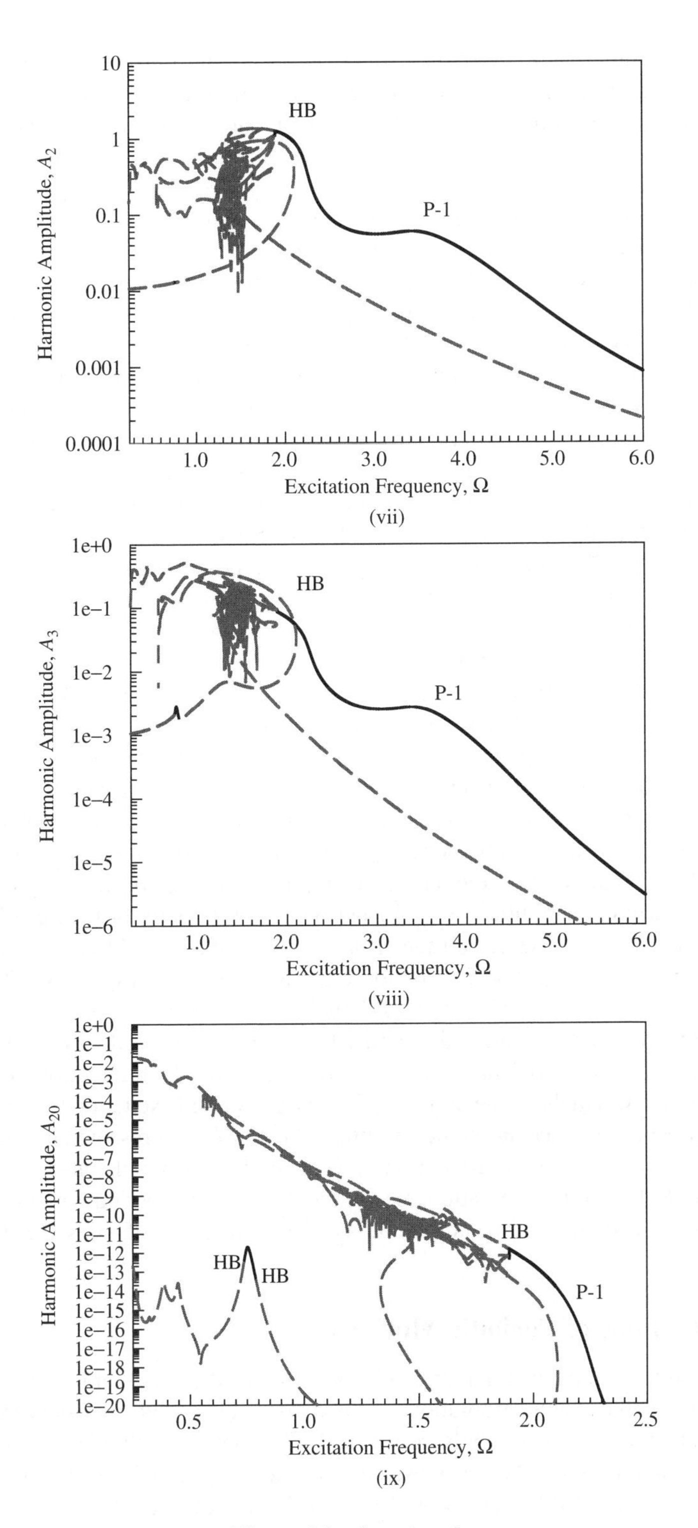

Figure 6.1 (*continued*)

similar to the harmonic amplitude $A_{1/8}$ for period-8 motion only herein. In Figure 6.1(v), the harmonic amplitude $A_{1/2}$ varying with excitation frequency is presented for period-2, period-4, and period-8 motions. The saddle-node and Hopf bifurcations of period-2 motion occur at $\Omega \approx 1.8957$ and $\Omega \approx 1.8955$. Such harmonic amplitude for period-1 motion is zero. The harmonic amplitudes $A_{k/m}$ ($\text{mod}(k,m) \neq 0, m = 8$) will not be presented to reduce illustrations. In Figure 6.1(vi), the primary harmonic amplitude A_1 versus excitation frequency is presented, and the Hopf bifurcation of period-1 motion occurs at $\Omega \approx 1.8955$. The amplitude peak is around $\Omega = 3.8$. To check the amplitude decrease, the harmonic amplitudes A_2 and A_3 versus excitation frequency are presented with common logarithm scale in Figure 6.1(vii),(viii), respectively. The harmonic amplitudes drop dramatically with increasing excitation frequency. However, for small excitation frequency, the harmonic amplitudes do not change too much. Thus, the variation of harmonic amplitude A_{20} with excitation frequency is presented in Figure 6.1(ix). Two Hopf bifurcations of period-1 motions are at $\Omega \approx 0.7355$ and 0.7625. The stable period-1 motion also exists $\Omega \in (0.7355, 0.7625)$.

The local view of the bifurcation tree of period-1 to period-8 motion is presented in Figure 6.2 in the range of $\Omega \in (1.885, 1.891)$. In Figure 6.2(i), the constant term $a_0^{(m)}$ versus excitation frequency is presented. The bifurcation tree of period-1 to period-8 motion is clearly observed. The Hopf bifurcation of period-1 motion gives the birth of the period-2 motion. The Hopf bifurcation of period-2 motion gives the birth of the period-4 motion, and the Hopf bifurcation of period-4 motion is the onset of period-8 motion. The Hopf bifurcation of period-8 motion can generate period-16 motion. In addition, the unstable period-1 to period-8 motions are presented. For this local view, the constant term is located in the range of $a_0^{(m)} \in (-0.65, -0.55)$. In Figure 6.2(ii), the harmonic amplitude $A_{1/8}$ varying with excitation frequency is presented for period-8 motion only. In the specific excitation range, the harmonic amplitude $A_{1/8} < 0.03$. In Figure 6.2(iii), the local view of the bifurcation tree of period-4 to period-8 motion is presented through the harmonic amplitude $A_{1/4}$. Such harmonic amplitude $A_{1/4}$ lies in the range of $A_{1/4} < 0.1$. In Figure 6.2(iv), the harmonic amplitude $A_{3/8} \in (0, 0.06)$ is presented that is similar to the harmonic amplitude $A_{1/8}$ for period-8 motion only. In Figure 6.2(v), the harmonic amplitude $A_{1/2}$ is presented for the bifurcation tree of period-2 to period-8 motion. The bifurcation structure from period-2 to period-8 motion is very clearly shown. The harmonic amplitude $A_{1/2} < 0.5$ is observed for period-2, period-4, and period-8 motions. To avoid abundant illustrations, the primary harmonic amplitude $A_1 < 0.5$ in the local view of bifurcation tree is presented in Figure 6.2(vi). The harmonic amplitude A_2 for the bifurcation tree of period-1 to period-8 motion lies in the range of $A_2 < 1.3$ in the prescribed excitation frequency range, as shown in Figure 6.2(vii). The harmonic amplitude $A_3 < 0.095$ is presented in Figure 6.2(viii), and the bifurcation tree for period-1 to period-8 motion is clearly illustrated. Finally, the harmonic amplitude A_{20} for the bifurcation tree of period-1 motion to period-8 motion is presented in Figure 6.2(ix). The range of the amplitude A_{20} lies in the range of $A_{20} < 2 \times 10^{-12}$.

6.3 Illustrations of Periodic Motions

The initial conditions and the initial time-delay values for $t \in (-\tau, 0)$ for numerical simulation are computed from the analytical solution. The numerical and analytical results are depicted by solid curves and red circular symbols, respectively. The big filled circular symbols are initial conditions and initial time-delay response values. The delay initial starting and delay initial final points are represented by acronyms D.I.S. and D.I.F., respectively.

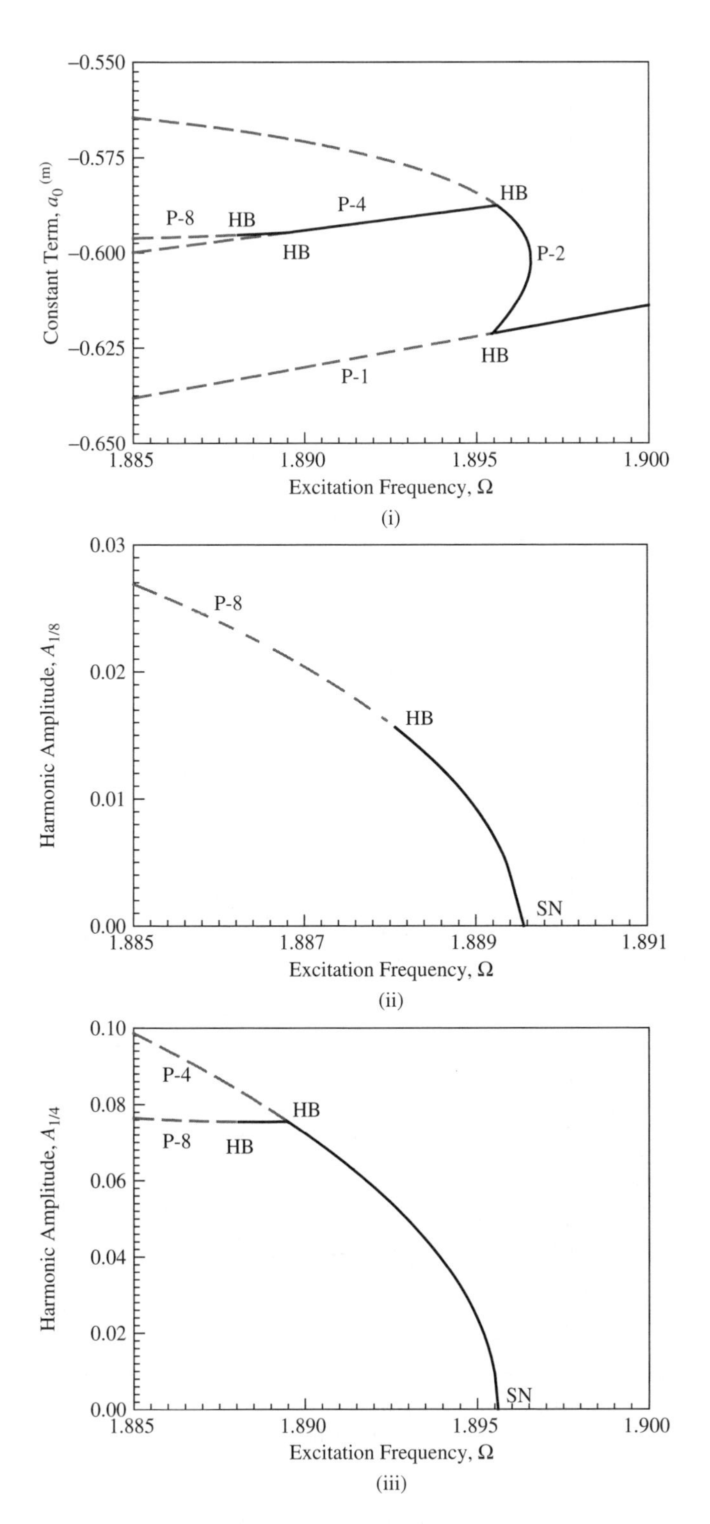

Figure 6.2 The zoomed view for the analytical prediction of period-1 to period-8 motions based on the 160 harmonic terms (HB160): (i) $a_0^{(m)}$ and (ii)–(ix) $A_{k/m}$ ($k = 1, 2, 3, 4, 6, 8, 12, \ldots, 24, 160, m = 8$). Parameters: ($\delta_1 = 0.05, \alpha_1 = 15.0, \alpha_2 = 5.0, \beta_1 = 5.0, Q_0 = 4.5, \tau = T/4$)

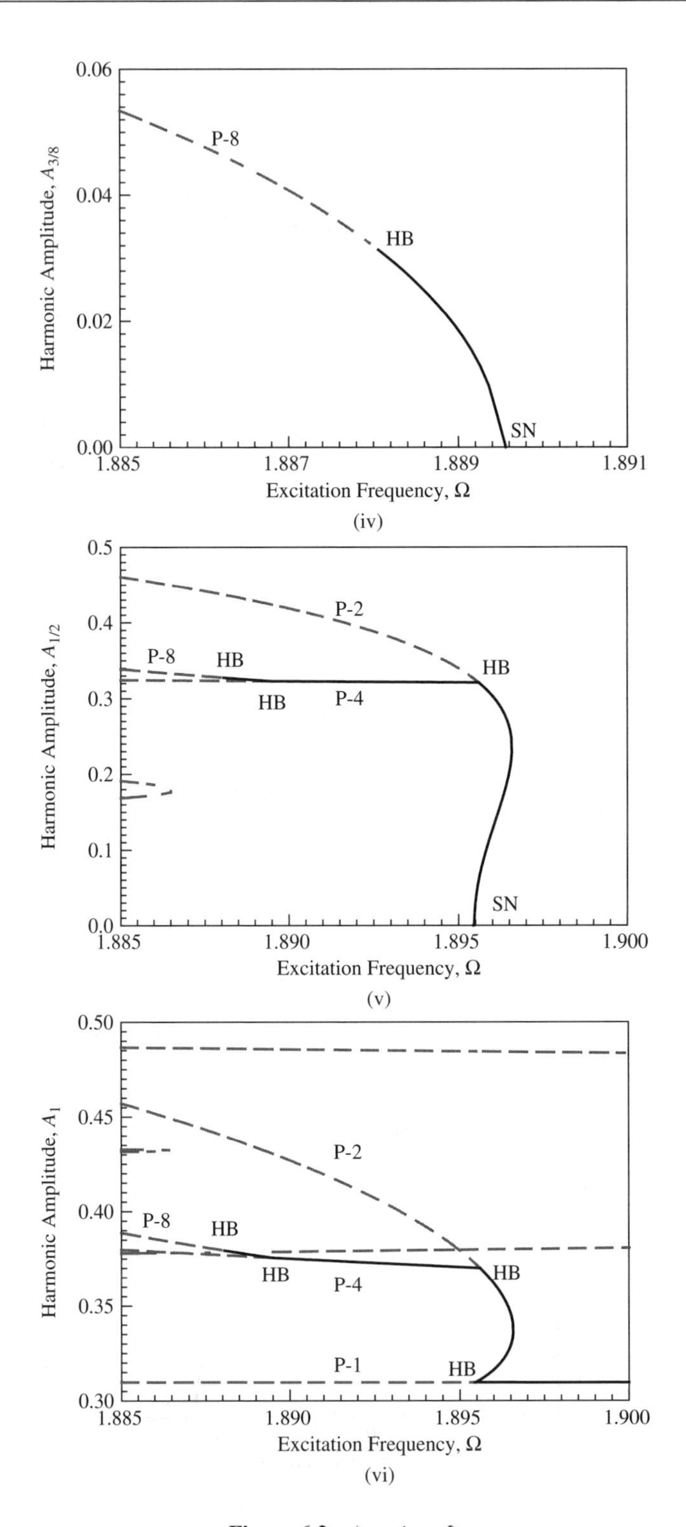

Figure 6.2 (*continued*)

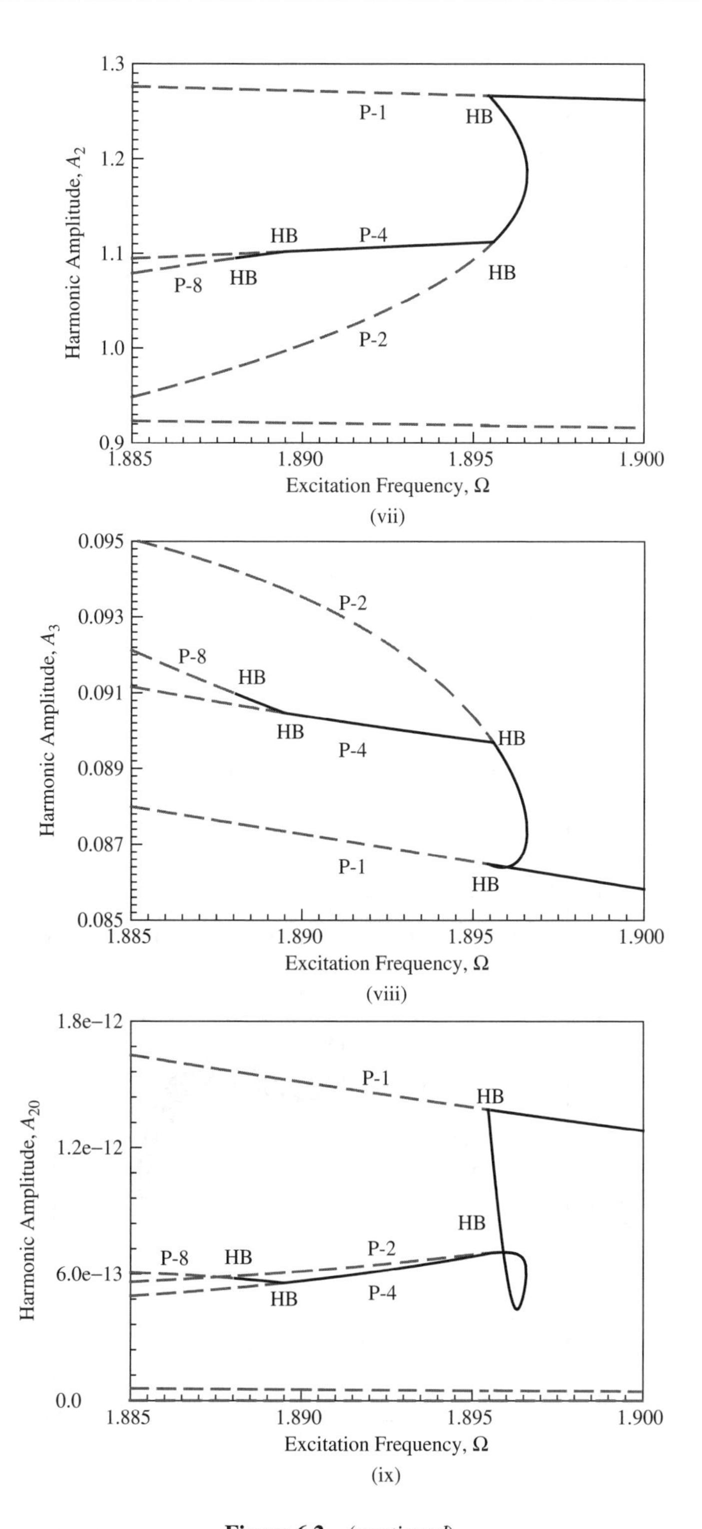

Figure 6.2 (*continued*)

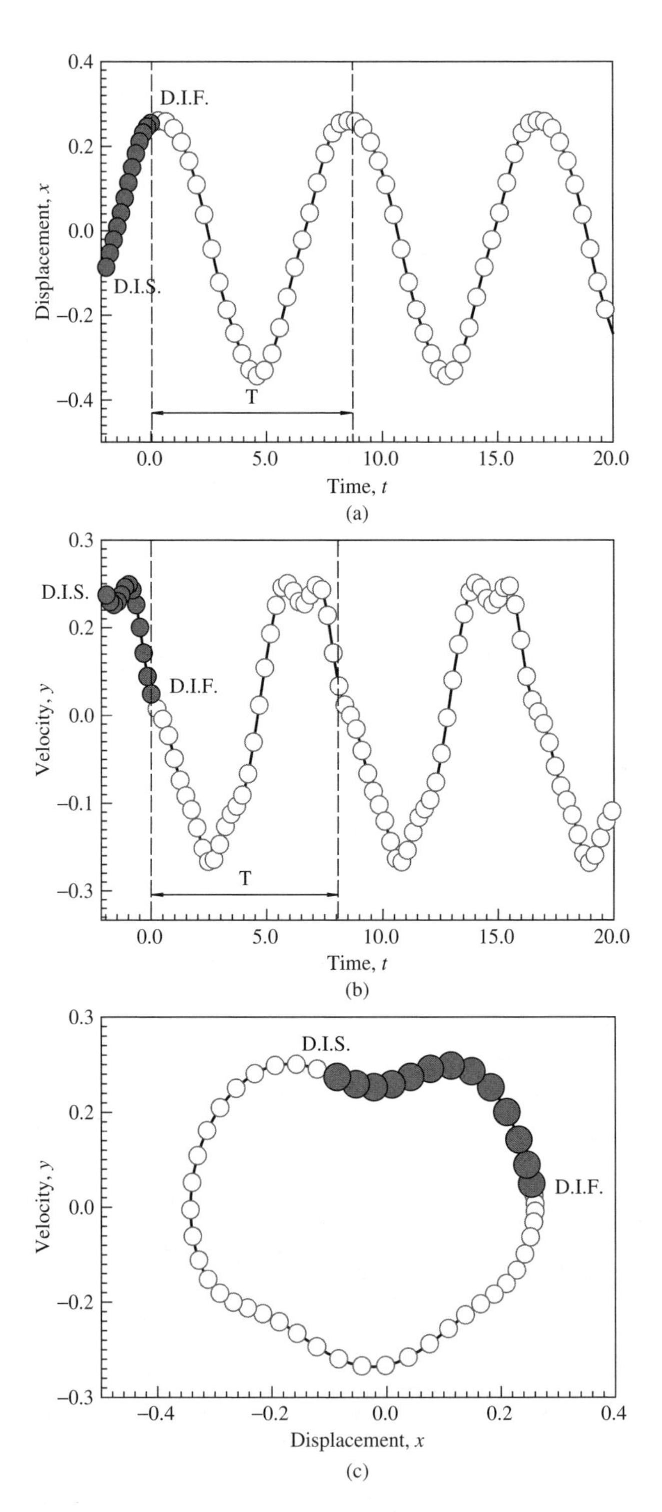

Figure 6.3 Analytical and numerical solutions of stable period-1 motion based on 48 harmonic terms (HB48): (a) displacement, (b) velocity, (c) phase plane, and (d) amplitude spectrum. Initial condition ($x_0 \approx 0.253793, \dot{x}_0 \approx 0.036826$). Parameters: ($\delta_1 = 0.05, \alpha_1 = 15.0, \alpha_2 = 5.0, \beta_1 = 5.0, Q_0 = 4.5, \tau = T/4$)

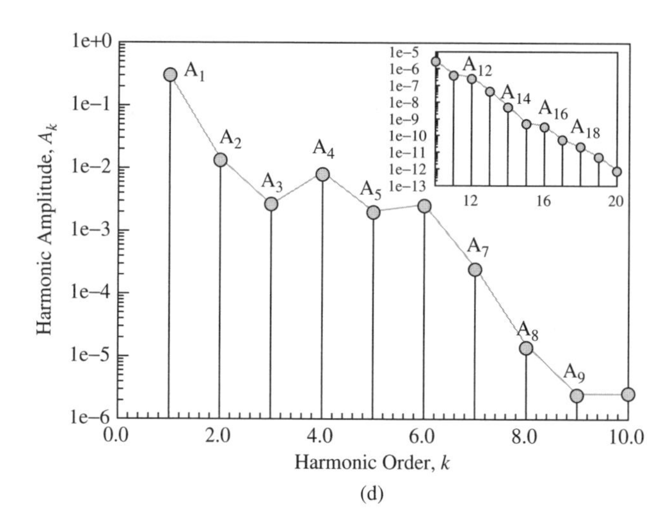

Figure 6.3 *(continued)*

As in Luo and Jin (2014), the displacement, velocity, trajectory and amplitude spectrum of stable period-1 motion for the time-delayed, quadratic nonlinear oscillator are presented in Figure 6.3 for $\Omega = 0.7665$ with initial condition ($x_0 \approx 0.253793, \dot{x}_0 \approx 0.036826$) with initial time-delayed responses. This analytical solution is based on 20 harmonic terms (HB20) in the Fourier series solution of period-1 motion. In Figure 6.3(a),(b), for over 100 periods, the analytical and numerical solutions of the period-1 motion in the time-delayed, quadratic nonlinear oscillator match very well. The initial time-delayed displacement and velocity are presented by the large circular symbols for the initial delay period of $t \in (-\tau, 0)$. In Figure 6.3(c), analytical and numerical trajectories match very well, and the initial time-delay responses in phase plane is clearly depicted. In Figure 6.3(d), the amplitude spectrum is presented. The quantity levels of the harmonic amplitudes are $a_0 \approx -0.022855$, $A_1 \approx 0.300206$, $A_2 \approx 0.013171$, $A_3 \sim 2.64 \times 10^{-3}$, $A_4 \sim 7.91 \times 10^{-3}$, $A_5 \sim 1.97 \times 10^{-3}$, $A_7 \sim 2.44 \times 10^{-4}$, $A_8 \sim 1.38 \times 10^{-5}$, $A_9 \sim 2.44 \times 10^{-6}$, $A_{10} \sim 2.59 \times 10^{-6}$, $A_{11} \sim 3.75 \times 10^{-7}$, $A_{12} \sim 2.38 \times 10^{-7}$, $A_{13} \sim 4.27 \times 10^{-8}$, $A_{14} \sim 4.8 \times 10^{-9}$, $A_{15} \sim 5.02 \times 10^{-10}$, $A_{16} \sim 3.24 \times 10^{-10}$, $A_{17} \sim 5.21 \times 10^{-11}$, $A_{18} \sim 2.02 \times 10^{-11}$, $A_{19} \sim 4.88 \times 10^{-12}$, $A_{20} \sim 7.32 \times 10^{-13}$. The harmonic amplitudes decrease with harmonic order non-uniformly. The main contributions for this periodic motion are from the primary and second harmonics. The truncated harmonic amplitude is $A_{20} \sim 10^{-13}$.

The trajectory and amplitude spectrum of stable period-1 motion for the time-delayed, quadratic nonlinear oscillator are presented in Figure 6.4 for $\Omega = 1.921$ and 5.52. The initial conditions are listed in Table 6.1 and initial time-delayed values are also computed from the analytical conditions. This analytical solution is based on 20 harmonic terms (HB20) in the Fourier series solution of period-1 motion. In Figure 6.4(a), analytical and numerical trajectories is presented for $\Omega = 1.921$, and the initial time-delay responses in the phase plane is illustrated, and this period-1 motion possesses two cycles. In Figure 6.4(b), the amplitude spectrum is presented. The main harmonic amplitudes are $a_0 \approx -0.579650$, $A_1 \approx 0.309288$, $A_2 \approx 1.240523$, $A_3 \approx 0.082741$, and $A_4 \approx 0.071660$. The other harmonic amplitudes are

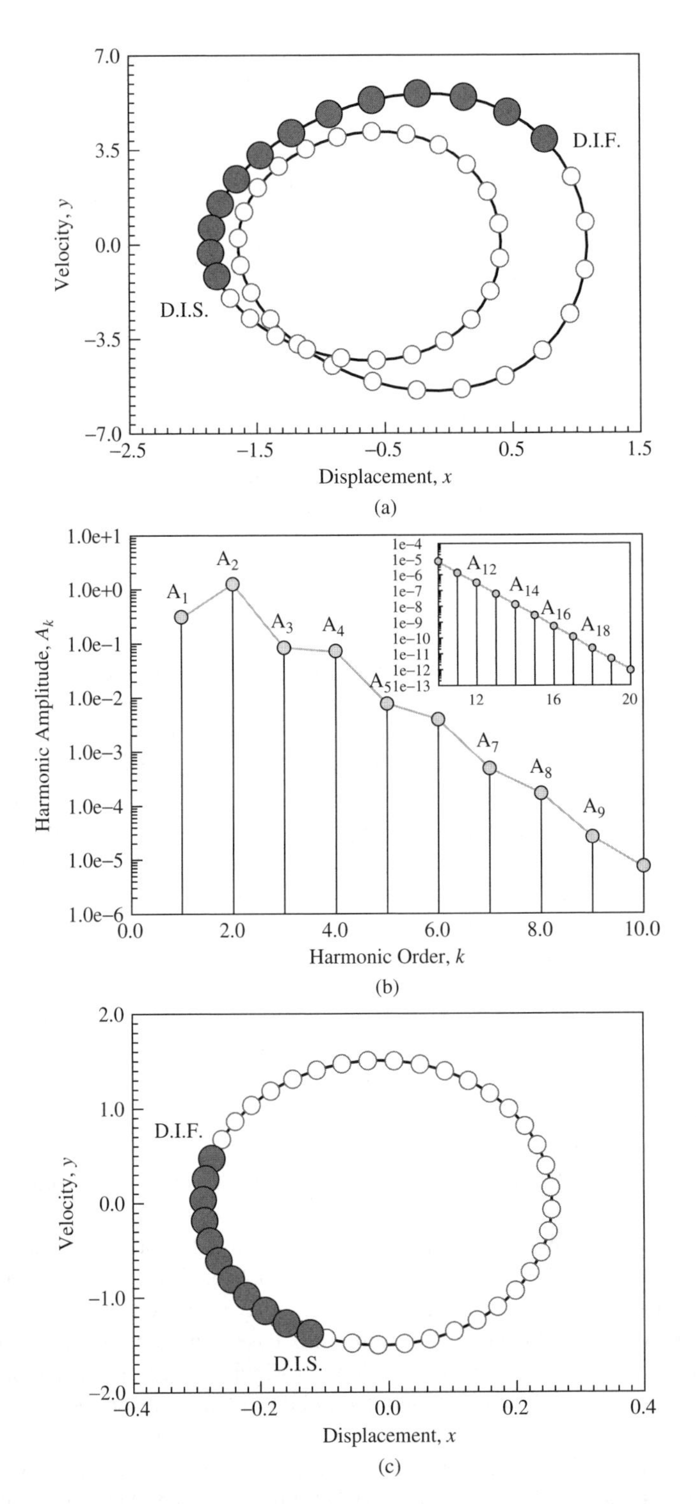

Figure 6.4 Phase plane and amplitude spectrums of period-1 motions: (a,b) $\Omega = 1.921$ with $(x_0 \approx 0.753207, \dot{x}_0 \approx 3.874847)$. (c,d) $\Omega = 5.52$ with $(x_0 \approx -0.275515, \dot{x}_0 \approx 0.468443)$. Parameters: $(\delta_1 = 0.05, \alpha_1 = 15.0, \alpha_2 = 5.0, \beta_1 = 5.0, Q_0 = 4.5)$

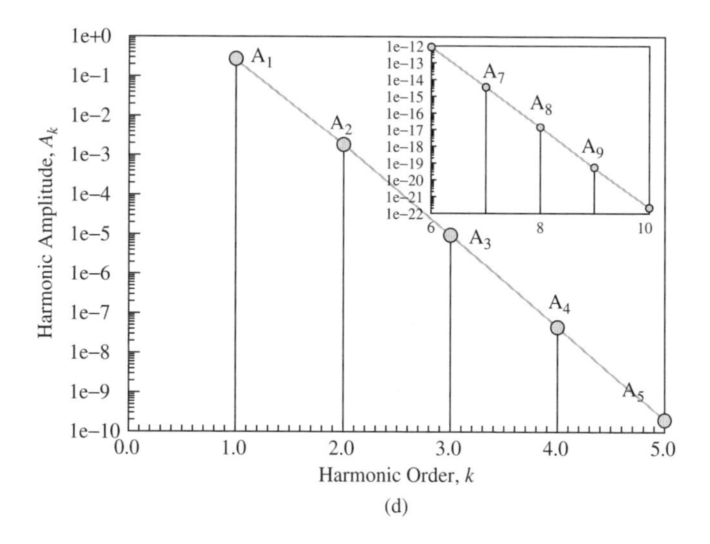

(d)

Figure 6.4 (*continued*)

Table 6.1 Input data for numerical illustrations
($\delta = 0.05, \alpha_1 = 10.0, \alpha_2 = 5.0, \beta_1 = 5.0, Q_0 = 4.5, \tau = T/4$)

Figure no.	Ω	Initial condition $(x_0, \dot{x}_0)$	Types	Harmonics terms
Figure 6.4(a),(b)	1.921	(−0.753207, 3.874847)	P-1	HB20 (stable)
Figure 6.4(c),(d)	5.52	(−0.275515, 0.468443)	P-1	HB20 (stable)
Figure 6.5(i),(ii)	1.8965	(0.258268, 4.712170)	P-2	HB40 (stable)
Figure 6.5(iii),(iv)	1.8920	(−0.788258, 0.116180)	P-4	HB80 (stable)
Figure 6.5(v),(vi)	1.88876	(−0.319273, 4.696123)	P-8	HB160 (stable)

$A_k \in (10^{-14}, 10^{-3})$ for $k = 5, 6, \ldots, 20$. In Figure 6.4(c), analytical and numerical trajectories with the initial time-delay values are presented for $\Omega = 5.52$. In Figure 6.4(d), the amplitude spectrum distribution is presented. The main harmonic amplitudes are $a_0 \approx -0.018739$ and $A_1 \approx 0.272493$. The other harmonic amplitudes are $A_k \in (10^{-46}, 10^{-3})$ for $k = 2, 3, \ldots, 20$.

The stable period-2, period-4, and period-8 motions are presented in Figure 6.5 at $\Omega = 1.8965, 1.8920, 1.88906$ for illustrations of complexity of periodic motions. The initial conditions for such stable periodic motions are listed in Table 6.1. In Figure 6.5(i), the analytical and numerical trajectories of a period-2 motion are presented. Such a period-2 motion possesses four cycles and the initial time-delay conditions are presented. The harmonic amplitude distribution is presented in Figure 6.5(ii). The main amplitudes of the period-2 motion in such time-delayed, nonlinear system are $a_0^{(2)} \approx -0.597643$, $A_{1/2} \approx 0.261850$, $A_1 \approx 0.345902$, $A_{3/2} \approx 0.299359$, $A_2 \approx 1.163761$, $A_{5/2} \approx 0.176459$, $A_3 \approx 0.087922$, $A_{7/2} \approx 0.035945$, $A_4 \approx 0.058968$, and $A_{9/2} \approx 0.018524$. The other harmonic amplitudes are $A_{k/2} \in (10^{-14}, 10^{-3})$ for $k = 10, 11, \ldots, 40$. The biggest contribution is from the harmonic term of $A_2 \approx 1.163761$ In Figure 6.5(iii), the analytical and numerical

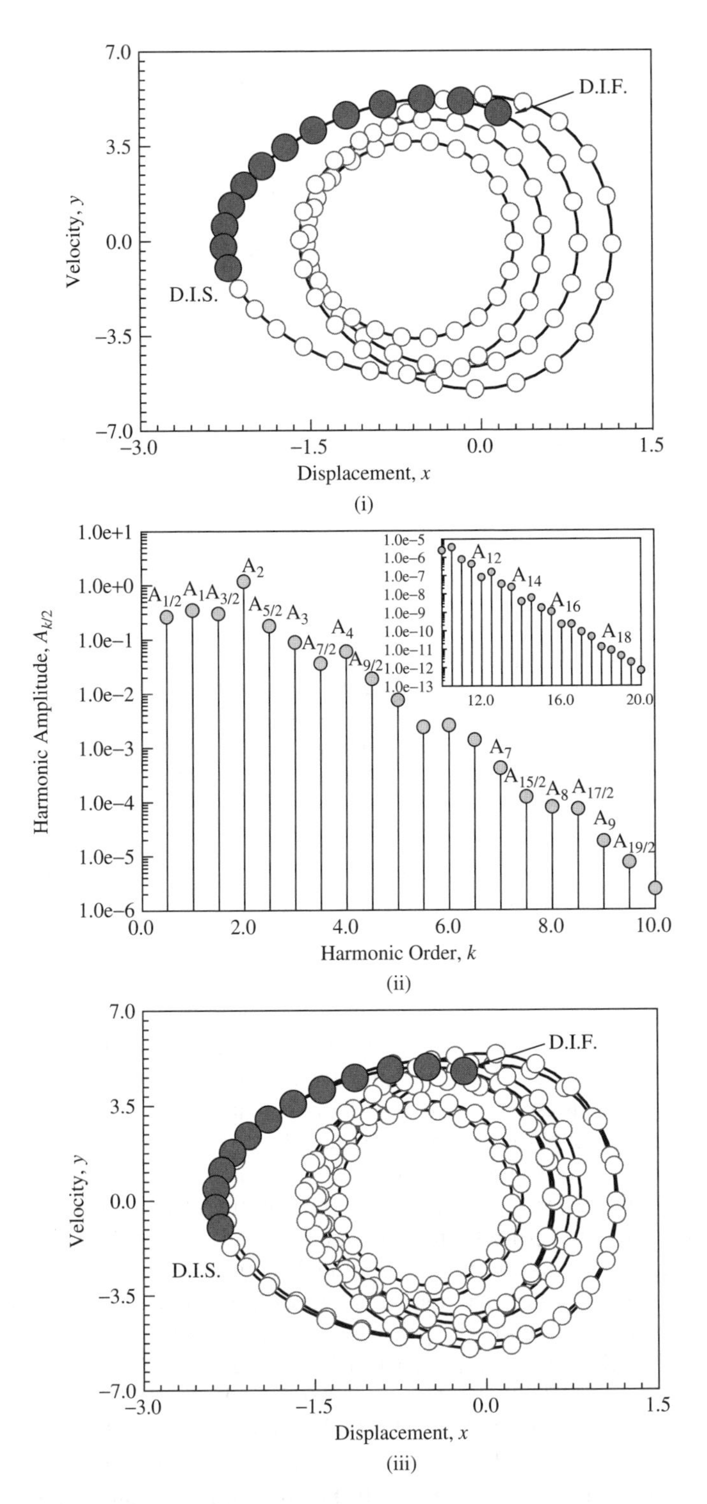

Figure 6.5 Phase plane and amplitude spectrums: (i,ii) period-2 motion ($\Omega = 1.8965$, $x_0 \approx 0.158268$, $\dot{x}_0 \approx 4.712170$, HB40), (iii,iv) period-4 motion ($\Omega = 1.8920$, $x_0 \approx -0.788258$, $\dot{x}_0 \approx 0.116180$, HB80), and (v,vi) period-8 motions ($\Omega = 1.88906$, $x_0 \approx 0.982777$, $\dot{x}_0 \approx 2.844490$, HB160). ($\delta_1 = 0.05, \alpha_1 = 15.0, \alpha_2 = 5.0, \beta_1 = 5.0, Q_0 = 4.5$)

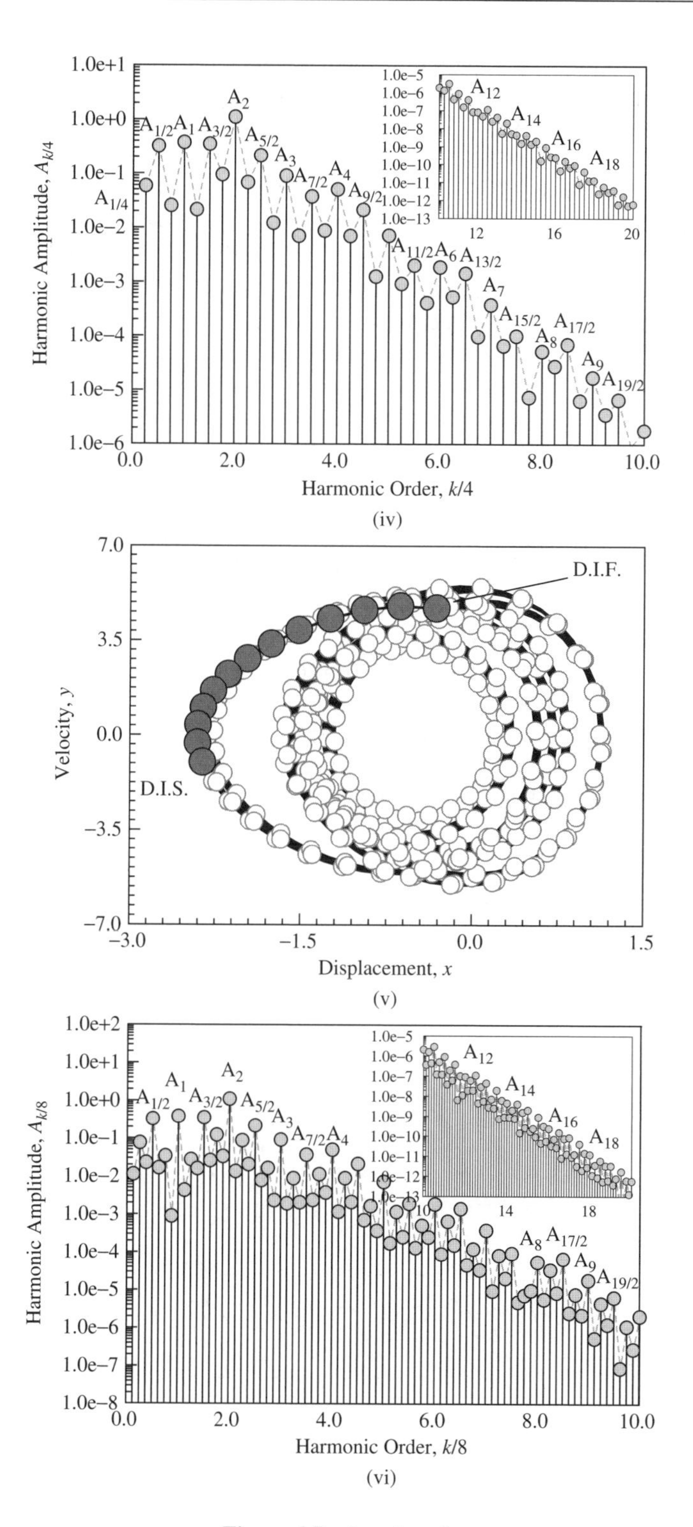

Figure 6.5 *(continued)*

trajectories of period-4 motion are presented. Such a period-4 motion possesses eight cycles and the initial time-delay conditions are presented. The harmonic amplitude distribution is presented in Figure 6.5(iv). The main amplitudes of the period-4 motion are $a_0^{(4)} \approx -0.591813$, $A_{1/4} \approx 0.058286$, $A_{1/2} \approx 0.322076$, $A_{3/4} \approx 0.025289$, $A_1 \approx 0.373248$, $A_{5/4} \approx 0.021254$, $A_{3/2} \approx 0.351173$, $A_{7/4} \approx 0.094394$, $A_2 \approx 1.106125$, $A_{9/4} \approx 0.067732$, $A_{5/2} \approx 0.214359$, $A_{11/4} \approx 0.012157$, $A_3 \approx 0.090130$, $A_{13/4} \approx 7.042438\text{E-}3$, $A_{7/2} \approx 0.037581$, $A_{15/4} \approx 8.784526\text{E-}3$, $A_4 \approx 0.050681$, $A_{17/4} \approx 7.035358\text{E-}3$, $A_{9/2} \approx 0.021354$, and $A_{19/4} \approx 1.263319\text{E-}3$. The other harmonic amplitudes are $A_{k/4} \in (10^{-14}, 10^{-3})$ for $k = 20, 21, \ldots, 80$. The biggest contribution of the period-4 motion is from the harmonic amplitude of $A_2 \approx 1.106125$. In Figure 6.5(v), the analytical and numerical trajectories of a period-8 motion are presented. Such a period-8 motion possesses 16 cycles and the initial time-delay conditions are presented. The harmonic amplitude spectrum is presented in Figure 6.5(vi). The main harmonic amplitudes of a period-8 motion are $a_0^{(8)} \approx -0.595049$, $A_{1/8} \approx 0.011233$, $A_{1/4} \approx 0.075460$, $A_{3/8} \approx 0.022571$, $A_{1/2} \approx 0.325151$, $A_{5/8} \approx 0.016434$, $A_{3/4} \approx 0.033788$, $A_{7/8} \approx 8.864442\text{E-}4$, $A_1 \approx 0.377481$, $A_{9/8} \approx 4.220962\text{E-}3$, $A_{5/4} \approx 0.027066$, $A_{11/8} \approx 0.015783$, $A_{3/2} \approx 0.351642$, $A_{13/8} \approx 0.025705$, $A_{7/4} \approx 0.122130$, $A_{15/8} \approx 0.033176$, $A_2 \approx 1.098604$, $A_{17/8} \approx 0.013382$, $A_{9/4} \approx 0.086979$, $A_{19/8} \approx 0.020426$, $A_{5/2} \approx 0.215422$, $A_{21/8} \approx 7.792078\text{E-}3$, $A_{11/4} \approx 0.016277$, $A_{23/8} \approx 2.312084\text{E-}3$, $A_3 \approx 0.090725$, $A_{25/8} \approx 1.920124\text{E-}3$, $A_{13/4} \approx 8.823134\text{E-}3$, $A_{27/8} \approx 2.050977\text{E-}3$, $A_{7/2} \approx 0.036823$, $A_{29/8} \approx 2.370049\text{E-}3$, $A_{15/4} \approx 0.011261$, $A_{31/8} \approx 3.737412\text{E-}3$, $A_4 \approx 0.049915$, $A_{33/8} \approx 1.169501\text{E-}3$, $A_{17/4} \approx 8.909312\text{E-}3$, $A_{35/8} \approx 2.117430\text{E-}3$, $A_{9/2} \approx 0.021195$, $A_{37/8} \approx 7.018696\text{E-}4$, $A_{19/4} \approx 1.663909\text{E-}3$, and $A_{39/8} \approx 3.704522\text{E-}4$. The other harmonic amplitudes are $A_{k/8} \in (10^{-14}, 10^{-3})$ for $k = 40, 41, \ldots, 160$. The biggest contribution is still from the harmonic amplitude of $A_2 \approx 1.098604$.

References

Arnold, V.I. (1964) Instability of dynamical systems with several degrees of freedom. *Soviet Mathematics Doklody*, **5**, 581–585.
Birkhoff, G.D. (1913) Proof of Poincare's geometric theorem. *Transactions on American Mathematical Society*, **14**, 14–22.
Birkhoff, G.D. (1927) *Dynamical Systems*, American Mathematical Society, New York.
Bogoliubov, N. and Mitropolsky, Y. (1961) *Asymptotic Methods in the Theory of Nonlinear Oscillations*, Gordon and Breach, New York.
Carr, J. (1981) *Applications of Center Manifold Theory, Applied Mathematical Science 35*, Springer-Verlag, New York.
Cartwright, M.L. and Littlewood, J.E. (1945) On nonlinear differential equations of the second order I. The equation $\ddot{y} - k(1 - y^2)\dot{y} + y = b\lambda k \cos(\lambda t + \alpha)$, k large. *Journal of London Mathematical Society*, **20**, 180–189.
Cartwright, M.L. and Littlewood, J.E. (1947) On nonlinear differential equations of the second order II. The equation $\ddot{y} + kf(y)\dot{y} + g(y, k) = p(t) = p_1(t) + kp_2(t)$, $k > 0\, f(y) \geq 1$. *Annals of Mathematics*, **48**, 472–494.
Coddington, E.A. and Levinson, N. (1955) *Theory of Ordinary Differential Equations*, McGraw-Hill, New York.
Chirikov, B.V. (1979) A universal instability of many-dimensional oscillator systems. *Physics Reports*, **52**, 263–379.
Coppola, V.T. and Rand, R.H. (1990) Averaging using elliptic functions: approximation of limit cycle. *Acta Mechanica*, **81**, 125–142.
Duffing, G. (1918) *Erzwunge Schweingungen bei veranderlicher eigenfrequenz*, F. Viewig u. Sohn, Braunschweig.
Fatou, P. (1928) Suré le mouvement d'un systeme soumis á des forces a courte periode. *Bulletin de la Société Mathématique*, **56**, 98–139.
Greenspan, B.D. (1981) Bifurcations in periodically forced oscillations: subharmonics and homoclinic orbits. PhD thesis. Center for Applied Mathematics, Cornell University.
Greenspan, B.D. and Holmes, P.J. (1983) Homoclinic orbits, subharmonics and global bifurcations in forced oscillations, In *Nonlinear Dynamics and Turbulence*, G. Barenblatt, G. Iooss, and D.D. Joseph (eds.) London, Pitman, pp. 172–214.
Guckenheimer, J. and Holmes, P. (1983) *Nonlinear Oscillations, Dynamical Systems, and Bifurcations of Vector Fields*, Springer-Verlag, New York.
Hartman, P. (1964) *Ordinary Differential Equations*, Wiley, New York (2nd ed. Birkhauser, Boston Basel Stuttgart, 1982).
Hayashi, C. (1964) *Nonlinear oscillations in Physical Systems*, McGraw-Hill Book Company, New York.

Toward Analytical Chaos in Nonlinear Systems, First Edition. Albert C. J. Luo.
© 2014 John Wiley & Sons, Ltd. Published 2014 by John Wiley & Sons, Ltd.

Holmes, P.J. (1979) A nonlinear oscillator with strange attractor. *Philosophical Transactions of the Royal Society A*, **292**, 419–448.

Holmes, P.J. and Rand, D.A. (1976) Bifurcations of Duffing equation; An application of catastrophe theory. *Quarterly Applied Mathematics*, **35**, 495–509.

Hu, H.Y., Dowell, E.H. and Virgin, L.N. (1998) Resonance of harmonically forced Duffing oscillator with time-delay state feedback. *Nonlinear Dynamics*, **15**(4), 311–327.

Hu, H.Y. and Wang, Z.H. (2002) *Dynamics of Controlled Mechanical Systems with Delayed Feedback*, Springer, Berlin.

Insperger, T. and Stepan, G. (2011) *Semi-Discretization for Time-Delay Systems: Stability and Engineering Applications*, Springer, New York.

Kao, Y.H., Wang, C.S. and Yang, T.H. (1992) Influences of harmonic coupling on bifurcations in Duffing oscillator with bounded potential wells. *Journal of Sound and Vibration*, **159**, 13–21.

Krylov, N.M. and Bogolyubov, N.N. (1935) *Methodes approchees de la mecanique non-lineaire dans leurs application a l'Aeetude de la perturbation des mouvements periodiques de divers phenomenes de resonance s'y rapportant*. Kiev, Academie des Sciences d'Ukraine (in French).

Lagrange, J.L. (1788) *Mecanique Analytique*, **2** vol., Edition Albert Balnchard, Paris.

Leung, A.Y.T. and Guo, Z. (2012) Bifurcation of the periodic motions in nonlinear delayed oscillators. *Journal of Vibration and Control*. Doi: 10.1177/1077546312464988

Levinson, N. (1948) A simple second order differential equation with singular motions. *Proceedings of the National Academy of Science of the United States of America*, **34**(1), 13–15.

Levinson, N. (1949) A second order differential equation with singular solutions. *Annals of Mathematics, Second Series*, **50**(1), 127–153.

Liu, L. and Kalmar-Nagy, T. (2010) High-dimensional harmonic balance analysis for second-order delay-differential equations. *Journal of Vibration and Control*, **16**(7–8), 1189–1208.

Luo, A.C.J. (1995) Analytical modeling of bifurcations, chaos and multifractals in nonlinear dynamics. PhD Dissertation. University of Manitoba, Winnipeg, Manitoba.

Luo, A.C.J. (2008) *Global Transversality, Resonance, and Chaotic Dynamics*, World Scientific, Singapore.

Luo, A.C.J. (2012a) *Continuous Dynamical Systems*, Higher Education Press/L&H Scientific, Beijing/Glen Carbon.

Luo, A.C.J. (2012b) *Regularity and Complexity in Dynamical Systems*, Springer, New York.

Luo, A.C.J. (2013) Analytical solutions for periodic motions to chaos in dynamical systems with/without time-delay. *International Journal of Dynamics and Control* **1**(4), 330–359.

Luo, A.C.J. and Han, R.P.S. (1997) A quantitative stability and bifurcation analyses of a generalized Duffing oscillator with strong nonlinearity. *Journal of Franklin Institute*, **334B**, 447–459.

Luo, A.C.J. and Han, R.P.S. (1999) Analytical predictions of chaos in a nonlinear rod. *Journal of Sound and Vibration*, **227**(3), 523–544.

Luo, A.C.J. and Han, R.P.S. (2001) The resonance theory for stochastic layers in nonlinear dynamical systems. *Chaos, Solitons and Fractals*, **12**, 2493–2508.

Luo, A.C.J. and Huang, J. (2012a) Approximate solutions of periodic motions in nonlinear systems via a generalized harmonic balance. *Journal of Vibration and Control*, **18**, 1661–1671.

Luo A.C.J. and Huang J. (2012b) Analytical dynamics of period-m flows and chaos in nonlinear Systems, *International Journal of Bifurcation and Chaos* **22**(4): Article No. 1250093 (29 pp).

Luo, A.C.J. and Huang, J.Z. (2012c) Analytical routes of period-1 motions to chaos in a periodically forced Duffing oscillator with a twin-well potential. *Journal of Applied Nonlinear Dynamics*, **1**, 73–108.

Luo, A.C.J. and Huang, J.Z. (2012d) Unstable and stable period-*m* motions in a twin-well potential Duffing oscillator. *Discontinuity, Nonlinearity and Complexity*, **1**, 113–145.

Luo, A.C.J. and Huang, J.Z. (2013a) Analytical solutions for asymmetric periodic motions to chaos in a hardening Duffing oscillator. *Nonlinear Dynamics*, **72**, 417–438.

Luo, A.C.J. and Huang, J.Z. (2013b) Analytical period-3 motions to chaos in a hardening Duffing oscillator. *Nonlinear Dynamics*, **73**, 1905–1932.
Luo, A.C.J. and Huang, J.Z. (2013c) An analytical prediction of period-1 motions to chaos in a softening Duffing oscillator. *International Journal of Bifurcation and Chaos*, **23**(5), Article No: 1350086 (31 pages).
Luo, A.C.J. and Huang, J.Z. (2014) Period-3 motions to chaos in a softening Duffing oscillator. *International Journal of Bifurcation and Chaos*, **24**(3), Article No: 1430010 (26 pages).
Luo, A.C.J. and Jin, H.X. (2014) Bifurcation trees of period-*m* motions in a periodically forced, time-delayed, quadratic nonlinear oscillator. *Discontinuity, Nonlinearity, and Complexity*, **3**, 87–107.
Luo, A.C.J. and Laken, A.B. (2013) Analytical solutions for period-m motions in a periodically forced van der Pol oscillator. *International Journal of Dynamics and Control*, **1**, 99–115.
Luo, A.C.J. and Yu, B. (2013a) Analytical solutions for stable and unstable period-1 motions in a periodically forced oscillator with quadratic nonlinearity. *ASME Journal of Vibrations and Acoustics*, **135**, Article No: 034505 (5 pp).
Luo, A.C.J. and Yu, B. (2013b) Complex period-1 motions in a quadratic nonlinear oscillator. *Journal of Vibration and Control*. Doi: 10.1177/1077546313490525
Luo, A.C.J. and Yu, B. (2013c) Period-*m* motions and bifurcation trees in a periodically forced, quadratic nonlinear oscillator. *Discontinuity, Nonlinearity, and Complexity*, **2**(3), 265–288.
MacDonald, N. (1995) Harmonic balance in delay-differential equations. *Journal of Sounds and Vibration*, **186**(4), 649–656.
Marsden, J.E. and McCracken, M.F. (1976) *The Hopf Bifurcation and Its Applications, Applied Mathematical Science 19*, Springer-Verlag, New York.
Melnikov, V.K. (1962) On the behavior of trajectories of system near to autonomous Hamiltonian systems. *Soviet Mathematics-Doklady*, **3**, 109–112.
Melnikov, V.K. (1963) On the stability of the center for time periodic perturbations. *Transaction Moscow Mathematical Society*, **12**, 1–57.
Minorsky, N. (1962) *Nonlinear Oscillations*, Van Nostrand, New York.
Nayfeh, A.H. (1973) *Perturbation Methods*, John Wiley & Sons, Inc., New York.
Nayfeh, A.H. and Mook, D.T. (1979) *Nonlinear Oscillation*, John Wiley & Sons, Inc., New York.
Peng, Z.K., Lang, Z.Q., Billings, S.A. and Tomlinson, G.R. (2008) Comparisons between harmonic balance and nonlinear output frequency response function in nonlinear system analysis. *Journal of Sound and Vibration*, **311**, 56–73.
Poincare, H. (1890) Sur les equations de la dynamique et le probleme de trios corps. *Acta Mathematica*, **13**, 1–270.
Pioncare, H. (1892) *Les Methodes Nouvelles de la Méchanique Célèste*. Gauthier-Villars, Paris.
Poincare, H. (1899) *Les Methods Nouvelles de la Mechanique Celeste*, vol. **3**, Gauthier-Villars, Paris.
van der Pol, B. (1920) A theory of the amplitude of free and forced triode vibrations. *Radio Review*, **1**, 701–710, 754–762.
van der Pol, B. and van der Mark, J. (1927) Frequency demultiplication. *Nature*, **120**, 363–364.
Smale, S. (1967) Differentiable dynamical systems. *Bulletin of the American Mathematical Society*, **73**, 747–817.
Stepan, G. (1989) *Retarded Dynamical Systems*, Longman, Harlow.
Sun, J.Q. (2009) A method of continuous time approximation of delayed dynamical systems. *Communications in Nonlinear Science and Numerical Simulation*, **14**(4), 998–1007.
Stoker, J.J. (1950) *Nonlinear Vibrations*, John Wiley & Sons, Inc., New York.
Tlusty, J. (2000) *Manufacturing Processes and Equipment*, Prentice-Hall, Upper Saddle River, NJ.
Ueda, Y. (1980) Explosion of strange attractors exhibited by the Duffing equations. *Annuals of the New York Academy of Science*, **357**, 422–434.
Wang, C.S., Kao, Y.H., Huang, J.C. and Gou, Y.H. (1992) Potential dependence of the bifurcation structure in generalized Duffing oscillators. *Physical Review A*, **45**, 3471–3485.
Wang, H. and Hu, H.Y. (2006) Remarks on the perturbation methods in solving the second order delay differential equations. *Nonlinear Dynamics*, **33**, 379–398.

Index

Toward Analytical Chaos in Nonlinear Systems, First Edition. Albert C. J. Luo.
© 2014 John Wiley & Sons, Ltd. Published 2014 by John Wiley & Sons, Ltd.